Jordan Rotheiser
Joining of Plastics

Jordan Rotheiser

Joining of Plastics

Handbook for Designers and Engineers

2nd Edition

HANSER

Hanser Publishers, Munich

Hanser Gardner Publications, Inc., Cincinnati

The Author:
Jordan Rotheiser, 3075 University Avenue, Highland Park, IL 60035, USA

Distributed in the USA and in Canada by
Hanser Gardner Publications, Inc.
6915 Valley Avenue, Cincinnati, Ohio 45244-3029, USA
Fax: (513) 527-8801
Phone: (513) 527-8977 or 1-800-950-8977
www.hansergardner.com

Distributed in all other countries by
Carl Hanser Verlag
Postfach 86 04 20, 81631 München, Germany
Fax: +49 (89) 98 48 09
www.hanscr.de

The use of general descriptive names, trademarks, etc. in this publication, even if the former are not especially identified, is not to be taken as a sign that such names, as understood by the Trade Marks and Merchandise Marks Act, may accordingly be used freely by anyone.
While the advice and information in this book are believed to be true and accurate at the date of going to press, neither the authors nor the editors nor the publisher can accept any legal responsibility for any errors or omissions that may be made. The publisher makes no warranty, express or implied, with respect to the material contained herein.

Library of Congress Cataloging-in-Publication Data

Rotheiser, Jordan.
    Joining of plastics : handbook for designers and engineers / by Jordan
Rotheiser.— 2nd ed.
        p. cm.
    ISBN 1-56990-354-9 (hardcover)
1. Plastics. 2. Joints (Engineering) 3. Sealing (Technology) I. Title.
    TA455.P5R675 2004
    668.4—dc22

                          2004000939

Bibliografische Information Der Deutschen Bibliothek
Die Deutsche Bibliothek verzeichnet diese Publikation in der Deutschen Nationalbibliografie; detaillierte bibliografische Daten sind im Internet über < http://dnb.ddb.de > abrufbar.

ISBN 3-446-22454-8

©Carl Hanser Verlag, Munich 2004
Production Management: Oswald Immel
Typeset: Alden Group, UK
Cover Concept: Marc Müller-Bremer, Rebranding, München, Germany
Coverdesign: MCP - Susanne Kraus GbR, Holzkirchen, Germany
Printed and bound by Druckhaus "Thomas Müntzer", Bad Langensalza, Germany

Dedication

The author wishes to dedicate this book to his wife Gail, without whose help and encouragement it could not have been written – and to his family from whom he was absent so many hours.

Foreword

The Society of Plastics Engineers is pleased to sponsor *Joining of Plastics Handbook for Designers and Engineers* by Jordan I. Rotheiser. Mr. Rotheiser is a practicing plastics engineer and industrial designer, with more than 35 years of experience in the design of plastic products and a well respected member of SPE. He has held numerous seminars on assembly methods for plastics. His concise book will help the reader to determine the most cost-effective joining method for any given application.

SPE, through its Technical Volumes Committee, has long sponsored books on various aspects of plastics. Its involvement has ranged from identification of needed volumes and recruitment of authors to peer review and approval and publication of new books.

Technical competence pervades all SPE activities, not only in the publication of books, but also in other areas such as sponsorship of technical conferences and educational programs. In addition, the Society publishes periodicals including *Plastics Engineering, Polymer Engineering and Science, The Journal of Injection Molding Technology, Journal of Vinyl & Additive Technology* and *Polymer Composites* as well as conference proceedings and other publications, all of which are subject to rigorous technical review procedures.

The resource of some 36,000 practicing plastics engineers, scientists, and technologists has made SPE the largest organization of its type worldwide. Further information is available from the Society at 14 Fairfield Drive, Brookfield, Connecticut 06804, USA.

Michael A. Cappelletti
Executive Director
Society of Plastics Engineers

Technical Volumes Committee:
Robert C. Portnoy, Chairperson
Exxon Chemical Company

Reviewed by:
Dr. Billy Chow
IBM Corporation

Preface

When plastics first emerged as a viable material from which products could be manufactured, parts were designed and assembled in the same manner as those of existing materials. In time, the unique characteristics of this family of materials and the processes used in their manufacture became known. Creative designers and engineers soon recognized that plastic products could be marketed at substantially reduced price over those made of traditional materials.

A large portion of the reduced costs achieved through the use of plastics in the design of products is attained through a reduction in the number of parts required and the efficiency attained by many of the joining methods used with these materials. However, it is more complicated to design with plastics than with metals because the physical properties of plastics are significantly affected by changes in temperature and chemical environment within the normal range of usage. Furthermore, the available joining methods vary with the material and with the processing method used to create the parts.

In his design and engineering practice over the years, the author has observed that the relationship between the assembly methods, the materials, and the plastics manufacturing processes is generally neglected in the available references. Rather, it is left for the readers to discover through experience. In this book, the assembly method limitations for a given molding process can be found in Chapter 6, "Assembly Method Selection by Process." Chapter 5, entitled "Assembly Method Selection by Material," will provide the methods that can be used with a given plastic resin along with the acceptable fitment tolerances where available. The assembly processes also have size limitations, which are addressed in the respective chapters.

The theoretical ultimate in assembly method efficiency is the complete elimination of all joining operations in the creation of a one-piece product. Part reduction not only eliminates assembly labor, it reduces the purchasing, inspecting, warehousing, capital requirements, and piece part costs as well. While the ultimate goal is rarely achieved, a significant reduction in the number of parts used can often be attained. Most often, this objective is attempted by combining design elements in the process currently being used to manufacture the product. This sometimes results in a part so complicated that it is more expensive to make than the ones it replaced — with their assembly cost included. Many of the author's most successful cost reduction programs have involved changes in both process and material, and he felt it important that this approach be included in the book. There is also a table to provide guidance to the range of product size attainable with each process.

The book also contains sections on the design of assemblies for disassembly and recycling. In addition, it provides the basic design for manufacturability fundamentals necessary to create parts which are not warped or distorted such that they cannot be efficiently assembled. Concurrent engineering practices are further developed to a level we refer to as the "holistic design of plastic parts."

Following the design chapters, there are the assembly method chapters which basically constitute most of the balance of the book. Reference value is enhanced by a full chapter devoted to each of the 14 principal fastening and joining methods used to assemble plastic products today. This is very much a "how to" book, with a great deal of hard-to-find detailed design information and a large number of illustrations. It is intended to be both a handy desk reference and a design guide.

In the accelerated pace of today's design environment, engineers rarely have the time to read at leisure. The author often finds himself reaching for a reference book and looking for the shortest path to the information needed. When he scans the table of contents of a book, he is looking to see if it is the kind of book he can use in that manner. He believes other engineers do the same, so this book is designed to accommodate the practice. In Chapter 1, "Rapid Guidelines for the Assembly of Plastics and the Efficient Use of This Handbook," the reader can scan the various joining methods and determine which ones are most likely solutions for the problem at hand. He or she can then go on to read only those chapters. To further this objective, each of the assembly chapters begins with a list of the advantages and disadvantages of that particular assembly method. The author hopes this effort results in quick and appropriate joining solutions for his readers.

Highland Park, Illinois   Jordan Rotheiser

Acknowledgments

The author would like to first thank Glenn Beall of Glenn Beall Plastics Ltd., who introduced him to the world of plastics many years ago, who provided much reference material, and from whom he has learned a great deal about plastics over the years. He further wishes to express special appreciation to the following gentlemen who took the time and trouble to provide peer reviews of the chapters pertaining to their areas of special expertise:
Paul Burleigh (Forward Technology Industries, Inc.)
Mark Caldwell (Sonics and Materials, Inc.)
Jeffrey Franz (Branson Ultrasonics Corp.)
David A. Grewell (The Ohio State University)
Michael Luehr (Hermann Ultrasonics)
Herbert Mikeworth (Dukane Corp.)
Kishor Mehta (Bayer)
Jim Nordgren (3M)
Don Schewe (Forward Technology Industries, Inc.)
Dr. Zan Smith (Ticona)
Michael Topping (Ashland Chemical)

In addition to those listed in the reference section of this book, the author further wishes to express his appreciation to those who provided information and assistance in obtaining information used in this book:
J. Andrew Besuyen (Branson Plastic Joining)
John Bottelle (Emhart Helicoil)
Steve Chookazian (Ashland Chemical)
Ed Collins (Kamweld Products Co., Inc.)
Shawn Dalton (Service Tectonics Corp.)
Russell DiLuciano (V and A Process)
Jonathan Gorbold (Ameritherm Inc.)
Steve Ham (Steve Ham Plastics)
Peter Hebert (Forward Technology Industries, Inc.)
Steven A. Kocheny (Leister Technologies, LLC)
Junusz Lachowski (Sonics and Materials)
Tom Hoyer (Tom Hoyer and Associates)
Nicholas Nagurny (Montell North America)
Robert T. Ruffini (Fluxtrol Inc.)
Jerry Zybko (Leister Technologies, LLC)

Finally, the author wishes to express his appreciation to those whose work is listed in the reference section of this book and to the many others whose efforts in the plastics industry over the years have contributed to this book in anonymity.

Contents

Introduction

Ever since early man first figured out how to attach a spearhead to a tree limb, assembly has been one of mankind's principal endeavors. Initial efforts consisted mainly of tying the parts together, largely with leaves, roots, vines, or strips of rawhide. Quality became an issue the first time a spearhead came off a hunter's shaft in the middle of a struggle.

Other methods of joining soon followed. Man learned that some materials stuck to each other, and mortar and adhesives came into use. Eventually he discovered that parts fitting tightly together needed no adhesives at all. This led to the development of dowels and, eventually, screws.

Somewhere along the line, he discovered that metals could be more readily shaped if heated. Taken to a high enough temperature, they would actually weld together. By the nineteenth century, all the means of assembling traditional materials were in place.

Then came plastics.

Here was a new group of materials; at once both wondrous and frightening. Initial applications were simply as substitutes for difficult to obtain natural materials. To this day, many plastics applications fall into this category. It was many years before engineers began to recognize these materials for their own unique properties. This understanding is important, for, without it, it is impossible to achieve their fullest utilization. The most outstanding of these properties are as follows.

1. *Transparency with flexibility*. The traditional transparent material is glass. However, this material is very brittle and breaks readily upon impact. It is also quite stiff and cannot be readily bent without cracking. While glass can be formed, the process requires very high temperatures.

A significant number of plastic materials in the styrenic, polycarbonate, acrylic, cellulosic, and polyolefin families offer varying degrees of transparency with flexibility and ductility. In fact, acrylics are capable of offering even greater transparency than the finest lead glass. These materials can be formed at relatively low temperatures and by a group of processes unique to the plastics industry. They can permit shapes that are impossible to duplicate with glass processes. Not only do these attributes provide product design opportunities, they vastly increase the palette of assembly techniques available to the fabricator.

2. *Integral coloring*. Painting and finishing of traditional materials are highly developed technologies. Indeed, on some occasions (e.g., when an identical color match is required) they are used on plastics as well. However, painting is an expensive process and one in which consistency is difficult to maintain. Nearly all the plastics materials and processes permit a vast spectrum of integral coloring. Not only is there substantial saving in cost, but surface abrasions do not remove the paint to reveal a different colored substrate. Thus, over time, a product improvement is provided.

Integral coloring has some impact on assembly considerations as well. Color matching between parts is accomplished readily when the parts are of the same material and with varying degrees of precision when they are of dissimilar materials. Multipart injection molding permits parts of different colors or materials to be molded together, thus eliminating an assembly operation. Coextrusion permits two-color extrusions, which can be used alone, as in striped soda straws, or in combination with another process, such as blow molding or thermoforming. For example, a package might be thin-gauge thermoformed from a coextruded sheet such that the cover is one color and the base is another. Such a design would eliminate attaching the assembly operation of the cover to the base. A clear material could be substituted for one of the colors, thus creating a package with an opaque base and a clear top.

3. *Corrosion resistance.* Plastics simply do not corrode. Aside from the obvious benefits for outdoor applications, there are assembly advantages as well, since galvanic action between metal components is not a concern.

4. *Highly complex integral shapes.* Short of hand carving out of a solid block, none of the processing methods associated with traditional materials can come close to creating the kind of complex integral shapes that the plastics processing methods can accomplish. In addition, most of the wastage from cuttings associated with traditional materials is avoided. From an assembly standpoint, the use of plastics permits the combination of many parts into one, thus eliminating many assembly operations altogether. Furthermore, changes in wall thickness can create some variation in part stiffness.

5. *Extreme adaptability.* When the volumes warrant, the chemical composition of plastics often allow modification to achieve specific characteristics. Thus, the most economical set of properties can be combined to optimize the material utilization. For example, the stiffness of a given plastic might be reduced just enough to permit the arm of a device to deflect adequately to perform a given function. This, in turn, could eliminate an entire mechanism composed of rigid arms and springs, thereby saving a whole series of assembly operations.

6. *Weight reduction.* With the exception of a few woods, such as balsa, plastics generally weigh far less than natural materials. This advantage has been the driving force for many applications. For example, increased use of plastics has enabled automobile manufacturers to reduce the weight of each vehicle by approximately 25% in recent years, with corresponding improvements in fuel economy. In some cases, the weight reduction associated with the use of plastics can be excessive and weight must be added to a product to provide the proper ergonomic "feel" required for comfortable use.

7. *Insulation qualities.* Plastics can provide both thermal and electrical insulating properties. Many plastics provide both, permitting the elimination of parts and assembly operations. A plastic housing for an electric drill is an excellent example of such an application. These savings may be negated to some degree when additional measures must be employed to provide for EMI/RFI (electromagnetic interference/radio-frequency interference) shielding or heat dissipation.

8. *New assembly techniques.* While flame welding cannot be recommended, the traditional methods of assembly such as press fits, adhesive joining, and the use of fasteners are readily available for plastics. In addition, parts made of plastic can also be snap-fitted, staked, ultrasonically welded, induction-welded, hot-die-welded,

hot-gas-welded, spin-welded, vibration-welded, and solvent-welded. Threads and threaded inserts can be molded right into the part or added as a secondary operation. New processes, such as focused infrared welding, are under development.

These capabilities of plastics have been among the driving forces in the success of the plastics industry. They have played a large part in the enhancement of our standard of living. While it is still possible to purchase an automobile manufactured by traditional hand methods, the cost of such vehicles is 10 to 15 times that of the average car. Few can afford to pay that price.

While everyone is eager to enjoy the benefits of cost reductions achievable through the use of plastics, it is important to recognize that one cannot take an object designed to be made of metal, for example, and simply convert its manufacture to plastic. Most engineers schooled in traditional engineering institutions find it difficult to fully understand the peculiarities of the plastics medium. Thus, their projects encounter development problems that could be avoided, or they fail to take full advantage of the capabilities of the medium. The most difficult to comprehend are as follows:

1. *The scope of the materials known as plastics.* Plastics are a broad category of materials that can be divided into some three dozen families, each of which may have hundreds, if not thousands of individual compounds. While these may seem vaguely similar to the casual observer, they can be truly unlike one another. Polycarbonate is no more like polyethylene than steel is like tin or oak is like balsa. Some are elastomers like rubber and others are as strong as iron. Therefore, anyone intending to work with plastics should acquire a reasonable understanding of the scope of properties for each of the families. It is not necessary to be knowledgeable about all the individual compounds, since the bulk of them fall in the category of specialized versions developed to compete for a particular market. It is sufficient to have enough general knowledge of the customary applications and relative physical, thermal and electrical properties for each family to know where to look for the precise properties needed for a given application. Familiarity with the specifics of the compounds within many of the families improves with experience.

Note that we use the term "compound" when referring to the individual materials. That is because most of them contain additives beyond the basic polymer. Among these are fillers, colorants, reinforcements, antioxidants, ultraviolet light inhibitors, and lubricants. Variations in the additives are one of the principal reasons for the availability of so many compounds. The term "resin" is also in common use.

2. *The general nature of plastics.* Traditional materials such as glass, stone, and brick tend to fail in a highly brittle fashion. They have a stress–strain curve that is nearly vertical and almost no strain to speak of. Metals and woods are less brittle but still tend to approach their ultimate strengths before they start to exhibit significant strain. In contrast, plastics generally exhibit a great deal of strain before reaching their ultimate strength. Their curves tend to be relatively low, with virtually no flat portion. Therefore, a Young's modulus is difficult to come by, and a secant modulus must be used as a substitute.

Low cost plastics used in large quantities, often referred to as "commodity plastics," tend to have strength and stiffness properties far below those of metals. Therefore, geometrical configuration and wall thickness play a much larger role in determining the structural integrity of the product. Consequently, these characteristics become significant in the assembly aspects of the products. The more expensive plastics, which offer higher physical properties and generally are referred as "engineering plastics," also exhibit this behavior, although in a less pronounced fashion.

3. *The thermal characteristics of plastics.* Generalizing again, plastics' properties change markedly with changes in temperature. Of course, so do the properties of metals; but they do so at such high temperatures that this characteristic is not relevant for most applications. The properties of plastics, however, change at temperatures in the range of normal human living conditions. Data sheets usually list physical properties at 72 °F. However, properties such as tensile strength and stiffness can drop off noticeably by the time the temperature has reached 100 °F. As temperatures drop, ductility drops noticeably; indeed, plastics normally thought of as relatively ductile at room temperature may actually be quite brittle when the temperature drops below freezing. Experienced engineers learn to take these characteristics into account and to use them to advantage when designing assemblies of plastic parts. One part can be heated to fit over another and then cooled to form a tight fitment, for example, or, conversely, one part can be cooled in a freezer to permit it to be fitted into another part. Upon returning to room temperature, the parts are inseparable.

Failure to properly account for the difference in the coefficient of linear thermal expansion (CLTE) between two mating parts has been the undoing of many a plastic assembly. In general, the difference in CLTE between most plastics and most metals can be quite pronounced. This can lead to significant changes in the amount of stress on a given joint over the functional temperature range. The result may be stress cracks or disassembly of the product.

Finally, there is the matter of thermal degradation. Plastics lose their physical properties with prolonged exposure to heat. At 72 °F, this can take years. However, at elevated temperatures, degradation can occur in minutes. Unfortunately, each time a thermoplastic is elevated, some damage occurs. Hence, a material can be reground and reused only a limited number of times. For some critical applications, only fresh resin, known as "virgin," can be used. This characteristic must be taken into account when one is designing with plastics. In certain cases, some form of heat shielding is required.

4. *Flammability.* Flammability is defined as relative ease of ignition and ability to withstand combustion. Materials behave quite differently when exposed to open flame, and standards for flammability are difficult to establish. While a few polymers do not support their own combustion, it is fair to state that in general, plastics burn — most of them, quite readily. Flame retardants can be used, but they are expensive and affect other properties as well. If fire is a major risk, or if the application entails the possible presence of an open flame, the flammability of the selected resin must be carefully examined and the use of shielding considered.

5. *Ultraviolet light.* Ultraviolet light causes or catalyzes chemical degradation in many plastics. The result is photooxidation, which leads over time to the loss of color, transparency, and physical properties. Some polymers exhibit a natural resistance to ultraviolet rays. Others require barrier coatings or additives, such as ultraviolet stabilizers and antioxidants, for prolonged life in environments characterized by the presence of high concentrations of ultraviolet light (i.e., outdoors).

6. *Cost*. Plastics are generally thought of as inexpensive. In reality, on a price per pound basis, they are generally more costly than competing materials. Often, far more expensive. However, the competing materials are normally processed by methods that result in large amounts of unusable scrap. Conversely, most plastics processes utilize all, or nearly all, of the material, with little or no scrap. In addition, the processes enable the creation of forms that combine multiple parts into one. This, plus the light weight of plastics, results in a low total product cost. Thus, plastics are actually expensive materials used in a very economic fashion.

7. *The chemical characteristics of plastics*. It is easy to forget that plastics are essentially chemical in nature and, consequently, most of them are significantly affected by chemical exposures and environments. This can be a desirable quality, since many plastics can be joined by solvent welding. Unfortunately, it can also lead to disaster when the potential for chemical exposure is not taken into account. Cleaning solutions have led to the downfall of many a plastics application. However, various greases, oils, acids, bases, gases, and other chemicals have taken their toll as well. Many of these exposures occur in the field far from the eyes of the manufacturer, hence are blind to the engineer. The prudent designer must investigate the potential chemical exposures carefully to avoid waking up one day to reports of failures in the field.

The author recalls a project early in his career where a medical product made of an ABS was introduced to the market. Since the product was gas-sterilized, little thought was given to effect of alcohol on this particular ABS (which is no longer on the market). Among the users of the product, however, were nurses who had been trained to wipe everything with alcohol. The combination of alcohol and molding stresses turned out to be disastrous for the product, and reports of cracks soon began to filter back to the manufacturer. Fortunately, it was possible to switch to another ABS and the product was salvaged.

This story had a happy ending. Others have not. However, with the proper knowledge of plastic assembly properties and the exercise of due diligence, the engineer need not be unduly concerned with the hazards of using plastics.

1 Rapid Guidelines for Joining of Plastics and Efficient Use of This Handbook

1.1 Efficient Use of This Handbook

The author of this book is a consultant who charges a fee based on the time required to perform the work. Often, by the time he is called in, the project is already overdue and over budget. Therefore, he must use his time efficiently and be prepared to justify the hours spent. Presumably, at the accelerated pace of today's business environment, other engineers are working under the same pressures. Therefore, *this book is designed to provide the engineer with the shortest path possible to the selection and implementation of the appropriate plastics assembly method.*

Initial direction can be established by referring to Section 1.2, "Rapid Guidelines for Assembly of Plastics." This is a cursory review of the various assembly methods, which the engineer can use to determine those that merit further investigation relative to the project at hand. Chapters addressed to each process are provided in alphabetical order commencing with Chapter 7, "Adhesive and Solvent Joining." Effort has been devoted to providing detailed practical design information wherever possible. Each chapter begins with the advantages and disadvantages of the process. Lists of suppliers for these processes with their addresses and phone numbers follow each chapter where appropriate.

There are two other quick reference chapters. Chapter 5, "Assembly Method Selection by Material," has two purposes. One is to permit the engineer to quickly check a suggested material (they are listed in alphabetical order) for the assembly methods appropriate for it. A reader who needs to know if it is feasible to solvent seal polyethylene, would look here (it is not feasible). The other is to allow the reader to scan the materials in a search for those that might be of interest for a given application. To this end, the chapter includes a brief summary of its property ranges (those most pertinent to assembly), general characteristics, and examples of some of its applications.

The other quick reference chapter is Chapter 6, "Assembly Method Selection by Process." It is designed to permit the engineer to quickly check a plastic process under evaluation for its assembly considerations. An engineer who needs to know if it is possible to create an efficient ultrasonic energy director with the blow molding process, would look here (it is not possible). The chapter also briefly describes the principal processes used to manufacture plastic products, to permit the reader to quickly scan for those that might be of value to the suggested application.

There are three design chapters devoted to the essential considerations for the assembly of plastics. Chapter 2, "Designing for Efficient Assembly," addresses the problem of providing parts for assembly that are not out of tolerance or distorted to the point of making assembly difficult or impossible. It also discusses industry design practices and methods of designing parts with looser tolerances which cost less to produce. This is the basic "blocking and tackling," which will be invaluable to engineers who only occasionally design plastic parts, but somewhat redundant to those who live in the plastics world. However, it can provide a useful "refresher course" even for them. A "Plastic Product Design for Assembly Checklist" is provided at the end of this chapter.

Chapter 3, "Cost Reduction in Assembly," raises the issue of concern to everyone involved in the manufacture of products today. Micro, macro, fastener reduction, and holistic approaches to cost reduction in plastic parts are addressed.

The increasing affluence of the world and the efficiency with which it is producing plastic product combined with the dramatic drop in available landfills can result in nothing less than increasing stress on reuse of resources. Chapter 4, "Design for Disassembly and Recycling," is devoted to this topic which all plastics engineers will need to deal with sooner or later. Better sooner than later.

1.2   Rapid Guidelines for Assembly of Plastics

The following very general guidelines are intended to provide direction for further reference, not solutions.

1.2.1   Adhesives (Chapter 7)

1.2.1.1   Liquids: Solvent-Based, Water-Based, and Anaerobic Adhesives

Reputation First choice for sealing thermosets, dissimilar plastics, and plastics to other materials that cannot be sealed by other means. Extra cost item. Must compete with fasteners and snap fits for simple joining of these materials. Can be used to reinforce screws and inserts. (Inserts are usually installed in thermosets with adhesive reinforcement if they are not molded-in.)

Description Solvent- and water-based liquid adhesives, available in a wide number of bases (e.g., polyester, vinyl) in one- or two-part form, fill bonding needs ranging from high speed lamination to one-of-a-kind joining of dissimilar plastics parts. Solvents provide more bite, but cost much more than similar-base, water-type adhesives. Anaerobics are a group of adhesives that cure in the absence of air, with a minimum amount of pressure required to effect the initial bond. Adhesives are used for almost every type of plastic.

Advantages Good repeatability; can join dissimilar materials. Easy to apply; adhesives available to fit most applications.

Limitations Adjustable speed (varies with adhesive, but can be slow). Surface preparation may be required. Shelf and pot life often limited. Solvents may cause pollution problems; water-based adhesives not as strong; anaerobics toxic.

Processing limitations Application techniques range from simply brushing on to spraying and roller coating-lamination for very high production. Adhesive application techniques, often similar to decorating equipment.

Anaerobics are generally applied a drop at a time from a special bottle or dispenser.

1.2.1.2 Mastics

Reputation First choice for sealing applications.
Description Highly viscous single- or two-component materials, which cure to a very hard or flexible joint depending on adhesive type.
Advantages Do not run when applied.
Limitations Shelf and pot life often limited. Not good for polyolefins.
Processing considerations Often applied via a trowel, knife, or gun-type dispenser; one-component systems can be applied directly from a tube. Various types of roller coaters are also used.

1.2.1.3 Hot Melts

Reputation Best adhesives for polyolefins. Good for tacking prior to final joining and high speed production.
Description 100% solid adhesives that become flowable when heat is applied. Often used to bond continuous flat surfaces.
Advantages Fast application; clean operation.
Limitations Hot melts cannot produce structural joints.
Processing considerations Hot melts are applied at high speeds via heating the adhesive, then extruding (actually squirting) it onto a substrate, roller coating, using a special dispenser or roll to apply dots, or simply dipping.

1.2.1.4 Pressure-Sensitive Adhesives

Reputation Quick bond, but light duty; used for adhesive tapes.
Description Tacky adhesives used in a variety of commercial applications. Often used with polyolefins.
Advantages Flexible.
Limitations Bonds not very strong.
Processing considerations Generally applied by spray; bonding is effected by light pressure.

1.2.2   Fasteners and Inserts (Chapter 8)

Reputation The method to go to for reliable openability, unusual shapes, and limited volumes. Extra cost item – must compete with snap fits for injection-molded applications. Excellent for dissimilar materials. These and adhesives are the principal assembly means for thermosets.

Description Self-tapping screws, machine screws, threaded inserts, rivets, and a variety of specialty fasteners designed particularly for plastics. Devices are made of metal or plastic. Type selected will depend on how strong the end product must be; appearance factors. Often used to join dissimilar plastics or plastics to non plastics.

Advantages Adaptable to many materials; low to medium costs; can be used for parts that must be disassembled.

Disadvantages Some have limited pullout strength; molded-in inserts may result in stresses.

Processing considerations Nails and staples are applied by simply hammering or stapling. Other fasteners may be inserted by drill press, ultrasonics, air or electric gun, or hand tool. Special molding (i.e., molded-in-hole) may be required.

1.2.3   Hinges (Chapter 9)

Reputation The method to use for repeated reopenability.

Description Traditional means of assembly; however, the use of plastics permits hinge components to be molded directly into the part. One-piece integral hinges are possible in some materials.

Advantages One-piece integral hinges require only closing as an assembly operation. Most other plastic hinges require no additional parts.

Disadvantages In general, design must be carefully done and additional tooling costs are involved.

Processing considerations No additional processing is required; however, molding cycles may be altered.

1.2.4   Hot Plate/Hot Die/Fusion and Hot Wire/Resistance Welding (Chapter 10)

Reputation These processes weld nearly any thermoplastic but have highest operating cost. Compete for applications with vibration welding, which has lower operating cost, but higher equipment cost. The techniques to go to when lower cost systems cannot be used.

Description Mating surfaces are heated against a hot surface, allowed to soften sufficiently to produce a good bond, then clamped together while bond sets. Applicable to rigid thermoplastics.

Advantages Can be very fast (4–10 s in some cases); strong bonds. Excellent repeatability. Flash provides additional bonding area.

Disadvantages Stresses may occur in bonding area. Problems with high-temperature-resistant materials. Continuous heating of platen results in high operating cost. Considerable flash.

Processing considerations Simple soldering guns and hot irons are used; relatively simple hot plates attached to heating elements up to semiautomatic hot plate equipment. Clamps needed in all cases.

1.2.5   Hot Gas Welding (Chapter 11)

Reputation The method to use for thermoplastic welds of very large parts, unusual shapes, and field repairs.

Description Welding rod of the same material being joined (largest application is vinyl) is softened by hot air or nitrogen as it is fed through a gun that is softening part surface simultaneously. Rod fills in joint area and cools to effect a bond.

Advantages Strong bonds, especially for large structural shapes. Equipment is inexpensive.

Disadvantages Relatively slow; weld is not aesthetically pleasing without substantial expenditure on finishing. Commercial welding rod available for limited number of polymers.

Processing considerations Requires a hand gun, special welding tips, an air source, and a welding rod. Regular hand gun speeds run 6 in./min; high speed handheld tool boosts this to 48 to 60 in./min.

1.2.6   Induction Welding (Chapter 12)

Reputation Excellent hermetic seals in most materials, but insert is extra cost item. Unusual in providing a welded joint that can be x-rayed for quality assurance and reopened.

Description A ferromagnetic insert, metal insert, or screen is placed between the parts to be welded and energized with an electromagnetic field. As the insert heats, the parts around it melt, and, when cooled, form a bond. For most thermoplastics.

Advantages Provides rapid heating of solid sections to reduce chance of degradation. Can x-ray joint to check weld. Reopenable for disassembly.

Disadvantages Extra cost of preformed electromagnetic insert. May cause stress at bond because metal embedded in plastic.

Processing considerations High frequency generator, heating coil, and inserts (generally 0.02–0.04 in. thick). Hooked up to automated devices, speeds are high (1–5 kW used). Work coils, water cooling for electronics, automatic timers, and multiple position stations may also be required.

1.2.7   Insert Molding (Chapter 13)

Reputation The highest strength method of emplacing threaded inserts; particularly desirable for thermosets. The only way to create embediments, internal structural support and some decorative effects. Extra part cost. Sufficiently troublesome from a molding standpoint to be avoided if possible.

Description The insert is placed in the mold and the part is molded around it.

Advantages Very strong – the insert can be removed only by destructive methods.

Disadvantages Inserts disrupt the molding cycle and create high levels of molded-in stress, which can lead to long-term failure. They cost money, contaminate the waste stream, and must be removed for recycling.

Processing considerations Can be molded in several processes. The inserts are placed on pins or in holes in the mold and the mold closes on them. The part is then formed or molded around the part capturing it within.

1.2.8   Multipart Molding (Chapter 13)

Reputation The most indelible means of molding a second color. The only way to make some designs such as pen barrels with soft sections for gripping comfort. Expensive tooling.

Description A form of insert molding in which the insert is first molded as part of the process.

Advantages Combines design freedom and permanence with high production capability.

Limitations Has very high tooling cost.

Processing considerations The first part is molded and the second part is molded over or under it. This requires special molding equipment that is limited in availability.

1.2.9   Press Fits/Force Fits/Interference Fits/Shrink Fits (Chapter 14)

Reputation Heavy duty – good way to assemble a plastic gear to a steel shaft. Light duty – excellent for pen caps.

Description One part is pressed or forced into the other such that the force resulting from the interference keeps them in place.

Advantages Can assemble dissimilar materials without tools or additional materials. Permanent or reopenable depending on design.

Disadvantages Creates high levels of hoop stress, which can cause failure in time due to creep, cold flow, stress relaxation, thermal relaxation, or moisture absorption. Process and shape limitations.

Processing considerations Easily accomplished with a minimum of tools. Light duty is readily automated.

1.2.10   Solvent Joining (Chapter 7)

Reputation Quick and inexpensive method of joining small parts made of suitable polymers. Also useful for fabrication of prototype or low volume production runs.

Description Solvent softens the surface of an amorphous thermoplastic; mating takes place when the solvent has completely evaporated. Bodied cement with small percentage of parent material can give more workable cement, fill in voids in bond area. Cannot be used for polyolefins and acetal homopolymers.

Advantages Strength, up to 100% of parent materials, easily and economically obtained with minimum equipment requirements.

Disadvantages Long evaporation times required; solvent may be hazardous; may cause crazing in some resins.

Processing considerations 'Equipment ranges from hypodermic needle or just a wiping medium to tanks for dip and soak. Clamping devices are necessary, and air dryer is usually required. Solvent recovery apparatus may be necessary or required. Processing speeds are relatively slow because of drying times. Equipment costs are low to medium.

1.2.11 Snap Fits (Chapter 15)

Reputation Inexpensive way to assemble parts once the tooling has been paid for.

Description One part is pressed into the other beyond its snap fitment, which restricts its ability to be reopened.

Advantages Can assemble dissimilar materials without tools or additional materials. Permanent or reopenable depending on design.

Disadvantages Process limitations, shape limitations, and high tooling cost. Can fail in time owing to thermal expansion and moisture absorption. Creep, cold flow, stress relaxation, and environmental exposures can cause snap fits under load to fail in time.

Processing considerations Easily accomplished with a minimum of tools and readily automated.

1.2.12 Spin Welding (Chapter 16)

Reputation Quick and inexpensive way to permanently join round thermoplastic shapes.

Description Parts to be bonded are spun at high speed, developing friction at the bond area; when spinning stops, parts cool in fixture under pressure to set bond. Applicable to most rigid thermoplastics.

Advantages Very fast (as low as 1–2 s); strong bonds. Uses no additional parts and can join materials and shapes which ultrasonic welding cannot weld.

Disadvantages Bond area must be circular. Difficult to register.

Processing considerations Basic apparatus is a spinning device, but sophisticated feeding and handling devices are generally incorporated to take advantage of high speed operation.

1.2.13 Staking/Swaging/Peening/Cold Heading/Cold Forming (Chapter 17)

Reputation Quick, inexpensive way to join two parts of dissimilar materials, if one is a thermoplastic, with no additional materials.

Description One part is fitted to the thermoplastic part, which has a design detail that can be formed over to capture the first part.

Advantages Can assemble parts of dissimilar materials rapidly with loose tolerances and no additional materials such as adhesives or fasteners. Permanent assembly but easily disassembled for recycling.

Disadvantages Tendency toward recovery of original shape, shape limitations, poor appearance, and limitations on use of fillers.

Processing considerations The parts are fitted to each other, and the tool forms the plastic. Equipment is relatively inexpensive.

1.2.14 Threads – Molded in (Chapter 18)

Reputation Mainly for bottle caps.
Description Screw threads are molded directly into the part.
Advantages Strongest possible thread; no additional thread-forming operation, and thread can be modified to suit plastic requirements.
Disadvantages Expensive and slow running molds.
Processing considerations Basic molding operation, but runs slower because time is required to strip or unthread the molds.

1.2.15 Threads – Tapped (Chapter 18)

Reputation For some precision threaded situations and reopenable low volume applications.
Description Hole is drilled or molded and then tapped as in metals.
Advantages Most precise location possible. Reopenable; can assemble dissimilar materials; familiar technology; shape and process freedom.
Disadvantages Thread form not well suited to plastics. Subject to creep, cold flow, thermal expansion, stress relaxation, and environmental effects. Slow to assemble.
Processing considerations Operator must be aware of specific speeds and feeds appropriate for each polymer.

1.2.16 Ultrasonic Welding (Chapter 19)

Reputation Inexpensive workhorse method of permanently welding small to medium size injection-molded thermoplastic parts.
Description High frequency sound vibrations transmitted by a metal horn generate friction at the bond area of a thermoplastic part, melting plastics just enough to permit a bond.
Advantages Strong bonds for most thermoplastics; energy efficient and fast. Uses no additional materials and can be handled immediately.
Disadvantages Size and shape limited for continuous welds. Does not weld all thermoplastics well. Limited application beyond injection-molded parts.

Processing considerations Converter to change electrical into mechanical energy is required, along with stand and horn to transmit energy to part. Rotary tables and high speed feeder can be incorporated.

1.2.17   Vibration Welding (Chapter 20)

Reputation Next place to go if ultrasonic or spin welding is not applicable.

Description Melting of plastic is achieved by friction in the joint area under defined pressure.

Advantages Greater shape and material freedom than ultrasonic or spin welding, but still makes permanent welds at high production rates.

Disadvantages High equipment cost and imprecise alignments. Cannot capture delicate components which could be damaged by vibrations.

Processing considerations Welding is horizontal with frequencies between 120 and 300 Hz. The whole part is physically moved.

Processing considerations Welding is horizontal with frequencies between 120 and 300 Hz. The whole part is physically moved.

1.2.18   Welding with Lasers (Chapter 21)

Reputation The technique to consider when the application requires non-contact, extremely precise, flash-free welding beyond the capability of the traditional methods. It is the only technique that can weld elastomers.

Description Laser welding is infrared welding using lasers as the energy source. Lasers are used directly as the heat source for staking and butt welding. However, the technique that makes it unique is Through Transmission Infra-red Laser Welding in which the laser passes through the laser-transparent upper part to the surface of the laser absorbent lower part and the weld occurs at the interface of the two parts.

Advantages The most precise of all the welding methods with heat confined to a minimal internal area. No marring of the surface and no flash or loose particulates.

Disadvantages Although progress in these two areas has been dramatic, material limitations and high equipment cost continue to limit the number of applications.

Processing Considerations Special laser welding equipment is required. Parts made from virtually all the thermoplastics processes can be laser welded.

1.3   Assembly Methods Selection by Size

Size is one criterion often forgotten in the preliminary selection of an assembly method. Most methods can cope with very small parts where the problem of handling of the parts determines the minimum size for which the technique be used. However, there are some limits in that direction. The maximum size is another matter. In many cases, it is controlled by the maximum size of the equipment. In others, it is limited

Table 1-1   Application Size Ranges for Assembly Techniques

	Smallest application	largest application
Adhesives joining	No limit (1)	No limit (1)
Fasteners and inserts	No limit (1)	No limit (1)
Hot die	25.4 mm dia. (1 in. dia.)	508 × 1829 mm (20 in. × 72 in.)
Hot gas	2.3 × 2.3 mm (0.09 in. × 0.09 in.)	No limit (1)
Induction heating	No limit (1)	127 × 508 mm (5 in. × 20 in.) 25.4 × 2032 mm (1 in. × 80 in.)
Insert molding	No limit (1)	No limit (2)
Laser welding	2 mm (0.079 in.) × 5 mm (0.197 in.)	250 mm (9.843 in.) × 400 mm (15.748 in.)
Multi-part molding	No limit (2)	No limit (2)
Press fits	0.80 mm dia. (0.031 in. dia.)	No limit (2)
Snap fits	1.0 mm (0.040 in.)	No limit (2)
Solvent welding	0.80 mm dia. (0.031 in. dia.)	No limit (1)
Spin welding	1.27 mm dia. (0.50 in. dia.)	356 mm dia. (20 in. dia.)
Staking	0.80 mm dia. (0.031 in. dia.)	No limit (2)
Swaging	0.80 mm dia. (0.031 in. dia.)	No limit (2)
Threads – tapped and Molded-in	m4 × 0.8 (6-32)	No limit (1)
Ultrasonic welds	6.35 mm × 6.35 mm (0.25 in. × 0.25 in.)	254 mm × 305 mm (10 in. × 12 in.)
Inserts	2.5 × 0.45 (4-40)	No limit (1)
Spot welds	0.80 mm dia. (0.031 in. dia.)	No limit (1)
Stake head	1.60 mm dia. (0.63 in. dia.)	No limit (2)
Vibration:		
Linear	25.4 mm × 25.4 mm (1 in. × 1 in.)	1016 mm × 1829 mm (40 in. × 72 in.)
Orbital	25.4 mm × 25.4 mm (1 in. × 1 in.)	304.8 mm dia. (12 in. dia.)

(1) Restricted only by handling limitations
(2) Restricted only by molding press limitations

only by the limitations of handling large pieces for joining. Some assembly methods are limited more by the molding process that will be needed to make its details than by any other factor. The mere fact that the application in question may fall within the extreme limits of the process should not lead to the presumption that the largest equipment will be affordable or readily available. A quick guide to the approximate limits of the assembly processes is provided in Table 1-1. It should be noted that there may be some disagreement, as to the precise sizes since the author's resources were not in complete agreement.

1.4   Assembly Methods Selection by Joining Time

A comparison of welding time for some of the welding techniques is provided in Fig. 1-1. It is difficult to rate assembly methods by joining time because this parameter is highly application sensitive. Furthermore, methods like press fits and snap fits have no processing time at all; only the time required for handling must be considered. Indeed, handling of the parts is an often overlooked aspect of design, as well as assembly process selection, because it is assumed that the handling time will not vary between processes. This is not true and the handling time for an adhesive joining will often be far greater than that of a method like snap fitting.

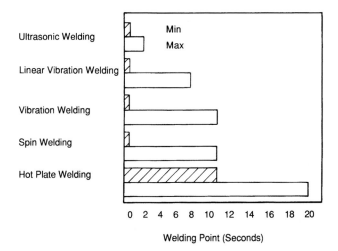

Figure 1-1   Welding method time comparison

2   Designing for Efficient Assembly

2.1   Avoiding Part Distortion

The most common problem encountered in plastic part assembly is parts that do not fit
together properly. In most cases, this problem is due to parts that are distorted or out of
tolerance. It is solvable with careful molding and skillfully designed parts, assemblies,
and molds.

While inadequate cooling before ejection can cause part distortion, the problem is
fundamentally due to nonuniform cooling of the part following injection of the
material into the mold. This can result from a number of conditions centering
around the tool design and molding parameters. These topics are beyond the scope
of this book, although a discussion on the fundamentals is included in the chapter
on tooling. However, the author recommends the following books for a more in-
depth discussion of the topic.
Injection Molds and Molding, Joseph B. Dym, Van Nostrand Reinhold
Plastics Mold Engineering, J. H. DuBois and W. I. Pribble, Van Nostrand Reinhold

In many cases the design itself is the cause of distortion. Nonuniform cooling will result
from nonuniform wall thicknesses as thinner sections cool and set up before thicker ones.
Figure 2-1a illustrates wall sections that are the height of poor practice. To the left there is
a thick corner section that will clearly require more time to set up than the thin center
section. To the right is a thicker section, that will also require more time to cool. Clearly,
the part will not be received as drawn. The actual molded result will exhibit voids,
distortion, and sink, as illustrated in Fig. 2-1b. All these are symptoms of high levels
of molded-in stress. Assembly to a mating part will be nearly impossible.

There is a rule the engineer can follow to avoid the problem just described. It is as
follows:

> Wall thickness variations greater than 25% will exhibit high levels of molded-
> in stress, resulting in sinks, voids, and distortion.

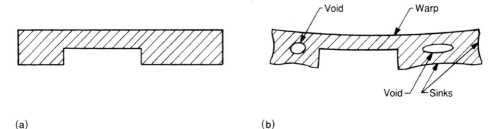

(a)   (b)

Figure 2-1   Distortion due to nonuniform cooling: (a) part as drawn and (b) part as molded

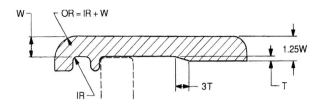

Figure 2-2 Improved design

This statement applies principally to thermoplastics, of which those with high post-molding shrinkage rates are somewhat more vulnerable to this condition than the others. Fillers will reduce the effect of this phenomenon much as they affect shrinkage. Structural foam parts will not experience this condition, and it is significantly reduced in thermosets.

Even when compliance with this rule is maintained, the change in wall thickness should be gradual as shown in Fig. 2-2. When the mold is filled, the flow of material should be from thick section to thin section.

Presumably, the thick wall on the left side of Fig. 2-1 was initially dictated by the need to locate a component. That component is illustrated by the dotted line in Fig. 2-2, where the thick corner has been modified to a uniform wall and a locating rib. The thick wall section has been reduced to the prescribed maximum. Thus the function has been maintained and the source of the nonuniform cooling condition has been eliminated.

2.2   Inside Corner Stress

Elimination of internal stress due to nonuniform cooling does not remove all the stresses resulting from the part contour. Sharp inside radii are another source of such stress. Figure 2-3a illustrates the stress pattern resulting from a sharp corner. Figure 2-3b shows how the stress is distributed in a rounded corner. This effect can be seen when a clear part is observed under polarized light.

The effect of corner radius on stress concentration is clearly illustrated by the curve plotted in Fig. 2-4. Note that the curve rises sharply at the 0.25 level. Therefore, an R/WT ratio of 0.25 should be regarded as a minimum. Note, also, that the curve flattens out beyond the 0.8 ratio. *A ratio of 0.5 should be regarded as optimum.* If the two walls are not equal, the thinner wall may be used for this computation. Reinforced materials are better able to withstand stress than unreinforced plastics. In that case,

Figure 2-3 Stress patterns: (a) corner stress concentration and (b) relieved stress

(a)   (b)

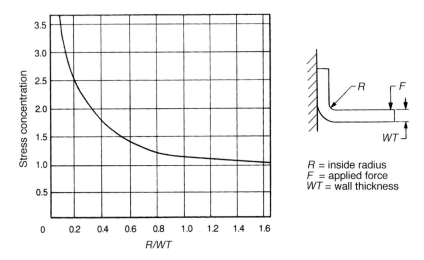

R = inside radius
F = applied force
WT = wall thickness

Figure 2-4   Relationship of inside corner radius to stress concentration factor

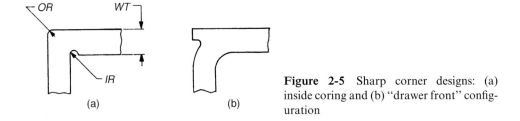

(a)

(b)

Figure 2-5 Sharp corner designs: (a) inside coring and (b) "drawer front" configuration

the curve of Fig. 2-4 can be disregarded to some extent. The inside radius, however, should never be less than 0.50 mm (0.020 in.).

Aesthetic requirements sometimes require sharp outside corners. In those cases, the designs illustrated in Fig. 2-5 can be used to approximate this appearance.

2.3   Ribs and Bosses

The use of longitudinal and box ribbing significantly increases the strength of the part. A method for the application of the basic design rules to the design of ribs is illustrated in Fig. 2-6, where a circle has been drawn at the intersection of the rib and the wall. The circle has three points of contact: at the nominal wall and at each inside radius at the base of the rib. The diameter (D) of this circle must not exceed 1.25 times the nominal wall (W). The same device may be used for the intersection of two ribs, adjacent bosses, a rib and a boss, etc. Care must be taken to identify the thickest condition correctly. For example, in the case of the intersection of two ribs, the circle must be drawn through the diagonal radii.

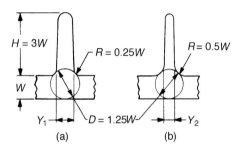

Figure 2-6 The 25% rule for ribs:
(a) $R = 0.25W$ and (b) $R = 0.5W$

In the design of ribs or bosses, whose bases must meet the same criteria, competing factors must be considered. Increasing the inside radius to reduce the stress concentration factor results in less material thickness available for the rib. Figure 2-6a illustrates a rib constructed with an inside radius of $0.25W$. Figure 2-4 indicates that the stress concentration factor for that inside radius is about 2.3. To reduce the stress concentration factor to 1.5, the inside radius would need to be increased to $0.5W$ (Fig. 2-6b). That results in a wall thickness at the base (Y_2) of the rib of $0.5W$, a significant reduction from the original thickness (Y_1). Obviously the rib wall thickness will provide less strength with less material. However, the overall strength provided by the rib may be greater because the internal stress has been reduced. If more strength is required, additional ribs are a better solution than a rib with high levels of molded-in stress.

There is an additional consideration in the design of ribs and bosses: the thinning down of the rib due to draft. Not only does this result in further thinning of the rib, it could result in a rib thickness at the top that is too thin to fill or one that is so thin it can be filled only at elevated temperature and pressure. In short, such thinning could be the driving force in determining the molding conditions, the nature of which increases the likelihood of distorted parts. For example, suppose that the value of W in Fig. 2-6b is placed at 2.5 mm (0.100 in). Then radius r would be 1.25 mm (0.050 in.), resulting in a rib thickness of 1.25 mm (0.050 in.) at its base. The recommended maximum rib height is three times the nominal wall thickness (more will not significantly increase strength). For a rib height H of 7.5 mm (0.300 in.) and a recommended draft of 2°/side, the rib would have a thickness of 0.86 mm (0.34 in.). That would be difficult to fill in the mold with a viscous material, particularly if the rib were located at a point distant from the gate and with a variety of contours in the flow path, which would cause pressure drops.

2.4   Draft

The part shrinks down on the core as it cools. Draft is used to help the part break free of that core. Although the temptation may be present, **do not give serious consideration to the elimination of draft.** If the draft angle is reduced, the locking force becomes greater, increasing the amount of pressure required to remove the part from the tool. The molding cycle will need to be extended to allow the part to cool enough to be ejected. The part must cool until it is rigid enough to withstand the pressure of ejection without

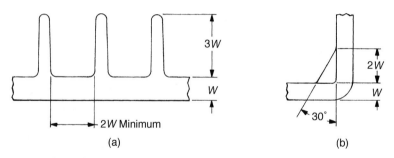

(a)   (b)

Figure 2-7   Ribs and gussets: (a) multiple ribbing and (b) gussets

distortion. This can be 50 to 80% of the molding cycle. The greater the draft, the lower the force necessary to eject the part, and the sooner ejection is possible.

A reduction in draft angle to one degree is a reasonable compromise between these competing demands for most materials. Unfortunately, it is too easy to allow one degree to become a habit, thus ignoring the price it is extracting.

The locking effect described above is increased with ribs because they provide additional gripping surfaces. Multiple ribs are worse; however, they are preferable to a rib without draft or a thick one with high levels of molded-in stress. Figure 2-7a demonstrates the minimum spacing between ribs of twice the nominal wall thickness. It is always desirable to attach ribs to a wall. This permits the gas (air) that is in the mold to escape the rib cavity recess. Freestanding ribs can result in gas traps.

Figure 2-7b illustrates the recommended configuration for gussets. A height greater than $2W$ will not significantly increase the strength. The same spacing rule which applies to ribs $(2 \times W)$ would apply for gussets.

A general guideline of 2°/side is desirable for outside walls as well. The effect of draft is demonstrated in Fig. 2-8. Note the flat angle in section A. Obviously it would have no difficulty separating from core or cavity.

Section B illustrates a 2°/side draft angle. It will separate from the mold, but considerably greater force will be necessary. However, compared to section C, which has no draft, it will be readily ejected. Without draft, section C will require tremendous force to overcome the effects of shrinkage on the core. The part may be too distorted to assemble in addition to needing more cycle time. Also there will be drag marks on the part where there is inadequate draft.

Draft must be present on all surfaces perpendicular to the parting line.

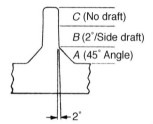

Figure 2-8   Effects of draft

There is, however, one exception to this rule with regard to the leading edge of a locating rib. This is described in Section 2.7.1.

It should be noted that the 2°/side direction represents an increase from the angle recommended for many years, namely, 1°/side. Several reasons account for this change. First, molding cycles have decreased, and there is strong economic pressure to eject parts faster than previously. Second, consumer expectations have been elevated by competition. Parts with slight scuff or drag marks from too little draft on the tool are no longer acceptable. Additional draft is preferable to painting the part to achieve an acceptable finish, an alternative imposed by competition in some cases. Finally, the combining of parts in the ongoing effort to eliminate assembly operations has resulted in components of greater complexity. Such parts have more detail, which increases the gripping effect of the part in the mold.

When texture is applied to the outside wall of a part, it results in tiny undercuts in the surface. To accommodate this surface, additional draft must be provided. While the draft allowance can vary from pattern to pattern and from one engraver to another, the most generally accepted rule is:

> Allow 1.5°/side plus 1.5° for each 0.001 in. depth of engraving.

Thus, for a texture depth of 0.025 mm (0.003 in.), allow a draft of 6°/side. The draft allowance should be approved by the engraver. When the amount of draft so determined interferes with critical requirements, the engraver may be able to feather the engraving to some extent. On some cases, less draft is acceptable. Regardless of the number of shops that manufacture molds for the product, all the engraving should be performed by the same engraver (in the same location), to ensure the greatest uniformity of engraving.

2.5   Shrinkage

Basic physics describes the phenomena whereby objects expand with heat and contract upon cooling. An object not constrained in some fashion behaves in exactly that way. When a plastic part cools in a mold, however, it is constrained by the mold. Usually, the core of the mold constrains the part, although such part details as external ribs can result in the cavity constraining the part.

The molten plastic begins to cool the instant it enters the mold. When the mold opens and the cavity side is removed, the part is exposed to the air and cooling commences in earnest. As the part cools, it stiffens. When it has done so enough to withstand the stress of ejection, it is removed from the mold.

At this point, the part is no longer constrained. This is the stage at which most of the shrinkage takes place, hence the term "post molding shrinkage." The majority of this shrinkage (75–95%) occurs within the first 2 h. By the end of a week, nearly all of it has taken place. Crystalline thermoplastics, however, can take a year to reach final equilibrium.

This phenomenon must be considered when post molding operations such as machining or assembly are planned. As a general rule, it is wise to wait 24 h before

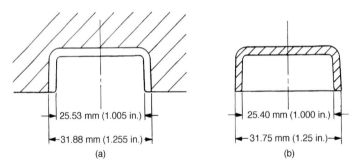

Figure 2-9   Postmolding shrinkage: (a) core and cavity and (b) finished part

performing any machining operations more precise then edge or gate trimming. The same guideline applies to assembly operations. However, in some cases, it can be an advantage to assemble two parts while one is still hot, provided the hot part is the female fitment. Cooling will then result in a firm joint as the outside part cools and shrinks down on the inside part.

Postmolding shrinkage figures for resins are readily available from the manufacturer. Generally, the moldmaker will add an allowance for shrinkage to the dimension. That is the number to which the mold core and cavity are built. Figure 2-9a illustrates the dimensions to which the mold for the part shown in Fig. 2-9b would be built. Those numbers were arrived at in this fashion.

$$(1 + a)b = c \tag{2-1}$$

where a = shrinkage rate

b = part dimension

c = mold dimension

A multiplier is created by adding the shrinkage rate to the number. Thus, a shrinkage rate of 0.005 in. per inch (a) would result in a multiplier of 1.005. Therefore: for a part dimension of 25.4 mm (1.000 in.), the mold dimension would be

1.005 × 25.4 mm = 25.53 mm

or

1.005 × 1.000 in. = 1.005 in.

Shrinkage rates are often indicated as a range. Usually the lower figure applies to thin walls and the higher figure to thicker walls. Shrinkage rates for amorphous thermoplastics tend to be low; however, those for crystalline materials can be large and can have an enormous range. Some polyethylene resins have shrinkage rates ranging from 0.38 mm/mm (0.015 in./in.) to 1.27 mm/mm. (0.050 in./in.)

Shrinkage rates have a direct relationship to assembly considerations with respect to material selection and joint design. Once the parts have been designed and the mold constructed, engineers' materials selection options are limited to resins with like shrinkage factors. Thus, it would be impossible to switch from a polyethylene with a shrinkage of 0.75 mm/mm (0.030 in./in.) to a polystyrene with a shrinkage of 0.10 mm/mm (0.004/5 in./in.). The part dimensions would be substantially different.

Of course, this can be used to advantage. If one of the parts is a little large and the mold is difficult or impossible to alter, a version of the material with a slightly higher shrinkage could be used. Conversely, a material with a slightly lower shrinkage or with a filler added could solve the problem if the parts were too small. Molding conditions can also have a significant affect on shrinkage.

2.6   Fitments

2.6.1   Drawing Conventions for Plastic Assembly

It is important for designers and engineers to understand the legal aspects of engineering drawings. A contract in the form of a purchase order that is let for the tooling for the new part generally refers to the engineering drawing as the source for the dimensions and tolerances to which the tooling will be made. Thus, that drawing becomes part of the contract, and the toolmaker is obligated to incorporate every detail described on the drawing within the tolerance limits provided. The toolmaker cannot legally protest that he is entitled to additional money because of details or tolerances he failed to notice when he quoted the mold. Conversely, he has the right to demand additional fees and production time for details or tolerances the designer or engineer failed to include in the drawing when it was originally released for tooling.

It therefore behooves all involved in the design and development process to pay engineering drawings the respect due a legal document.

Efforts to reduce the amount of time required to bring a new product to market have brought great pressures to bear on the design community. Unfortunately, in some cases this has resulted in drawing shortcuts that have led to the production of parts that do not fit together. The consequence of this circumstance is usually mold revisions, which are both costly and time consuming. In some cases, the mold cannot be salvaged and a complete new one must be built. Legal disputes may ensue.

One of the most serious of such ill-advised shortcuts is the failure to draw draft angles and corner radii. Both these fundamental features of plastic parts have a significant effect on their assembly and on the cost of their tooling. Designers and engineers from other disciplines, however, tend to underestimate the importance of these basics. The problem is exacerbated when the drawing is inconvenient for the moldmaker to check because it contains few or no dimensions. When a three-dimensional computer-aided design program is used, the file will be transferred directly to the machining center and the mold will be built exactly as the object is drawn.

Figure 2-10a illustrates the type of drawing just referred to. If such a drawing contains notes that state "Allowable draft 1°/side" and "1.5 mm (0.059 in.) inside radius permitted," it is a recipe for disaster. The use of the words "allowable" and "permitted" gives the moldmaker every right to build a mold that will make a part just as shown in the drawing. In fact, it would be less costly for the moldmaker to make the mold exactly that way. It costs money to add radii and draft walls. For that matter, the moldmaker

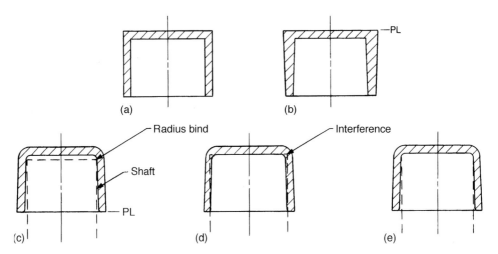

Figure 2-10   Drawing effects: (a) part drawn without radii or draft, (b) part made with parting line at top, (c) part made with parting line at bottom, (d) potential for interference based on mating part without draft consideration, and (e) fitment with draft considered

could easily have made it as shown in Fig. 2-10b with the parting line at the top. The drawing provides that freedom. Best intentions notwithstanding, anything that is not on the drawing is not in the contract. Molds have actually been made just as shown in these drawings.

A prudent policy to eliminate any possible misunderstandings is to follow this rule:

The drawing must represent the part exactly as its shape is intended.

Thus, the part that was depicted in Fig. 2-10a should actually be represented as it is shown in Fig. 2-10c. Both draft and radii are drawn into the part, clearly showing some of the potential fitment problems. The dashed line demonstrates the interference that would occur if the mating part were made with a square corner.

The dashed lines in Fig. 2-10d demonstrate the interference that would occur if the mating part were made without consideration for draft. If the stated dimension were taken as the maximum number in the draft, the fitment would have turned out as in Fig. 2-10d. If the stated dimension were taken as the minimum number in the draft, the fitment would have turned out as in Fig. 2-10e.

There are so many computer-aided design (CAD) programs in general use that it simply is not economically feasible for moldmakers to maintain seats for them all. Therefore, each selects the one or two best suited to his own operation. Drawings provided in other programs are converted to the one at hand. When a CAD file is provided that can be used directly in the moldmaker's system, much of the responsibility for the accuracy of the mold shifts from the moldmaker to the CAD designer and his company. In such cases the CAD designer must exhibit great care lest walls without draft and radiused corners appear in the mold. He or she must also be aware that two levels of design inspection have been removed from the system: the checker and the moldmaker.

2.6.2   Importance of Tolerancing for Assembly

If football can be called a "game of inches," then assembly must be called a "game of thousandths of inches." The difference between a force fit, solvent weld, or other type of joint that provides adequate strength and one that fails can literally be just a very few thousandths of an inch or hundredths of a millimeter.

A vast fortune in extra mold costs and mold revisions has resulted from improper tolerancing. In some cases, projects have been canceled because of a failure in joining that could be directly traced to poorly executed tolerances. In other cases, large multi-cavity molds have been routinely operated at partial capacity because all the cavities could not be kept functioning within tolerance.

Many of these problems can be solved with better dimensioning practice. Once the object is drawn as it is intended to appear, it becomes apparent that careful control of fitment details is critical to the success of the joining.

Dimension A in Fig. 2-11 is one that does not actually exist either in the mold or in the part. While it can be measured in the mold before the radius is added to the core, it is useful only as a reference dimension because it does not exist in the part. The first point that can be measured in either the finished mold or the part is the tangent to the radius, dimension B. This distance can be significantly different from dimension A. For a draft angle of $2°$/side and an outside radius of 3.18 mm (0.125 in.), dimension B would be 0.23 mm (0.009 in.) greater than dimension A. That is sufficient to invalidate a tolerancing scheme. Furthermore, it could cause a close fitting part to be incorrectly deemed in or out of tolerance by quality assurance personnel.

CAD programs will provide the dimension at the tangent. For those who wish to determine this location manually, the computations are as follows:

$$\angle\beta = 90° - \angle\alpha \tag{2-2}$$

$$z = R\tan\tfrac{1}{2}\beta \tag{2-3}$$

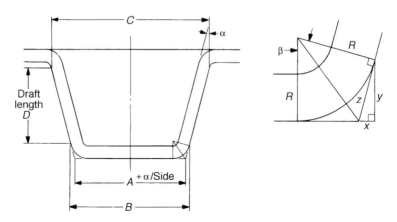

Figure 2-11   Fitment tolerancing

$$x = z \sin \alpha \tag{2-4}$$

$$y = z \cos \alpha \tag{2-5}$$

where $\angle \alpha$ = draft angle

$\quad\quad \angle \beta$ = construction angle

$\quad\quad z$ = construction hypotenuse

$\quad\quad x$ = horizontal offset

$\quad\quad y$ = vertical distance to tangent

One of the problems encountered in the practical application of this principle is the tendency of moldmakers to presume that the dimension is meant to be to the intersect even when it is indicated to the tangent. This is a result of the widespread practice of dimensioning the intersect instead of the tangent. Thus it behooves the designer to show very clearly where the dimension is taken to, even to the point of providing both dimensions.

The dimension at the other end of the part, dimension C, is of equal importance and the same condition with regard to a square corner would also apply. Its location is determined by equation 2-6. The generally accepted drafting rules state that if a dimension and the related angle are toleranced, the dimension at the other end must be referenced. However, if the fitment requires close control, it may be better to tolerance the dimensions at each end and permit the angle to be referenced.

$$x' = D \tan \alpha \tag{2-6}$$

where x' = horizontal offset of wall

$\quad\quad \alpha$ = draft angle

$\quad\quad D$ = length of the drafted wall

This practice prevents the problems that result when a toleranced end is carefully controlled and the designer fails to check the range of dimensions at the other end resulting from the angle tolerance. For example, for an angle tolerance of $\pm \frac{1}{2}°$ and a height D of 25.4 mm (1 in.), dimension C could vary by 0.43 mm (0.017 in.) easily enough to get a fitment in trouble.

2.6.3   Special Drafting Practices for Plastics

One drafting practice that is somewhat unique to the plastics industry is the designation for draft. It is usually specified as $+\alpha°$/side and placed on the dimension taken from that point. Thus, the designation $+2°$/side indicates that a 2° draft is intended to increase from the point of the dimension so indicated. Conversely, the designation $-2°$/side would indicate that the draft decreases from the point dimensioned by 2°/side.

In addition, several notes commonly used on plastics drawings are not typically found in drawings for other industries.

1. *Each cavity must contain an identification number – location to be approved.* This refers to the common practice of using multiple cavities, which will vary in dimension and should be identified by a small number or letter located where it will not interfere with a fitment or the appearance of the part. Identification is necessary to locate a cavity that is not producing acceptable parts.

2. *Gate location must be approved.* A number of plastics processes require a gate through which the mold is filled. Neither the molder nor the moldmaker has a thorough understanding of the function of the part or its mating fitments. Therefore, the designer or engineer must approve the location of the gate to ensure that it does not interfere with the function or the appearance of the part. That statement notwithstanding, the processor should be given all the freedom possible to permit the production of the part in the most economical manner.

3. *Weld line must not be visible.* Parts from gated processes will contain a weld line (knit line) that will be located, depending on the contour of the part, at a point approximately opposite the gate. Multiple gates will result in multiple weld lines. There will also be a weld line around each core pin, since the material must flow around that pin to fill the mold. Unless the hole is drilled, a hole and a boss will always have a core pin and, therefore, a weld line. Cored holes should not be positioned less than 1.5 to 2.0 mm (0.059 to 0.079 in.) from the edge of an injection molded part; or two diameters of the largest cored hole from the edge of a compression molded part.

   The temperature and pressure of the melt will determine the amount of weld achieved at the weld line. "Open weld line" is the term used to indicate that no weld at all has been achieved. Such a condition always results in a reject part. In some cases, a partial weld is achieved. This will result in a visible weld line much weaker than the surrounding material. As indicated above, this is also a reject condition. A good weld line will not be noticeable, although a slight line may be visible on a polished surface if closely inspected. This can be acceptable; however, testing of weld lines to establish acceptable limits may be necessary. This may require the words "limit samples" to be added to note 3. Even a good weld, however, is somewhat weaker than the surrounding material, and that is where failure is most likely to occur. Critical areas that cannot withstand any loss of strength will need to be flagged with the notation "No weld line permitted in this area" so the gating can be located accordingly. Methods of dealing with bosses in regard to boss cracking due to self-tapping screws are discussed in Chapter 8, Fasteners and Inserts.

4. *Maximum allowable flash is .XXX in.* Flash is the tiny amount of plastic that fills the crevice between the mold components. It is almost always present in a part molded under optimum conditions. The amount varies according to the process. To demand no flash brands the designer as a neophyte and may lead to costly post-molding trimming operations, which are truly unnecessary. A far better practice is to design the part so the fitments do not occur on the parting line. Failing that, the next best approach is to design the fitment to allow for a reasonable amount of flash, which will avoid an increase in the cost of the part.

5. *Maximum allowable mismatch is .XXX in.* "Mismatch" generally refers to the alignment between the upper and lower parts of the mold at the parting line. However, it can be construed to refer to the alignment between any two mold components anywhere on the core or cavity. To avoid disputes over whether any misalignment is or is not included in the tolerance, maximum allowable mismatch should be so indicated. Misalignment between core and cavity results in a variation in the wall thickness between the two sides. Furthermore, since the cavity space is different between the two sides of the mold, the flow of the molten plastic in the mold will be altered, potentially affecting the strength and location of a weld line at a hole or boss. Misalignment between core and cavity is also controlled by placing a tolerance on the wall thickness.

6. *Part must be flat within .XXX mm (or within .XXX mm/mm).* It has been said that there is no such thing as an absolutely flat or perfectly round plastic part. Perhaps not, but it is certainly within reason to set limits on how much out of flat or out of round a part can be. Plastics processes are cycle sensitive. Without such limits, the process can go out of control and proper fitments can no longer be guaranteed.

7. *Material is to be [name of manufacturer] [exact number of resin]. Part is to include XX% additive [name of manufacturer] [exact number of additive]. No substitutions permitted without written authorization.* The material, which has been tested and approved, should be clearly indicated and no substitutions permitted. That is because there is no such thing as a competing material exactly the same as the original material. Patents prevent that. Thus, another material will behave differently in some respects and will have somewhat different properties. Depending on the application, such variations may or may not impact the performance of the product, with potential legal implications.

   Also, the material supplier may not be the actual resin producer. In fact, the supplier may use several sources. His practice must be ascertained, and resin from all possible sources tested for critical applications.

   In his exercise of due diligence, the engineer will test additional resin grades until at least two can be approved (three is better). This provides the molder with some flexibility to respond to availability and pricing fluctuations thus keeping costs under control. The vague term "or equivalent" should be avoided at all times, since it is subject to interpretation, which in turn can result in disputes.

8. *XX% regrind acceptable.* "Regrind" is the term used to refer to sprues, runners, and rejected parts that have been reground so they can be mixed in with the virgin material and run through the molding machine again. The use of regrind is one of the characteristics of thermoplastics that has made them economically competitive by enabling nearly 100% material utilization and solves the problem of in-process recycling. However, the material suffers some degradation when it is raised to elevated temperatures. Thus, regrind reduces the physical properties of the material. However, the degree of drop varies between materials and according to molding conditions.

   Obviously, the higher the amount of regrind permitted, the greater the drop in physical properties. However, this diminution is magnified by the multiplier effect which takes place. For example, if 20% regrind is used, the batch will contain 4%

(20% of 20%) that has been through twice, 0.8% that has been through three times, 0.16% that has been through four times, etc. Readily visible signs of degradation are an increase in brittleness and a yellowing or darkening of color.

9. *Ejector locations must be approved.* The designer must bear in mind that neither the molder nor the moldmaker is intimately familiar with the product or its application. These suppliers will select ejector pin locations based on mold construction and part ejection criteria, and their choices may conflict with fitment or other features in the part. Therefore, the locations selected must be approved by the designer.

Be particularly alert to the location of ejector pins on ribs. Ribs are a logical place for ejector pins because of their gripping effect on the core of the mold. However, often the rib is too thin to support any but the smallest diameter ejector pins, the type that break often and need to be replaced. The moldmaker can solve this problem by using a larger diameter pin. The larger diameter pin will require a circle of material to push on. This is known as an ejector pin pad and is illustrated in Fig. 2-12. If made too large, however, it can lead to a sink on the outside surface. (*Ref. 2.3, Ribs and Bosses*)

Before agreeing to this location, the designer should check that the pin pad or recess will not interfere with any fitments nor create an unacceptable sink mark at the intersection with the outside wall.

Figure 2-12   Ejector pin pad

10. *SPI Finish No. X on all outside surfaces except as noted. SPI Finish No. Y on all inside surfaces.* The pattern, location, and depth of finish should be clearly indicated across the area of intended application. Exterior surfaces must meet appearance requirements, which may include sections that vary from the widely used SPI finishes. Interior finishes must be polished just enough to avoid the need to use increased force for ejection due to rough, unpolished interior surfaces, which create tiny undercuts that cause the part to adhere to the core.

11. *Tolerances ±.XXX if not otherwise specified.* When properly employed, this note, or a tolerance box, is a convenience that can save considerable drafting time. Unfortunately, more often than not, it is not properly employed and it leads to excessively tight tolerances. That is because it is easier for the designer to use the overall tolerance than to work out the tolerance for each location. When the range of tolerances used is great, it may be more useful to use the note "Dimensions reference if not toleranced." ("Reference" means for informational purposes only – usually abbreviated REF.)

Excessively tight tolerances increase mold and molding costs considerably. The moldmaker must build the tool to tolerances one-third those indicated on the drawing, the balance being reserved for molding variations. A common blanket tolerance is ±0.13 mm (±0.005 in.). That means the moldmaker is building those dimensions to a tolerance of ±0.043 mm (±0.0017 in.). That is careful, precision work. It also means slow, expensive work, which often unnecessarily lengthens the amount of time required to build the tool. For his part, the molder must slow the molding cycle to meet tight tolerances. Clearly, a large portion of the cost of a part is in the tolerancing.

Unfortunately, casual requests for ultra-tight tolerances are so widespread that moldmakers and molders have become accustomed to asking engineers to indicate which tolerances they really want held. The establishment of so-called critical dimensions undermines the validity of the entire tolerancing system. All tolerances should be held, and none should be asked for that are not necessary. All parties to the transaction should bear in mind that the drawing is part of a contract. In the event of a disaster that leads to legal action, what is written on that drawing is what will count the most.

The foregoing reminder of the legal standing of engineering drawings is not to be interpreted to mean that a deviation from the drawing tolerance cannot be approved. Tolerances created on paper before the part is molded sometimes turn out to be excessively tight when the actual part is available. This is particularly true because there is no inexpensive method of predetermining the actual rigidity of the part in advance of manufacture. Therefore, the actual part may flex more than anticipated, and greater tolerances may be acceptable. When parts are accepted with deviations from the contract drawing, a written record should be retained and the drawing should be altered to reflect the newly approved tolerance.

Regardless of how specified, the objective remains the same: the parts must fit together readily and stay together within acceptable parameters.

2.6.4   Procedure for Establishing Tolerances

Figure 2-13a illustrates a typical male/female fitment. Its tolerances will be developed as in Fig. 2-13b. Presuming a desired dimension of 25.40 mm (1 in.) for the male fitment, the trial tolerance of ±0.13 mm (±0.005 in.) is applied. That will result in a high side of 25.53 mm (1.005 in.) and a low side of 25.27 mm (0.995 in.). Since the starting point is the male fitment, the high side will establish the minimum clearance. If a clearance of 0.05 mm (0.002 in.) is determined to be the most desirable, the lowest acceptable dimension for the female fitment becomes 25.58 mm (1.007 in.). If the same tolerance of ±0.13 mm (±0.005 in.) is retained, the nominal dimension for the female fitment is 25.70 mm (1.012 in.) and the high side becomes 25.83 mm (1.017 in.).

Referring to the lowest acceptable dimension for the male fitment of 25.27 mm (0.995 in.), the maximum clearance becomes 0.56 mm (0.022 in.) If there were two such fitments on the part, as illustrated in Fig. 2-13c, and a center-to-center distance of 76.20 ± 0.13 mm (3.000 ± 0.005 in.) between them were to be applied, an additional

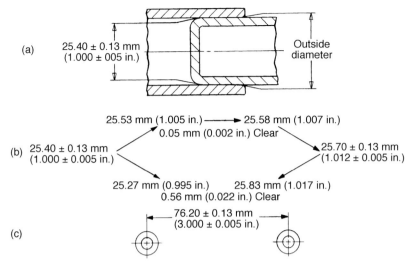

Figure 2-13   Tolerancing procedure

clearance of 0.25 mm (0.010 in.) would have to be added for a total clearance of 0.81 mm (0.032 in.). That would be an unacceptably loose fit for many applications. Yet, a tolerance of ±0.13 mm (±0.005 in.) on a 25.40 mm (1 in.) dimension is quite tight for many high shrinking plastics. For the 76.20 mm (3 in.) dimension, that would be an extremely tight tolerance indeed.

The immediate reaction of the neophyte designer is to tighten the tolerances, thereby dramatically increasing the cost of both the part and the mold. Such a design change might even result in a respecification to a much more costly material. That, in effect, is fighting the material and the process instead of working with it. Experienced plastics engineers learn to use the inherent advantages of plastics to devise ways of fitting parts together using looser and less costly tolerances.

2.7   Design Practices for Looser Tolerances in Plastics

To design for the loosest tolerances compatible with desired performance, the first principle entails taking advantage of the properties plastics can provide. The most important one in this respect is the ability to alter the rigidity of the material such that one of the mating parts can be more rigid than the other. This property allows the more rigid of the two parts to cause the other part to conform to its contour, as illustrated in Fig. 2-14. The greater the difference in rigidity, the easier it is to accomplish this. Normally, the internal fitment is the more rigid. However, it is possible to specify the higher rigidity for the outer filaments – although usually there is less latitude.

If both parts must be made of materials similar in rigidity, the same effect can sometimes be achieved by thinning the walls of one. The difference in rigidity will be far less,

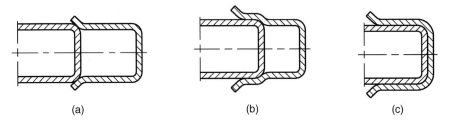

(a)   (b)   (c)

Figure 2-14   Rigid to soft fitment: (a) starting point, (b) during assembly, and (c) assembled

however. In some cases, heat will achieve the same result. In solvent joining, the solvent will sometimes soften the parts enough to perform the same function.

2.7.1   Three-Point Location

The next principle to establish with respect to tolerance design is that of three-point location. This is demonstrated in Fig. 2-15a, where the cylinder is completely constrained by three points of location around its perimeter. That is all that is required. Figure 2-15b illustrates the same three-point location applied to a rectangular object. With this configuration, rotational motion is constrained, as is vertical movement. Horizontal motion, however, is not limited. Figure 2-15c demonstrates that with the addition of a point of contact at each end, all motion is constrained.

> What has been demonstrated is the minimum contact necessary to locate an object.

However, the minimum may not be sufficiently robust to withstand stresses such as severe drop tests. Additional ribs may be necessary.

> The leading edge of a locating rib does not need to be drafted if the sides of the rib have draft.

Mating parts often have straight sides. When it is neither desirable nor practical to use the crush-rib principle, straight walls would obviously provide an increased bearing surface. However, undrafted walls are extremely difficult to remove from a mold and very costly as a result. A rib can be readily molded with a straight leading edge, provided its side walls are drafted, as shown in Fig. 2-15c.

2.7.2   Hollow Bosses

When the part to be fitted is located away from the walls of the mating part, the temptation is to use a freestanding pin or "floating" rib. Such a pin, however, may leave a sink mark in the outer surface. The rib becomes a gas trap in the mold, which leaves a burn mark. What

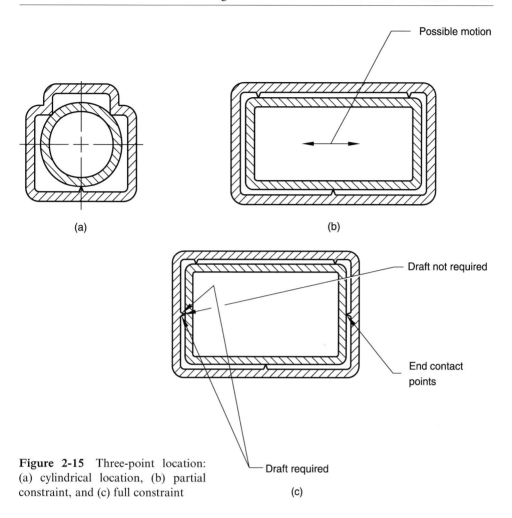

Figure 2-15 Three-point location: (a) cylindrical location, (b) partial constraint, and (c) full constraint

happens is that the molten plastic enters the mold pushing the air that is already in the cavity on ahead of it. There is no place for the gas (air) to escape in a freestanding rib, and it becomes trapped at the end of the rib away from the gate and burns under heat and pressure. The pin, if large enough, can have an ejector pin under it. Vents are customarily provided around ejector pins and on the parting line for the gas to escape.

When the rib is attached to a wall, the gas is chased down the rib to the wall and on down to the vent at the parting line. One way to avoid that circumstance is to use locator bosses instead of freestanding ribs, as illustrated in Fig. 2-16. Core pins which can have vents ground on their sides create the center opening. These pins can be vented to the base of the cavity block and on out to the parting line. Since they are stationary, these pin vents can plug with dirt and debris, making their cleaning a maintenance item. If the number of cavities is low enough to warrant hand removal, or if two-stage ejection can be used, ejector pins can replace the core pins. (That is necessary

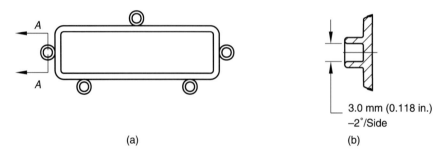

3.0 mm (0.118 in.)
−2°/Side

(a)   (b)

Figure 2-16 Hollow locator bosses: (a) plan view and (b) section *A–A* (enlarged)

because the cored hole prevents the part from falling free.) Otherwise, a more costly ejector sleeve is an option.

2.7.3   Crush Ribs

The advantages of the crush-rib type of location to plastic parts is that these points of contact can be accomplished with hollow bosses or ribs. These devices are much more controllable in plastic parts than are outer walls; moreover, they are less costly to alter and can be adjusted with little effect on the part weight or wall thickness. Most important, they permit the mating part to be located while also maintaining uniform wall thicknesses and without increasing the wall for locating purposes.

Ribs, in particular, can be designed to be deformable. Such ribs are known as "crush ribs" and can be used in a variety of ways. The amount of deformation possible is dependent on the material; however, thinning the wall and bringing the edge of the rib to a point can also be effective. Figure 2-17a illustrates such a rib in cross section. As the mating part is emplaced, the tip of the rib is crushed over. Being very thin at the tip, it is readily deformed in this way. For example, a 2° angle on the rib would vary the fitment dimension by 0.035 in. per side over a 1-in. depth. This would permit a considerable increase in the tolerance. The end view of the object shown in Fig. 2-17b illustrates how the ribs can be used for centering as well as for the fitment. That is

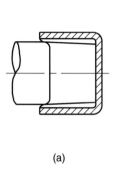

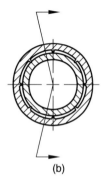

(a)   (b)

Figure 2-17 Crush ribs: (a) cross-section and (b) end view

more easily accomplished by using four ribs instead of three because four ribs are easier to adjust.

Unfortunately, precise determination of shrinkages and the amount of deformation are very difficult to accomplish. Therefore, the prudent approach is to "close in" on the fitment. When one is working with a single cavity or a prototype mold, the usual practice is to start with the targeted dimensions on the loose side of the intended fitment. For a simple location, the dimensions would be set to just clear. For an interference fit, a light contact would be targeted. The final adjustment of the ribs can be based on the results of the first shots. This is best accomplished following the "steel safe" practice. That means cavity dimensions should start to the small side so they can be increased by removing steel instead of welding and machining, which is far more expensive. Core dimensions should be to the large side for the same reason.

2.7.4   Flexible Ribs

Greater tolerance range can be gained from flexible ribs such as the one illustrated in Fig. 2-18a. Whereas the crush rib is a fairly stiff support with limited ability to accept a range of matting fitments, this design permits a much larger range. That, of course, depends on the material in use, however this device can be employed even with relatively stiff materials such as polycarbonate. Figure 2-18b illustrates this design as assembled.

A common version of the flex concept is the ring found on closures, which has long been used to provide a bottle seal. In this form, illustrated in Fig. 2-18c, it is often referred to as a "crab claw." The reader is cautioned that specific applications of this concept may be covered by patents. The flex concept can also be used as a gripping

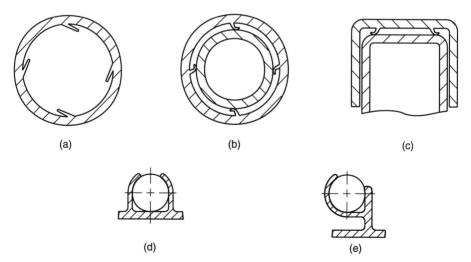

(a)   (b)   (c)

(d)   (e)

Figure 2-18   Flexible ribs: (a) as molded, (b) assembled, (c) crab claw, (d) two arm flex, and (e) one arm flex

device, as illustrated in Figs. 2-18d and 2-18e. Both arms flex in the design in Fig. 2-18d, whereas only the left arm is designed to flex in the configuration in Fig. 2-18e.

2.7.5   Inside/Outside Fitments

Another approach to wider tolerances is the inside/outside fitment as illustrated in Fig. 2-19. If the male part is on the large side, the inner fitment is engaged as illustrated in Fig. 2-19a. Figure 2-19b demonstrates the effect when the male part is on the small side, causing the outer fitment to be engaged. If the materials are sufficiently flexible, a wedging effect will occur as shown in Fig. 2-19c.

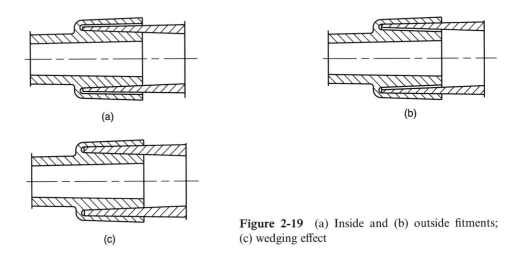

(a)

(b)

(c)

Figure 2-19   (a) Inside and (b) outside fitments; (c) wedging effect

2.7.6   Step Fitments

Another way to utilize the wedge concept involves the step fitment, a concept derived from the traditional laboratory hose fitting. Figure 2-20a illustrates the external version, and the internal variety is shown in Fig. 2-20b.

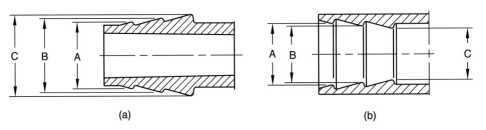

(a)

(b)

Figure 2-20   Step fitments: (a) external and (b) internal

2.8   More Relaxed Tolerances for Large Parts

The preceding devices are principally intended for small part locations. However, the most severe tolerance problems usually occur with large parts. There, the 0.25 mm/mm (0.010 in./in.) tolerance common with processes such as thermoforming and rotational molding can present significant fitment problems.

2.8.1   Drill in Place

A common solution for low volume applications is to simply *locate two parts together and drill the holes that need to be located through both of them at one time*. Thus, the need to hold costly tight location tolerances to the holes is completely eliminated. This feature is particularly useful when both parts are molded as one part, common with rotational and blow molding, separated by a section to be cut away between them. In that case, they will both be on the high or low side of the tolerance at the same time. The drawback to this approach is the loss of interchangeability. Should a replacement to one of the parts be required, it must be shipped without the holes, which must be drilled in place in the field.

2.8.2   Oversize Hole with Washer

One approach to the problem of non-interchangeability is to make one of the holes sufficiently oversize to accommodate the large tolerances, as illustrated in Fig. 2-21a. The difficulty with this approach is that while it deals with medium tolerances with standard large washers, custom washers must be obtained to cope with very wide tolerances, since these result in very large oversize holes.

2.8.3   Criss Cross Slots

One way to cope with the wide tolerances required of plastics is the use of slots placed perpendicular to each other. This practice, illustrated in Fig. 2-21b, accommodates dimensional variations in either direction and can cope with larger tolerance variations

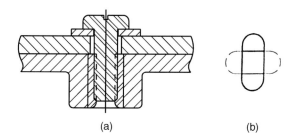

Figure 2-21   (a) Oversize hole with washer and (b) crisscross slots

(a)   (b)

than the oversize hole, without the need for custom washers. While no more expensive to mold in place than the round holes used in that method, it is a more costly post-molding operation.

2.8.4   Separation of Functions

Figure 2-22a represents two parts held together and located with self-tapping screws. To achieve this objective, tolerances must be held on the bosses for both the locating and self-tapping functions. This can lead to excessively tight tolerances that are expensive to maintain.

Figure 2-22b illustrates the principle of separation of functions. If locating ribs are used, the holes for the screws can be clearance. The boss for the self-tapping screw no longer is located to a tight tolerance; thus the molder can concentrate on the integrity of the boss weld line itself. The locating function has been relocated to another site and can be adjusted independently without interfering with the strength of the boss.

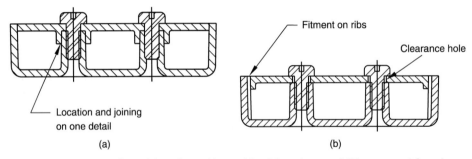

(a)   (b)

Figure 2-22   Separation of functions: (a) combined functions and (b) separated functions

2.8.5   Corner Clearance

Plastic parts are often stiffest, at their corners leading to interference in the corners, a condition known as "corner bind" (Fig. 2-23a). This is a common problem in plastic part assembly. The usual practice of making the inside part with a smaller radius than the outside part exacerbates this problem, particularly if the parts are out of tolerance. It is readily avoided by the following rule:

> The outside radius of the inner part must be greater than the inside radius of the outer part.

Referring to Fig. 2-23b, that means that radius R_1 must be greater than radius R_2. That will leave a gap between the two radii and avoid a corner bind.

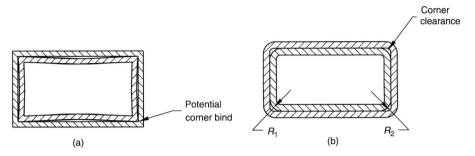

Figure 2-23   Corner clearance criteria: (a) corner bind and (b) corner clearance

2.9   **Semidovetail Joint**

The last bit of distortion may be impossible to eliminate without incurring costs impossible to recover under competitive conditions. The semidovetail joint, illustrated in an overhang version of Fig. 2-24a, is probably the most common joint used in plastic assemblies, as it provides fitment and location. However it can also be a source of problems. First, it requires a 50% drop in wall weight, which causes distortion due to nonuniform cooling. Second, it can result in notches at the inside radii which are stress concentrations and can be particularly dangerous for notch-sensitive plastics. The greater the outside wall draft angle above the joint, the more significant this combination of conditions can be.

While the very nature of the semidovetail configuration makes it extremely difficult to alleviate these conditions, certain steps can be taken. The first is to make the walls uniform (if appearance is not an issue). The second is to make the draft angle the minimum necessary for the particular finish being used until a point clear of the joint is reached. (If there is etching on the outside wall, feathering can be used to reduce this angle.) The next step is to make the wall thickness at the joint 25% greater than the nominal wall thickness, the maximum permissible without distortion. The major portion of that increase would go to the outer wall of the joint, to minimize the visible distortion. Finally, the inside radii is increased to reduce the notch sensitivity problem – remembering, of course, that rim radius of the mating part must be increased even more to avoid corner contact at that point. For large parts, ribs can be spaced along the wall to control inward movement of the inner portion of the joint.

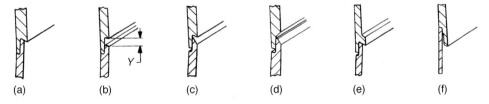

Figure 2-24   Minimizing the effect of misalignment on appearance: (a) semidovetail with overhang, (b) separation (slot), (c) half v-groove, (d) full v-groove, (e) bead, and (f) skirt

2.10   Minimizing the Effect of Misalignment on Appearance

Between the effects of distortion, gate scars, and ejector marks, it is often extremely difficult or costly to create a perfect edge. Fortunately, the human eye is not truly able to discern slight deviations if they are separated by a relatively small amount of space. Figure 2-24 illustrates a variety of ways to utilize this phenomenon. The version in Fig. 2-24a is essentially a semidovetail joint with an overhang to hide misalignment. The same result can be achieved with a separation, as shown in Fig. 2-24b: the separation (Y) can be effective if it is as small as 1.57 mm (0.062 in.); anything more only helps the effect. The half v-groove and full v-groove versions of the slot concept are represented in Figs. 2-24c and 2-24d. The bead version in Fig. 2-24e is sometimes used as a bumper on conveyors for automatic assembly. The last design, a variety known as a "skirt" design, is illustrated in Fig. 2-24f. The lower edge of the skirt can be highly contoured, while the actual contact takes place on the *internal fitment*.

2.11   The Plastic Product Design for Assembly Checklist

Studies indicate that the time required to conduct a plastic product design program is often doubled by reworks. While these can be due to the occasional failure of an experimental concept or a change in objective, the bulk are attributable to unforeseen engineering phenomena. The author has observed countless project disasters due to the failure to fully identify *all* the parameters of the project in advance. The checklist provided in Fig. 2-25 is designed to unearth all the requirements of a project in advance so they may be dealt with. Surprises are expensive. The reader is encouraged to use this checklist as the basis for customized checklists.

2.12   Testing

The accelerated pace of plastic product development programs has reduced the time available for actual testing of the finished product. Yet this is the only way to be certain the product will perform according to specification. The author recommends the following guidelines for the establishment of a test program.

1. *Duplicate actual conditions.* It is easy to become so involved in the minutiae of test procedures that one loses sight of the whole purpose of the test, which is to determine whether the product will perform successfully in practice. Therefore the test procedure must reproduce actual conditions as closely as possible.
2. *Test completed assemblies made from parts produced by production processes.* Anything short of this is a waste of time and resources. The individual parts will not behave as the completed assembly would, nor will handmade samples.
3. *Test for temperature extremes first.* In the author's experience, plastic assemblies are most likely to fail as a result of thermal expansion, loss of strength or rigidity at

elevated temperatures, or thermal contraction or reduced impact strength at low temperatures. If a product passes these trials, the likelihood is strong that it will pass the remainder of the tests.

4. *Time related tests.* The next greatest hurdles to overcome are the time related tests. Whether called creep, cold flow, or stress relaxation, these effects can be disastrous to a plastic product assembly, and there is no reliable method of correlating an accelerated test program to product life. Be alert to the influences of thermal cycling, chemical attack, and ultraviolet light on the time effects.

Plastic Product Design for Assembly Checklist
Notes:
1. Omitted presumes "does not apply."
2. Include conditions encountered in use, cleaning, shipping, assembly, testing and decorating.
3. Overspecification results in unnecessary cost.

Part name and description of application: _____

_____

A. Physical Limitations
Length: _____ Width: _____ Height: _____

Weight: _____ Density _____

Mating parts: _____

Mating fitments: _____

B. Mechanical Requirements
Functional (dynamic) life: _____

Nonfunctional (static) life: _____ Shelf life: _____

Tensile strength: _____ Type: _____ Duration: _____

Flexural strength: _____ Type: _____ Duration: _____

Compressive strength: _____ Type: _____ Duration: _____

Flexural modulus (stiffness): _____ Hardness: _____

Maximum allowable creep: _____ Deflection: _____

Impact strength: Room temp. _____ High temp. _____ Low temp. _____

Shear strength: _____ Abrasion resistance: _____

C. Environmental Limitations
Chemical resistance: Continuous _____ Intermittent _____

          Occasional _____

Temperature: Maximum + duration _____ Minimum + duration_____

          Operating + duration _____

Ultraviolet light: _____ Water immersion: _____

Moisture vapor transmission: _____ Radiation: _____

Flammability: _____

Figure 2-25 Checklist: plastic product design for assembly

D. Electrical Requirements

Volume resistivity: _____   Surface resistivity: _____

Dielectric constant: _____   Dissipation factor: _____

Dielectric loss: _____   Arc resistance: _____

Electrical conductance: _____   Microwave transparency: _____

EMI/RFI shielding: _____

E. Appearance Requirements

Surface finish: Inside (SPI #) _____   Outside (SPI #) _____

Texture no. _____   Depth _____

Color: _____   Match to: _____

Color maintenance: _____

Transparency: _____   Translucency: _____

Metallizing: _____

Decoration: _____

Identification: Model _____   Production date _____

Recycling _____   Instructions _____

Warnings _____   Nameplates _____

Manufacturing limits: Flash _____   Mismatch _____

Gate Location _____

Ejector location _____

F. Assembly Requirements

Parts to be assembled to, method prepared for (screws, solvents etc.), and type (permanent, serviceable, occasionally reopenable, watertight, hermetic)

1. _____

2. _____

3. _____

4. _____

G. Other Design Parameters

Anticipated volume: Annual _____   Order _____

Anticipated production date: Start _____   Volume _____

Legal exposure: Liability _____

Patents _____

Regulatory approvals required: 1. _____   2. _____   3. _____

4. _____   5. _____   6. _____

Specifications: Military _____   Building codes _____

Foreign production: _____

Foreign sales areas: _____

Figure 2-25   Continued

H. List All Tests to Be Performed on This Product

1. _____   2. _____

3. _____   4. _____

5. _____   6. _____

7. _____   8. _____

9. _____   10. _____

I. Additional Comments:

_____

_____

_____

_____

_____

_____

Figure 2-25 Continued

3   Cost Reduction In Assembly

3.1   Introduction

The plastics processes continuously change as manufacturers improve their products. Competition exists not only among the producers in each means of assembly, but among the different methods themselves.

As long as the author has been involved in plastic product design, there has been an effort to reduce assembly costs by combining parts. For years, this effort was largely confined to small injection-molded applications, which could take advantage of the range of capabilities of a given process. The author can recall a project in the 1960s in which a 27-part ratchet was combined into a one-piece moldment, as well as the first time an electrical bobbin was molded in one piece. Indeed, part reduction has been one of the major driving forces in the expansion of the plastics industry.

This chapter deals with four fundamental approaches to part reduction: micro, macro, fastener elimination, and holistic design. In a micro method, the basic design concept is retained, but the individual parts within it are examined for possible cost savings. The macro approach is to rethink the entire concept. Fastener elimination has been a driving force in the assembly of plastic parts in recent years. Holistic design encompasses knowledge in the total aspect, including all the disciplines necessary to take design to a higher level.

3.2   The Micro Approach to Part Reduction

A full process review is not executed in the micro approach. However, some reevaluation of processes should be undertaken. Combining parts usually leads to more complex parts. Processes such as blow molding, cold press molding, compression molding, extrusion, filament winding, layup, pultrusion, resin transfer molding, and thermoforming do not favor highly complex parts. Others, such as casting and machining, do not lend themselves to complex parts very efficiently. Combining those parts might force a change to a process like transfer or injection molding.

The combined part might result in one with a different set of physical characteristics than the original parts. Therefore, combining parts could also force a change in material. This, in turn, could drive a change in process.

For the micro approach, the first step is to determine the fundamental functions of the parts to be considered for combining. The following are the types of fundamental function we refer to:

1. *Product function* Elements of parts that are essential to the function of the product.
2. *Protection from external forces* Shielding or other protection of components from external forces.

3. *Environmental protection* Protection of internal components from the environment.
4. *Location* Location of one part to another.
5. *Shock absorption* Protection of internal components from shock and vibration damage
6. *Mechanical control* Linkage and drive functions.
7. *Support* Structural support of other components.

During this exercise, a component may be discovered for which there is no apparent function. Such "vestigial" parts were once quite commonly found in part reduction programs because product elements had been changed to plastic from other materials by inexperienced designers who did not understand that there was no need for the part with the new design. One example that comes to mind was a venting device to accommodate the condensate that accumulated on the surface of a steel component. That housing had been changed to a plastic, which had insulating qualities that prevented it from "sweating" as its metal predecessor had. The vent, however, had remained in place long after the change until its function was finally questioned. One cautions, however, against removing such a part unless its original function has been identified. The component might have a function that simply is not immediately obvious.

Having identified the basic functions of the parts in question, one must identify the secondary functions the parts may have. Then the parts are examined for whatever nonessential components may remain. When those have been noted and removed, the process of combining the parts can commence.

3.2.1   Combining Parts Through Materials

Plastics materials offer the engineer a wide palette of properties, which can be particularly useful in combining parts. In addition, because of their chemical composition, plastics often can be modified for specific characteristics. Thus, the most economical combination of properties can be combined to optimize the material utilization.

A significant number of plastic materials offer varying degrees of flexibility and ductility both with and without transparency. They can be formed at relatively low temperatures and by a group of processes unique to the plastics industry. These can permit shapes that are impossible to duplicate with traditional processes. Not only do these attributes provide product design opportunities, they vastly increase the palette of assembly techniques available to the fabricator.

In particular, the ability to combine strength with stiffness that varies with the wall thickness can be very useful. This might affect assembly, for example, when it is possible to reduce the stiffness of a given plastic just enough so that the arm of a device can deflect adequately to perform a given function. This, in turn, could eliminate an entire mechanism composed of rigid arms and springs, thereby saving a whole series of assembly operations, as demonstrated by the spring-loaded ratchet in Fig. 3-1. The arms of the ratchet were successfully converted to integral components by utilizing the ability of acetal to vary its stiffness in one part through a variation in wall thickness.

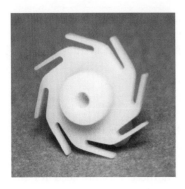

Figure 3-1   Acetal ratchet

With the arms integrated into the main part, 24 parts were eliminated: the arm, the spring, and the pin for each of the eight arms of the ratchet.

Another example of this type of part elimination is the incorporation of a seal ring into a cap. This practice, which is now commonplace, is illustrated in Fig. 3-2a. It replaced the two-part assembly, illustrated in Fig. 3-2b, which combined a metal cap with a separate seal liner.

For a guide to the general characteristics of plastic materials and the assembly methods they can accommodate, see Chapter 5, "Assembly Methods by Material."

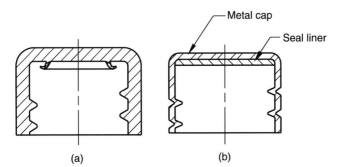

(a)   (b)

Figure 3-2   Lined caps: (a) cap with integral seal ring and (b) metal cap with separate liner

3.2.2   Combining Parts Through Processes

One of the principal advantages offered by plastics is the capability of its processes to produce highly complex parts. The volume process with the greatest capability in this respect is injection molding.

The mold, the broad availability of materials, and their extreme controllability are the keys to the success of this process. Moldmakers have succeeded in constructing incredibly complex molds with side cores coming in from virtually any angle conceivable. Collapsing cores are no longer a rarity, and with special equipment, even thermosets can be injection-molded. With combinations of heat and pressure, molders have been able to successfully fill these molds and create parts that combine many parts into one with accompanying savings in assembly costs (although the tooling can be very costly). The widest possible material availability has created great opportunities for product designers.

A number of applications, however, attempt to do so much that they fail to achieve their objectives. Some parts simply become too complicated to be properly molded. Furthermore, the molds may be too costly to be written off in a reasonable length of time or to maintain at an acceptable quality level. They may simply be too complex to keep a profitable number of cavities operating within tolerance.

3.3   The Macro Approach to Part Reduction

The "micro" approach to part reduction is often a struggle to find a better way to put two parts together. Among product designers and engineers, there is sometimes a need to stand back and see the project from a greater perspective. The way to achieve deep cuts in assembly costs is to substantially reduce the number of assembly operations required. In short, to challenge the entire concept of how the product is to be manufactured. We refer to this as the "macro" approach to part reduction.

This concept is not brand new. Most boat hulls are now made in one piece. One manufacturer (Lotus) has been building a car body in two half-body moldments for some time. Perhaps the twenty-first century will see entire car bodies molded in one shot. One does not need to be clairvoyant to discern the direction in which we are headed.

Here are some examples of this type of thinking.

The barrel, which was a successful storage container for many years, is a design composed of staves, hoops, and nails. In construction, the staves needed to be softened with steam. No matter how efficiently the barrel maker created that steam, no matter how little he paid his workers, no matter how hard he squeezed his nail source, he could never compete with a drum blow-molded in one piece.

Blow molding is not the only way that large storage containers have been manufactured in one piece. They have been rotationally molded, structural foam molded, and reaction injection molded as well. Smaller sizes can be injection molded, and shallow shapes can be thermoformed. It is simply a matter of determining the optimum process for a given set of shape, physical property, and volume requirements.

In recent years, Komatsu undertook a parts reduction program for the exterior parts of their mini wheel loaders. Their conventional method required 298 steel plate parts. With RIM selected as the appropriate plastic process for this conversion, they managed to reduce the number of parts to 62. The largest reduction came in the hood, where 140 steel plate parts were replaced by 14 RIM parts.

Other processes offer similar examples. Lay-up and spray-up have almost completely replaced other methods in the manufacture of large boat hulls. For smaller sizes, these processes must themselves compete with hulls made by a variety of processes. Small boat hulls have been thermoformed, rotationally molded, resin transfer molded, and even compression molded.

Hood and fender units for truck tractors have been made by the spray-up, SMC compression and reaction injection molding processes. Automobile panels, except for the largest sizes, are now largely plastic as well. As we referred to earlier, Lotus manufactures an entire car body in two parts. They assemble the two halves with adhesives much like is commonly done with toy models. They use their VARI process, a variation on what is known in the United States as resin transfer molding (RTM).

Other macro approaches to part reduction in the automotive industry include the General Motors' Super Plug (Fig. 3-3), an interior door module that consolidates as many as 61 parts into a single gas-assisted, injection molded unit for a substantial cost and weight savings. The material used is 30% glass-fiber-reinforced poly-carbonate/polyester resin.

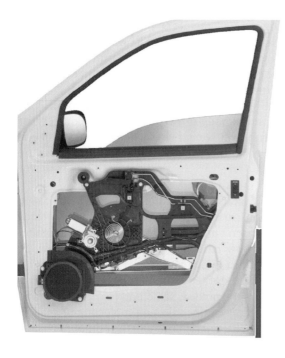

Figure 3-3 General Motors' Super Plug door eliminates 60 parts (Courtesy of Delphi Division of General Motors and GE Appliances)

Referring to the major appliance industry, the Plastics Industry Mobilization (a joint initiative with the American Plastics Council and the Society of the Plastics Industry) reports that "Parts consolidation, molded complex shapes and ease of fabrication through the use of plastics result in a 15% to 20% savings per application in product cost and often as much as 50% less manufacturing investment to fabricate major structures. For instance, the Carry Cool® air conditioner (Fig. 3-4) combines 17 metal parts into a one-piece molded thermoset plastic."

Structure is a key element in the macro approach. Many of the parts eliminated in the preceding examples were the stiffeners, hardware, and fasteners used to hold the

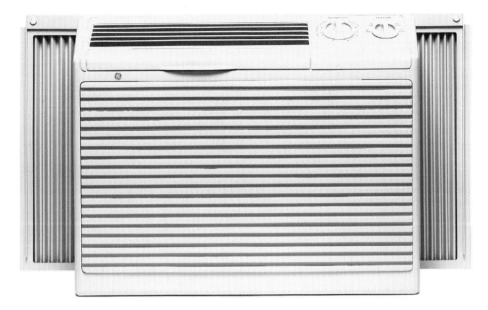

Figure 3-4   General Electric Carry Cool® air conditioner reduces 17 parts to 1 (Courtesy of General Electric)

original assembly together. Several assembly methods are capable of providing this type of integral structure.

The thermoset molding processes gain their strength from the glass fiber reinforcement placed within. Other processes, like injection molding, can also employ glass fiber reinforcement. These methods can mold structural elements made of steel directly into the part. A housing made by injection molding is illustrated in Fig. 3-5a. That process can mold integral ribs, but it has size and tool cost limitations. However, the structural foam process (Fig. 3-5b) can offer stiffening ribs with substantially lower mold costs – provided one can resist the temptation to put a lot of expensive detail into the part. The gas-assisted, injection-molded version (Fig. 3-5c) has a similar structure, except that a void replaces the foam.

The hollow processes like rotational molding, industrial blow molding, and twin-sheet thermoforming have the capability to provide a different form of structure.

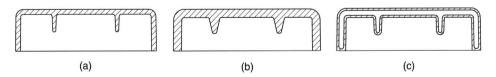

(a)   (b)   (c)

Figure 3-5   Molded-in structural elements: (a) injection molding, (b) structural foam, and (c) gas-assisted injection molding

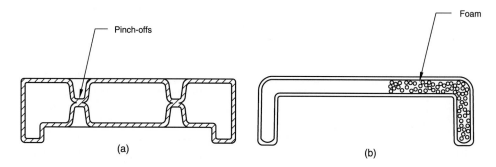

Figure 3-6   Structural elements in hollow parts: (a) design with pinch-offs and (b) foam-filled hollow part

Figure 3-6a illustrates what is known as a "pinch-off," in which the two outer walls touch each other. This process can leave a visible mark, and thus it may be preferable to stiffen the structure with foam, as shown in Fig. 3-6b.

Perhaps the ultimate in sophisticated hollow processes is actually a modern injection molding variant on an ancient process used by sculptors, the "lost core process." The air intake manifold in Fig. 3-7 won the Grand Award for the Automotive Division of the Society of Plastics Engineers for the most innovative use of plastic

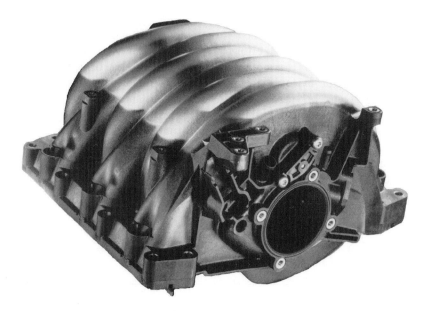

Figure 3-7   One-piece manifold manufactured by lost core process (Courtesy of MANN + HUMMEL Automotive)

technologies in the 1995 automotive model year. It reduced the number of parts required in the assembly from 118 to 33, while providing additional features like snap fit attachments. A 5-horsepower increase in performance was gained as a result of better air flow, while reductions were achieved in costs, packaging space, and weight. There are actually several competing varieties of this concept, but the basic idea is to use a core that is prevented by undercuts from being withdrawn from the moldment in the normal manner but can be dissolved or melted out, according to the technology selected.

Clearly, we, as designers and engineers, must get out of the ruts we find ourselves in when we try to take every application into the one or two plastics processes we are most comfortable with. We must clear the cobwebs from our minds and open them to the wide range of processes available.

The key principle in process selection is to establish the most efficient means of manufacture. To do so, one must take into account the investment cost as well as the piece part cost. For example, injection molding might offer the lowest piece part price for a given project, but if only a few thousand parts would be made each year, it probably would not be possible to save enough to pay the difference in mold cost over structural foam.

Decision making would be tremendously simplified if we could determine a given volume for each process beyond which the additional investment could be justified by the reduction in piece part cost. Actually, if all parts were identical in shape and size, that might be possible. Since they are not, each designer and engineer must learn through experience with the type of parts he or she works with. As a rough guide, however, one may presume that the larger and/or more complex the part, the greater the investment will be. As the investment grows, greater volume will be needed to pay the difference with lower piece part costs within an acceptable period. That period also varies considerably between companies.

A further consideration lies in the size capability of each of the processes. Chapter 6, "Assembly Method Selection by Process," indicates the size range each process is capable of (Table 6-1).

3.3.1   Multiple Material Processing

Multiple material processing permits an even wider range of properties to be achieved with a given molding. It can be used to mold an elastomer over a rigid form, an inner sandwich of recycled polymer, foam, or another resin, to mate two different colors of a material, or an opaque and transparent combination. These combinations can eliminate one or more finishing or assembly operations.

Coextrusion, coinjection molding, and multipart injection molding are the processes that currently offer multiple material processing. Their relative benefits to the assembly of plastic parts are discussed in the following sections. Coextrusion and coinjection molding are discussed further in Chapter 6, "Assembly Method Selection by Process." Additional information on insert and multipart injection molding can be found in Chapter 13, "Insert and Multipart Molding."

3.3.2   Coextrusion

Coextrusion can be used to create either a laminate with one material over the other or one with the two materials side by side. In a typical laminate application, the major portion of the coextrusion is one material and a very thin coat, known as a cap sheet, of another material laid over it, as shown in Fig. 3-8. The principal use of this technique is to provide a barrier coating for the primary layer, whose properties are needed. This is typically done with ABS covered with an AS cap coat, the ABS being vulnerable to ultra-violet light. The sheet is then used for thermoforming. In this fashion, the need for an expensive painting operation on the finished part is eliminated.

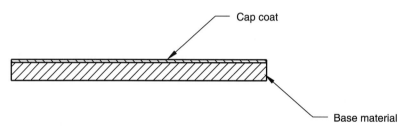

Figure 3-8   Layered coextrusion

Coextrusion of this sort is also used to create the parison used in industrial blow molding. The finished part, in this case, is a fuel tank with a barrier liner.

In another form of coextrusion, the materials are extruded side by side. An example of this kind of application is illustrated in Fig. 3-9. This is a package thin-gauge-thermoformed from a coextruded sheet such that the cover is clear and the base is opaque. The coextruded material (Fig. 3-9a) is then formed as in Fig. 3-9b; the finished package is depicted in Fig. 3-9c. The assembly operation of the cover to the base is thereby eliminated and, possibly, some hardware as well. Material of another color could be

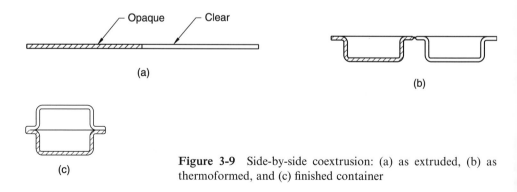

Figure 3-9   Side-by-side coextrusion: (a) as extruded, (b) as thermoformed, and (c) finished container

substituted for the transparent resin, thus creating a package with a different colored base and top.

3.3.3   Coinjection Molding

A similar concept is used to eliminate finishing operations in the manufacture of structural foam articles, a process that typically leaves a poor finish, requiring secondary operations. By molding a surface coating on the part, the secondary steps are eliminated, with significant savings as a result. This technique can also be used to replace the inner material with one of lower cost.

3.3.4   Multipart or Two-Color Injection Molding

The process originally created to mold the letters in key caps, often called two-color injection molding, permits parts of different materials or colors of a material to be molded together. Since the second material is molded into the first, it is more resistant to abrasion, and therefore more durable, than a decoration applied to the surface. Two-color product housings have also been made this way, thus eliminating the assembly operation. Other applications include clear lenses in opaque housings and pen barrels with rubber-like gripping surfaces. An example of a two-color moldment is shown in Fig. 3-10.

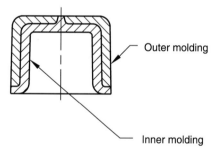

Figure 3-10   Multipart or two-color molding

3.4   Elimination of Fasteners

Fasteners are parts and, as such, they cost money to purchase and inventory. They also take time to install, thereby adding to the labor cost. Furthermore, they contaminate the waste stream, making recycling more time-consuming and costly: plastic cannot be recycled with metal in it, therefore any metal must be removed before the plastic can be reprocessed. The cost of doing this can exceed the value of the reclaimed

material. For all these reasons, a wholesale reduction in the use of fasteners in assembly
has been a major trend in recent years.

3.4.1   Multiple Parts per Fastener

The number of fasteners in a given assembly can be substantially reduced, though not
entirely eliminated, by using one fastener to assemble several parts. Two versions of this
concept are illustrated in Fig. 3-11. In a reopenable version of this concept (Fig. 3-11a)
parts A, B, and C are placed over one another with their holes in line and secured by a
bolt and threaded insert. The boss on part C is used for location. This assembly uses a
metal insert (D) to provide threads strong enough to withstand repeated reopening of
the screw (E). This is the traditional approach. However, the author has observed an
increase in the use of self-tapping screws for reopenable applications, the increased
risk being driven by economic pressures.

   The design illustrated in Fig. 3-11b is a permanent assembly in that the fastener must
be destroyed to open it. A post both locates parts A, B, C, and D to part E and provides
a place for a push-on fastener (F). This concept can be used for both thermoplastic and
thermoset parts. Chapter 8, "Fasteners and Inserts," discusses these devices in greater
depth.

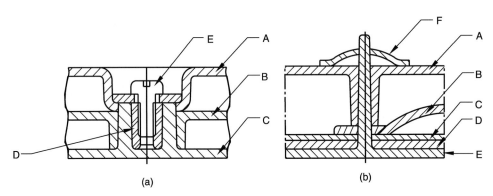

(a)   (b)

Figure 3-11   Multiple parts per fastener: (a) three-part reopenable assembly and (b) five-part
permanent assembly

3.4.2   Press and Snap Fits

Fasteners can be replaced entirely through the use of press and snap fits. Both can be
designed to create either permanent or reopenable assemblies. The press fit in Fig. 3-12a
is intended to be permanent. The stresses need to be carefully calculated to establish the
proper dimensions to provide enough interference to keep the parts in position, yet not
so much that cracking will result. The snap fit (Fig. 3-12b) is intended to be reopenable
because the snap can be pressed in through the hole in the side. If that hole were to be

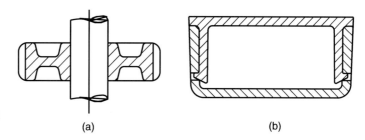

Figure 3-12(a) Press fit
and (b) snap fit   (a)   (b)

closed, it would be a permanent-type snap fit. The tooling for snap fits is generally much
more costly than the tooling for a similar part with a press fit. For further information
on either of these methods, the reader is referred to Chapter 14, "Press Fits/Force Fits/
Interference Fits/Shrink Fits," and Chapter 15, "Snap Fits."

3.4.3   Integral Hinges

Hinges molded directly into the part reduce the number of fasteners. They also permit
two parts to be molded as one, reducing part and inventory costs as well. When being
used to replace an existing hinge assembly, the integral hinge eliminates the two hinge
components (assuming they were not already molded in), the hinge pin, and the fasteners needed to attach them to the hinged parts. In addition, the use of integral hinges
saves the assembly time required to put the unit together.

There are, however, assembly costs that can be saved even when hinging between
two parts is not required. Any two parts, such as the top and bottom of an enclosure,
can be molded as one piece through the use of integral hinges. Figure 3-13 illustrates
what is known as the "living hinge," so called because, when properly manufactured,
it can withstand numerous flexures. The hinge is shown in the "as-molded" position in
Fig. 3-13a and in the "closed" position in Fig. 3-13b. Polypropylene is the principal
material used for living hinges, however there have been applications in other flexible
polymers. Living and other integral hinge designs are discussed in Chapter 9,
"Hinges."

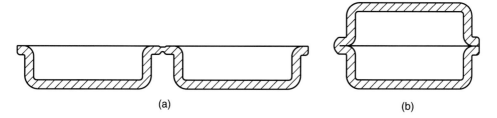

                    (a)   (b)

Figure 3-13   Living hinge: (a) as molded and (b) closed

3.4.4   Combining Fastener Elimination Concepts

The ideas expressed for the elimination of fasteners are often most effective taken in combination. Figure 3-14 illustrates a two-part integral hinge that has been combined with a snap fit. Either end can be snapped open or hinged with no additional parts. In this case, the parts are symmetrical, so only one mold is needed.

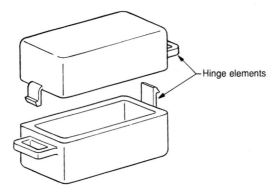

Figure 3-14   Two-part integral hinge with snap fit

On a larger scale, a number of fasteners and even snap fits have been eliminated in the assembly in Fig. 3-15. This is accomplished by providing slide-in features

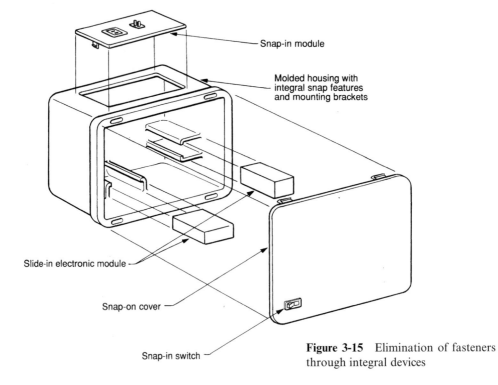

Figure 3-15   Elimination of fasteners through integral devices

for the internal parts, which are held immobile by the cover, which is, itself, a snap-in part. Snap fits have practically become standard for the assembly of electrical components.

3.5   Holistic Design

Traditionally, plastic part design has been segregated from the assembly, materials, processing, and tooling disciplines. Each of these aspects is pursued independently by individuals representing different disciplines. First the designer designs the part, then the moldmaker builds the mold, and the processor molds the part. In truth, these aspects are not discrete; rather, they are so totally interrelated that any attempt to discuss one without considering the others will fall short of the level necessary to compete in a world marketplace. When these elements are taken together as part of the design procedure, a greater degree of sophistication is achieved. Thus, the whole is greater than the sum of its parts and the result is "holistic design."

A holistic approach to design is, of necessity, more time-consuming and more costly in the design phase. First, the designer must know more about the materials, molds, processes, and assembly methods than would otherwise be necessary. This does not mean that in addition to being an expert designer, he or she must be able to recite the properties of thousands of resins, build a mold, operate every variety of processing equipment, and build an automated assembly machine. However, a reasonable familiarity with the nuances of those elements is necessary.

3.5.1   The Overall Design Considerations

For a good example of the interrelationships entailed in holistic design, consider a low cost, high volume, threaded thermoplastic bottle cap of the type that might be used for the soda industry (Fig. 3-16). This product contains an integral assembly device (screw threads) and a seal liner as well. The body of the cap must be sufficiently robust to withstand the forces used to emplace the liner and those developed by the screw thread, both initially and over the shelf life of the product. In addition, it must not loosen when exposed to elevated temperatures at any time during

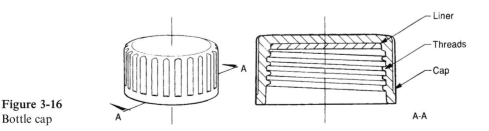

Figure 3-16
Bottle cap

production, shipping, or storage; nor may it crack as a result of stress relaxation or contraction at low temperatures. Furthermore, the issue of moisture vapor transmission may need to be examined.

Next, there is the matter of the thread design. The bottle thread will have a predetermined standard design for a bottle neck finish of a given size. For the cap thread, however, there is some design freedom in the exact dimensions, selection of radii, and determination of whether the thread will be complete or intermittent.

Thus far, this seems like a straightforward project. One might assume that the designer needs to determine what the loads will be and then establish the nominal wall thickness and thread form. With the addition of a seal ring, the design is ready to go out for quotes to the molder and the moldmaker. A design so arrived at, however, won't go very far in the marketplace. The tooling investment will be high because of the requirement for a very large unscrewing mold with many cavities. The molding cost of the cap will be excessive as well, because of the extended cycle needed to unscrew the caps on each cycle – all this because the fine points of the process and the tool design were not considered in the design of the part. (*Note*: The author does not wish to imply that there are no applications in which that type of cap would be appropriate. There are. This simply is not one of them.)

Both the tool cost and the part cost could be substantially reduced if the requirement for unscrewing were eliminated. This can be done with a stripping mold; that is, it must be possible to strip the thread off the core. However, the internal threads of the cap will form an undercut in the mold. Thus, they could either smash over or break off during ejection. Nevertheless, if the internal threads of the cap are designed correctly, if the right material is selected, if the mold is built properly and if the molding conditions are just right, the threads can be stripped off the core without the need for an unscrewing mechanism. This will result in a faster molding cycle, and the mold will be less expensive. It will also permit more cavities in a given platen area, which could allow the use of a smaller molding machine, with a lower operating cost. Thus, careful consideration of all the process and tooling elements can result in a bottle cap with a significant competitive edge over other closures. If, however, the designer had considered only the design elements of the cap, the manufacturer might have wound up with a less profitable product.

3.5.2   The Thread Design

Let us examine the thread in the bottle cap of Fig. 3-16 more closely. If it is continuous, it will provide hoop strength, which will stiffen the wall of the cap. That, in turn, will make it more difficult to expand the cap enough to stretch over the core and will resist the stripping of the thread. If the cap cannot stretch sufficiently, the thread will either mash over (if ejected hot enough) or shear off (if ejected cold), thus creating a need for an unscrewing mechanism. An intermittent thread, which would not have that additional strength, would make the cap easier to expand to facilitate removal from the core.

Unfortunately, what benefits ejection detracts from the properties of the finished cap. An intermittent thread will be weaker than a continuous one of the same depth. Now it would appear that the designer's choice becomes one of designing the cap with a continuous thread of shallow depth or with an intermittent thread. The decision will not be that simple.

Radii are also significant in this equation. Obviously, a round thread requires far less force to eject from the core than a squared-off buttress thread. It also has much less surface contact with the mating thread, so it does not exhibit the same holding power. To select the radii for the new thread, therefore, a compromise between these two conflicting requirements must be found. Lest one opt for the round thread too quickly, it should be pointed out that round threads possess a greater ratio of base width to height than buttress threads, which would likely result in a sink behind the thread as a result of nonuniform cooling. Often, a compromise is attained with a modified buttress thread to reduce the ratio of the thread base to the wall thickness.

Another factor determining the ease with which the cap can expand enough for the threads to strip off the core is the thickness of the wall itself. If that wall is kept to a minimum, its strength is reduced, and consequently its resistance to stripping. A thin wall also requires less material. If the wall is too thin, however, the nonuniform cooling condition will again be present, with its resulting sink mark, wherever there is thread. This, in turn, can be masked somewhat by vertical ribs in some applications. If that is not an acceptable solution and sink marks cannot be tolerated, the wall thickness will have to be increased until the sink no longer appears. That will result in a much stronger wall which will be more difficult to stretch over the core for stripping.

3.5.3   The Processing Considerations

Clearly, the stiffness of the material selected will also have a pronounced effect on this equation. However, the issue of whether to favor piece part benefits over processing advantages will arise, since the stiffer material will create a stronger cap while making ejection from the mold more difficult. In addition, the melt flow rate will be a factor because an increased ability to flow through the cylinder, sprue, runner system, and part details at lower temperatures will shorten the processing cycle time. Less heat into the process means less heat has to be removed before the part can set up enough to withstand the forces of ejection. This is also a cost reduction because the processor must pay to remove heat from the process as well as to put it in.

Besides providing greater throughput, which reduces the machine rate component of the part cost, a decrease in cycle time, if sufficient, can permit production goals to be met with fewer cavities. In turn, this could lead to greater tolerance control and the use of a smaller molding machine with a lower machine rate. Of course, all these blessings have a price. The tariff for increasing the melt flow is a drop in physical properties.

3.5.4   The Tooling Considerations

The ability to flow through the sprue and runner system can also be affected by the design of the mold. A carefully designed runner system with corners having generous radii will reduce the heat loss in the melt en route to the cavities. Cold slug wells, placed in the runner at a point just before the molten plastic enters the cavity, will remove the cooled tip (cold slug) of the melt so it will not impede flow into the mold. Melt flow programs can be used to develop the optimum flow path.

Next, there is the matter of part ejection. The least costly method of ejecting a piece part from the mold is the use of ejector pins. However, the concentrated loading on the moldment that is created means that the cooling cycle must be long enough for the part to become sufficiently stiff to withstand the forces of ejection. A stripper plate costs more, but it spreads the ejection forces uniformly around the edges of the cap. The cap can then be ejected in a softer state. This can shorten the cooling period significantly, leading to a potential 35% reduction in the cycle time.

Equally important, a stripper plate provides more freedom for the part designer in selecting the wall thickness and for the processor in setting the correct cycle. Such flexibility will help the processor find the precise point at which the thread is set up enough to retain its shape for ejection but the wall is soft enough for the part to expand over the core and be ejected.

3.5.5   Execution

Manufacturers usually base new versions of products on existing designs if such exist. Owing to the inherent inaccuracies of the computerized engineering programs available at the time of writing and the data to use in them, the precise dimensions cannot be reliably predetermined. The drawing will, therefore, be made with "steel-safe" dimensions: that is, the cavity dimension will be on the small side and the core dimensions will be on the large side, resulting in a wall thickness slightly thinner than the anticipated final thickness. The dimensions can then be adjusted by trial and error by simply removing steel from the mold.

Normally, a single cavity pre-production mold is built. There are a variety of methods for producing such a tool. That which is most appropriate for a given application will vary. A bottle cap will require a mold that fully duplicates the proposed production mold in gating, cooling, and ejection systems. The more the pre-production mold varies from the production mold design, the less reliable the results of the preliminary trials will be in determining the optimum combination of material, molding cycle, wall thickness, and details for the final design. The pre-production mold can also be used to produce the many hundreds of parts required for the final adjustments to the automatic assembly equipment.

A mold like this is, for all practical purposes, a one-cavity production mold. Consequently, it is expensive to modify. Therefore, it is often desirable to build one of soft steel, which can be easily, and inexpensively, altered. This is particularly useful when certain design details are likely to require considerable modification.

Light-duty snap and press fitments are examples of such applications, since they often involve finding the right "feel," which is difficult to calculate.

3.5.6   Toward Holistic Design

The amount of knowledge required can be considerable and difficult for individuals to acquire. Universities cannot impart it because students simply are not there long enough. Even postgraduate work does not suffice, since its focus is too narrow for the scope of this endeavor. For the present, this is knowledge that must be acquired while in the workplace and through industry seminars.

This level of sophistication is attained today in some industries through the use of design teams composed of individuals from various disciplines. At its best, this system can be nearly as effective as one individual in possession of the broad spectrum. At worst, it can degenerate into a group of individuals narrowly protecting their own sectors. In most cases, it is somewhere between these extremes.

Regardless of how they are achieved, the benefits of the holistic approach to design are necessary to maintain a competitive advantage in a world market.

4   Design for Disassembly and Recycling

4.1   Introduction

Each year municipalities in the United States dispose of approximately 320 billion pounds of solid waste. In 1984, 80% of that total went into landfills, 9% was incinerated, and 11% was recycled. Unfortunately, the availability of landfills is rapidly diminishing. Between 1980 and 1990, the number of landfills in the United States diminished by one-third. Despite vast improvements in landfill technology, no one is interested in having such facilities created nearby. Thus, the availability of landfills is certain to continue to decline, with consequent rises in the cost of disposing of solid waste, which will have to be transported ever greater distances for disposal.

At some point, solid waste disposal will become so expensive that governing bodies will seek respite. While there is little debate over the desirability of recycling, members of the general public are inclined to consider the problem solved once they have separated their newspapers and some containers into recycling bins. The cost of garbage collection will certainly rise and there will, no doubt, be screams of agony as consumers discover that they must pay $100 to dispose of an old dishwasher. In Europe, governments have already turned to industry with product "take back" legislation. There they have discovered that the "tomorrow" of dealing with the problem is at hand.

In the United States, the function, market, and cost priorities have elbowed disassembly and recycling to a secondary position. The evidence suggests that the issues of recycling and disassembly will soon rise to a higher level of priority. Indeed, in some areas, there is already pressure in this direction.

When the recycling of automobiles was first proposed, someone figured out that to completely disassemble and sort the materials for one model year's worth of cars sold in the United States, several thousand containers would have to be spread over acres of land, and engineers would be needed to do the work! Frankly, this pessimistic estimate was not far wrong. For a product as complex as an automobile, the problem of disassembling and removing the component materials is quite formidable. Many different resins are used. Even parts made of the same resin will contain different additives and have different coatings.

Nonetheless, the bulk of the volume of the product is readily salvageable. According to research from the Ford Motor Company, more than 94% of junked automobiles went to a dismantling center in 1994, and 75% of the content was recycled.

An infrastructure exists for the recycling of automobile components, and steps are being taken to implement it. The Society of Automotive Engineers (SAE) and the American Society for Testing and Materials (ASTM) have developed marking systems for plastics (ASTM D1972-94) similar to the system of the International Organization for Standardization (ISO).

Individually, the automobile companies have initiated programs on their own. Ford makes new splash shields from the polypropylene recycled from old batteries, taillamp housings from salvaged bumpers and grille opening reinforcement panels, and headliners and luggage rack side rails from recycled PET.

Most business and household products are composed of a far more modest number of parts, and significant progress can be made toward achieving total or near-total recyclability.

4.2   Design for Disassembly

Design for disassembly is not a new concept. Products with serviceable components have always been designed this way – in fact, it was more widespread in the days before repairs became so expensive that it became cheaper for the consumer to dispose of most small products and replace them with new models than to have the original units repaired.

In those days, it principally meant designing the housings with metal screws. If the housings were to be opened repeatedly, metal inserts were used. Self-tapping or thread-forming screws would suffice for housings not expected to be opened often. Frequently opened components, like battery doors, were snap-fit together for convenient access.

These practices are still widespread and for the very same reasons. What is new is the adoption of these practices for parts that are not intended to be reopened. However, the term "design for disassembly" as used today usually refers to the disassembly associated with recycling. In this case, the objective is to free the part from mating components in an uncontaminated fashion so it can be isolated with other parts of the same material. Since there is no intent to reassemble, damage incurred during disassembly is of no concern.

The biggest problems confronting the disassembly and recycling process are those of economics. Corporate good intentions notwithstanding, if the cost of the recycled material exceeds that of virgin resin offering the same properties, only legislative mandates will induce industry to move toward such recycling. Therefore, it is necessary to disassemble the product in the most efficient manner possible.

4.2.1   Reopenable Assemblies

The most obvious contaminants are metal fasteners, since plastic parts containing them cannot be readily recycled. One technique is to partially eliminate such fasteners by substituting self-tapping or thread-forming screws for machine screws and metal inserts. Then there are only the screws to remove, and they are easier to get out than inserts. This approach has the disadvantage of limiting the number of times a product can be disassembled, however, because threads of these types are readily stripped by excessive torque. Once the product is out of the factory, it is impossible to control the application torque on these types of screws.

Another approach is to eliminate metal fasteners altogether through the use of press and snap fits. Press fits have limitations because they cannot have a great deal of holding power if they are to be readily disassembled. The difference in holding power can vary greatly with just a few thousandths of an inch in interference. Thus, such fitments can be tricky to use. The reader is referred to Chapter 14, "Press Fits/ Force Fits/Interference Fits/Shrink Fits," for more details.

Snap fits are easier to control and can be designed with a through hole such as that illustrated in Fig. 4-1a for easy disassembly, or with a closed recess for a clean external appearance. In the latter case, a slot in the bottom, as illustrated in Fig. 4-1b, can provide access for a screwdriver blade, which will activate the snap finger.

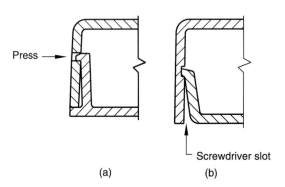

Press →

Screwdriver slot

(a)   (b)

Figure 4-1   Snap fits: (a) through hole type and (b) tool activated

There are some concerns regarding the use of snap fits. First, both the snap finger and the recess it fits will require tooling significantly more expensive than is needed for the same part without these details. Second, the maintenance cost is greater for these types of molds. Next, the mechanisms required to make these elements will require more space in the mold, which could cause either a reduction in the number of cavities that can be run in a given molding machine or call for a larger press, with a higher operating rate. These factors, plus the longer cycle associated with such mechanisms, could increase the processing cost. Finally, there is some material in the snaps themselves which adds to the material cost. Chapter 15, "Snap Fits," presents a more extensive discussion of this topic.

When metal inserts cannot be avoided, they can at least be designed in a fashion that promotes the most efficient handling for recycling. Molded-in inserts are the most difficult to remove and should be avoided. Inserts made of steel can be readily removed from the regrind with magnets, but they can damage the grinder during the regrinding operation prior to their removal.

As illustrated in Fig. 4-2, an insert can also be removed by heating it with a hot probe until the plastic around it softens enough for it to be pulled out. This process is not recommended for plastics with high melting points, since its efficiency is highly dependent on the melting point and strength of the material. In most cases, faster methods can be found.

Some inserts are readily removed with tools. A tool called the Easy Out is illustrated in Fig. 4-3. It is used to wrench inserts of various types from low to medium strength

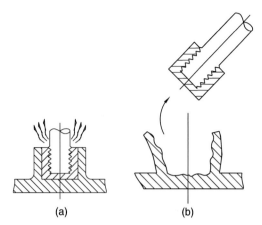

Figure 4-2 Hot probe removal: (a) hot probe melting plastic and (b) insert removal

(a)   (b)

plastics. It can handle screws up to 3/8 in. diameter. Other tools used for this purpose are the "T" wrench and the power impact wrench. Inserts that can be removed in this way include coiled wire inserts, press-in inserts, double-threaded inserts, and any made of a material soft enough for this method to be effective.

There are two types of expansion insert: those with a center disk, which forces the insert to expand (Fig. 4-4a), and those that use the screw to force expansion (Fig. 4-4b). The latter are less expensive, but sometimes fall out of their own accord when the screw is removed. They are quite easy to remove, however.

The type of expansion insert shown in Fig. 4-4a is more difficult to remove. Inserts such as these, molded-in inserts and those that are emplaced with adhesives, are often used for thermosets. When they are in bosses, they can sometimes be broken out, since such bosses are often "resin-rich" (short of glass reinforcement) and therefore quite brittle. If this is not possible, they must be drilled out, a slow process that leaves metal chips that must be cleaned up to prevent contamination of the plastic to be recycled.

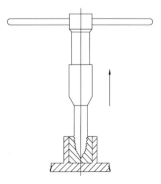

Figure 4-3 Insert removal with Easy Out

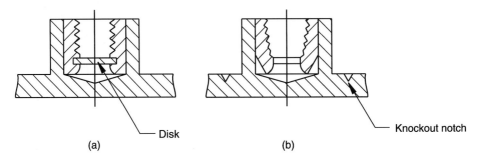

(a)   Disk   (b)   Knockout notch

Figure 4-4 Expansion inserts: (a) disk expansion insert and (b) screw expansion insert and knockout notch

The breakout approach can be used for thermoplastic inserts as well. In fact, notches such as illustrated in Fig. 4-4b can also be used for bosses, stakes, ultrasonic joints, or any other type of joining detail. This concept is similar to the "knockouts" used for optional holes in metal electrical boxes. Depending on the material used, it may be possible to "knock out" the welded area with a hammer in the same fashion. Unfortunately, notches also create weak points in the part which detract from its function. Another device used for this purpose is the hole saw (Fig. 4-5), which removes the boss and a circle of its surrounding material.

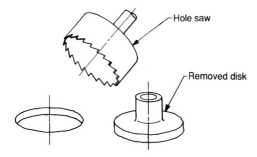

Hole saw

Removed disk

Figure 4-5 Hole saw removal

In some cases, the insert or screw can be placed in a through hole, such as that illustrated in Fig. 4-6, and pressed out from behind.

There are other fasteners besides inserts to be concerned with. Self-tapping and drive screws, of course, can be unscrewed, although some are more economically pressed out if they are in through holes. Cap fasteners, such as illustrated in Fig. 4-7a, can be wrenched off with pliers. The push-on fastener shown in Fig. 4-7b can be pried up with a screw driver and twisted off with pliers or simply broken off with its stud. The spring-loaded version shown in Fig. 4-7c can be removed by pressing down the upper segment to relax the tooth grip on the stud. The self-threading nuts illustrated in Fig. 4-7d have a single tooth form that cuts a coarse thread when pressed on during assembly. They can be removed with pliers or a socket wrench, depending on the shape of their heads. The latter can also be removed

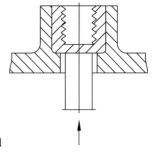

Figure 4-6 Press-out method

with an electric or pneumatic screwdriver. These and push-on versions can also be twisted off with pliers.

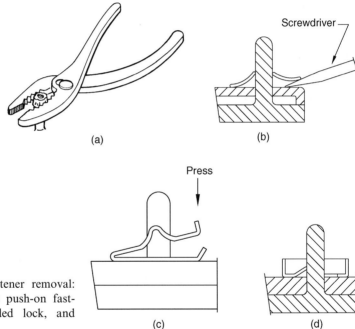

Figure 4-7 Stud fastener removal: (a) plier twist-off, (b) push-on fasteners, (c) spring-loaded lock, and (d) self-threading

4.2.2 Permanent Assemblies

Permanent assemblies include all the welding methods plus stakes and adhesive joints. Stakes are the easiest to deal with because it is usually possible to knock their heads off with a chisel or to pry them off as illustrated in Fig. 4-8a. If there is adequate clearance, they can be knocked off from the back side with a hammer.

If the stake is a flat-head type, where the surface is flush with the surface of the part, or if there is not enough space to get under the head of the stake, then a wedge, like that

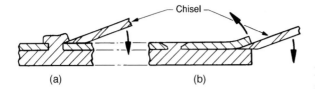

(a)   (b)

Figure 4-8   Removal of: (a) headed stakes and (b) flush stakes

illustrated in Fig. 4-8b, can be driven between the two parts and they can be separated in that fashion.

Snap fits can also be designed to be permanent rather than reopenable. If the snaps are too strong to be broken off, the area around the snaps must also be designed with a weak segment that would allow them to be broken out much like the "knockouts" described earlier and shown in perspective in Fig. 4-9.

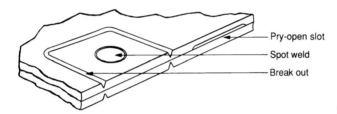

Pry-open slot
Spot weld
Break out

Figure 4-9   Spot weld opening

It is far more difficult to separate parts that have been welded together, although the level of difficulty depends on the location of the welds and their strength. Intermittent welds like those often created by ultrasonic energy directors or spot welds like the one shown in Fig. 4-9 are most vulnerable to a wedge at points between the welds. A pry-open slot for this purpose is also illustrated in that figure.

More robust welds may need to be cut away with a band or circular saw. That is probably the best approach to adhesive joints. Both they and induction/electro-magnetic welds involve the use of contaminating third materials that substantially reduce the value of the recycled resin because it becomes a "mixed material recycled material." In the case of the adhesive joint, it is the adhesive itself. In electromagnetic induction welding, there is an insert containing ferrite particles between the two parts. The latter joint is one of the easiest to disassemble because the joint can be remelted if the cost of the time and equipment can be justified.

Solvent welds are also easier to deal with. Since solvent welds are most commonly used to join parts of the same material, it may not be necessary to disassemble them at all. The best approach is to establish whether solvent welds indeed were used before attempting to disassemble the parts. If the parts must be disassembled, they can sometimes be softened enough for disassembly by exposing them to the same solvents used to assemble them. If not, they should be treated the same way as other welded joints.

4.3   Design for Recycling

The design of a product involves a wide variety of considerations, of which recycling is only one. The following discussion presumes that recycling is the primary concern, all others being secondary. In this way, the designer can quickly rate the recyclability of the assembly method under evaluation, for reference along with the other considerations.

4.3.1   Simplification

Part reduction was discussed in Chapter 3 as a means of cost reduction. However, it has disassembly importance as well, since it is clearly less expensive to disassemble fewer parts. Furthermore, the fewer parts there are, the less likely it is that the product will break down and require service. If it does fail, the repair will cost less because less time will be needed to effect it. Thus, a simpler design is not only cost-effective, it is environmentally responsible as well.

4.3.2   Assembly Method Selection

From a recyclability standpoint, the author recommends the following assembly method selection hierarchies. This presumes that it is not possible to make the two parts of the same material such that no disassembly would be required.

4.3.2.1   Reopenable Methods

1. *Press fit* Easy to disassemble, often without heat or tools. However, it can require very tight tolerances and be too easy to disassemble. If the two materials have significantly different coefficients of linear thermal expansion (e.g., a plastic gear on a metal shaft), heat can be used to disassemble them.
2. *Snap fit* Often thought of as a permanent assembly method, it can be readily reopenable if so designed. Increases the cost of tooling, maintenance, and processing, however, it saves the cost of a fastener.
3. *Self-tapping, thread forming, drive screw, or sheet metal nut* Saves the cost of the opposite threaded fastener, its assembly, and its disassembly. However, it can be opened only a limited number of times, the drive screw least of all.
4. *Machine screw with simple expansion insert* This is the easiest of the threaded inserts to remove. Unfortunately, it will sometimes fall out of its own accord, once the screw is removed, and be lost.
5. *Machine screw with helical wire, double-threaded insert, or onsert* These are the easiest of the female fasteners to remove, but they either do not have the strength and durability of a full metal insert and their installation is time-consuming.
6. *Machine screw with molded-in or postmolded insert* It does not matter which way these inserts are emplaced; they are all difficult to remove. Molded-in inserts are usually more difficult to remove simply because they are often custom-made for a

specialized requirement. As such, they can have odd shapes or be more flimsy than standard inserts. When fasteners are required, their number can be reduced through careful design as illustrated in the previous chapter.

4.3.2.2   Permanent Methods

By definition, the permanent assembly techniques are all difficult to disassemble. Therefore the hierarchy for these methods is much shorter.

1. *Press or snap fit* Both these methods can be designed to be permanent, although the press fit may require expansion of one part with heat to accomplish. However, they can also be designed to be readily disassembled with limited difficulty.
2. *Staking* In most cases, the head of the stake can be located such that a tool can be used to pry it off. Failing that, the parts can be designed to be readily pried apart.
3. *Hot gas, hot die, solvent, ultrasonic, and vibration welding* Products made by all these processes are difficult to disassemble, the level of difficulty varying with the structure and material. When possible, design the assemblies so that the weld can be broken out or the parts can be pried apart as illustrated in Fig. 4-9.
4. *Adhesives and induction/electromagnetic* Products joined by these means are more difficult to disassemble than those above in that they are contaminated with foreign matter, which must also be removed.

4.3.3   Material Selection

A great deal of disassembly can be avoided through material selection with recycling in mind. First, those materials must be identified that can meet the product's function and cost parameters. Once that has been accomplished, the recycling concerns can be considered in arriving at the final selection.

All plastic materials can be recycled in some fashion. When selecting a material that is to be recycled, the designer must establish the level of recyclability of the resin that has otherwise met the function and cost criteria. The thermosets, for example, have very limited recyclability because they cannot be remelted and reused. Current recycling of thermosets calls for the material to be shredded, glass fibers to be separated out, and the remaining resin to be mixed with calcium carbonate and ground to a fine powder. The powder, referred to as a "composite filler," is mixed with virgin resin at a level of around 7% by weight to form a new compound. Thermosets have also been chopped up and used as filler for a variety of applications.

There are more options for the thermoplastics. To begin with, if they can be reground and recycled through the same process, the processor has the opportunity to recycle sprues, runners, cut-offs, and scrap parts. This is one aspect of recycling that has proven to have current economic benefits. Provided the returned parts are not contaminated and the same material is still in use, this also provides the manufacturer with a means for dealing with returned parts.

The SPI has a "Clean Sweep" program to help companies identify internal scrap that can be reused, establish collection methods, and find a market for the reprocessed material. Under this program, Ticona reclaims and sells acetal, polyester, nylon 6/6, polyphenylene sulfide, and liquid crystal polymers. General Electric has a program for repurchasing scrap polycarbonate for resale, mixed with virgin. Many materials must be mixed with virgin to have reliable physical properties and be reused in this fashion.

In addition, there are commercial markets for some post-consumer recycled resins. Polyethylene terephthalate, the material used in soda bottles, is known to be recycled into sheeting, textile fibers, films, and bottles. High density polyethylene, the milk bottle material, is made into traffic barriers, detergent bottles, drainage pipes, lumber, pallets, flower pots, toys, and boat gears. Polypropylene is recycled for packaging, closure, automotive, and medical applications.

The designer concerned with creating a product that can be recycled must select a resin that has an established market for the recycled material and an organized method of collecting, disassembling, and sorting the product for reuse. To interest the commercial market, a plastic must be available in a significant and steady supply. Note that all the polymers for which applications were listed in the preceding paragraph are commodity resins produced in large volume.

If a resin is to be used for which such a market does not exist, it will be necessary to create a market. This is a less formidable task than it might seem if the resin is a high value engineering resin and the volume is adequate. Look first to the resin supplier, who may have a recycling program, or at least people on staff who know what is taking place in the market for their recycled resin. To be successful, a recycling program must be carefully planned. One cannot simply manufacture a product made of a material that could be recycled and presume that it will be recycled if one is to be regarded as serious about recycling.

4.3.4   Additives

Plastics materials are often referred to as "compounds" instead of resins because, as used, they contain a great deal more than the basic resin. Additives are used to enhance the properties of the material and its processing. However, these additives can adulterate the material sufficiently to limit its value as a recycled material. The following is a partial list of such additives.

- Antioxidants
- Antistatic agents
- Colorants (dyes and pigments)
- Coupling agents
- Fillers/extenders (calcium carbonate, talc, kaolinite, alumina trihydrate, feldspar, silica)
- Flame retardants
- Foaming agents
- Heat stabilizers
- Impact modifiers

- Lubricants
- Organic peroxides
- Plasticizers
- Preservatives
- Property enhancers (mica, spheres)
- Reinforcements (carbon, glass, aramid)
- Stabilizers
- Ultraviolet inhibitors

Although these additives add substantial benefit to the material, they limit its acceptance to the high value recycled material stream. Filled and reinforced compounds are being recycled; however, each specific material under consideration needs to be investigated with respect to its ability to be recycled. When the total cost picture is evaluated, it may be more desirable to go to a resin with higher properties that can be used in an unadulterated form.

4.3.5   Contaminants

There are many ways for a product to become contaminated once it leaves the manufacturer and is, therefore, uncontrollable. Usually these are sources of contamination, dirt or grease, and the recycler deals with them through washing protocols. However, some contaminants are within the control of the manufacturer.

Paint and metallic coatings represent the type of coating that can make it difficult to reclaim and recycle a material. While one cannot deny the necessity to provide EMI/RFI shielding, enhance the external appearance of the product, or protect it from ultraviolet light through the use of color or metallics, there are choices to be made. An integral color, such as a pigment, may be an acceptable alternative, depending on the intended use of the recycled resin. Color can be eliminated from internal parts completely, thereby making them more desirable from a recycling standpoint.

Lubricants also contaminate the recycled material stream. Two types of external lubricant are in use. One is the variety used in the function of the product, such as that which is often found on bearings and the like. The other is the mold release used in the processing of the part. While it may be difficult to remove the former, the latter can be reduced or eliminated through the use of greater draft angles. These reduce the force necessary to strip the part off its core, thus eliminating the need for a mold release. As added ecological and economic benefits, increased draft angles can reduce the cycle time, which reduces energy usage as well as lowering the machine cost.

4.3.6   Material Reduction

There are several ways to reduce the amount and number of materials to be recycled. By using materials sufficiently alike to be recycled together, the number of parts to be disassembled can be reduced. This can have an effect on the assembly aspects as well, since similar materials usually have melting points close together. This characteristic facilitates the assembly techniques that utilize heat for welding. Of the materials that can be solvent-welded, those that are chemically close to each other lend themselves most easily to the process. Although it is always advantageous to eliminate a disassembly operation, it is particularly desirable with these assembly methods because products so manufactured are more difficult to disassemble.

Sometimes the number of materials used can be reduced by combining parts. Variations in geometry, derived carefully by a designer mindful of molded-in stress limitations, can alter the strength and stiffness of a component such that the need for an additional part of another material can be eliminated. The stiffness of a given material can also be altered through the use of ribbing or foaming. Thus, one component could be made of solid resin by injection molding and the other might be structural foam molded out of the same material.

When the number of materials can be no further diminished, it may be possible to reduce the amount of material used through a technique called "thinwalling." While essentially a cost reduction technique, it also has recycling and ecological benefits because, in addition to reducing the amount of resin used, it can reduce the cycle time during processing. Two criteria must be met for effective thinwalling: strength and moldability. The wall thickness that meets the strength requirements of the part can be determined by manual methods, however it is best done with finite element analysis. A mold filling analysis program will help determine the minimum wall thickness that will fill. While new, high flow rate versions of many resins have been

Code		Material
1	PETE	Polyethylene terephthalate (PET)
2	HDPE	High density polyethylene
3	V	Vinyl/polyvinyl chloride (PVC)
4	LDPE	Low density polyethylene
5	PP	Polypropylene
6	PS	Polystyrene
7	Other	All other resins and layered multimaterial

Figure 4-10 Plastics recycling code for use in packaging applications: Examples of container code system for plastic bottles. The stand alone bottle code is different from standard industry identification to avoid confusion with registered trademarks (From Rosato, *Plastics Encyclopedia and Dictionary*, p. 151.)

developed to accommodate this practice, there is significant danger that high molding temperatures and cavity packing pressures will result in distortion and high levels of molded-in stress.

4.3.7   Identification and Disassembly Instructions

Regardless of which material and assembly technique is used, the designer must recognize that this will not be immediately obvious to the disassembler. A sticker bearing printed instructions should be affixed to the interior of the main housing. Directions on opening the main housing should be etched into the mold on the bottom part. Material identification should be in the mold for all the parts according to one of the internationally recognized systems.

For packaging use (only), the Society of the Plastics Industry has produced the Plastics Recycling Code, illustrated by the example in Fig. 4-10.

The balance of consumer and industrial products are identified for recycling under the SAE, ISO, and ASTM standards, which are regarded as quite similar. The ASTM standard is designated D1972-94 and is available from:

American Society for Testing and Materials (ASTM)
100 Barr Harbor Drive
P.O. Box C700
West Conshohocken, PA 19428-2959
Telephone: (610) 832-9585
Facsimile: (610) 832-9555
www.astm.org

5 Assembly Method Selection by Material

5.1 Thermoplastics Versus Thermosets

This is a book about the assembly of plastic products, and most of this chapter is concerned with the assembly aspects of the plastics materials. However, a knowledge of material fundamentals is necessary to understand these relationships.

There are two fundamental types of plastic: thermoplastics and thermosets. Thermosets undergo a chemical change known as "cross-linking" when they are heated to their melting temperatures and cannot be returned to their original state. Scrap generated in processing cross-linked materials cannot be reused.

Thermoplastics melt but do not undergo a chemical change when they are heated. That means they can be reheated and returned to their original state. As long as they are not contaminated, sprues, runners, trimmings and scrap parts can be reground and reintroduced to the processing cycle. In this fashion, material usage efficiencies approaching 100% can be reached.

Unfortunately, thermoplastics do not completely return to their original state. An infinitesimally small amount of degradation does take place. That means that the material does not have quite the same physical properties the second time around as it did originally. If the material is reground and run through an injection molding machine again, it will produce the part. However, the part will not be quite as strong as a part made from virgin material; in addition, it will be more brittle and slightly discolored.

Will the degradations in quality just described affect the performance of the part? That depends on a number of things. Some resins are more vulnerable to such degradation than others. Then there are questions of temperature: How high was it, and how long was the melt exposed to it? Finally, there is the matter of the safety factor, if any, included in the original design and the percentage of regrind that was permitted. A relatively unaffected resin with a small percentage (5 or 10%) of regrind will be imperceptible in a part that has a substantial safety factor and is not exposed to high temperatures in its normal life. The other extreme is a highly affected polymer, which might have a 50% regrind content, a minimal safety factor, and daily exposure to elevated temperatures. A part made of such a resin will undoubtedly suffer a substantial drop in physical properties that could result in a failure before too much time has passed. In varying degrees depending on the particular resin's vulnerability, ultraviolet light exacerbates the problem.

Between these extremes there is a gray area. Limiting the amount of regrind and the temperature exposure will result in a reduced rate of deterioration. It is difficult to tell how great the reduction will be without extensive testing over an extended period of time. At the time of this writing, there is no accurate method of predetermining exactly when a given part will fail. With an adequate design safety factor, the part could survive for years before experiencing failure. (Where the continuous use temperatures are available, they are provided in the tables for individual materials found in Section 5.4.1.)

"Heat history" is the term applied to the accumulated degradation just described. The resin will encounter exposure to heat when it is first manufactured as well as when it is processed. Some processes add more heat to "history" than others. Thermoforming, for example, is a process that produces parts with a rather high heat history. Having been exposed to elevated temperatures in the formation of the compound, the material is transformed into sheet, usually by extrusion, which entails additional exposure. It is then thermoformed, which gives it another dose of elevated heat.

Thermoforming produces a high percentage of scrap (offal) because the part must be cut out of the sheet. The resulting blank can be reprocessed as scrap and usually is, thus incurring additional heat history. If a 50% regrind is permitted, that means that the second time around, 25% has been through both processes twice. The third time, 12.5% of the part has been through three times, and so on. One could readily conclude that it is unwise to use regrind when thermoforming. However, efficiency in material utilization requires the use of regrind. Fortunately for thermoformers, the molded in stress associated with this process is usually insignificant. That tends to compensate for the additional heat history.

A number of the assembly processes also add to the heat history of the material. Thus, the determination of the best process for a given application should consider not only the heat applied during the assembly process, but that which the product has encountered up to that point.

5.2   Amorphous Versus Semicrystalline Thermoplastics

There are the two basic varieties of thermoplastics: amorphous and semicrystalline. Rosato's *Plastics Encyclopedia and Dictionary* defines an amorphous plastic as

> the type in which the molecular chains exist in random coil conformation (such as water-boiled spaghetti). There is no regularity of structure as there is with crystalline thermoplastics. The structure of an amorphous plastic is characterized by the absence of a regular three-dimensional arrangement of molecules or subunits of molecules extending over distances that are large compared to atomic dimensions; there is no close packing long-range order. However, due to the close packing in the condensed state, certain regularity of structure exists on a local scale, denoted as a short-range order.

Rosato defines semicrystalline plastics as those which

> have their molecules arranged in a very regular repeating lattice structure, so precise that every atom of the polymer molecule must recur at very specific points in the repeat structure. While no material is completely crystalline (typically less than 80%), very regular polymers can be mostly crystalline with only small amorphous areas remaining between crystallites. Technically it is more accurate to refer to these plastics as semi-crystalline but in general the industry refers to them as crystalline.
> [For simplicity, this book will henceforth do the same.]

Basically, the individual molecules tend to pack into a neat orderly crystalline arrangement of high density, sharp melting point, and low solubility. Their stability at low temperatures and energy levels is demonstrated by the heat of fusion of enthalpy which is liberated during the formation of the crystal, and which must be supplied in order to melt and disperse the molecules out of the crystal. At higher temperatures and energy levels, however, vibrations and Browninan motions of the molecules tend to drive them apart, destroying the crystals and dispersing them into a highly random and disordered state containing many degrees of freedom. The balance between these two factors determines the melting point of each particular plastic.

There is far more to the topic. Since, however, this is a book on assembly, the author must refer the reader to other sources to learn more about the nature of amorphous plastics. This book must concern itself with the different properties associated with amorphous and crystalline plastics which affect assembly. (The type of thermoplastic is listed next to the name of each polymer in Section 5.4.1.)

5.2.1   Postmolding Shrinkage

First, the amorphous plastics will, as a group, tend to have fairly low postmolding shrinkage rates that do not vary a great deal with the direction of flow. The crystalline materials will tend to have much higher shrinkage factors (which are reduced when they contain fillers) that may vary considerably with direction of flow.

Shrinkage rates affect the processor's ability to maintain tight tolerances. The fact that a shrinkage rate is low and directionally uniform means that it is more predictable as far as the moldmaker and the molder are concerned. Thus they are better able to hold tight tolerances with an amorphous material than with a crystalline material. This is important to know when selecting a material for a tight tolerance assembly.

Crystalline thermoplastics tend to be more compressible than amorphous thermoplastics. That helps the designer overcome some of the effects of their higher shrinkage rates because they permit the use of compressible fitments. Fillers and reinforcements, particularly glass fibers, will reduce both the shrinkage and the compressibility of crystalline thermoplastics. The stress crack resistances of crystalline thermoplastics are generally higher than those of amorphous thermoplastics, thus making the former a better choice for press fitments that will be under continuous pressure. (Shrinkages for the various plastics are listed in Section 5.4.1.)

5.2.2   Coefficient of Linear Thermal Expansion

The coefficients of linear thermal expansion (CLTEs) for amorphous plastics will, as a group, be lower than those for crystalline thermoplastics. There are two important assembly considerations associated with this fact.

The first concerns fitments and is particularly worrisome when materials with significantly different rates of thermal expansion are fitted together. If the materials are welded, the welds will be stressed beyond the values inherent in the basic product design. If the joint is a mechanical one, the part with the greater expansion rate will obviously expand more than the other part, causing the fitment to weaken. The larger the product, the worse this problem will be.

The other effect of the differences in CLTE has to do with the use of heat in the joining process, particularly in the assembly of mechanical joints. One of the secrets of efficient assembly is to heat the outer, or female, part so that it fits easily over the male part. This is more readily accomplished with parts made of materials with higher rates of thermal expansion. (The CLTE for each material is listed in Section 5.4.1.)

5.2.3   Weldability

Basically, thermoplastics are weldable. However, they vary considerably in degree, and some are so difficult to weld that it is, not feasible to weld them commercially. (Where available, weldabilities of the different polymers and welding processes are listed in Section 5.4.1.)

5.2.4   Solvent Sealability

The crystalline thermoplastics, as a group, are much more resistant to solvents than the amorphous polymers. Consequently, crystalline thermoplastics are not commercially solvent welded. Most amorphous thermoplastics are readily solvent-weldable, usually to parts made of the same material. However, some dissimilar amorphous resins can be welded to each other. (Solvent sealability for each polymer is listed in Section 5.4.1.)

5.3   Thermosets

As their name implies, thermosets take a permanent set with the application of heat and pressure. Cross-linking takes place between the molecular chains, and they cannot be returned to their original state. When additional heat and pressure are applied, there is degradation in the form of charring and burning. Thermosets cannot be reground and reused. The only commercial way to recycle them at the time of writing is to chop them up and use them for fillers in resins and roadways and the like. (There is a method under development to extract the glass fibers from reinforced materials for reuse.)

The assembly effects are significant when thermosets are used. Moreover, not all the assembly methods that use heat are available – which means that heat distortion will not be a problem either. Thus thermoset parts can be handled while they are still hot from molding or in a very hot environment.

In general, the engineer receives greater physical and thermal properties for the money spent with thermosets than with thermoplastics. However, that statement does not give the complete picture, since thermosets do not provide the opportunity for complete utilization of material; indeed, they have more limited processing opportunities and much more limited availability of adhesive- or fastener-free assembly. (Thermoset plastics are so designated next to their name in Section 5.4.1.)

5.4   Assembly Method by Material

There are a number of criteria to use in the selection of a plastic material for a given application. This chapter deals with assembly considerations. In Section 5.4.1 the materials are listed in alphabetical order for easy reference. Next to each material name is the type of plastic it is. The information on properties is very general, since it must cover the range for the hundreds of grades of polymers made possible by the availability of a broad spectrum of alloys and additives. Note that the range presented does not include values for filled resins: the presence of mineral or glass would substantially alter these figures. This information is presented to provide direction; it is not to be used in engineering formulas. The reader is advised to obtain the specific properties for the resin grade under evaluation from the manufacturer.

An alphabetical list of plastic material abbreviations is provided in Table 5-1.

Table 5-1   Alphabetical list of plastic material abbreviations

ABA	acrylonitrile-butadiene-acrylate
ABS	acrylonitrile-butadiene-styrene copolymer
ACS	acrylonitrile-chlorinated PE-styrene
AES	acrylonitrile-ethylene-propylene-styrene
AMMA	acrylonitrile-methyl methacrylate
AN	acrylonitrile
APET	amorphous polyethylene terephthalate
APP	atactic polypropylene
ASA	acrylic-styrene-acrylonitrile
BMC	bulk molding compounds
BOPP	biaxially-oriented polypropylene
BR	butadiene rubber
BS	butadiene styrene rubber
CA	cellulose acetate
CAB	cellulose acetate butyrate
CAP	cellulose acetate propionate
CFC	chlorofluorocarbons
CN	cellulose nitrate
COC	cyclo-olefin copolymer
COP	copolyester
COPA	copolyamide
COPE	copolyester
CP	cellulose propionate
CPE	chlorinated polyethylene
CPET	crystalline polyethylene terephthalate

Table 5-1   Alphabetical list of plastic material abbreviations *Continued*

CPP	cast polypropylene
CPVC	chlorinated polyvinyl chloride
CR	chloroprene rubber
CTA	cellulose triacetate
DAM	diallyl maleate
DAP	diallyl phthalate
DCPD	dicyclopentadiene
DETDA	diethyltoluenediamine
DMT	dimethyl ester of terephthalate
EBA	ethylene butyl acrylate
EC	ethyl cellulose
ECTFE	ethylene-chlorotrifluorethylene copolymer
EEA	ethylene-ethyl acrylate
EMA	ethylene-methyl acrylate
EMAC	ethylene-methyl acrylate copolymer
EMPP	elastomer modified polypropylene
EnBA	ethylene normal butyl acrylate
EP	epoxy resin
EP	ethylene-propylene
EPDM	ethylene-propylene terpolymer rubber
EPM	ethylene-propylene rubber
EPS	expandable polystyrene
ESI	ethylene-styrene copolymer
ETE	engineering thermoplastic elastomer
ETFE	ethylene-tetrafluoroethylene copolymer
EVA(C)	polyethylene-vinyl acetate
EVOH	polyethylene-vinyl alcohol copolymer
FEP	fluorinated ethylene propylene copolymer
FPVC	flexible polyvinyl chloride
FRP	fiber reinforced plastic
GMT(P)	glass mat reinforced thermoplastic
GPPS	general purpose polystyrene
GRP	glass fiber reinforced plastic
HCFC	hydrochlorofluorocarbon
HCR	heat-cured rubber
HDPE	high-density polyethylene
HFC	hydrofluorocarbon
HIPS	high-impact polystyrene
HNP	high nitrile polymer
LCP	liquid crystal polymer
LDPE	low-density polyethylene
LLDPE	linear low-density polyethylene
LP	low-profile resin
MBS	methacrylate-butadiene-styrene
MC	methyl cellulose
MDI	methylene diphenylene diisocyanate
MF	melamine formaldehyde
MMA	methyl methacrylate
MPE	metallocene polyethylenes
MPF	melamine-phenol-formaldehyde
MPR	melt-processalbe rubber
NBR	nitrile rubber

Table 5-1   Alphabetical list of plastic material abbreviations *Continued*

NR	natural rubber
OPET	oriented polyethylene terephthalate
OPP	oriented polypropylene
OSA	olefin-modified styrene-acrylonitrile
PA	polyamide
PAEK	polyaryletherketone
PAI	polyamide imide
PAN	polyacrylonitrile
PB	polybutylene
PBAN	polybutadiene-acrylonitrile
PBI	polybenzimidazole
PBN	polybutylene naphthalate
PBS	polybutadiene styrene
PBT	polybutylene terephthalate
PC	polycarbonate
PCD	polycarbodiimide
PCT	polycyclohexylenedimethylene terephthalate
PCTA	copolyester of CHDM and PTA
PCTFE	polychlorotrifluoroethylene
PCTG	glycol-modified PCT copolymer
PE	polyethylene
PEBA	polyether block polyamide
PEC	chlorinated polyethylene
PEDT	3,4 polyethylene dioxithiophene
PEEK	polyetheretherketone
PEI	polyether imide
PEK	polyetherketone
PEKEKK	polyetherketoneetherketoneketone
PEN	polyethylene naphthalate
PES	polyether sulfone
PET	polyethylene t2222erephthalate
PETG	PET modified with CHDM
PF	phenol formaldehyde
PFA	perfluoroalkoxy resin
PI	polyimide
PIBI	butyl rubber
PMDI	polymeric methylene diphenylene diisocyanate
PMMA	polymethyl methacrylate
PMP	polymethylpentene
PO	polyolefins
POM	polyacetal
PP	polypropylene
PPA	polyphthalamide
PPC	chlorinated polypropylene
PPE	polyphenylene ether, modified
PPO	polyphenylene oxide
PPS	polyphenylene sulfide
PPSU	polyphenylene sulfone
PS	polystyrene
PSU	polysulfone
PTFE	polytetrafluoroethylene
PU	polyurethane

Table 5-1   Alphabetical list of plastic material abbreviations *Continued*

PUR	polyurethane
PVC	polyvinyl chloride
PVCA	polyvinyl chloride acetate
PVDA	polyvinylidene acetate
PVDC	polyvinylidene chloride
PVDF	polyvinylidene fluoride
PVF	polyvinyl fluoride
PVOH	polyvinyl alcohol
RHDPE	recycled high density polyethylene
RPET	recycled polyethylene terephthalate
SI	silicone plastic
SAN	styrene acrylonitrile copolymer
SB	styrene butadiene copolymer
SBC	styrene block copolymer
SBR	styrene butadiene rubber
SMA	styrene maleic anhydride
SMC	sheet molding compound
SMC-C	SMC-continuous fibers
SMC-D	SMC-directionally oriented
SMC-R	SMC-randomly oriented
TEO	thermoplastic elastomeric olefin
TLCP	thermoplastic liquid crystal polymer
TMC	thick molding compound
TP	thermoplastic
TPE	thermoplastic elastomers
TPO	thermoplastic olefins
TPU	thermoplastic polyurethane
TPV	thermoplastic vulcanizate
TS	thermoset
UF	urea formaldehyde
UHMW	ultrahigh molecular weight
ULDPE	ultralow-density polyethylene
UP	unsaturated polyester resin
UR	urethane
VA(C)	vinyl acetate
VC	vinyl chloride
VDC	vinylidene chloride
VLDPE	very low-density polyethylene

The first property listed is the heat deflection temperature. In addition to providing a comparative idea of the temperature resistance of the resin, the value listed should give an idea of the weldability of the material, since polymers that melt at lower temperatures are more economically welded than those with high melt temperatures. In comparing plastics, one must be certain that the loading (66 psi or 264 psi) is the same. In many cases a maximum use temperature is also supplied. This is a more practical gauge from an application standpoint.

The next property provided is the tensile strength (at yield). This, too, is presented as a comparative value. Bear in mind that this property will change dramatically for many polymers within what might be regarded as the normal operating range.

The flexural modulus will serve as a rough gauge of the polymer's ability to be used for hinges plus snap and press fits. The mold shrinkage provides a guide to the materials ability to hold tolerances. The coefficient of linear thermal expansion describes a resin's dimensional stability with respect to temperature. The listing goes on to discuss the other properties of the material, and there is a sampling of some of its applications. It then proceeds to rate the material's ability to deal with the various assembly methods. If no reliable information could be found for a given assembly method/material pair, that method is omitted from the material's listing.

Where available, the 1991 revision of the applicable SPI table of standards for molding tolerances is designated at the end of the material listing. These tables are presented as Figs. 5-1 to 5-20 at the end of the chapter.

Many considerations in addition to those listed below must be borne in mind when selecting an assembly method for a material. Before making a final selection, the reader should consult the chapter treating the material of interest.

5.4.1   Properties and Assembly-Related Data for Selected Materials

Acetal (Polyoxymethylene: POM), Crystalline Thermoplastic

	Homopolymer	Copolymer
Heat deflection temperature (°F @ 66 psi)	338	316
(°F @ 264 psi)	255	230
Maximum use temperature (°F, no load)	195	212
Tensile strength (psi)	9700–10,000	8,800
Flexural modulus (psi $\times$ 10^3 @ 73 °F)	380–490	370–450
Mold shrinkage (in./in.)	0.018–0.025	0.020
Coefficient of linear thermal expansion (in./in./°F $\times$ 10^{-6})	28–62	34–61

Additional Properties
Very high tensile strength, toughness, stiffness and dimensional stability; high abrasion resistance, chemical resistance and resistance to creep under load. Excellent resistance to vibration fatigue. Low coefficient of friction and high melt point. Glass-filled acetal provides higher stiffness, lower creep, higher heat deflection temperature, better arc resistance, and greater dimensional stability.

Homopolymer
Advantages include being strongest and stiffest of unreinforced thermoplastics, good resistance to bending under load, resistance to organic compounds, low moisture absorption (good stability), low wear, and friction characteristics superior to other thermoplastics.

Copolymer
Toughness decreases with grades offering increased flowability. Maximum tensile strength, rigidity, and heat deflection temperature with 25% glass. Special grades for repeated food contact, low friction, and low wear. One of the most creep-resistant crystalline thermoplastics, with long-term retention of properties. Can handle pH values from 4 to 14 and can be immersed in common solvents, lubricants, and gas.

Can retain properties in long term water immersion to 180 °F (air to 220 °F) and has good fatigue endurance.

Applications
Uses include growth in applications traditionally made of zinc, brass, and aluminum castings and some steel stampings, such as automobile fuel system components, seat belts, and steering columns. Other uses include bearings, plumbing fixture components, hardware items, butane lighter housings, zippers, mechanical couplings, gears, and pulleys.

Assembly Methods
- Adhesives: Ref. Chapter 7, Adhesive and Solvent Joining.
- Fasteners: Acetals have low drive torque and high fail torque and are well suited to fasteners. Special design thread-rolling screws and push-in thread-forming screws are recommended for high holding power with low relaxation characteristics and low tendency toward stress cracking. Fasteners may also be ultrasonically installed.
- Hinges: Good.
- Inserts: Ultrasonic inserts are good, and this is the preferred method. Heat installation is also good. Press-in, self-tapping, and helical coil inserts can also be used. Molded-in inserts are not recommended because acetal has a high shrinkage rate of the material.
- Press Fits: Good.
- Snap Fits: Excellent.
- Solvents: Not recommended.
- Staking/Swaging: Heat, good; Hot air/cold staking, poor. Ultrasonics, fair to good.

Weldability
- Hot die/fusion: Fair to good, weld strength up to 90% of the strength of the material.
- Hot gas: Fair, 20–30% of the strength of the material.
- Induction/electromagnetic: Fair to good.
- Spin welding: Fair to good, 50–70% of the strength of the material.
- Ultrasonic welding: Near field, good; Far field, fair to good; Spot welding, fair; Swaging, fair to good.
- Vibration welding: Good to excellent; Difficult with shear joint.

Figure 5-1 presents the SPI standards for molding tolerances for POM.

Acrylonitrile–Styrene–Acrylate (ASA), Amorphous Thermoplastic

Heat deflection temperature (°F @ 66 psi)	200–210
(°F @ 264 psi)	180–200
Tensile strength (psi)	4000–7500
Flexural modulus (psi $\times$ 10^3 @ 73 °F)	200–341
Mold shrinkage (in./in.)	0.004–0.008
Coefficient of linear thermal expansion (in./in./°F $\times$ 10^{-6})	33
Behaves very much like ABS, but has superior resistance to outdoor weathering	

Applications
Outdoor applications found in construction (e.g. gutters), mailboxes, flower pots and window trim. Recreation uses include outdoor furniture, boats, spas, and pool equipment. Used for exterior auto parts such as grilles and trim.

Assembly Methods
- Adhesives: Ref. Chapter 7, Adhesive and Solvent Joining.
- Fasteners: Fair to good; special design and push-in thread-forming screws can be used. Fasteners may also be ultrasonically installed.
- Inserts: Heat, good. Ultrasonics, good; inserts can be molded in.
- Solvents: Acetone, glacial acetic acid, *N*-methylpyrrolidone, *O*-dichlorobenzol.
- Staking/Swaging: Heat, good. Hot air/cold staking, good. Ultrasonics, fair to good.

Weldability
- Hot die/fusion: Fair to good.
- Hot gas: Fair to good.
- Induction/electromagnetic: Good to excellent.
- Spin welding: Good.
- Ultrasonic welding: Near and far field, excellent. Spot welding, fair.
- Vibration welding: Fair to good

Acrylic (Polymethylmethacrylate: PMMA), Amorphous Thermoplastic

Heat deflection temperature (°F @ 66 psi)	175–225
(°F @ 264 psi)	165–210
Maximum use temperature (°F, no load)	130–230
Tensile strength (psi)	7000–10,500
Flexural modulus (psi $\times$ 10^3 @ 73 °F)	325–460
Mold shrinkage (in./in.)	0.001–0.004 (flow)
	0.002–0.008 (trans.)
Coefficient of linear thermal expansion (in./in./°F $\times$ 10^{-6})	28–50

Additional Properties
Acrylic offers high optical clarity and brilliant transparent colors as well as resistance to outdoor weathering. It is also a good electrical insulator and is known for its hard surface (Rockwell hardness M85–105). Impact resistance can be improved with PVC modification.

Applications
Used in lighting applications, auto lenses, dials, and instrument panels; nameplates, windows, appliance panels, knobs, housings, and telephone touch buttons.

Assembly Methods
- Adhesives: Ref. Chapter 7, Adhesive and Solvent Joining.
- Fasteners: Drive and fail torques, clamp loads, and tensile pullout strengths are high. Initial material relaxation loads are low. Notch sensitive and vulnerable to cracking.

Trilobular thread-forming screws preferred, push-in types feasible in certain cases. Fasteners may be ultrasonically installed.
- Hinges: Integral not possible; two-piece hinges may require heat to assemble.
- Inserts: Ultrasonic inserts are good to excellent (fair to good for acrylic/PVC) and are the preferred method. Heat-installed inserts are also good to excellent. Press-in, self-tapping, and helical coil inserts may be used. Inserts can be molded in.
- Solvents: Ethylene dichloride, methylene chloride, vinyl trichloride, and glacial acetic acid. Can produce joint strengths up to 60% of the strength of the material.
- Staking/swaging: Heat, good to excellent. Hot air/cold staking, good. Ultrasonics, fair to good; good for acrylic/PVC.

Weldability
- Hot die/fusion: Excellent; weld strength up to 80% of the strength of the material; can also weld to ABS, polycarbonate and SAN in some cases.
- Hot gas: Welds with PVC rod (check with rod manufacturer).
- Induction/electromagnetic: Excellent; welds to itself as well as ABS, PC, SAN, and PS in some cases.
- Spin welding: Good.
- Ultrasonic welding: Near field, good to excellent, same for acrylic/PVC; far field, fair to good (energy director recommended). Spot welding, good; also welds to ABS, Polyphenylene Oxide, PC, and SAN in some cases.
- Vibration welding: Fair to good; can also weld to some versions of ABS, polyphenylene oxide, and polycarbonate.

Figure 5-2 presents the SPI standards for molding tolerances for PMMA.

Acrylonitrile Butadiene Styrene (ABS), Amorphous Thermoplastic

Heat deflection temperature (°F @ 66 psi)	210–225
(°F @ 264 psi)	190–206
Maximum use temperature (°F, no load)	140–210
Tensile strength (psi)	4000–7500
Flexural modulus (psi $\times$ 10^3 @ 73 °F)	179–375
Mold shrinkage (in./in.)	0.004–0.009
Coefficient of linear thermal expansion (in./in./°F $\times$ 10^{-6})	44–61

Additional Properties
Combines high rigidity and impact strength depending on formulation and has excellent finishing characteristics. Provides good combination of flame retardancy, relatively good heat resistance, and overall chemical resistance. ABS is the material of choice for most plating applications, and special grades are available for this purpose. ABS does not adhere to mold surfaces and can be molded with reduced draft ($\frac{1}{2}$° per side).

Disadvantages of ABS are solubility in esters, ketones, and some chlorinated hydrocarbons, vulnerability to molded in stress (which adversely affects chemical resistance), and poor resistance to ultraviolet light (requires additives to retard loss in physicals properties, color, and gloss).

ABS can be tailored to end product requirements in the following manner: Acrylonitrile improves heat stability, chemical resistance, and aging resistance. Butadiene increases low temperature retention, toughness, and impact resistance. Styrene adds luster, rigidity, and processing ease.

ABS can be alloyed with other polymers such as PVC or polycarbonate. Methyl methacrylate can be added as fourth monomer to provide transparency. Thus modified, it can attain 75 to 80% light transmission.

Applications
Drain and pipe fittings, appliance housings, automobile kick panels and headlight housings, refrigerator door liners, luggage, camper tops, sporting goods, and power tool housings. Most widely used engineering thermoplastic in the world.

Assembly Methods
- Adhesives: Ref. Chapter 7, Adhesive and Solvent Joining.
- Fasteners: Values are dependent on butadiene content. Tensile pullout strengths, torque values and material relaxation qualities improve with reduced butadiene content. Resistance to cracking improves with greater butadiene content, as low levels may produce crazing on screw entry, which requires counterboring to control. Special thread-rolling screws and push-in thread-forming screws are recommended. Thread-cutting screws are acceptable. Plastic machine screws are good – Unified Coarse (UNC) threads are recommended. Rivets and spring clips are good. Fasteners may be ultrasonically installed.
- Hinges: Fair to good.
- Inserts: Ultrasonic inserts are excellent and the preferred method. Heat installation is also excellent, but slower in most cases. Press-in self-tapping and helical coil inserts can also be used. Ultrasonic insertion is good with polycarbonate-modified ABS and fair with PVC-modified ABS. Molded-in inserts are good but should be preheated to 95 to 120 °C (200–250 °F) prior to molding.
- Press Fits: Good.
- Snap Fits: Good.
- Solvents: Methyl ethyl ketone (MEK), methyl isobutyl ketone (MIBK), methylene chloride, acetone, ethylene dichloride, perchlorethylene, tetrahydrofuran, trichlorethylene. Can produce joint strengths up to 60% of the strength of the material.
- Staking/swaging: Heat, good to excellent. Hot air/cold staking, excellent. Ultrasonics, excellent; good for ABS/PC and ABS/PVC.

Weldability
Note: ABS is somewhat hygroscopic and may need to be dried before some heat based processes.
- Hot die/fusion: Good, weld strengths up to 80% of the strength of the material. Can also weld to acrylic, polycarbonate and SAN in some cases.
- Hot gas: Good with some blends; weld strengths up to 70% of the strength of the material.

- Induction/electromagnetic: Excellent; also welds to acrylic, PC, and SAN in some cases.
- Spin welding: Good to excellent
- Ultrasonic welding: Near field, excellent, good for ABS/PC and ABS/PVC; far field, good (energy director recommended), fair for ABS/PC and ABS/PVC. Spot welding, excellent, good for ABS/PC and ABS/PVC. Also welds to PVC and SAN in some cases.
- Vibration welding: Good, can also weld to some versions of polystyrene, acrylic, polyphenylene oxide, polycarbonate, and styrene acrylonitrile.

Figure 5-3 presents the SPI standards for molding tolerances for ABS.

Allyl: See Diallyl Phthalate

Cellulosics: Cellulose Acetate (CA), Cellulose Acetate Butyrate (CAB), Cellulose Acetate Proprionate (CAP), Ethyl Cellulose (EC), Crystalline Thermoplastics

	Cellulose Acetate	Cellulose Acetate Butyrate	Cellulose Proprionate	Ethyl Cellulose
Heat deflection temperature				
(°F @ 66 psi)	120–209	130–227	147–250	115–190
(°F @ 264 psi)	111–195	113–202	111–228	115–190
Maximum use temperature (°F, no load)	140–220	140–220	155–220	115–185
Tensile strength (psi)	1900–9000	2600–6900	2000–7800	2000–7800
Flexural modulus (psi $\times 10^3$ @ 73 °F)	1200–4000	90–300	120–350	
Mold shrinkage (in./in.)	0.003–0.010	0.003–0.009	0.003–0.009	0.005–0.009
Coefficient of linear thermal expansion (in./in./°F $\times 10^{-6}$)	44–100	61–94	61–94	55–111

Additional Properties
The cellulosics have similar properties that vary in degree. They offer high toughness, strength, stiffness, and hardness, brilliant transparent colors, and toughness at low temperature. They are easy to clean and provide good weather resistance.

Applications
Cellulose acetate is mainly an extrusion (film and sheet) material, but injection applications include premium toys, tool handles, appliance housings, shields, lenses, and eyeglass frames. Cellulose acetate butyrate is used for pen barrels, steering wheels, tool handles, machine guards, and skylights. Cellulose acetate propionate applications include lighting fixtures, safety goggles, motor covers, brush handles, face shields, and steering wheels. Ethyl cellulose is used for flashlight cases, fire extinguisher parts and electrical appliance parts.

Assembly Methods
- Adhesives: Ref. Chapter 7, Adhesive and Solvent Joining.
- Fasteners: High torque values, high tensile pullout strengths, and low material relaxation.

Oversize pilot holes required to avoid stress cracking with thread-forming screws. Thread-cutting screws preferred. Fasteners may be ultrasonically installed.
- Inserts: Ultrasonic inserts are good to excellent and the preferred method. Heat installation is also excellent, but slower in most cases. Press-in, self-tapping, and helical coil inserts can also be used. Inserts may be molded in.
- Solvents: Chloroform, methylene dichloride, acetone, ethyl acetate, methyl ethyl ketone, methyl cellosolve. Cellulose acetate butyrate is also solvent in cyclohexanone, ethylene dichloride, and a solution of 80% toluene and 20% ethanol. Can produce strengths equivalent to the strength of the material.
- Staking/swaging: Heat, Good. Hot air/cold staking, fair. Ultrasonics, fair to good.

Weldability
- Hot die/fusion: Limited; weld strengths up to 80% of the strength of the material in some cases.
- Hot gas: Good; weld strengths up to 90% of the strength of the material.
- Induction/electromagnetic: Good to excellent.
- Spin welding: Good to excellent.
- Ultrasonic welding: Near field, poor to fair; far field, very limited (energy director recommended). Spot welding, poor to fair.
- Vibration welding: Good; weld strengths up to 80% of the strength of the material.

Figure 5-4 presents the SPI standards for molding tolerances for cellulosics.

Diallyl Phthalate (DAP), Thermoset

	Glass Filled	Mineral Filled
Heat deflection temperature (°F @ 264 psi)	330–550	320–550
Maximum use temperature (°F, no load)	300–400	300–400
Tensile strength (psi)	6000–11,000	5000–8000
Flexural modulus (psi $\times 10^3$ @ 73 °F)	1200–1500	1000–1400
Mold shrinkage (in./in.)	0.0005–0.005	0.002–0.007
Coefficient of linear thermal expansion (in./in./°F $\times 10^{-6}$)	5.6–20	5.6–23

Additional Properties
Diallyl phthalate offers exceptional dimensional stability (most stable of all thermosets), excellent electrical properties and heat and chemical resistance. It has very low moisture absorption and hard tough surfaces.

Applications
Critical electronic and television components, military and aerospace components plus furniture laminates. This is a high cost material used mainly for applications requiring high reliability over long term adverse conditions.

Assembly Methods
- Adhesives: Ref. Chapter 7, Adhesive and Solvent Joining.
- Fasteners: Good; thread-cutting screws recommended. Thread-forming trilobular screws can be used for single applications. Pilot holes recommended. Push-in thread-forming screws not recommended. Can withstand soldering of electrical inserts. Ultrasonic installation is not possible.
- Inserts: Press-in type with adhesives or molded-in inserts recommended. Can also use helical coil, self-tapping inserts and expansion inserts.
- Staking/swaging: Not applicable.

Weldability: Does not weld.

Figure 5-5 presents the SPI standards for molding tolerances for DAP.

Epoxy (EP), Thermoset

	Mineral-filled Bisphenol	Glass-filled SMC	Carbon-filled SMC
Heat Deflection Temperature (°F @ 264 psi)	225–500	250–500	250–500
Maximum use temperature (°F, no load)	300–500	300–500	300–500
Tensile strength (psi)	4000–10,800	20,000–35,000	40,000–50,000
Flexural modulus (psi $\times$ 10^3 @ 73 °F)	1400–2000	2000–3000	5000
Mold shrinkage (in./in.)	0.002–0.010	0.001	0.002
Coefficient of linear thermal expansion (in./in./°F $\times$ 10^{-6})	11–33	7	2

Additional Properties
Outstanding physical and electrical strengths and adhesive qualities. Very low shrinkage during molding. Superior dimensional stability under adverse environments. Capable of very thin [1.5 mm (0.060 in.)], SMC wall thicknesses.

Applications
Epoxies are used mainly in adhesives, molds, and coatings. Product applications include ignition coils, high voltage insulators, bushings, switch gear, and semiconductors. Used for casting and potting applications.

Assembly Methods
- Adhesives: Ref. Chapter 7, Adhesive and Solvent Joining.
- Fasteners: Thread-cutting screws preferred. Push-in screws are not recommended. Ultrasonic installation is not possible.
- Inserts: Press-in type with adhesive reinforcement and molded-in inserts recommended. Can also use helical coil, self-tapping, and expansion inserts.
- Staking/swaging: Not applicable.

Weldability: Does not weld.

Figure 5-6 presents the SPI standards for molding tolerances for EP.

Melamine Formaldehyde (MF), Thermoset

	Cellulose filled	Glass filled
Heat deflection temperature (°F @ 264 psi)	350–390	375–400
Maximum use temperature (°F, no load)	250	300–400
Tensile strength (psi)	5000–13,000	5000–10,500
Flexural modulus (psi $\times$ 10^3 @ 73 °F)	1100	–
Mold shrinkage (in./in.)	0.005–0.015	0.001–0.006
Coefficient of linear thermal expansion (in./in./°F $\times$ 10^{-6})	22–25	8.3–15.6

Additional Properties
Good electrical properties and high torque retention, surface hardness, and heat resistance. Highest colorability of all thermoset materials. Fillers used to improve hardness and strength.

Applications
Dinnerware, buttons, toilet seats, knobs, handles, ashtrays, food utensils, mixing bowls, electrical switches, and military applications.

Assembly Methods
- Adhesives: Ref. Chapter 7, Adhesive and Solvent Joining.
- Fasteners: Low drive torques with thread-cutting screws. High tensile pullout strengths and practically no relaxation, but susceptible to cracking at the edge of the hole. Type BT standard thread-cutting screws are recommended. Special design thread-rolling screws can sometimes be used with assistance of screw manufacturer's laboratory.
- Inserts: Press-in type with adhesive reinforcement recommended. Molded-in inserts feasible due to low shrinkage rate. Can also use helical coil, expansion, and self-tapping inserts. Ultrasonic and heat installation not possible.
- Solvents: Not applicable.
- Staking/swaging: Not applicable.

Weldability: Does not weld.

Figure 5-7 presents the SPI standards for molding tolerances for MF.

Nylon: See Polyamide

Phenolic (Phenol-Formaldehyde: PF), Thermoset

	Wood flour filled	Glass filled
Heat deflection temperature (°F @ 264 psi)	300–370	300–600
Maximum use temperature (°F, no load)	300–350	350–550
Tensile strength (psi)	5000–9000	7000–18,000
Flexural modulus (psi $\times$ 10^3 @ 73 °F)	1000–1200	1150–3300
Mold shrinkage (in./in.)	0.004–0.009	0.001–0.004
Coefficient of linear thermal expansion (in./in./°F $\times$ 10^{-6})	16.7–25	4.4–19

Additional Properties

Phenolic is a low cost plastic with an excellent combination of high physical strength, high temperature resistance, scratch resistance, good dimensional stability, electrical properties, and chemical resistance. It is weak and brittle without fillers and has very poor colorability.

Applications

Electrical devices, pulleys, commutators, pumps, closures, and automotive ignition components.

Assembly Methods

- Adhesives: Ref. Chapter 7, Adhesive and Solvent Joining.
- Fasteners: Low drive torques with thread-cutting screws. High tensile pullout strengths and practically no relaxation, but very susceptible to cracking at the edge of the hole. Type BT standard thread-cutting screws are recommended. Special design thread-rolling screws can sometimes be used for single insertions with assistance of manufacturer. Not suitable for push-in fasteners. Ultrasonic installation is not possible.
- Inserts: Molded-in inserts feasible due to low shrinkage rate. Press-in type inserts with adhesive reinforcement recommended for light-duty applications. Can also use helical coil expansion and self-tapping inserts. Heat, hot air, and ultrasonic inserts not applicable.
- Staking/swaging: Not applicable.

Weldability: Does not weld.

Figure 5-8 presents the SPI standards for molding tolerances for PF.

Polyamide (Nylon: PA), Crystalline Thermoplastic

	Nylon 6	Nylon 6/6
Heat deflection temperature (°F @ 66 psi)	300–375	360–474
(°F @ 264 psi)	140–155	150–220
Maximum use temperature (°F, no load)	180–250	180–300
Tensile strength (psi)	6000–24,000	9000–12,000
Flexural modulus (psi $\times$ 10^3 @ 73 °F, 50% RH)	140	185
Mold shrinkage (in./in.)	0.003–0.015	0.007–0.018
Coefficient of linear thermal expansion (in./in./°F $\times$ 10^{-6})	44–46	44

Other grades of nylon, as indicated by their chemical structures, are available with varying properties.

Additional Properties

Generally, nylons are characterized by high toughness and wear resistance, and low coefficient of friction, good electrical strength and impact strength, and excellent chemical resistance. They are extremely hygroscopic, and property values can drop substantially with increasing moisture content. Nylon has very low viscosity at elevated temperatures, allowing it to mold mechanical parts with very thin sections.

Applications
Gears, bearings, and other antifriction parts. Automotive uses include exhaust canisters, transmission covers, headlamp housings, fender extensions, cowl vents, exterior parts, and underhood mechanical components. Electrical applications include plugs, connectors, computer parts, and bobbins. Other applications include springs, drapery hardware, brush bristles, bike wheels, mallet heats, and hot comb and brush handles.

Assembly Methods
- Adhesives: Ref. Chapter 7, Adhesive and Solvent Joining.
- Fasteners: Torque values and tensile pullout strengths will be high when dry, dropping as moisture levels increase. Material relaxation and resistance to stress cracking are good. Special design thread-rolling and push-in thread-forming screws are excellent, however push-in fasteners good but should be ultrasonically installed if glass-fiber reinforcement is used. Material manufacturer also recommends type A, B, and AB standard thread-forming screws. Can mold strippable thread.
- Inserts: Ultrasonic inserts are good and the preferred method. Heat installation is also excellent, but slower in most cases. Press-in, self-tapping, and helical coil inserts can also be used. Molded-in inserts are acceptable, however be alert to stresses resulting from the high shrinkage rate of the material. Installation should be performed immediately after molding before moisture can be absorbed. Otherwise, parts must be kept dry with desiccants or redried before installation can proceed.
- Press Fits: Excellent.
- Snap Fits: Excellent; parts may be too brittle for snap fit applications immediately after molding and, in such cases, require remoisturing.
- Solvents: Aqueous phenol, calcium chloride in alcohol, formic acid. Solvent welding is not recommended as a production assembly method.
- Staking/swaging: Heat, good. Hot air/cold staking, fair. Ultrasonics, fair to good.

Weldability
- Hot die/fusion: Fair to good; nylon 6 better with noncontact welding; weld strengths up to 90% of the strength of the material.
- Hot gas: Unsatisfactory.
- Induction/electromagnetic: Good to excellent.
- Spin welding: Good for nylon 6/6, excellent for nylon 6.
- Ultrasonic welding: Near field, Good, shear joint recommended; far field, fair to good. Spot welding, fair to good.
- Vibration welding: Good to excellent; weld strengths up to 70% of the strength of the material. Can also weld to polyphenylene oxide in some cases.

Figure 5-9 presents the SPI standards for molding tolerances for PA.

Polyamide-Imide (PA-I), Amorphous Thermoplastic

Heat deflection temperature (°F @ 264 psi)	532
Maximum use temperature (°F, no load)	450–500
Tensile strength (psi)	22,000
Flexural modulus (psi × 10^3 @ 73 °F)	730
Mold shrinkage (in./in.)	0.006–0.0085
Coefficient of linear thermal expansion (in./in./°F × 10^{-6})	17

Additional Properties
Polyamide-imide offers a combination of high service temperature with low coefficient of linear thermal expansion. It retains strength and impact resistance at high temperatures and has low coefficient of friction. It is hygroscopic, but its chemical resistance is good.

Applications
Extreme service applications in aerospace. Automotive underhood and transmission applications.

Assembly Methods
- Adhesives: Ref. Chapter 7, Adhesive and Solvent Joining.
- Fasteners: Thread-cutting, self-tapping screws may be used, but can be difficult to emplace due to the toughness of the material. Tapped holes are recommended.
- Inserts: Ultrasonic inserts recommended. Other postmolded inserts can be used. Molded-in inserts are recommended. Inserts should be preheated to mold temperature.
- Press Fits: Good press fits are attainable due to excellent resistance to creep.
- Snap Fits: Cannot strip undercuts. Non-undercut snap fits acceptable except for graphite fiber reinforced grades.

Weldability
- Induction/electromagnetic: Good.
- Spin welding: Fair to good.
- Ultrasonic welding: Near field, good; far field, fair
- Vibration welding: Good.

Polyarylsulfone (PASU), Amorphous Thermoplastic

Heat deflection temperature (°F @ 264 psi)	525
Maximum use temperature (°F, no load)	500
Tensile strength (psi)	13,000
Flexural modulus (psi × 10^3 @ 73 °F)	330–400
Mold shrinkage (in./in.)	0.007–0.008
Coefficient of linear thermal expansion (in./in./°F × 10^{-6})	17–27

Additional Properties
Polyarylsulfone offers high long-term stability, creep resistance, stress crack resistance and electrical properties. It can withstand heat of soldering.

Applications
Electrical components and printed circuit boards. Extreme service environment applications.

Assembly Methods
- Adhesives: Ref. Chapter 7, Adhesive and Solvent Joining.
- Fasteners: Thread-cutting self-tapping screws acceptable.
- Snap Fits: Excellent.
- Solvents: *N,N*-Dimethylformamide, *N*-methylpyrrolidone, and *O*-dichlorobenzol.

Weldability
- Hot die/fusion: Fair
- Spin welding: Good.
- Ultrasonic welding: Near field, fair, far field, fair.

Polycarbonate (PC), Amorphous Thermoplastic

Heat deflection temperature (°F @ 66 psi)	273–287
(°F @ 264 psi)	265–280
Maximum use temperature (°F, no load)	250
Tensile strength (psi)	9100–10,500
Flexural modulus (psi $\times 10^3$ @ 73 °F)	330–340
Mold shrinkage (in./in.)	0.005–0.007
Coefficient of linear thermal expansion (in./in./°F $\times 10^{-6}$)	38

Additional Properties
Rigid, transparent, and tough, Polycarbonate has excellent outdoor dimensional stability and exhibits very low creep under load. It is an excellent engineering material, but it is vulnerable to greases and oils.

Applications
Automotive uses include tail and side marker lights, headlamp support fixtures, instrument panels, trim strips, exterior body components. It is also used in traffic light housings, optical lenses, glazing, and signal lenses. Food uses include returnable milk containers and microwave ovenware, mugs, ice cream dishes, food storage containers, microwave oven applications, and water cooler bottles. Other applications are intravenous drug delivery and blood processing equipment, appliance and tool housings, and telephone, T.V., boat and conveyor components.

Assembly Methods

- Adhesives: Ref. Chapter 7, Adhesive and Solvent Joining.
- Fasteners: Excellent torque and tensile pullout values, approaching those of metals with longer engagements. Material relaxation is low; lower for reinforced polycarbonate. Thread-cutting screws are recommended. Special thread-rolling screws may be used if molding stress does not exceed 1200 psi, as thread-rolling screws will add 300 psi. Otherwise, stress cracking may occur. Push-in thread-forming screws are not generally recommended unless ultrasonically installed. This caveat does not apply to structural foam polycarbonate. Copper and aluminum rivets can be used; however use washers or shoulder rivets and avoid overcompression. Avoid molding in V-threads owing to notch sensitivity.
- Inserts: Ultrasonic inserts are good and the preferred method, but parts must be dry. Molded-in inserts can be used in certain cases. Heat installation is good to excellent, but slower. Self-tapping and helical coil inserts can also be used. Press-in inserts are not recommended. Molded-in inserts are feasible, however one resin manufacturer does not recommend their use with unfilled polycarbonate. Larger sizes may require preheating to 177 to 204 °C (350°–400 °F). Avoid molded-in inserts for applications requiring thermocycling. Boss caps can be used for light duty applications.
- Press Fits: Excellent.
- Snap Fits: Good to excellent.
- Solvents: Ethylene dichloride, methylene chloride, ethyl acetate, glacial acetic acid, and methyl cellosolve. Weld strengths up to 60% of adjacent walls.
- Staking/swaging: Heat, good. Hot air/cold staking, excellent. Ultrasonics, good.

Weldability

- Hot die/fusion: Fair to good; weld strengths up to 80% of the strength of the material. Also welds to ABS, acrylic, and polysulfone in some cases.
- Hot gas: Fair to good; weld strengths up to 50% of the strength of the material.
- Induction/electromagnetic: Good to excellent; also welds to ABS, acrylic, PS, polysulfone, and SAN in most cases.
- Spin welding: Good to excellent.
- Ultrasonic welding: Near field, good to excellent; far field, fair to good (energy directors recommended). Spot welding, good. Also welds to acrylic, polyphenylene oxide, PC and PVC in some cases.
- Vibration welding: Good; weld strengths up to 50% of the strength of the material. Can also weld to some versions of ABS and acrylic.

Figure 5-10 presents the SPI standards for molding tolerances for PC.

Polydicyclopentene (pDCPD), Thermoset

Heat deflection temperature (°F @ 264 psi)	248
Tensile strength (psi)	6,700
Flexural modulus (psi $\times 10^3$ @ 73 °F)	284
Mold shrinkage (in./in.)	0.009
Coefficient of linear thermal expansion (in./in./°F $\times 10^{-6}$)	49

Additional Properties
Polydicyclopentene offers a combination of high rigidity and impact strength with excellent finishing characteristics. It is a liquid molding resin that allows part sizes up to $11\,m^2$ ($120\,ft^2$) outer surface area in low pressure rated molds. Part thicknesses can be varied. Adhesion to paint and structural adhesives is excellent, and the product can be chrome plated. A flame-retardant grade is available.

Applications
Typical uses are heavy truck hoods, roof fairings, bumpers, fenders, consoles and engine support cradles, as well as video games, medical instruments, computers, copiers, and electronic housings. Recreational applications include snowmobile hoods and components, golf cart body parts, and personal watercraft covers.

Assembly Methods
- Adhesives: Ref. Chapter 7, Adhesive and Solvent Joining.
- Fasteners: Thread-forming self-tapping screws preferred.
- Inserts: Molded-in inserts recommended. Self-tapping and press-in inserts with adhesives usable for light duty applications. Can also use expansion inserts.

Polyester, Crystalline thermoplastic (PBT: Polybutylene terephthalate; PET: Polyethlene terephthalate)

	PBT	PET
Heat deflection temperature (°F @ 66 psi)	310	240
(°F @ 264 psi)	130	185
Maximum use temperature (°F, no load)	280	175
Tensile strength (psi)	8200–8700	7000–10,500
Flexural modulus (psi $\times$ 10^3 @ 73 °F)	330–400	350–450
Mold shrinkage (in./in.)	0.009–0.022	0.002–0.030
Coefficient of linear thermal expansion (in./in./°F $\times$ 10^{-6})	33–53	36

Additional Properties
Thermoplastic polyesters offer good tensile strength, toughness, impact resistance, frictional, chemical and electrical properties.

Applications
- PBT: Electrical connectors, coil bobbins, light sockets, and other electronic components are typical applications. Automotive uses include hardware and underhood components. Also used for industrial conveyor and hardware parts.
- PET: This polymer has been used as a metal replacement for motor housings and furniture frames. Switches, relays and sensors are some of its electrical applications.

Assembly Methods
- Adhesives: Ref. Chapter 7, Adhesive and Solvent Joining.
- Fasteners: Low material relaxation. Tensile pullout strength of 1000 psi and

fail/drive ratios from 3:1 to 5:1 with special design thread-rolling screws. Notch sensitivity requires special care in boss design. Push-in thread-forming screws acceptable with or without ultrasonic installation. Standard thread-cutting screws can be used with polyesters except structural foam. Tapped threads feasible (use carbide tools with 30% glass reinforced polyester).

- Inserts: Ultrasonic inserts are good and the preferred method. Heat installation is good to excellent, but slower in most cases. Press-in, self-tapping, and helical coil inserts can also be used. Special grades available for molded-in inserts.
- Press Fits: Good to excellent.
- Snap Fits: PBT; Good to excellent. PET; Good to excellent. Special grades available for snap fits.
- Staking/swaging: Heat, Good; use heat staking temperatures of 218 to 288 °C (425°–550 °F). Hot air/cold staking, fair. Ultrasonics, fair to good.

Weldability
- Hot die/fusion: Fair.
- Hot gas: Not satisfactory.
- Induction/electromagnetic: Good to excellent.
- Spin welding: Fair to good.
- Ultrasonic welding: Near field, fair to good (shear joint recommended). Far field, fair. Spot welding, fair.
- Vibration welding: PET and PBT can weld to themselves.

Figure 5-11 presents the SPI standards for molding tolerances for TPPE.

Polyester, Unsaturated (UP), Thermoset[*]

	SMC	BMC	TMC
Heat deflection temperature (°F @ 264 psi)	400–500	320–400	
Maximum use temperature (°F, no load)	300–350	300–350	
Tensile strength (psi)	7000–25,000	3000–6000	7000–10,500
Flexural modulus (psi $\times 10^3$ @ 73 °F)	1000–2200	1400	1550
Mold shrinkage (in./in.)	0.0005–0.004	0.0003–0.004	0.0005–0.0035
Coefficient of linear thermal expansion (in./in./°F $\times 10^{-6}$)	7.5–14.0	11	–

[*]Thermoset polyester is usually identified by the method of glass fiber reinforcement: SMC is sheet molding compound, BMC is bulk molding compound, and TMC is thick molding compound.

Additional Properties
Thermoset polyesters provide good dimensional stability, electrical properties, physical properties, and coloring ability, as well as ease of handling and rapid curing. A wide variety of properties are available with modification. This polymer is brittle without fillers and forms hard, stiff, tough surfaces.

Applications
This material is the workhorse of the reinforced plastics industry, almost always with glass filler content greater than 12%. Widespread usage for boats, truck hoods, air deflectors, and other body components. Also used for bath fixtures and chemical tanks.

Assembly Methods
- Adhesives: Ref. Chapter 7, Adhesive and Solvent Joining.
- Fasteners: Good; type BT thread-cutting screws acceptable, but produce low torque values. Special design thread-rolling screws recommended if used properly. (Consult manufacturer.) Virtually no material relaxation, and tensile pullout strength can surpass that of screw. Pilot holes required to avoid crazing. "Hi-Lo" threaded screws recommended by SMC Automotive Alliance. Rivets can be used. U- and J-nuts effective. Tapping and stud plates often used. Push-in thread-forming screws not recommended. Ultrasonic installation is not possible.
- Inserts: Press-in type with adhesive reinforcement and expansion inserts recommended. Molded-in inserts feasible, but require large-diameter bosses. Can also use helical coil and self-tapping inserts. Ultrasonic or heat installation not possible.
- Staking/swaging: Not applicable.

Weldability: Does not weld.

Polyetheretherketone (PEEK), Crystalline Thermoplastic

Heat deflection temperature (°F @ 264 psi)	320
Maximum use temperature (°F, no load)	480
Tensile strength (psi)	10,200–15,000
Flexural modulus (psi $\times$ 10^3 @ 73 °F)	560
Mold shrinkage (in./in.)	0.011
Coefficient of linear thermal expansion (in./in./°F $\times$ 10^{-6})	<302 °F: 22–26
	>302 °F: 60

Additional Properties
Polyetheretherketone has high strength and chemical resistance. It is steam sterilizable and able to withstand hostile high temperature, chemical, and fire-risk environments.

Applications
Automotive transmission, suspension, braking, and engine components. Chemical pumps, valves, and flow meter parts. Electronic, aerospace, and business machine applications.

Assembly Methods
- Adhesives: Ref. Chapter 7, Adhesive and Solvent Joining.
- Fasteners: Thread-cutting self-tappng screws applicable.
- Inserts: Heat, excellent. Ultrasonics, excellent.
- Staking/swaging: Heat, good. Ultrasonics, good.

Weldability
- Hot die/fusion: Fair
- Induction/electromagnetic: Not applicable.
- Spin welding: Good.
- Ultrasonic welding: Near field, excellent; far field, good. Spot welding, good.
- Vibration welding: Good.

Polyetherimide (PEI), Amorphous Thermoplastic

Heat deflection temperature (°F @ 66 psi)	405–410
Maximum use temperature (°F, no load)	338
Tensile strength (psi)	14,000
Flexural modulus (psi × 10^3 @ 73 °F)	480
Mold shrinkage (in./in.)	0.005–0.007
Coefficient of linear thermal expansion (in./in./°F × 10^{-6})	26–31

Additional Properties
Polyetherimide has excellent tensile strength and heat resistance, and good electrical properties, machinability, and resistance to ultraviolet light and radiation. It offers low water absorption and unusual ductility for a high strength material with impact resistance to −4 °C (−40 °F).

Applications
Automotive parts, aerospace components, and steam-sterilizable medical equipment. Heat- and corrosion-resistant fluid and air handling parts. Microwave cookware and coextruded packaging.

Assembly Methods
- Adhesives: Ref. Chapter 7, Adhesive and Solvent Joining.
- Fasteners: Low to moderate drive torque and high fail torque values combined with excellent tensile pull out strengths and resistance to vibration for special design thread-rolling screws. Low material relaxation. Counterbore of pilot holes recommended to counteract notch sensitivity of the material. Push-in threaded fasteners should be ultrasonically inserted only. Molded-in threads feasible (truncate V-threads). Tapped threads acceptable.
- Inserts: Ultrasonic inserts are excellent and the preferred method. Molded-in inserts are good in some cases with minimal knurls and simple design. Heat excellent, but slower in most cases.
- Snap Fits: Excellent.
- Solvents: Excellent with use of clamping. Use methylene chloride with a 1 to 5% resin component.
- Staking/swaging: Heat, good. Ultrasonics, good.

Weldability
- Hot die/fusion: Good.
- Induction/electromagnetic: Good.

- Spin welding: Good.
- Ultrasonic welding: Near field, excellent; far field, good. Spot welding, good.
- Vibration welding: Welds to itself.

Figure 5-12 presents the SPI standards for molding tolerances for PEI.

Polyethersulfone (PES), Amorphous Thermoplastic

Heat deflection temperature (°F @ 264 psi)	397
Maximum use temperature (°F, no load)	356
Tensile strength (psi)	9800–13,800
Flexural modulus (psi $\times$ 10^3 @ 73 °F)	348–380
Mold shrinkage (in./in.)	0.006–0.007
Coefficient of linear thermal expansion (in./in./°F $\times$ 10^{-6})	31

Additional Properties
Polyethersulfone has excellent tensile strength, electrical properties, and chemical resistance. It is platable, but mildly hygroscopic. It must be pigmented for exterior applications.

Applications
Premium material for high heat aerospace, automotive, chemical, and electrical components. Also used for pumps, valves, and medical applications.

Assembly Methods
- Adhesives: Ref. Chapter 7, Adhesive and Solvent Joining.
- Fasteners: Very high fail torque values and tensile pullout strengths. Very low material relaxation with outstanding elastic recovery. Tough material; it is resistant to stress cracking which improves with moisture absorption. Excellent results with use of special design thread-rolling screws, however trials required to determine optimum hole sizes. Ultrasonic installation required for push-in fasteners.
- Inserts: Ultrasonic installation preferred. Heat installation requires very high temperatures. Thread-cutting, press-in, and helical coil inserts applicable. Inserts may be molded in.
- Solvents: Good bonds possible with dichloromethane.
- Staking/swaging: Heat, good. Ultrasonics, fair.

Weldability
- Hot die/fusion: Fair.
- Hot gas: suitable for large parts.
- Induction/electromagnetic: Good.
- Spin welding: Good.
- Ultrasonic welding: Near field, fair to good; shear joints required; far field, fair. Spot welding, fair.

Polyethylene (PE), Crystalline Thermoplastic, Except for Cross-Linked PE, Which Is a Thermoset

	Low–Medium Density	High Density
Heat deflection temperature (°F @ 66 psi)	104–120	140–196
(°F @ 264 psi)	90–105	110–130
Maximum use temperature (°F, no load)	180–212	175–250
Tensile strength (psi)	1200–4550	3200–4500
Flexural modulus (psi $\times$ 10^3 @ 73°F)	35–48	145–225
Mold shrinkage (in./in.)	0.015–0.050	0.015–0.040
Coefficient of linear thermal expansion (in./in./°F $\times$ 10^{-6})	56–122	32.8–61.1

Additional Properties

Polyethylene is a member of the polyolefin family and is mainly a low cost material whose physical properties increase with its' density. Low-density polyethylene (LDPE) is tough and flexible. High density polyethylene (HDPE) is strong, stiff and resistant to creep and heat. It has good physical and electrical strengths, toughness, good chemical resistance and flexibility. Polyethylene is readily processed by most processing techniques.

Applications

Used for toys, lids, closures, packaging, rotationally molded tanks, and medical apparatus. Other applications are pipe, gas tanks, large containers, institutional seating, luggage, outdoor furniture, pails, containers, and housewares. Polyethylene is the workhorse of the rotational molding industry.

Assembly Methods

- Adhesives: Ref. Chapter 7, Adhesive and Solvent Joining.
- Fasteners: LDPE produces low tensile pullout strengths, low drive and fail torques; it is recommended for light-duty applications only. Excellent resistance to cracking, but initial material relaxation can reach 50%. Standard thread-forming screws recommended, but special design thread-rolling screws effective for HDPE. Push-in thread-forming screws can also be used for light duty applications.
- Inserts: Ultrasonic inserts preferred for HDPE. Heat installation is excellent. Press-in and self-tapping inserts can also be used. All types have low holding power in unreinforced polyethylene. Inserts may be molded in in some cases.
- Snap Fits: Excellent.
- Solvents: Not appropriate.
- Staking/swaging: Heat, excellent. Hot air/cold staking, good. Ultrasonics, fair.

Weldability

- Hot die/fusion: Good to excellent; weld strengths can approach those of the strength of the material.
- Hot gas: Good to excellent (use inert gas); weld strengths up to 80% of the strength of the material.

- Induction/electromagnetic: Good to excellent; also welds to polypropylene in some cases.
- Spin welding: Good.
- Ultrasonic welding: Near field, poor to good (with 15 kHz); far field, poor to good (with 15 kHz). Spot welding, good.
- Vibration welding, fair to good; weld strengths up to 90% of the strength of the material. Can also weld to polypropylene in some cases.

Figures 5-13 and 5-14 present the SPI standards for molding tolerances of LPDE and HDPE, respectively.

Polyimide (PI), Amorphous Thermoplastic

Heat deflection temperature (°F @ 264 psi)	460–680
Maximum use temperature (°F, no load)	500
Tensile strength (psi)	10,500–17,100
Flexural modulus (psi $\times 10^3$ @ 73 °F)	360–500
Mold shrinkage (in./in.)	0.008
Coefficient of linear thermal expansion (in./in./°F $\times 10^{-6}$)	25–31

Additional Properties
Polyimide provides excellent heat resistance, impact strength, wear resistance, and electrical and physical properties. It has inherent lubricity but offers limited weather resistance.

Applications
High performance electrical, aerospace, nuclear, aircraft, business machine, and military components. Also used for industrial hydraulic equipment, jet engines, automobiles, recreation vehicles, machinery, pumps valves, and turbines.

Assembly Methods
- Adhesives: Ref. Chapter 7, Adhesive and Solvent Joining.
- Fasteners: High initial clamp loads, substantial tensile pullout strengths, moderate drive-torque, and high fail-torque values. Notch sensitivity requires counterboring of pilot holes. High resistance to creep. Special design thread-rolling screws effective; push-in thread-forming fasteners should be limited to light-duty applications. Caution should be exercised in the use of tapped holes and thread-cutting screws. Screws can be made of this material.
- Inserts: Press-in type with adhesive reinforcement or expansion inserts recommended. Self-tapping inserts can be used with caution. Ultrasonic and heat installation not recommended. Inserts may be molded in.

Weldability
- Hot die/fusion: Fair.
- Induction/electromagnetic: Good.
- Spin welding: Fair.
- Ultrasonic welding: Fair.

Polymethylpentene (PMP), Crystalline Thermoplastic

Heat deflection temperature (°F @ 66 psi)	180–190
(°F @ 264 psi)	120–130
Maximum use temperature (°F, no load)	275
Tensile strength (psi)	2300–2500
Flexural modulus (psi $\times$ 10^3 @ 73 °F)	70–190
Mold shrinkage (in./in.)	0.016–0.021
Coefficient of linear thermal expansion (in./in./°F $\times$ 10^{-6})	36

Additional Properties
Polymethylpentene offers high transparency, good electrical properties, and chemical resistance. It maintains physical properties to levels near its melting point.

Applications
Used for laboratory and medical ware (syringes, connectors, hollow ware, disposable curettes), lighting, lenses, food (freeze to cooking range), and liquid level and flow indicators.

Assembly Methods
- Adhesives: Ref. Chapter 7, Adhesive and Solvent Joining.
- Inserts: Heat, excellent. Ultrasonics, excellent.
- Staking/swaging: Heat, excellent. Ultrasonics, Fair.

Weldability
- Induction/electromagnetic: Good.
- Ultrasonic welding: Near field, fair to good; far field, poor to fair. Spot welding, good.

Polyphenylene Oxide (PPO), Amorphous Thermoplastic

Heat deflection temperature (°F @ 66 psi)	230–279
(°F @ 264 psi)	212–265
Maximum use temperature (°F, no load)	175–220
Tensile strength (psi)	6800–9600
Flexural modulus (psi $\times$ 10^3 @ 73 °F)	325–400
Mold shrinkage (in./in.)	0.005–0.008
Coefficient of linear thermal expansion (in./in./°F $\times$ 10^{-6})	18–43

Additional Properties
Polyphenylene oxide provides high dimensional stability, toughness, chemical resistance, excellent electrical properties and low moisture absorption. It offers the highest resistance to creep under loads.

Applications
Automobile dashboards, electrical connectors, grilles, and wheel covers. Also used for hot water pumps, underwater components, shower heads, appliances, and electrical and appliance housings.

Assembly Methods
- Adhesives: Ref. Chapter 7, Adhesive and Solvent Joining.
- Fasteners: Excellent torque, tensile pullout, and material relaxation values – particularly when reinforced. Molded-in stress can create boss problems. Excellent application for special design thread-forming screws, particularly for foamed parts; with glass reinforcement, however, boss cracking can result for some of these screws. Highly stressed bosses may be better suited to thread-cutting screws for nonfoamed material. Resin manufacturer recommends standard ASA type T (type 23) or ASA type BT (type 25) thread-cutting screws. Boss caps and push-in screws are good for light-duty applications. Rivets acceptable, but can create high stresses on installation. Aluminum and shoulder rivets preferred. Tapped holes acceptable.
- Hinges: Two-piece integral hinges excellent.
- Inserts: Ultrasonic inserts recommended. Heat insertion excellent; press-in, self-tapping, and helical coil inserts applicable. Use molded-in inserts for highest pullout strengths.
- Snap Fits: excellent.
- Solvents: Chloroform, ethylene dichloride, methylene chloride, toluene, perchlorethylene, dichloromethane, methyl ethyl ketone, monochlorobenzene, chlorobenzene, trichloroethylene, xylene, dimethyl benzene, toluol and tetrahydrofuran.
- Staking/swaging: Heat, excellent. Hot air, good. Ultrasonics, excellent.

Weldability
- Hot die/fusion: Good.
- Induction/electromagnetic: Good to excellent; also welds to polystyrene in some cases.
- Spin welding: Good to excellent
- Ultrasonic welding: Near field, good to excellent (energy director recommended); far field, poor to good (with 15 kHz). Spot welding, fair to excellent. Also welds to acrylic, PC, PS, and SAN in some cases.
- Vibration welding: Easy to medium with shear joint. Can also weld to some versions of ABS, acrylic, and nylon.

Figure 5-15 presents the SPI standards for molding tolerances for PPO.

Polyphenylene Sulfide (PPS), Crystalline Thermoplastic

Heat deflection temperature (°F @ 66 psi)	390
(°F @ 264 psi)	275
Maximum use temperature (°F, no load)	500
Tensile strength (psi)	7000–12,500
Flexural modulus (psi × 10^3 @ 73 °F)	550–600
Mold shrinkage (in./in.)	0.006–0.014
Coefficient of linear thermal expansion (in./in./°F × 10^{-6})	15–27

Additional Properties
Polyphenylene sulfide has high mechanical, impact, thermal stability, heat deflection, and dielectric properties with good flame and chemical resistance. Depending on molding cycle, it can cross-link to become a thermoset.

Applications
Mainly used for electrical (connectors, coil forms, bobbins), mechanical (chemical processing equipment and pumps, including submersibles), and automotive (underhood) applications.

Assembly Methods
- Adhesives: Ref. Chapter 7, Adhesive and Solvent Joining.
- Fasteners: High drive torque, fail torque, and tensile pullout strengths. Very low initial load relaxation. Notch sensitive; pilot holes must be counterbored similar to thermoset resins to reduce stress cracking tendency. Special design thread-rolling screws and thread-cutting screws recommended (standard type BF or BT). Use pilot hole sizes for phenolformaldehyde thermosets in *Machinery's Handbook*. Threads can be molded in or tapped. Polyphenylene sulfide can be successfully assembled using semi-tubular rivets. Push-in thread-forming screws not recommended unless inserted ultrasonically.
- Inserts: Ultrasonic inserts are the preferred method. Heat installation excellent. Can also use self-tapping, helical coil, expansion and press-in inserts with adhesive reinforcement. Molded-in inserts provide highest pullout strength and can be used except where repeated heat cycling can be expected.
- Press Fits: Excellent.
- Snap Fits: Good.
- Solvents: No known solvents below 204 °C (400 °F). Not recommended.
- Staking/swaging: Heat, fair to excellent (application sensitive). Hot air/cold staking, good. Ultrasonics, poor to good.

Weldability
- Hot die/fusion: Fair
- Induction/electromagnetic: Good.
- Spin welding: Good.

- Ultrasonic welding: Near field, good; far field, fair to good (shear joint recommended). Spot welding, fair to good.
- Vibration welding: Good.

Polyphthalamide (PPA), Crystalline Thermoplastic*

Heat deflection temperature (°F @ 264 psi)	248
Tensile strength (psi, dry)	9,000–11,000
(psi, 50% RH)	7,500–9,000
Flexural modulus (psi × 10^3 @ 73 °F)	320–380
Mold shrinkage (in./in.)	0.015–0.022

* Some polymers are amorphous.

Additional Properties
Polyphthalamide is basically a crystalline aromatic nylon that combines the high strength and stiffness of nylon with the heat deflection characteristics of polyphenylene sulfide. Like nylon, it is hygroscopic. It is typically used with glass reinforcement, which results in higher values for most of the properties listed above.

Applications
PPA is intended for automotive, electrical and industrial applications.

Assembly Methods
- Adhesives: Ref. Chapter 7, Adhesive and Solvent Joining.
- Fasteners: Thread-forming, self-tapping screws are recommended for unfilled grades; thread-cutting screws for glass-reinforced resins. For optimum strip-out torque, the pilot hole diameter should be the pitch diameter of the screw. Threads can be molded in. Torque values and tensile pullout strengths drop with increase in RH.
- Inserts: Ultrasonic inserts are good and the preferred method. Heat installation is also excellent, but slower in most cases. Press-in, self-tapping, and helical coil inserts can also be used. Molded-in inserts are acceptable, but be alert to the high shrinkage rate of the material. Installation should be performed immediately after molding, before moisture can be absorbed. Otherwise, parts must be kept dry with desiccants or redried before installation can proceed.
- Press Fits: Excellent.
- Snap Fits: Excellent; parts may be too brittle for snap fit applications immediately after molding (if so, remoisturing would be rquired).
- Solvents: Solvent welding is not recommended as a production assembly method.
- Staking/swaging: Heat, good. Hot air/cold staking, fair. Ultrasonics, fair to good.

Weldability
- Hot die/fusion: Good; best results in trials obtained with hot plate temperature of 330 °C (626 °F), a clamping pressure of 207 kPa (30 psi), a weld time of 40 seconds,

and a hold time of 20 seconds. Bond strength was comparable to the strength of the material itself.

- Induction/electromagnetic: Good to excellent.
- Spin welding: Good to excellent, depending on application.
- Ultrasonic welding: Near field, good (shear joint recommended); far field, poor to fair. Spot welding, fair to good.
- Vibration welding: Excellent; weld strengths up to the strength of the parent material. Good results obtained with weld times as short as 0.6 second at pressures down to 2.20 MPa (320 psi).

Polypropylene (PP), Crystalline Thermoplastic

	Homoploymer	Copolymer
Specific gravity	0.900–0.910	0.890–0.905
Izod impact strength (ft-lb/in.)	0.4–1.0	1.0–20.0
Heat deflection temperature (°F @ 66 psi)	200–250	185–220
(°F @ 264 psi)	125–140	120–140
Maximum use temperature (°F, no load)	225–300	190–250
Tensile strength (psi)	4500–6000	4000–4500
Flexural modulus (psi $\times$ 10^3 @ 73 °F)	170–250	130–200
Mold shrinkage (in./in.)	0.010–0.025	0.010–0.025
Coefficient of linear thermal expansion (in./in./°F $\times$ 10^{-6})	45–55	38–53

Additional Properties
Polypropylene is a very low cost material with excellent resistance to stress or flex cracking. It has good chemical, grease, heat, and scratch resistance. It can be electroplated and has a very low specific gravity. It offers good impact strength and toughness (best with the copolymer). It becomes brittle at low temperatures starting at freezing (particularly the homopolymer).

Applications
Used for tubs, agitators, dispensers, pump housings, and filters in appliances, and in automotive components (fan shrouds, fan blades, ducts, housings, batteries, door panels, trim on glove boxes, seat frames, louvers, and seatbelt retractor covers). Also medical, luggage, toy, and houseware applications.

Assembly Methods
- Adhesives: Ref. Chapter 7, Adhesive and Solvent Joining.
- Fasteners: The homopolymer and the copolymer have similar fastener characteristics. Polypropylene produces low drive torques, fail torques, and tensile pullout strengths in unfilled versions. Longer engagement preferred. Excellent resistance to stress cracking, but material relaxation can reach 50%. Standard and special design thread-forming screws recommended, with somewhat higher stripping and pullout values for the latter. Push-in thread-forming screws effective. Molded-in threads possible, rounded coarse threads recommended. Ultrasonic installation feasible in certain instances.

- Hinges: Principal material for living hinges.
- Inserts: Low holding power of unfilled polypropylene must be taken into consideration. Heat insertion excellent. Ultrasonic inserts fair to good depending on application. Press-in inserts can also be used, as can self-tapping and helical coil inserts. Molded-in inserts not recommended owing to high shrinkage rate. If molded-in inserts are necessary, they should be preheated.
- Snap Fits: Excellent.
- Solvents: None.
- Staking/swaging: Heat, excellent. Hot air/cold staking, good. Ultrasonics, excellent.

Weldability
- Hot die/fusion: Good to excellent; weld strengths up to 80% of the strength of the material.
- Hot gas: Good; weld strengths up to 80% of the strength of the material.
- Induction/electromagnetic: Good to excellent. Also welds to polybutylene and PE in some cases.
- Spin welding: Good.
- Ultrasonic welding: Near field, fair to poor; far field, very poor. Spot welding, good to excellent
- Vibration welding: Good; weld strengths up to 90% of the adjacent walls. Can also be welded to some versions of polyethylene.

Figure 5-16 presents the SPI standards for molding tolerances for PP.

Polystyrene (PS), Amorphous Thermoplastic

	Crystal (GP)	High Impact
Heat deflection temperature (°F @ 66 psi)	155–204	165–200
(°F @ 264 psi)	169–202	170–205
Maximum use temperature (°F, no load)	150–170	140–175
Tensile strength (psi)	5200–7500	1900–6200
Flexural modulus (psi × 10^3 @ 73 °F)	380–490	160–390
Mold shrinkage (in./in.)	0.004–0.007	0.004–0.007
Coefficient of linear thermal expansion (in./in./°F × 10^{-6})	28–46	25

Additional Properties
Polystyrene is very easy to process and low in cost, with good rigidity and dimensional stability. It has low moisture absorption, glossy surface, and good clarity and is easy to decorate. General-purpose (GP) styrene is brittle without a modifier; butadiene is added to improve impact resistance. Clear is not resistant to ultraviolet light. Weather exposure discolors the material and reduces its strength. Improvement can be gained with pigments (finely disbursed carbon black). Limit outdoor use to applications to those in which parts can be replaced or exposure is intermittent. Polystyrene is available in heat resistant, UV-light-resistant, and flame-retardant grades.

Applications
- GP: Home furnishings (mirror and picture frames and moldings); housewares (personal care products, flower pots, toys, cutlery, bottles, combs, disposables such as tumblers, dishes, and trays). Consumer electronics (cassettes, reels, and housings) and medical uses (sample collectors, petri dishes, test tubes).
- Impact grade (with flame retardants): Used for TV, smoke detector and small appliance housings.

Assembly Methods
- Adhesives: Ref. Chapter 7, Adhesive and Solvent Joining.
- Fasteners: Drive and fail torque values are in the medium to high range, and relatively high clamp loads can be tolerated. Very slight material relaxation in unfilled grades; less in filled grades. Strong boss support and large ratios of boss diameter to pilot hole diameter required to avoid boss cracking. Special design thread-rolling screws effective; push-in thread-forming screws satisfactory. Use impact grades for mechanical fasteners. Fasteners can be ultrasonically installed.
- Inserts: Heat installation excellent. Ultrasonic inserts good for both GP and high impact polystyrenes. Press-in, self-tapping, and helical coil inserts feasible. Inserts can be molded in in some cases.
- Snap Fits: Poor for crystal polystyrene, fair for high impact polystyrene.
- Solvents: Ethylene dichloride, methyl ethyl ketone, methylene chloride, toluene, cyclohexanone, dioxane, ethyl acetate, glacial acetic acid, methyl Cellosolve, *N*-methyl pyrrolidone, *O*-dichlorobenzol, perchlorethylene, and xylene. Chlorinated hydrocarbons mar its surface and cause failure under load. Aliphatic and aromatic hydrocarbons dissolve polystyrene. Solvent welding is widely used to asemble polystyrene parts. Weld strengths range up to 60% of the strength of the material.
- Staking/swaging: Heat, Good. Hot air/cold staking – Excellent. Ultrasonics, Fair for both GP and High Impact Styrene.

Weldability
- Hot die/fusion: Good to excellent, GP weld strengths up to 80% of the strength of the material; high impact strengths up to 90% of the strength of the material. Can also weld to polyphenylene oxide and SAN in some cases.
- Hot gas: Fair to good; weld strengths up to 50% of the strength of the material. (Bubbles, blows around, gets stringy.)
- Induction/electromagnetic: Good to excellent; also welds to acrylic, PC, polyphenylene oxide. and SAN in some cases.
- Spin welding: Good.
- Ultrasonic welding: Near field, excellent (good to excellent for high impact styrene); far field, excellent for GP styrene, fair to good for high impact styrene (energy director recommended). Spot welding, fair for both GP and high impact styrene.
- Vibration welding: Good; weld strengths up to 60% of the strength of the material. Can weld to some ABS and SAN resins.

Figure 5-17 presents the SPI standards for molding tolerances for PS.

Polysulfone, (PSU) Serving the manufacturing industry utilizing agile technology
Amorphous Thermoplastic

Heat deflection temperature (°F @ 66 psi)	358
(°F @ 264 psi)	345
Maximum use temperature (°F, no load)	300
Tensile strength (psi)	10,200
Flexural modulus (psi $\times$ 10^3 @ 73 °F)	390
Mold shrinkage (in./in.)	0.006–0.007
Coefficient of linear thermal expansion (in./in./°F $\times$ 10^{-6})	31

Additional Properties
Polysulfone offers good transparency, high mechanical strengths, heat resistance, and
electrical strengths. It has unusual resistance to strong mineral acids and alkalies and
retention of properties on heat aging. Its weatherability is poor without coating.

Applications
High temperature automotive, office machine, consumer electronics, appliance, and
medical applications.

Assembly Methods
- Adhesives: Ref. Chapter 7, Adhesive and Solvent Joining.
- Fasteners: High tensile pullout, drive, and fail torque values. Very low initial material
  relaxation, but polymer tends to be notch sensitive, which can lead to stress cracking
  if hole and boss are not carefully designed. Special design thread-rolling screws
  acceptable. Push-in thread-forming screws not recommended unless installed ultrasonically. Tapped threads can be used.
- Inserts: Ultrasonic inserts recommended. Heat installation good. Press-in, self-tapping,
  molded in, and helical coil inserts are acceptable.
- Press Fits: Excellent due to low creep characteristics.
- Solvents: Methylene chloride, cyclohexanone, ethylene dichloride, methyl ethyl
  ketone, *O*-dichlorobenzol, toluene and xylene.
- Staking/swaging: Heat, good. Ultrasonics, fair.

Weldability
- Hot die/fusion: Fair to good; can also weld to polycarbonate in some cases.
- Hot gas: Satisfactory.
- Induction/electromagnetic: Good to excellent; also welds to PC in some cases.
- Spin welding: Good.
- Ultrasonic welding: Near field, good; far field, fair to good (shear joint recom-
  mended). Spot welding, fair to good.
- Vibration welding: Good.

Polyurethane (TPU), Thermoplastic

Heat deflection temperature (°F @ 66 psi)	115–275
(°F @ 264 psi)	158–260
Maximum use temperature (°F, no load)	190
Tensile strength (psi)	4500–9000
Flexural modulus (psi $\times$ 10^3 @ 73 °F)	235–310
Mold shrinkage (in./in.)	0.004–0.006

Additional Properties
Thermoplastic polyurethane is a rubberlike plastic with excellent adhesive properties. It combines the functional performance and properties of a themoset rubber with the processibility of a plastic.

Applications
Extensive use as a substitute for conventional thermoset rubbers in appliance, construction, food service and health care industries. Used in automotive industry in nontire applications.

Assembly Methods
- Adhesives: Ref. Chapter 7, Adhesive and Solvent Joining.
- Inserts: Molded-in inserts are effective, however they should be heated to 104 to 121 °C (220–250 °F) prior to molding.

Weldability
- Hot die/fusion: Fair.
- Hot gas: Unweldable.
- Induction/electromagnetic: Good to excellent.
- Spin welding: Fair.
- Ultrasonic welding: Near field, fair; far field, fair.
- Vibration welding: Fair.

Polyurethane (PUR), Thermoset

Heat deflection temperature (°F @ 264 psi)	190–200
Tensile strength (psi)	175–10,000
Flexural modulus (psi $\times$ 10^3 @ 73 °F)	10–100
Mold shrinkage (in./in.)	0.020
Coefficient of linear thermal expansion (in./in./°F $\times$ 10^{-6})	55–110

Additional Properties
Thermoset polyurethane is a rubberlike plastic with high toughness and resistance to tearing, abrasion, ozone, oxidation, fungi and humidity. It is available as flexible or rigid foam and as an elastomer.

Applications
- Flexible Foam: Bedding, furniture, carpet underlays, and transportation.
- Rigid Foam: Building insulation, refrigeration, packaging, insulation, and transportation.
- Elastomers: Shoe parts and reaction injection molding use in transportation.

Assembly Methods
- Adhesives: Ref. Chapter 7, Adhesive and Solvent Joining.
- Fasteners: Thread-forming self-tapping screws acceptable in some cases.
- Inserts: Molded-in inserts and press-in inserts combined with adhesive can be used. Inserts cannot be ultrasonically installed.
- Staking/swaging: Not applicable.

Weldability: Does not weld.

Polyvinyl Chloride (Vinyl) (PVC), Amorphous Thermoplastic

	Rigid	Flexible
Heat deflection temperature (°F @ 66 psi)	135–180	–
(°F @ 264 psi)	130–175	–
Maximum use temperature (°F, no load)	150–175	140–175
Tensile strength (psi)	5,000–8,000	1000–3500
Elongation at break (%)	40–80	200–400
Flexural modulus (psi $\times$ 10^3 @ 73 °F)	10,000–16,000	–
Mold shrinkage (in./in.)	0.001–0.003	0.008–0.035
Coefficient of linear thermal expansion (in./in./°F $\times$ 10^{-6})	5–10	7–25

Additional Properties
Polyvinyl chloride is available in a wide range of properties.
- Rigid: Rigid PVC is hard, tough, and difficult to process. It has excellent outdoor stability, electrical properties, and resistance to moisture and chemicals. It is self-extinguishing.
- Flexible: Rigid vinyl becomes flexible when additives are included in the formulation. Flexible vinyl is easier to process, but offers lower heat resistance, physical, and weathering properties. It provides the unusual combination of transparency with flexibility.

Applications
- Rigid: Used for pipe fittings, toys, dinnerware, sporting goods, shoe heels, credit cards, and electrical applications in appliances, television sets, and electrical boxes.
- Flexible: Used in intravenous medical equipment, floor and wall tiles, raincoats, and shower curtains, plus automotive dashboards, armrests, and headrests.

Assembly Methods
- Adhesives: Ref. Chapter 7, Adhesive and Solvent Joining.
- Fasteners: Flexible vinyls do not lend themselves well to fasteners owing to low tensile pullout and drive torque values coupled with high rates of material relaxation. Rigid vinyl yields moderate torque and strength values with good assembly load retention. Good stress crack resistance; however the most rigid of PVCs must be carefully designed. Standard thread-forming special-design thread-rolling and push-in thread-forming screws are excellent, however thread-cutting screws produce less stress. Fasteners may be ultrasonically installed.
- Inserts: Ultrasonic, excellent for rigid PVC; not feasible for flexible PVC. Heat installation excellent for rigid. Press-in, self-tapping, molded in, and helical coil inserts effective, best in rigid PVC.
- Snap Fits: Rigid PVC excellent for snap fits.
- Solvents: Acetone, cyclohexane, methyl ethyl ketone, tetrahydrofuran, cyclohexanone, methyl Cellosolve, and *O*-dichlorobenzol. Solvent welding is commonly used to assemble PVC parts. Weld strengths up to 70% of the strength of the material.
- Staking/swaging: Heat, good for rigid PVC. Hot air/cold staking, good for rigid PVC. Ultrasonics, fair to good for rigid, not feasible for flexible PVC.

Weldability
- Hot die/fusion: Fair to excellent; weld strengths up to 90% of adjacent walls for both rigid and flexible PVC. Special ventilation required.
- Hot gas: Good for rigid; weld strengths up to 70% of the strength of the material.
- Induction/electromagnetic: Good to excellent.
- Spin welding: Good.
- Ultrasonic welding: Near field, poor to fair for rigid PVC; far field, rigid PVC, poor to fair (energy director recommended), flexible PVC, not feasible. Spot welding, rigid PVC, fair, flexible PVC, poor. Also welds to ABS in some cases.
- Vibration welding: Fair to good; weld strengths up to 70% of the strength of the material.

Figure 5-18 and 5-19 present the SPI standards for molding tolerances of rigid and flexible PVC, respectively.

Styrene–Acrylonitrile (SAN), Amorphous Thermoplastic

Heat deflection temperature (°F @ 66 psi)	220–224
(°F @ 266 psi)	203–220
Maximum use temperature (°F, no load)	140–200
Tensile strength (psi)	10,000–11,900
Flexural modulus (psi × 10^3 @ 73 °F)	500–610
Mold shrinkage (in./in.)	0.003–0.005
Coefficient of linear thermal expansion (in./in./°F × 10^{-6})	36–38

Additional Properties
Styrene-acrylonitrile, a member of the styrenic group, is stronger and stiffer than poly-styrene with better chemical resistance, and more resistant to ultraviolet light than ABS.

Applications
Medical instruments and disposable intravenous equipment. Cigarette lighter and blender housings; overhead fan blades, faucet handles, and turntable covers.

Assembly Methods
- Adhesives: Ref. Chapter 7, Adhesive and Solvent Joining.
- Fasteners: High torque drive and tensile pullout strengths combined with moder-ately low load relaxation. Pilot hole must be counterbored to counteract tendency toward crazing on screw entry. Special design thread-rolling screws and push-in thread-forming screws excellent. Fasteners may be ultrasonically installed.
- Inserts: Ultrasonic recommended. Heat installation good. Press-in, self-tapping, molded in, and helical-coil inserts effective.
- Solvents: Acetone, glacial acetic acid, *N*-methylpyrrolidone, and *O*-dichlorobenzol. Solvent welding is often used to join SAN parts. Weld strengths up to 50% of the adjacent wall.
- Staking/swaging: Heat, good. Hot air/cold staking, good. Ultrasonics, fair.

Weldability
- Hot die/fusion: Good; also welds to ABS, acrylic, and polystyrene in some cases.
- Induction/electromagnetic: Good to excellent; also welds to ABS, acrylic, PC, and PS in some cases.
- Spin welding: Good.
- Ultrasonic welding: Near field, excellent; far field, excellent. Spot welding, fair. Also welds ABS, acrylic, polyphenylene oxide and polystyrene in some cases.
- Vibration welding: Good; can also weld to some versions of polystyrene, ABS and acrylic.

Figure 5-20 presents the SPI standards for molding tolerances for SAN.

Urea–Formaldehyde (UF), Thermoset

	Alpha-Cellulose Filled
Heat deflection temperature (°F @ 264 psi)	260–290
Maximum use temperature (°F, no load)	170
Tensile strength (psi)	5500–13,000
Flexural modulus (psi $\times$ 10^3 @ 73 °F)	1300–1600
Mold shrinkage (in./in.)	0.006–0.014
Coefficient of linear thermal expansion (in./in./°F $\times$ 10^{-6})	12–20

Additional Properties
Urea-formaldehyde is hard, rigid, solvent resistant, chip resistant, abrasion resistant, and self-extinguishing. It has excellent colorability and dimensional stability under load.

Applications
Military and industrial electrical applications.

Assembly Methods
- Adhesives: Ref. Chapter 7, Adhesive and Solvent Joining.
- Fasteners: Low drive torques with thread-cutting screws. High tensile pullout strengths and practically no relaxation, but susceptible to cracking at the edge of the hole. Type BT standard thread-cutting screws are recommended. Special design thread-rolling screws can sometimes be used with assistance of screw manufacturer's laboratory.
- Inserts: Press-in type with adhesive reinforcement recommended. Molded-in inserts feasible owing to low shrinkage rate. Can also use helical coil expansion and self-tapping inserts. Ultrasonic and heat installation not possible.
- Staking/swaging: Not applicable.

Weldability: Does not weld.

5.4.2   Adhesives

The reader will note the absence of the adhesives from the listings of Section 5.4.1. There was not enough space available to locate specific recommendations with each material for the many combinations and the variety of adhesives available. See instead, Chapter 7, Adhesives and Solvent Joining, which also gives suggested surface preparation methods.

5.4.3   Using the SPI Tables

The SPI tables are similar in basic format for all the materials that are covered, whether for thermoplastics or thermosets.

Two separate sets of values are given. The commercial values (Comm. ±) represent common production tolerances that can be achieved at the most economical level. The fine values (Fine ±) represent closer tolerances that can be held, but at a greater cost. The selection will depend on the application under consideration and the economics involved.

By referring to the hypothetical molded article and its cross section illustrated in the table, and by then using the applicable code number (e.g., *A* represents the diameter) in the first column and the exact dimensions as indicated in the second column, readers can find the recommended tolerances either in the chart at the top of the table or in the two columns underneath. (Note that the typical article shown in cross section in the tables may be of round, rectangular or another shape. Thus, dimensions *A* and *B* may be diameters or lengths.)

For example, an ABS part with a diameter (*A*) of 2 in. would show tolerances of ±0.004 in. (fine value) and 0.007 in. (commercial value). If the dimensions go up to 6 in., the tolerances would change to ±0.008 in. (fine) and 0.012 in. (commercial). If the dimension is greater than 6 in., however, the tolerance is increased by the amount indicated on the lines directly below the chart. Thus, if dimension *A* is 10 in., the tolerance (commercial) is ±0.024 in. (i.e., 0.012 in. for a 6-in. dimension plus 0.003 for each of the additional 4-in.).

Other tolerances for the various dimensions shown in the typical cross section are as indicated in the two columns running under the chart.

Special notes for users of the SPI charts:
1. The tolerances indicated in the tables for diameter and hole size do not include allowance for aging characteristics of the particular plastics material under consideration.
2. Tolerances are based on $\frac{1}{8}$-in. wall sections.
3. For depth, height, and bottom wall dimensions, parting line must be taken into consideration.
4. In terms of sidewall dimensions, flatness, and concentricity, part design should maintain a wall thickness as nearly constant as possible. Complete uniformity in this dimension is impossible to achieve.
5. In determining hole size depth and draft allowance per side, care must be taken that the ratio of the depth of a cored hole to the diameter does not reach a point that will result in excessive pin damage.
6. Values for fillets, ribs, and corners should be increased whenever compatible with desired design and good molding technique.
7. Where surface finish and color stability are concerned, customer and molder must come to an understanding as to what is necessary for the particular job under consideration before molding.

Standards & Practices **of Plastics Molders**	**Material** Polyoxymethylene (Acetal) (POM)

Note: The *Commercial* values shown below represent common production tolerances at the most economical level. The *Fine* values represent closer tolerances that can be held but at a greater cost. Any addition of fillers will compromise physical properties and alter dimensional stability. Please consult the manufacturer.

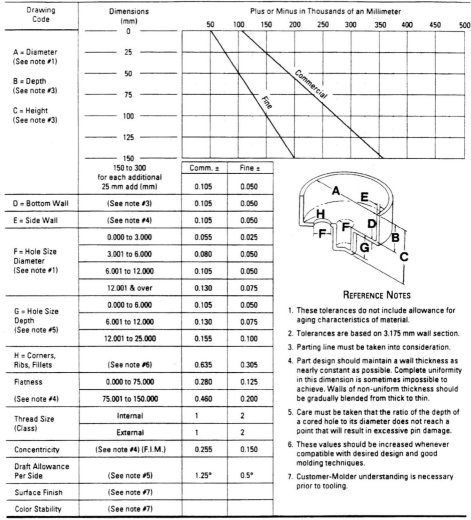

Drawing Code	Dimensions (mm)		
A = Diameter (See note #1)			
B = Depth (See note #3)			
C = Height (See note #3)			

Plus or Minus in Thousands of an Millimeter (chart: 50, 100, 150, 200, 250, 300, 350, 400, 450, 500 — curves labeled *Commercial* and *Fine*; dimensions 0, 25, 50, 75, 100, 125, 150)

		Comm. ±	Fine ±
	150 to 300 for each additional 25 mm add (mm)	0.105	0.050
D = Bottom Wall	(See note #3)	0.105	0.050
E = Side Wall	(See note #4)	0.105	0.050
F = Hole Size Diameter (See note #1)	0.000 to 3.000	0.055	0.025
	3.001 to 6.000	0.080	0.050
	6.001 to 12.000	0.105	0.050
	12.001 & over	0.130	0.075
G = Hole Size Depth (See note #5)	0.000 to 6.000	0.105	0.050
	6.001 to 12.000	0.130	0.075
	12.001 to 25.000	0.155	0.100
H = Corners, Ribs, Fillets	(See note #6)	0.635	0.305
Flatness	0.000 to 75.000	0.280	0.125
(See note #4)	75.001 to 150.000	0.460	0.200
Thread Size (Class)	Internal	1	2
	External	1	2
Concentricity	(See note #4) (F.I.M.)	0.255	0.150
Draft Allowance Per Side	(See note #5)	1.25°	0.5°
Surface Finish	(See note #7)		
Color Stability	(See note #7)		

REFERENCE NOTES

1. These tolerances do not include allowance for aging characteristics of material.
2. Tolerances are based on 3.175 mm wall section.
3. Parting line must be taken into consideration.
4. Part design should maintain a wall thickness as nearly constant as possible. Complete uniformity in this dimension is sometimes impossible to achieve. Walls of non-uniform thickness should be gradually blended from thick to thin.
5. Care must be taken that the ratio of the depth of a cored hole to its diameter does not reach a point that will result in excessive pin damage.
6. These values should be increased whenever compatible with desired design and good molding techniques.
7. Customer-Molder understanding is necessary prior to tooling.

Figure 5-1   Acetal (polyoxymethylene-POM)

Standards & Practices of Plastics Molders

Material
Polyoxymethylene
(Acetal) (POM)

Note: The *Commercial* values shown below represent common production tolerances at the most economical level. The *Fine* values represent closer tolerances that can be held but at a greater cost. Any addition of fillers will compromise physical properties and alter dimensional stability. Please consult the manufacturer.

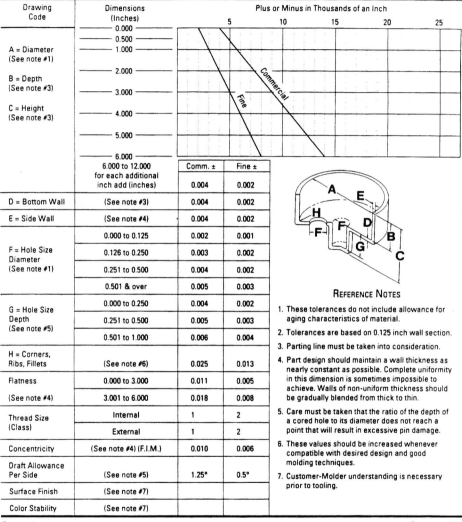

Drawing Code	Dimensions (Inches)	Comm. ±	Fine ±
A = Diameter (See note #1)			
B = Depth (See note #3)			
C = Height (See note #3)			
	6.000 to 12.000 for each additional inch add (inches)	0.004	0.002
D = Bottom Wall	(See note #3)	0.004	0.002
E = Side Wall	(See note #4)	0.004	0.002
F = Hole Size Diameter (See note #1)	0.000 to 0.125	0.002	0.001
	0.126 to 0.250	0.003	0.002
	0.251 to 0.500	0.004	0.002
	0.501 & over	0.005	0.003
G = Hole Size Depth (See note #5)	0.000 to 0.250	0.004	0.002
	0.251 to 0.500	0.005	0.003
	0.501 to 1.000	0.006	0.004
H = Corners, Ribs, Fillets	(See note #6)	0.025	0.013
Flatness	0.000 to 3.000	0.011	0.005
(See note #4)	3.001 to 6.000	0.018	0.008
Thread Size (Class)	Internal	1	2
	External	1	2
Concentricity	(See note #4) (F.I.M.)	0.010	0.006
Draft Allowance Per Side	(See note #5)	1.25°	0.5°
Surface Finish	(See note #7)		
Color Stability	(See note #7)		

REFERENCE NOTES

1. These tolerances do not include allowance for aging characteristics of material.

2. Tolerances are based on 0.125 inch wall section.

3. Parting line must be taken into consideration.

4. Part design should maintain a wall thickness as nearly constant as possible. Complete uniformity in this dimension is sometimes impossible to achieve. Walls of non-uniform thickness should be gradually blended from thick to thin.

5. Care must be taken that the ratio of the depth of a cored hole to its diameter does not reach a point that will result in excessive pin damage.

6. These values should be increased whenever compatible with desired design and good molding techniques.

7. Customer-Molder understanding is necessary prior to tooling.

Copyright   Revised 1991

Standards & Practices of Plastics Molders	Material
	Acrylic

Note: The *Commercial* values shown below represent common production tolerances at the most economical level. The *Fine* values represent closer tolerances that can be held but at a greater cost. Any addition of fillers will compromise physical properties and alter dimensional stability. Please consult the manufacturer.

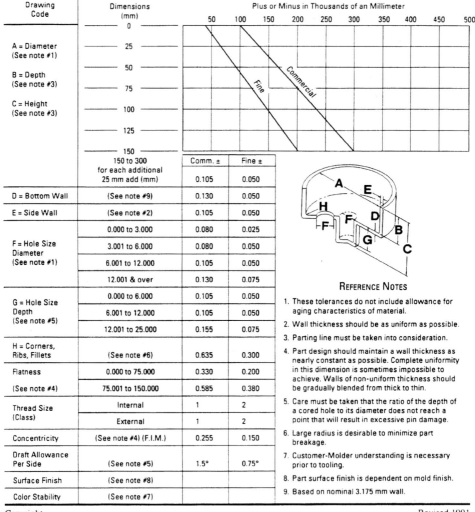

Drawing Code	Dimensions (mm)	Comm. ±	Fine ±
150 to 300 for each additional 25 mm add (mm)		0.105	0.050
D = Bottom Wall	(See note #9)	0.130	0.050
E = Side Wall	(See note #2)	0.105	0.050
F = Hole Size Diameter (See note #1)	0.000 to 3.000	0.080	0.025
	3.001 to 6.000	0.080	0.050
	6.001 to 12.000	0.105	0.050
	12.001 & over	0.130	0.075
G = Hole Size Depth (See note #5)	0.000 to 6.000	0.105	0.050
	6.001 to 12.000	0.105	0.050
	12.001 to 25.000	0.155	0.075
H = Corners, Ribs, Fillets	(See note #6)	0.635	0.300
Flatness (See note #4)	0.000 to 75.000	0.330	0.200
	75.001 to 150.000	0.585	0.380
Thread Size (Class)	Internal	1	2
	External	1	2
Concentricity	(See note #4) (F.I.M.)	0.255	0.150
Draft Allowance Per Side	(See note #5)	1.5°	0.75°
Surface Finish	(See note #8)		
Color Stability	(See note #7)		

A = Diameter (See note #1)

B = Depth (See note #3)

C = Height (See note #3)

REFERENCE NOTES

1. These tolerances do not include allowance for aging characteristics of material.
2. Wall thickness should be as uniform as possible.
3. Parting line must be taken into consideration.
4. Part design should maintain a wall thickness as nearly constant as possible. Complete uniformity in this dimension is sometimes impossible to achieve. Walls of non-uniform thickness should be gradually blended from thick to thin.
5. Care must be taken that the ratio of the depth of a cored hole to its diameter does not reach a point that will result in excessive pin damage.
6. Large radius is desirable to minimize part breakage.
7. Customer-Molder understanding is necessary prior to tooling.
8. Part surface finish is dependent on mold finish.
9. Based on nominal 3.175 mm wall.

Copyright

Revised 1991

Figure 5-2   Acrylic (polymethyl methacrylate – PMMA)

Standards & Practices of Plastics Molders

Material

Acrylic

Note: The *Commercial* values shown below represent common production tolerances at the most economical level. The *Fine* values represent closer tolerances that can be held but at a greater cost. Any addition of fillers will compromise physical properties and alter dimensional stability. Please consult the manufacturer.

Drawing Code	Dimensions (Inches)	Comm. ±	Fine ±
A = Diameter (See note #1) B = Depth (See note #3) C = Height (See note #3)	0.000 / 0.500 / 1.000 / 2.000 / 3.000 / 4.000 / 5.000 / 6.000		
	6.000 to 12.000 for each additional inch add (inches)	0.004	0.002
D = Bottom Wall	(See note #9)	0.005	0.002
E = Side Wall	(See note #2)	0.004	0.002
F = Hole Size Diameter (See note #1)	0.000 to 0.125	0.003	0.001
	0.126 to 0.250	0.003	0.002
	0.251 to 0.500	0.004	0.002
	0.501 & over	0.005	0.003
G = Hole Size Depth (See note #5)	0.000 to 0.250	0.004	0.002
	0.251 to 0.500	0.004	0.002
	0.501 to 1.000	0.006	0.003
H = Corners, Ribs, Fillets	(See note #6)	0.025	0.012
Flatness	0.000 to 3.000	0.013	0.008
(See note #4)	3.001 to 6.000	0.023	0.015
Thread Size (Class)	Internal	1	2
	External	1	2
Concentricity	(See note #4) (F.I.M.)	0.010	0.006
Draft Allowance Per Side	(See note #5)	1.5°	0.75°
Surface Finish	(See note #8)		
Color Stability	(See note #7)		

Plus or Minus in Thousands of an Inch (graph columns: 5, 10, 15, 20, 25) — curves labeled *Commercial* and *Fine*

REFERENCE NOTES

1. These tolerances do not include allowance for aging characteristics of material.

2. Wall thickness should be as uniform as possible.

3. Parting line must be taken into consideration.

4. Part design should maintain a wall thickness as nearly constant as possible. Complete uniformity in this dimension is sometimes impossible to achieve. Walls of non-uniform thickness should be gradually blended from thick to thin.

5. Care must be taken that the ratio of the depth of a cored hole to its diameter does not reach a point that will result in excessive pin damage.

6. Large radius is desirable to minimize part breakage.

7. Customer-Molder understanding is necessary prior to tooling.

8. Part surface finish is dependent on mold finish.

9. Based on nominal 0.125 inch wall.

Copyright

Revised 1991

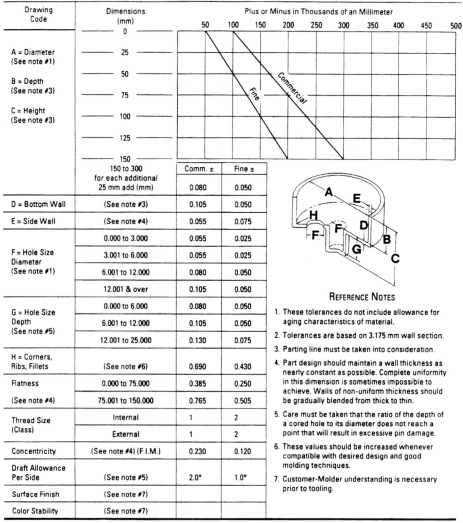

Standards & Practices of Plastics Molders		Material Acrylonitrile Butadiene Styrene (ABS)

Note: The *Commercial* values shown below represent common production tolerances at the most economical level. The *Fine* values represent closer tolerances that can be held but at a greater cost. Any addition of fillers will compromise physical properties and alter dimensional stability. Please consult the manufacturer.

Drawing Code	Dimensions (mm)	Plus or Minus in Thousands of an Millimeter
A = Diameter (See note #1) B = Depth (See note #3) C = Height (See note #3)	0 – 150	(graph: 50 100 150 200 250 300 350 400 450 500; curves Fine, Commercial)

	Dimensions (mm)	Comm. ±	Fine ±
	150 to 300 for each additional 25 mm add (mm)	0.080	0.050
D = Bottom Wall	(See note #3)	0.105	0.050
E = Side Wall	(See note #4)	0.055	0.075
F = Hole Size Diameter (See note #1)	0.000 to 3.000	0.055	0.025
	3.001 to 6.000	0.055	0.025
	6.001 to 12.000	0.080	0.050
	12.001 & over	0.105	0.050
G = Hole Size Depth (See note #5)	0.000 to 6.000	0.080	0.050
	6.001 to 12.000	0.105	0.050
	12.001 to 25.000	0.130	0.075
H = Corners, Ribs, Fillets	(See note #6)	0.690	0.430
Flatness	0.000 to 75.000	0.385	0.250
(See note #4)	75.001 to 150.000	0.765	0.505
Thread Size (Class)	Internal	1	2
	External	1	2
Concentricity	(See note #4) (F.I.M.)	0.230	0.120
Draft Allowance Per Side	(See note #5)	2.0°	1.0°
Surface Finish	(See note #7)		
Color Stability	(See note #7)		

REFERENCE NOTES

1. These tolerances do not include allowance for aging characteristics of material.
2. Tolerances are based on 3.175 mm wall section.
3. Parting line must be taken into consideration.
4. Part design should maintain a wall thickness as nearly constant as possible. Complete uniformity in this dimension is sometimes impossible to achieve. Walls of non-uniform thickness should be gradually blended from thick to thin.
5. Care must be taken that the ratio of the depth of a cored hole to its diameter does not reach a point that will result in excessive pin damage.
6. These values should be increased whenever compatible with desired design and good molding techniques.
7. Customer-Molder understanding is necessary prior to tooling.

Copyright   Revised 1991

Figure 5-3   Acrylonitrile butadiene styrene (ABS)

Standards & Practices of Plastics Molders	Material Acrylonitrile Butadiene Styrene (ABS)

Note: The *Commercial* values shown below represent common production tolerances at the most economical level. The *Fine* values represent closer tolerances that can be held but at a greater cost. Any addition of fillers will compromise physical properties and alter dimensional stability. Please consult the manufacturer.

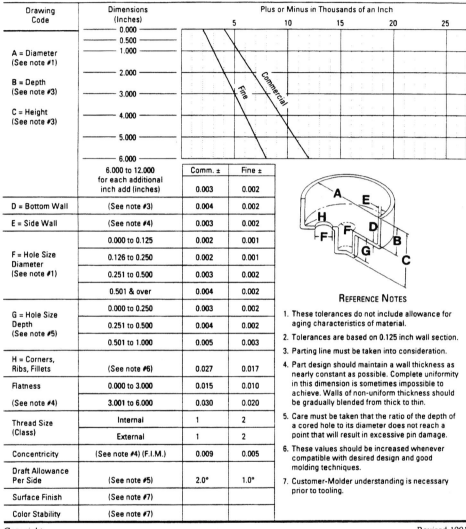

Drawing Code	Dimensions (Inches)		
A = Diameter (See note #1) B = Depth (See note #3) C = Height (See note #3)	0.000 0.500 1.000 2.000 3.000 4.000 5.000 6.000		
	6.000 to 12.000 for each additional inch add (inches)	Comm. ± 0.003	Fine ± 0.002
D = Bottom Wall	(See note #3)	0.004	0.002
E = Side Wall	(See note #4)	0.003	0.002
F = Hole Size Diameter (See note #1)	0.000 to 0.125	0.002	0.001
	0.126 to 0.250	0.002	0.001
	0.251 to 0.500	0.003	0.002
	0.501 & over	0.004	0.002
G = Hole Size Depth (See note #5)	0.000 to 0.250	0.003	0.002
	0.251 to 0.500	0.004	0.002
	0.501 to 1.000	0.005	0.003
H = Corners, Ribs, Fillets	(See note #6)	0.027	0.017
Flatness (See note #4)	0.000 to 3.000	0.015	0.010
	3.001 to 6.000	0.030	0.020
Thread Size (Class)	Internal	1	2
	External	1	2
Concentricity	(See note #4) (F.I.M.)	0.009	0.005
Draft Allowance Per Side	(See note #5)	2.0°	1.0°
Surface Finish	(See note #7)		
Color Stability	(See note #7)		

Plus or Minus in Thousands of an Inch: 5   10   15   20   25

REFERENCE NOTES

1. These tolerances do not include allowance for aging characteristics of material.

2. Tolerances are based on 0.125 inch wall section.

3. Parting line must be taken into consideration.

4. Part design should maintain a wall thickness as nearly constant as possible. Complete uniformity in this dimension is sometimes impossible to achieve. Walls of non-uniform thickness should be gradually blended from thick to thin.

5. Care must be taken that the ratio of the depth of a cored hole to its diameter does not reach a point that will result in excessive pin damage.

6. These values should be increased whenever compatible with desired design and good molding techniques.

7. Customer-Molder understanding is necessary prior to tooling.

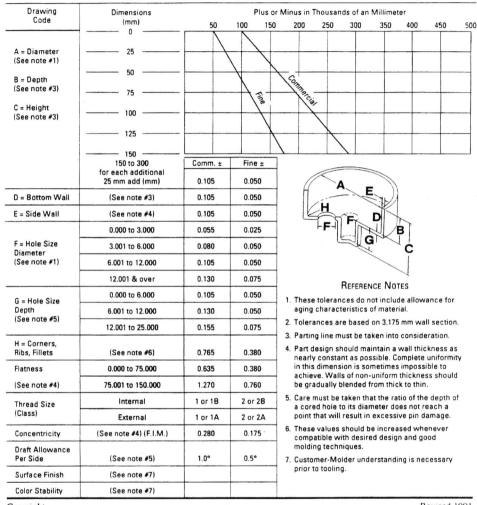

Standards & Practices of Plastics Molders

Material

Cellulosics

Note: The *Commercial* values shown below represent common production tolerances at the most economical level. The *Fine* values represent closer tolerances that can be held but at a greater cost. Any addition of fillers will compromise physical properties and alter dimensional stability. Please consult the manufacturer.

Drawing Code	Dimensions (mm)	Comm. ±	Fine ±
A = Diameter (See note #1)			
B = Depth (See note #3)			
C = Height (See note #3)			
150 to 300 for each additional 25 mm add (mm)		0.105	0.050
D = Bottom Wall	(See note #3)	0.105	0.050
E = Side Wall	(See note #4)	0.105	0.050
F = Hole Size Diameter (See note #1)	0.000 to 3.000	0.055	0.025
	3.001 to 6.000	0.080	0.050
	6.001 to 12.000	0.105	0.050
	12.001 & over	0.130	0.075
G = Hole Size Depth (See note #5)	0.000 to 6.000	0.105	0.050
	6.001 to 12.000	0.130	0.050
	12.001 to 25.000	0.155	0.075
H = Corners, Ribs, Fillets	(See note #6)	0.765	0.380
Flatness (See note #4)	0.000 to 75.000	0.635	0.380
	75.001 to 150.000	1.270	0.760
Thread Size (Class)	Internal	1 or 1B	2 or 2B
	External	1 or 1A	2 or 2A
Concentricity	(See note #4) (F.I.M.)	0.280	0.175
Draft Allowance Per Side	(See note #5)	1.0°	0.5°
Surface Finish	(See note #7)		
Color Stability	(See note #7)		

REFERENCE NOTES

1. These tolerances do not include allowance for aging characteristics of material.
2. Tolerances are based on 3.175 mm wall section.
3. Parting line must be taken into consideration.
4. Part design should maintain a wall thickness as nearly constant as possible. Complete uniformity in this dimension is sometimes impossible to achieve. Walls of non-uniform thickness should be gradually blended from thick to thin.
5. Care must be taken that the ratio of the depth of a cored hole to its diameter does not reach a point that will result in excessive pin damage.
6. These values should be increased whenever compatible with desired design and good molding techniques.
7. Customer-Molder understanding is necessary prior to tooling.

Copyright   Revised 1991

© The Society of the Plastics Industry, Inc.
1801 K Street, N.W., 600K
Washington, D.C. 20006-1301

Figure 5-4 Cellulosics: cellulose acetate (CA), cellulose acetate butyrate (CAB), cellulose acetate propionate (CA), ethyl cellulose (EC)

Standards & Practices of Plastics Molders	Material
	Cellulosics

Note: The *Commercial* values shown below represent common production tolerances at the most economical level. The *Fine* values represent closer tolerances that can be held but at a greater cost. Any addition of fillers will compromise physical properties and alter dimensional stability. Please consult the manufacturer.

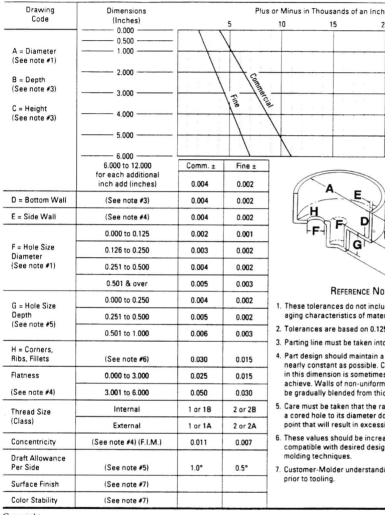

Drawing Code	Dimensions (Inches)			Plus or Minus in Thousands of an Inch			
		Comm. ±	Fine ±				
A = Diameter (See note #1)	0.000 / 0.500 / 1.000 / 2.000 / 3.000 / 4.000 / 5.000 / 6.000						
B = Depth (See note #3)							
C = Height (See note #3)							
	6.000 to 12.000 for each additional inch add (inches)	0.004	0.002				
D = Bottom Wall	(See note #3)	0.004	0.002				
E = Side Wall	(See note #4)	0.004	0.002				
F = Hole Size Diameter (See note #1)	0.000 to 0.125	0.002	0.001				
	0.126 to 0.250	0.003	0.002				
	0.251 to 0.500	0.004	0.002				
	0.501 & over	0.005	0.003				
G = Hole Size Depth (See note #5)	0.000 to 0.250	0.004	0.002				
	0.251 to 0.500	0.005	0.002				
	0.501 to 1.000	0.006	0.003				
H = Corners, Ribs, Fillets	(See note #6)	0.030	0.015				
Flatness	0.000 to 3.000	0.025	0.015				
(See note #4)	3.001 to 6.000	0.050	0.030				
Thread Size (Class)	Internal	1 or 1B	2 or 2B				
	External	1 or 1A	2 or 2A				
Concentricity	(See note #4) (F.I.M.)	0.011	0.007				
Draft Allowance Per Side	(See note #5)	1.0°	0.5°				
Surface Finish	(See note #7)						
Color Stability	(See note #7)						

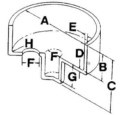

REFERENCE NOTES

1. These tolerances do not include allowance for aging characteristics of material.

2. Tolerances are based on 0.125 inch wall section.

3. Parting line must be taken into consideration.

4. Part design should maintain a wall thickness as nearly constant as possible. Complete uniformity in this dimension is sometimes impossible to achieve. Walls of non-uniform thickness should be gradually blended from thick to thin.

5. Care must be taken that the ratio of the depth of a cored hole to its diameter does not reach a point that will result in excessive pin damage.

6. These values should be increased whenever compatible with desired design and good molding techniques.

7. Customer-Molder understanding is necessary prior to tooling.

Copyright

Revised 1991

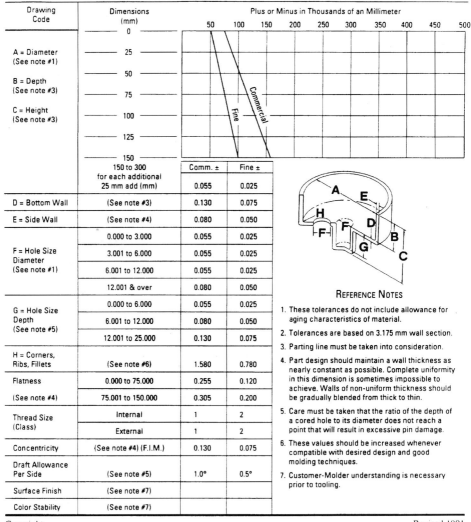

Standards & Practices of Plastics Molders		Material Diallylphthalate (DAP)

Note: The *Commercial* values shown below represent common production tolerances at the most economical level. The *Fine* values represent closer tolerances that can be held but at a greater cost. Any addition of fillers will compromise physical properties and alter dimensional stability. Please consult the manufacturer.

Drawing Code	Dimensions (mm)	Comm. ±	Fine ±
A = Diameter (See note #1) B = Depth (See note #3) C = Height (See note #3)	150 to 300 for each additional 25 mm add (mm)	0.055	0.025
D = Bottom Wall	(See note #3)	0.130	0.075
E = Side Wall	(See note #4)	0.080	0.050
F = Hole Size Diameter (See note #1)	0.000 to 3.000	0.055	0.025
	3.001 to 6.000	0.055	0.025
	6.001 to 12.000	0.055	0.025
	12.001 & over	0.080	0.050
G = Hole Size Depth (See note #5)	0.000 to 6.000	0.055	0.025
	6.001 to 12.000	0.080	0.050
	12.001 to 25.000	0.130	0.075
H = Corners, Ribs, Fillets	(See note #6)	1.580	0.780
Flatness (See note #4)	0.000 to 75.000	0.255	0.120
	75.001 to 150.000	0.305	0.200
Thread Size (Class)	Internal	1	2
	External	1	2
Concentricity	(See note #4) (F.I.M.)	0.130	0.075
Draft Allowance Per Side	(See note #5)	1.0°	0.5°
Surface Finish	(See note #7)		
Color Stability	(See note #7)		

Plus or Minus in Thousands of an Millimeter

REFERENCE NOTES

1. These tolerances do not include allowance for aging characteristics of material.

2. Tolerances are based on 3.175 mm wall section.

3. Parting line must be taken into consideration.

4. Part design should maintain a wall thickness as nearly constant as possible. Complete uniformity in this dimension is sometimes impossible to achieve. Walls of non-uniform thickness should be gradually blended from thick to thin.

5. Care must be taken that the ratio of the depth of a cored hole to its diameter does not reach a point that will result in excessive pin damage.

6. These values should be increased whenever compatible with desired design and good molding techniques.

7. Customer-Molder understanding is necessary prior to tooling.

Copyright   Revised 1991

Figure 5-5   Diallyl phthalate (DAP)

Standards & Practices of Plastics Molders	Material Diallylphthalate (DAP)

Note: The *Commercial* values shown below represent common production tolerances at the most economical level. The *Fine* values represent closer tolerances that can be held but at a greater cost. Any addition of fillers will compromise physical properties and alter dimensional stability. Please consult the manufacturer.

Drawing Code	Dimensions (Inches)	Plus or Minus in Thousands of an Inch
A = Diameter (See note #1)	0.000 / 0.500 / 1.000 / 2.000 / 3.000 / 4.000 / 5.000 / 6.000	Commercial / Fine lines (5, 10, 15, 20, 25)

	Dimensions	Comm. ±	Fine ±
A = Diameter (See note #1), B = Depth (See note #3), C = Height (See note #3)	6.000 to 12.000 for each additional inch add (inches)	0.002	0.001
D = Bottom Wall	(See note #3)	0.005	0.003
E = Side Wall	(See note #4)	0.003	0.002
F = Hole Size Diameter (See note #1)	0.000 to 0.125	0.002	0.001
	0.126 to 0.250	0.002	0.001
	0.251 to 0.500	0.002	0.001
	0.501 & over	0.003	0.002
G = Hole Size Depth (See note #5)	0.000 to 0.250	0.002	0.001
	0.251 to 0.500	0.003	0.002
	0.501 to 1.000	0.005	0.003
H = Corners, Ribs, Fillets	(See note #6)	0.062	0.031
Flatness	0.000 to 3.000	0.010	0.005
(See note #4)	3.001 to 6.000	0.012	0.008
Thread Size (Class)	Internal	1	2
	External	1	2
Concentricity	(See note #4) (F.I.M.)	0.005	0.003
Draft Allowance Per Side	(See note #5)	1.0°	0.5°
Surface Finish	(See note #7)		
Color Stability	(See note #7)		

REFERENCE NOTES

1. These tolerances do not include allowance for aging characteristics of material.

2. Tolerances are based on 0.125 inch wall section.

3. Parting line must be taken into consideration.

4. Part design should maintain a wall thickness as nearly constant as possible. Complete uniformity in this dimension is sometimes impossible to achieve. Walls of non-uniform thickness should be gradually blended from thick to thin.

5. Care must be taken that the ratio of the depth of a cored hole to its diameter does not reach a point that will result in excessive pin damage.

6. These values should be increased whenever compatible with desired design and good molding techniques.

7. Customer-Molder understanding is necessary prior to tooling.

Copyright   Revised 1991

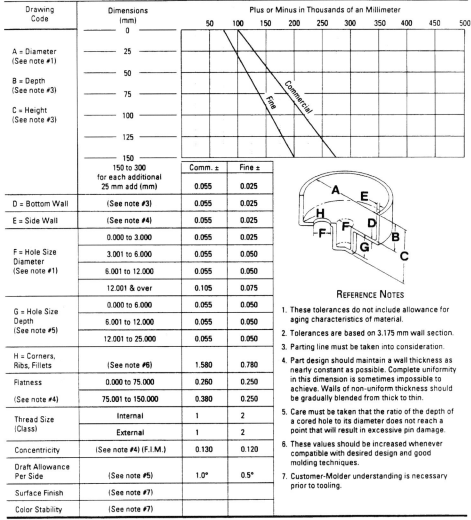

	Material Epoxy (EP)

Standards & Practices of Plastics Molders

Note: The *Commercial* values shown below represent common production tolerances at the most economical level. The *Fine* values represent closer tolerances that can be held but at a greater cost. Any addition of fillers will compromise physical properties and alter dimensional stability. Please consult the manufacturer.

Drawing Code	Dimensions (mm)	Comm. ±	Fine ±
A = Diameter (See note #1) B = Depth (See note #3) C = Height (See note #3)			
150 to 300 for each additional 25 mm add (mm)		0.055	0.025
D = Bottom Wall	(See note #3)	0.055	0.025
E = Side Wall	(See note #4)	0.055	0.025
F = Hole Size Diameter (See note #1)	0.000 to 3.000	0.055	0.025
	3.001 to 6.000	0.055	0.050
	6.001 to 12.000	0.055	0.050
	12.001 & over	0.105	0.075
G = Hole Size Depth (See note #5)	0.000 to 6.000	0.055	0.050
	6.001 to 12.000	0.055	0.050
	12.001 to 25.000	0.055	0.050
H = Corners, Ribs, Fillets	(See note #6)	1.580	0.780
Flatness	0.000 to 75.000	0.260	0.250
(See note #4)	75.001 to 150.000	0.380	0.250
Thread Size (Class)	Internal	1	2
	External	1	2
Concentricity	(See note #4) (F.I.M.)	0.130	0.120
Draft Allowance Per Side	(See note #5)	1.0°	0.5°
Surface Finish	(See note #7)		
Color Stability	(See note #7)		

REFERENCE NOTES

1. These tolerances do not include allowance for aging characteristics of material.
2. Tolerances are based on 3.175 mm wall section.
3. Parting line must be taken into consideration.
4. Part design should maintain a wall thickness as nearly constant as possible. Complete uniformity in this dimension is sometimes impossible to achieve. Walls of non-uniform thickness should be gradually blended from thick to thin.
5. Care must be taken that the ratio of the depth of a cored hole to its diameter does not reach a point that will result in excessive pin damage.
6. These values should be increased whenever compatible with desired design and good molding techniques.
7. Customer-Molder understanding is necessary prior to tooling.

Copyright   Revised 1991

© The Society of the Plastics Industry, Inc.
1801 K Street, N.W., 600K
Washington, D.C. 20006-1301

Figure 5-6   Epoxy (EP)

Standards & Practices of Plastics Molders	**Material** Epoxy (EP)

Note: The *Commercial* values shown below represent common production tolerances at the most economical level. The *Fine* values represent closer tolerances that can be held but at a greater cost. Any addition of fillers will compromise physical properties and alter dimensional stability. Please consult the manufacturer.

Drawing Code	Dimensions (Inches)		Plus or Minus in Thousands of an Inch
A = Diameter (See note #1)			
B = Depth (See note #3)			
C = Height (See note #3)			

		Comm. ±	Fine ±
	6.000 to 12.000 for each additional inch add (inches)	0.002	0.001
D = Bottom Wall	(See note #3)	0.002	0.001
E = Side Wall	(See note #4)	0.002	0.001
F = Hole Size Diameter (See note #1)	0.000 to 0.125	0.002	0.001
	0.126 to 0.250	0.002	0.002
	0.251 to 0.500	0.002	0.002
	0.501 & over	0.004	0.003
G = Hole Size Depth (See note #5)	0.000 to 0.250	0.002	0.002
	0.251 to 0.500	0.002	0.002
	0.501 to 1.000	0.002	0.002
H = Corners, Ribs, Fillets	(See note #6)	0.062	0.031
Flatness	0.000 to 3.000	0.010	0.010
(See note #4)	3.001 to 6.000	0.015	0.010
Thread Size (Class)	Internal	1	2
	External	1	2
Concentricity	(See note #4) (F.I.M.)	0.005	0.005
Draft Allowance Per Side	(See note #5)	1.0°	0.5°
Surface Finish	(See note #7)		
Color Stability	(See note #7)		

REFERENCE NOTES

1. These tolerances do not include allowance for aging characteristics of material.

2. Tolerances are based on 0.125 inch wall section.

3. Parting line must be taken into consideration.

4. Part design should maintain a wall thickness as nearly constant as possible. Complete uniformity in this dimension is sometimes impossible to achieve. Walls of non-uniform thickness should be gradually blended from thick to thin.

5. Care must be taken that the ratio of the depth of a cored hole to its diameter does not reach a point that will result in excessive pin damage.

6. These values should be increased whenever compatible with desired design and good molding techniques.

7. Customer-Molder understanding is necessary prior to tooling.

Copyright   Revised 1991

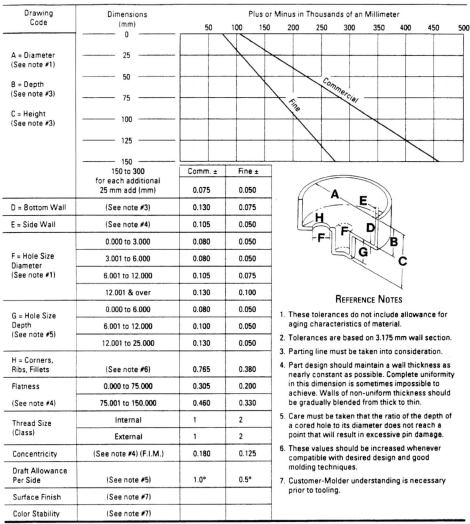

Standards & Practices of Plastics Molders			Material Melamine - Urea (MF-UF)

Note: The *Commercial* values shown below represent common production tolerances at the most economical level. The *Fine* values represent closer tolerances that can be held but at a greater cost. Any addition of fillers will compromise physical properties and alter dimensional stability. Please consult the manufacturer.

Drawing Code	Dimensions (mm)	Comm. ±	Fine ±
A = Diameter (See note #1)			
B = Depth (See note #3)			
C = Height (See note #3)			
	150 to 300 for each additional 25 mm add (mm)	0.075	0.050
D = Bottom Wall	(See note #3)	0.130	0.075
E = Side Wall	(See note #4)	0.105	0.050
F = Hole Size Diameter (See note #1)	0.000 to 3.000	0.080	0.050
	3.001 to 6.000	0.080	0.050
	6.001 to 12.000	0.105	0.075
	12.001 & over	0.130	0.100
G = Hole Size Depth (See note #5)	0.000 to 6.000	0.080	0.050
	6.001 to 12.000	0.100	0.050
	12.001 to 25.000	0.130	0.050
H = Corners, Ribs, Fillets	(See note #6)	0.765	0.380
Flatness	0.000 to 75.000	0.305	0.200
(See note #4)	75.001 to 150.000	0.460	0.330
Thread Size (Class)	Internal	1	2
	External	1	2
Concentricity	(See note #4) (F.I.M.)	0.180	0.125
Draft Allowance Per Side	(See note #5)	1.0°	0.5°
Surface Finish	(See note #7)		
Color Stability	(See note #7)		

Plus or Minus in Thousands of an Millimeter

REFERENCE NOTES

1. These tolerances do not include allowance for aging characteristics of material.

2. Tolerances are based on 3.175 mm wall section.

3. Parting line must be taken into consideration.

4. Part design should maintain a wall thickness as nearly constant as possible. Complete uniformity in this dimension is sometimes impossible to achieve. Walls of non-uniform thickness should be gradually blended from thick to thin.

5. Care must be taken that the ratio of the depth of a cored hole to its diameter does not reach a point that will result in excessive pin damage.

6. These values should be increased whenever compatible with desired design and good molding techniques.

7. Customer-Molder understanding is necessary prior to tooling.

Copyright   Revised 1991

Figure 5-7   Melamine formaldehyde (MF) and urea formaldehyde (UF)

Standards & Practices of Plastics Molders	Material Melamine - Urea (MF-UF)

Note: The *Commercial* values shown below represent common production tolerances at the most economical level. The *Fine* values represent closer tolerances that can be held but at a greater cost. Any addition of fillers will compromise physical properties and alter dimensional stability. Please consult the manufacturer.

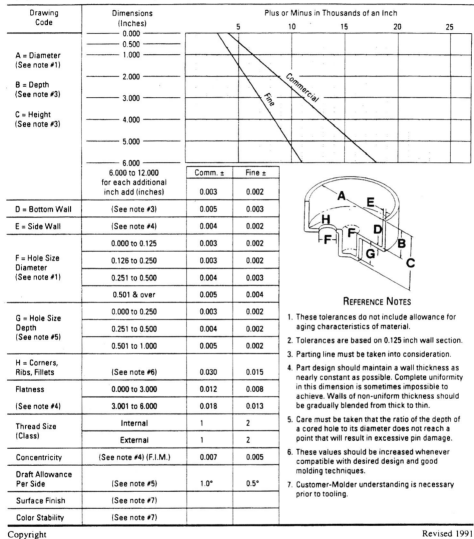

Drawing Code	Dimensions (Inches)	Plus or Minus in Thousands of an Inch		
		Comm. ±	Fine ±	
A = Diameter (See note #1) B = Depth (See note #3) C = Height (See note #3)	0.000–6.000			
	6.000 to 12.000 for each additional inch add (inches)	0.003	0.002	
D = Bottom Wall	(See note #3)	0.005	0.003	
E = Side Wall	(See note #4)	0.004	0.002	
F = Hole Size Diameter (See note #1)	0.000 to 0.125	0.003	0.002	
	0.126 to 0.250	0.003	0.002	
	0.251 to 0.500	0.004	0.003	
	0.501 & over	0.005	0.004	
G = Hole Size Depth (See note #5)	0.000 to 0.250	0.003	0.002	
	0.251 to 0.500	0.004	0.002	
	0.501 to 1.000	0.005	0.002	
H = Corners, Ribs, Fillets	(See note #6)	0.030	0.015	
Flatness	0.000 to 3.000	0.012	0.008	
(See note #4)	3.001 to 6.000	0.018	0.013	
Thread Size (Class)	Internal	1	2	
	External	1	2	
Concentricity	(See note #4) (F.I.M.)	0.007	0.005	
Draft Allowance Per Side	(See note #5)	1.0°	0.5°	
Surface Finish	(See note #7)			
Color Stability	(See note #7)			

REFERENCE NOTES

1. These tolerances do not include allowance for aging characteristics of material.

2. Tolerances are based on 0.125 inch wall section.

3. Parting line must be taken into consideration.

4. Part design should maintain a wall thickness as nearly constant as possible. Complete uniformity in this dimension is sometimes impossible to achieve. Walls of non-uniform thickness should be gradually blended from thick to thin.

5. Care must be taken that the ratio of the depth of a cored hole to its diameter does not reach a point that will result in excessive pin damage.

6. These values should be increased whenever compatible with desired design and good molding techniques.

7. Customer-Molder understanding is necessary prior to tooling.

Copyright   Revised 1991

Standards & Practices of Plastics Molders

Material
Phenol-Formaldehyde (PF)
(Phenolic) General Purpose

Note: The *Commercial* values shown below represent common production tolerances at the most economical level. The *Fine* values represent closer tolerances that can be held but at a greater cost. Any addition of fillers will compromise physical properties and alter dimensional stability. Please consult the manufacturer.

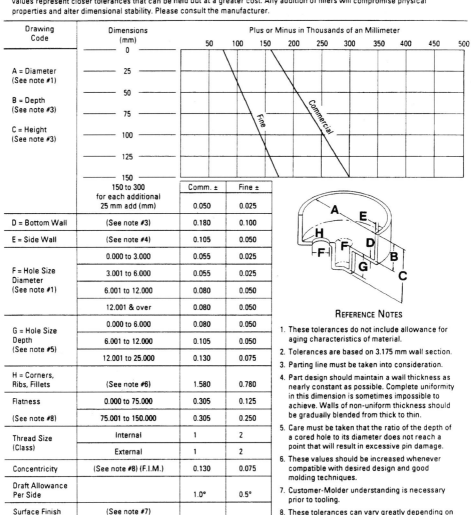

Drawing Code	Dimensions (mm)	Comm. ±	Fine ±
	150 to 300 for each additional 25 mm add (mm)	0.050	0.025
D = Bottom Wall	(See note #3)	0.180	0.100
E = Side Wall	(See note #4)	0.105	0.050
F = Hole Size Diameter (See note #1)	0.000 to 3.000	0.055	0.025
	3.001 to 6.000	0.055	0.025
	6.001 to 12.000	0.080	0.050
	12.001 & over	0.080	0.050
G = Hole Size Depth (See note #5)	0.000 to 6.000	0.080	0.050
	6.001 to 12.000	0.105	0.050
	12.001 to 25.000	0.130	0.075
H = Corners, Ribs, Fillets	(See note #6)	1.580	0.780
Flatness	0.000 to 75.000	0.305	0.125
(See note #8)	75.001 to 150.000	0.305	0.250
Thread Size (Class)	Internal	1	2
	External	1	2
Concentricity	(See note #8) (F.I.M.)	0.130	0.075
Draft Allowance Per Side		1.0°	0.5°
Surface Finish	(See note #7)		
Color Stability	(See note #7)		

Drawing Code:
A = Diameter (See note #1)
B = Depth (See note #3)
C = Height (See note #3)

Dimensions (mm): 0, 25, 50, 75, 100, 125, 150

Plus or Minus in Thousands of an Millimeter: 50, 100, 150, 200, 250, 300, 350, 400, 450, 500

REFERENCE NOTES

1. These tolerances do not include allowance for aging characteristics of material.
2. Tolerances are based on 3.175 mm wall section.
3. Parting line must be taken into consideration.
4. Part design should maintain a wall thickness as nearly constant as possible. Complete uniformity in this dimension is sometimes impossible to achieve. Walls of non-uniform thickness should be gradually blended from thick to thin.
5. Care must be taken that the ratio of the depth of a cored hole to its diameter does not reach a point that will result in excessive pin damage.
6. These values should be increased whenever compatible with desired design and good molding techniques.
7. Customer-Molder understanding is necessary prior to tooling.
8. These tolerances can vary greatly depending on method of molding and gate location.

Copyright   Revised 1991

Figure 5-8   Phenolic (phenyl-formaldehyde – PF)

Standards & Practices of Plastics Molders

Material
Phenol-Formaldehyde (PF)
(Phenolic) General Purpose

Note: The *Commercial* values shown below represent common production tolerances at the most economical level. The *Fine* values represent closer tolerances that can be held but at a greater cost. Any addition of fillers will compromise physical properties and alter dimensional stability. Please consult the manufacturer.

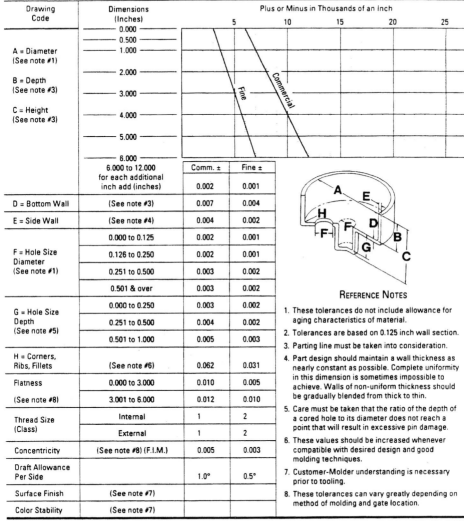

Drawing Code	Dimensions (Inches)	Comm. ±	Fine ±
A = Diameter (See note #1)			
B = Depth (See note #3)			
C = Height (See note #3)			
	6.000 to 12.000 for each additional inch add (inches)	0.002	0.001
D = Bottom Wall	(See note #3)	0.007	0.004
E = Side Wall	(See note #4)	0.004	0.002
F = Hole Size Diameter (See note #1)	0.000 to 0.125	0.002	0.001
	0.126 to 0.250	0.002	0.001
	0.251 to 0.500	0.003	0.002
	0.501 & over	0.003	0.002
G = Hole Size Depth (See note #5)	0.000 to 0.250	0.003	0.002
	0.251 to 0.500	0.004	0.002
	0.501 to 1.000	0.005	0.003
H = Corners, Ribs, Fillets	(See note #6)	0.062	0.031
Flatness	0.000 to 3.000	0.010	0.005
	3.001 to 6.000	0.012	0.010
Thread Size (Class)	Internal	1	2
	External	1	2
Concentricity	(See note #8) (F.I.M.)	0.005	0.003
Draft Allowance Per Side		1.0°	0.5°
Surface Finish	(See note #7)		
Color Stability	(See note #7)		

REFERENCE NOTES

1. These tolerances do not include allowance for aging characteristics of material.
2. Tolerances are based on 0.125 inch wall section.
3. Parting line must be taken into consideration.
4. Part design should maintain a wall thickness as nearly constant as possible. Complete uniformity in this dimension is sometimes impossible to achieve. Walls of non-uniform thickness should be gradually blended from thick to thin.
5. Care must be taken that the ratio of the depth of a cored hole to its diameter does not reach a point that will result in excessive pin damage.
6. These values should be increased whenever compatible with desired design and good molding techniques.
7. Customer-Molder understanding is necessary prior to tooling.
8. These tolerances can vary greatly depending on method of molding and gate location.

Copyright

Revised 1991

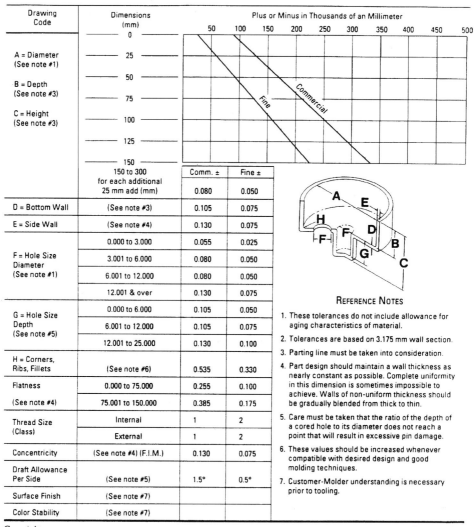

Standards & Practices of Plastics Molders	Material Polyamide (Nylon) (PA)

Note: The *Commercial* values shown below represent common production tolerances at the most economical level. The *Fine* values represent closer tolerances that can be held but at a greater cost. Any addition of fillers will compromise physical properties and alter dimensional stability. Please consult the manufacturer.

Drawing Code	Dimensions (mm)	Comm. ±	Fine ±
A = Diameter (See note #1) B = Depth (See note #3) C = Height (See note #3)			
	150 to 300 for each additional 25 mm add (mm)	0.080	0.050
D = Bottom Wall	(See note #3)	0.105	0.075
E = Side Wall	(See note #4)	0.130	0.075
F = Hole Size Diameter (See note #1)	0.000 to 3.000	0.055	0.025
	3.001 to 6.000	0.080	0.050
	6.001 to 12.000	0.080	0.050
	12.001 & over	0.130	0.075
G = Hole Size Depth (See note #5)	0.000 to 6.000	0.105	0.050
	6.001 to 12.000	0.105	0.075
	12.001 to 25.000	0.130	0.100
H = Corners, Ribs, Fillets	(See note #6)	0.535	0.330
Flatness (See note #4)	0.000 to 75.000	0.255	0.100
	75.001 to 150.000	0.385	0.175
Thread Size (Class)	Internal	1	2
	External	1	2
Concentricity	(See note #4) (F.I.M.)	0.130	0.075
Draft Allowance Per Side	(See note #5)	1.5°	0.5°
Surface Finish	(See note #7)		
Color Stability	(See note #7)		

REFERENCE NOTES

1. These tolerances do not include allowance for aging characteristics of material.
2. Tolerances are based on 3.175 mm wall section.
3. Parting line must be taken into consideration.
4. Part design should maintain a wall thickness as nearly constant as possible. Complete uniformity in this dimension is sometimes impossible to achieve. Walls of non-uniform thickness should be gradually blended from thick to thin.
5. Care must be taken that the ratio of the depth of a cored hole to its diameter does not reach a point that will result in excessive pin damage.
6. These values should be increased whenever compatible with desired design and good molding techniques.
7. Customer-Molder understanding is necessary prior to tooling.

Copyright   Revised 1991

Figure 5-9   Polyamide (nylon – PA)

Standards & Practices of Plastics Molders	Material Polyamide (Nylon) (PA)

Note: The *Commercial* values shown below represent common production tolerances at the most economical level. The *Fine* values represent closer tolerances that can be held but at a greater cost. Any addition of fillers will compromise physical properties and alter dimensional stability. Please consult the manufacturer.

Drawing Code	Dimensions (Inches)	Plus or Minus in Thousands of an Inch

A = Diameter (See note #1)

B = Depth (See note #3)

C = Height (See note #3)

Dimensions (Inches): 0.000, 0.500, 1.000, 2.000, 3.000, 4.000, 5.000, 6.000

Scale: 5, 10, 15, 20, 25 (Commercial, Fine)

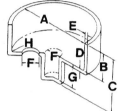

Drawing Code	Dimensions	Comm. ±	Fine ±
	6.000 to 12.000 for each additional inch add (inches)	0.003	0.002
D = Bottom Wall	(See note #3)	0.004	0.003
E = Side Wall	(See note #4)	0.005	0.003
F = Hole Size Diameter (See note #1)	0.000 to 0.125	0.002	0.001
	0.126 to 0.250	0.003	0.002
	0.251 to 0.500	0.003	0.002
	0.501 & over	0.005	0.003
G = Hole Size Depth (See note #5)	0.000 to 0.250	0.004	0.002
	0.251 to 0.500	0.004	0.003
	0.501 to 1.000	0.005	0.004
H = Corners, Ribs, Fillets	(See note #6)	0.021	0.013
Flatness	0.000 to 3.000	0.010	0.004
(See note #4)	3.001 to 6.000	0.015	0.007
Thread Size (Class)	Internal	1	2
	External	1	2
Concentricity	(See note #4) (F.I.M.)	0.005	0.003
Draft Allowance Per Side		1.5°	0.5°
Surface Finish	(See note #7)		
Color Stability	(See note #7)		

REFERENCE NOTES

1. These tolerances do not include allowance for aging characteristics of material.

2. Tolerances are based on 0.125 inch wall section.

3. Parting line must be taken into consideration.

4. Part design should maintain a wall thickness as nearly constant as possible. Complete uniformity in this dimension is sometimes impossible to achieve. Walls of non-uniform thickness should be gradually blended from thick to thin.

5. Care must be taken that the ratio of the depth of a cored hole to its diameter does not reach a point that will result in excessive pin damage.

6. These values should be increased whenever compatible with desired design and good molding techniques.

7. Customer-Molder understanding is necessary prior to tooling.

Copyright   Revised 1991

© The Society of the Plastics Industry, Inc.
1801 K Street, N.W., 600K
Washington, D.C. 20006-1301

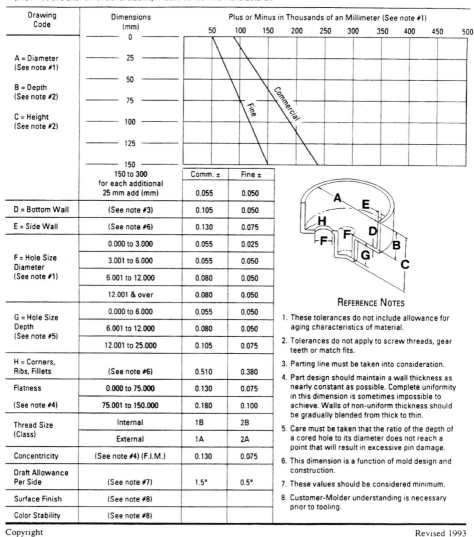

Standards & Practices of Plastics Molders		Material Polycarbonate (PC)

Note: The *Commercial* values shown below represent common production tolerances at the most economical level. The *Fine* values represent closer tolerances that can be held but at a greater cost. Any addition of fillers will compromise physical properties and alter dimensional stability. Please consult the manufacturer.

Drawing Code	Dimensions (mm)	Comm. ±	Fine ±
A = Diameter (See note #1) B = Depth (See note #2) C = Height (See note #2)	0 — 25 — 50 — 75 — 100 — 125 — 150 —		
	150 to 300 for each additional 25 mm add (mm)	0.055	0.050
D = Bottom Wall	(See note #3)	0.105	0.050
E = Side Wall	(See note #6)	0.130	0.075
F = Hole Size Diameter (See note #1)	0.000 to 3.000	0.055	0.025
	3.001 to 6.000	0.055	0.050
	6.001 to 12.000	0.080	0.050
	12.001 & over	0.080	0.050
G = Hole Size Depth (See note #5)	0.000 to 6.000	0.055	0.050
	6.001 to 12.000	0.080	0.050
	12.001 to 25.000	0.105	0.075
H = Corners, Ribs, Fillets	(See note #6)	0.510	0.380
Flatness (See note #4)	0.000 to 75.000	0.130	0.075
	75.001 to 150.000	0.180	0.100
Thread Size (Class)	Internal	1B	2B
	External	1A	2A
Concentricity	(See note #4) (F.I.M.)	0.130	0.075
Draft Allowance Per Side	(See note #7)	1.5°	0.5°
Surface Finish	(See note #8)		
Color Stability	(See note #8)		

Plus or Minus in Thousands of an Millimeter (See note #1)
50 100 150 200 250 300 350 400 450 500

Fine *Commercial*

REFERENCE NOTES

1. These tolerances do not include allowance for aging characteristics of material.

2. Tolerances do not apply to screw threads, gear teeth or match fits.

3. Parting line must be taken into consideration.

4. Part design should maintain a wall thickness as nearly constant as possible. Complete uniformity in this dimension is sometimes impossible to achieve. Walls of non-uniform thickness should be gradually blended from thick to thin.

5. Care must be taken that the ratio of the depth of a cored hole to its diameter does not reach a point that will result in excessive pin damage.

6. This dimension is a function of mold design and construction.

7. These values should be considered minimum.

8. Customer-Molder understanding is necessary prior to tooling.

Copyright   Revised 1993

© The Society of the Plastics Industry, Inc.
1801 K Street, N.W., 600K
Washington, D.C. 20006-1301

Figure 5-10   Polycarbonate (PC)

Standards & Practices	**Material**
of Plastics Molders	Polycarbonate (PC)

Note: The *Commercial* values shown below represent common production tolerances at the most economical level. The *Fine* values represent closer tolerances that can be held but at a greater cost. Any addition of fillers will compromise physical properties and alter dimensional stability. Please consult the manufacturer.

Drawing Code	Dimensions (Inches)	Plus or Minus in Thousands of an Inch (See note #1)	
		Comm. ±	Fine ±
A = Diameter (See note #1) B = Depth (See note #2) C = Height (See note #2)	0.000 / 0.500 / 1.000 / 2.000 / 3.000 / 4.000 / 5.000 / 6.000		
	6.000 to 12.000 for each additional inch add (inches)	0.002	0.002
D = Bottom Wall	(See note #3)	0.004	0.002
E = Side Wall	(See note #6)	0.005	0.003
F = Hole Size Diameter (See note #1)	0.000 to 0.125	0.002	0.001
	0.126 to 0.250	0.002	0.002
	0.251 to 0.500	0.003	0.002
	0.501 & over	0.003	0.002
G = Hole Size Depth (See note #5)	0.000 to 0.250	0.002	0.002
	0.251 to 0.500	0.003	0.002
	0.501 to 1.000	0.004	0.003
H = Corners, Ribs, Fillets	(See note #6)	0.020	0.015
Flatness (See note #4)	0.000 to 3.000	0.005	0.003
	3.001 to 6.000	0.007	0.004
Thread Size (Class)	Internal	1B	2B
	External	1A	2A
Concentricity	(See note #4) (F.I.M.)	0.005	0.003
Draft Allowance Per Side	(See note #7)	1.5°	0.5°
Surface Finish	(See note #8)		
Color Stability	(See note #8)		

REFERENCE NOTES

1. These tolerances do not include allowance for aging characteristics of material

2. Tolerances do not apply to screw threads, gear teeth or match fits.

3. Parting line must be taken into consideration.

4. Part design should maintain a wall thickness as nearly constant as possible. Complete uniformity in this dimension is sometimes impossible to achieve. Walls of non-uniform thickness should be gradually blended from thick to thin.

5. Care must be taken that the ratio of the depth of a cored hole to its diameter does not reach a point that will result in excessive pin damage.

6. This dimension is a function of mold design and construction.

7. These values should be considered minimum.

8. Customer-Molder understanding is necessary prior to tooling.

Revised 1993

© The Society of the Plastics Industry, Inc.
1801 K Street, N.W., 600K
Washington, D.C. 20006-1301

Standards & Practices of Plastics Molders

Material
Thermoplastic Polyester
(TPPE)

Note: The *Commercial* values shown below represent common production tolerances at the most economical level. The *Fine* values represent closer tolerances that can be held but at a greater cost. Any addition of fillers will compromise physical properties and alter dimensional stability. Please consult the manufacturer.

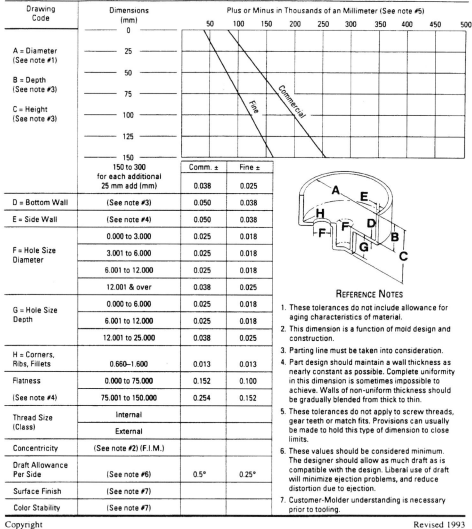

Drawing Code	Dimensions (mm)	Comm. ±	Fine ±
A = Diameter (See note #1)			
B = Depth (See note #3)			
C = Height (See note #3)			
	150 to 300 for each additional 25 mm add (mm)	0.038	0.025
D = Bottom Wall	(See note #3)	0.050	0.038
E = Side Wall	(See note #4)	0.050	0.038
F = Hole Size Diameter	0.000 to 3.000	0.025	0.018
	3.001 to 6.000	0.025	0.018
	6.001 to 12.000	0.025	0.018
	12.001 & over	0.038	0.025
G = Hole Size Depth	0.000 to 6.000	0.025	0.018
	6.001 to 12.000	0.025	0.018
	12.001 to 25.000	0.038	0.025
H = Corners, Ribs, Fillets	0.660–1.600	0.013	0.013
Flatness	0.000 to 75.000	0.152	0.100
(See note #4)	75.001 to 150.000	0.254	0.152
Thread Size (Class)	Internal		
	External		
Concentricity	(See note #2) (F.I.M.)		
Draft Allowance Per Side	(See note #6)	0.5°	0.25°
Surface Finish	(See note #7)		
Color Stability	(See note #7)		

REFERENCE NOTES

1. These tolerances do not include allowance for aging characteristics of material.
2. This dimension is a function of mold design and construction.
3. Parting line must be taken into consideration.
4. Part design should maintain a wall thickness as nearly constant as possible. Complete uniformity in this dimension is sometimes impossible to achieve. Walls of non-uniform thickness should be gradually blended from thick to thin.
5. These tolerances do not apply to screw threads, gear teeth or match fits. Provisions can usually be made to hold this type of dimension to close limits.
6. These values should be considered minimum. The designer should allow as much draft as is compatible with the design. Liberal use of draft will minimize ejection problems, and reduce distortion due to ejection.
7. Customer-Molder understanding is necessary prior to tooling.

Copyright

Revised 1993

Figure 5-11 Polyester (thermoplastic)

Standards & Practices of Plastics Molders

	Material
	Thermoplastic Polyester (TPPE)

Note: The *Commercial* values shown below represent common production tolerances at the most economical level. The *Fine* values represent closer tolerances that can be held but at a greater cost. Any addition of fillers will compromise physical properties and alter dimensional stability. Please consult the manufacturer.

Drawing Code	Dimensions (Inches)	Plus or Minus in Thousands of an Inch (See note #5)				
		5	10	15	20	25
A = Diameter (See note #1)	0.000 / 0.500 / 1.000 / 2.000 / 3.000 / 4.000 / 5.000 / 6.000					
B = Depth (See note #3)						
C = Height (See note #3)						

	Dimensions (Inches)	Comm. ±	Fine ±
	6.000 to 12.000 for each additional inch add (inches)	0.0015	0.001
D = Bottom Wall	(See note #3)	0.002	0.0015
E = Side Wall	(See note #4)	0.002	0.0015
F = Hole Size Diameter	0.000 to 0.125	0.001	0.0007
	0.126 to 0.250	0.001	0.0007
	0.251 to 0.500	0.001	0.0007
	0.501 & over	0.0015	0.001
G = Hole Size Depth	0.000 to 0.250	0.001	0.0007
	0.251 to 0.500	0.001	0.0007
	0.501 to 1.000	0.0015	0.001
H = Corners, Ribs, Fillets	0.025 to 0.062	0.005	0.005
Flatness	0.000 to 3.000	0.006	0.004
(See note #4)	3.001 to 6.000	0.010	0.006
Thread Size (Class)	Internal		
	External		
Concentricity	(See note #2) (F.I.M.)		
Draft Allowance Per Side	(See note #6)	0.5°	0.25°
Surface Finish	(See note #7)		
Color Stability	(See note #7)		

REFERENCE NOTES

1. These tolerances do not include allowance for aging characteristics of material.
2. This dimension is a function of mold design and construction.
3. Parting line must be taken into consideration.
4. Part design should maintain a wall thickness as nearly constant as possible. Complete uniformity in this dimension is sometimes impossible to achieve. Walls of non-uniform thickness should be gradually blended from thick to thin.
5. These tolerances do not apply to screw threads, gear teeth or match fits. Provisions can usually be made to hold this type of dimension to close limits.
6. These values should be considered minimum. The designer should allow as much draft as is compatible with the design. Liberal use of draft will minimize ejection problems, and reduce distortion due to ejection.
7. Customer-Molder understanding is necessary prior to tooling.

Copyright   Revised 1993

© The Society of the Plastics Industry, Inc.
1801 K Street, N.W., 600K
Washington, D.C. 20006-1301

| Standards & Practices of Plastics Molders | Material Polyetherimide (PEI) |

Note: The *Commercial* values shown below represent common production tolerances at the most economical level. The *Fine* values represent closer tolerances that can be held but at a greater cost. Any addition of fillers will compromise physical properties and alter dimensional stability. Please consult the manufacturer.

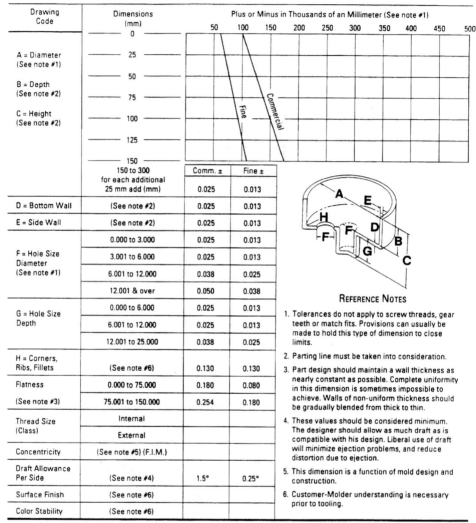

Drawing Code	Dimensions (mm)	Comm. ±	Fine ±
A = Diameter (See note #1)			
B = Depth (See note #2)			
C = Height (See note #2)			
	150 to 300 for each additional 25 mm add (mm)	0.025	0.013
D = Bottom Wall	(See note #2)	0.025	0.013
E = Side Wall	(See note #2)	0.025	0.013
F = Hole Size Diameter (See note #1)	0.000 to 3.000	0.025	0.013
	3.001 to 6.000	0.025	0.013
	6.001 to 12.000	0.038	0.025
	12.001 & over	0.050	0.038
G = Hole Size Depth	0.000 to 6.000	0.025	0.013
	6.001 to 12.000	0.025	0.013
	12.001 to 25.000	0.038	0.025
H = Corners, Ribs, Fillets	(See note #6)	0.130	0.130
Flatness	0.000 to 75.000	0.180	0.080
(See note #3)	75.001 to 150.000	0.254	0.180
Thread Size (Class)	Internal		
	External		
Concentricity	(See note #5) (F.I.M.)		
Draft Allowance Per Side	(See note #4)	1.5°	0.25°
Surface Finish	(See note #6)		
Color Stability	(See note #6)		

REFERENCE NOTES

1. Tolerances do not apply to screw threads, gear teeth or match fits. Provisions can usually be made to hold this type of dimension to close limits.

2. Parting line must be taken into consideration.

3. Part design should maintain a wall thickness as nearly constant as possible. Complete uniformity in this dimension is sometimes impossible to achieve. Walls of non-uniform thickness should be gradually blended from thick to thin.

4. These values should be considered minimum. The designer should allow as much draft as is compatible with his design. Liberal use of draft will minimize ejection problems, and reduce distortion due to ejection.

5. This dimension is a function of mold design and construction.

6. Customer-Molder understanding is necessary prior to tooling.

Copyright   Revised 1993

© The Society of the Plastics Industry, Inc.
1801 K Street, N.W., 600K
Washington, D.C. 20006-1301

Figure 5-12   Polyetherimide (PEI)

Standards & Practices of Plastics Molders

Material
Polyetherimide (PEI)

Note: The *Commercial* values shown below represent common production tolerances at the most economical level. The *Fine* values represent closer tolerances that can be held but at a greater cost. Any addition of fillers will compromise physical properties and alter dimensional stability. Please consult the manufacturer.

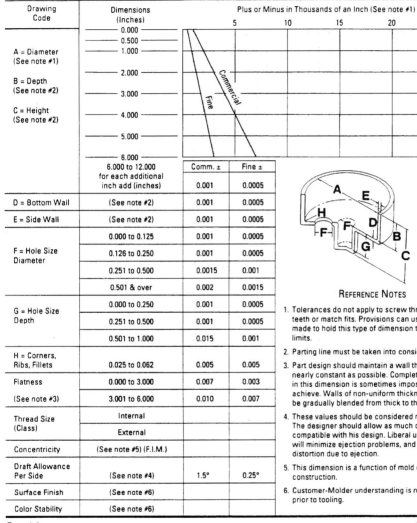

Drawing Code	Dimensions (Inches)	Comm. ±	Fine ±
A = Diameter (See note #1) B = Depth (See note #2) C = Height (See note #2)	0.000 / 0.500 / 1.000 / 2.000 / 3.000 / 4.000 / 5.000 / 6.000 (graph)		
	6.000 to 12.000 for each additional inch add (inches)	0.001	0.0005
D = Bottom Wall	(See note #2)	0.001	0.0005
E = Side Wall	(See note #2)	0.001	0.0005
F = Hole Size Diameter	0.000 to 0.125	0.001	0.0005
	0.126 to 0.250	0.001	0.0005
	0.251 to 0.500	0.0015	0.001
	0.501 & over	0.002	0.0015
G = Hole Size Depth	0.000 to 0.250	0.001	0.0005
	0.251 to 0.500	0.001	0.0005
	0.501 to 1.000	0.015	0.001
H = Corners, Ribs, Fillets	0.025 to 0.062	0.005	0.005
Flatness	0.000 to 3.000	0.007	0.003
(See note #3)	3.001 to 6.000	0.010	0.007
Thread Size (Class)	Internal		
	External		
Concentricity	(See note #5) (F.I.M.)		
Draft Allowance Per Side	(See note #4)	1.5°	0.25°
Surface Finish	(See note #6)		
Color Stability	(See note #6)		

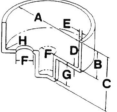

REFERENCE NOTES

1. Tolerances do not apply to screw threads, gear teeth or match fits. Provisions can usually be made to hold this type of dimension to close limits.

2. Parting line must be taken into consideration.

3. Part design should maintain a wall thickness as nearly constant as possible. Complete uniformity in this dimension is sometimes impossible to achieve. Walls of non-uniform thickness should be gradually blended from thick to thin.

4. These values should be considered minimum. The designer should allow as much draft as is compatible with his design. Liberal use of draft will minimize ejection problems, and reduce distortion due to ejection.

5. This dimension is a function of mold design and construction.

6. Customer-Molder understanding is necessary prior to tooling.

Copyright   Revised 1993

© The Society of the Plastics Industry, Inc.
1801 K Street, N.W., 600K
Washington, D.C. 20006-1301

Standards & Practices of Plastics Molders

Material
Low Density Polyethylene (LDPE)

Note: The *Commercial* values shown below represent common production tolerances at the most economical level. The *Fine* values represent closer tolerances that can be held but at a greater cost. Any addition of fillers will compromise physical properties and alter dimensional stability. Please consult the manufacturer.

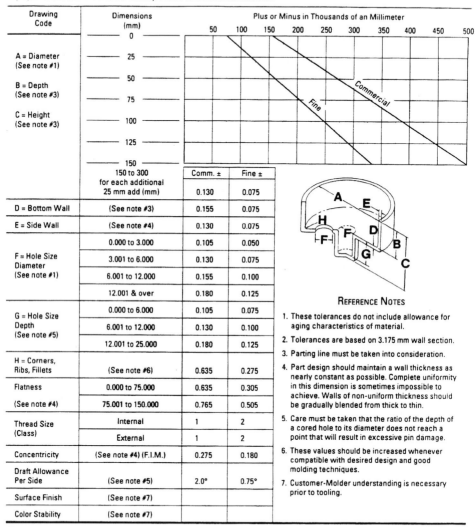

Drawing Code	Dimensions (mm)	Comm. ±	Fine ±
A = Diameter (See note #1) B = Depth (See note #3) C = Height (See note #3)	150 to 300 for each additional 25 mm add (mm)	0.130	0.075
D = Bottom Wall	(See note #3)	0.155	0.075
E = Side Wall	(See note #4)	0.130	0.075
F = Hole Size Diameter (See note #1)	0.000 to 3.000	0.105	0.050
	3.001 to 6.000	0.130	0.075
	6.001 to 12.000	0.155	0.100
	12.001 & over	0.180	0.125
G = Hole Size Depth (See note #5)	0.000 to 6.000	0.105	0.075
	6.001 to 12.000	0.130	0.100
	12.001 to 25.000	0.180	0.125
H = Corners, Ribs, Fillets	(See note #6)	0.635	0.275
Flatness	0.000 to 75.000	0.635	0.305
(See note #4)	75.001 to 150.000	0.765	0.505
Thread Size (Class)	Internal	1	2
	External	1	2
Concentricity	(See note #4) (F.I.M.)	0.275	0.180
Draft Allowance Per Side	(See note #5)	2.0°	0.75°
Surface Finish	(See note #7)		
Color Stability	(See note #7)		

REFERENCE NOTES

1. These tolerances do not include allowance for aging characteristics of material.

2. Tolerances are based on 3.175 mm wall section.

3. Parting line must be taken into consideration.

4. Part design should maintain a wall thickness as nearly constant as possible. Complete uniformity in this dimension is sometimes impossible to achieve. Walls of non-uniform thickness should be gradually blended from thick to thin.

5. Care must be taken that the ratio of the depth of a cored hole to its diameter does not reach a point that will result in excessive pin damage.

6. These values should be increased whenever compatible with desired design and good molding techniques.

7. Customer-Molder understanding is necessary prior to tooling.

Copyright   Revised 1991

Figure 5-13   Polyethylene, low density (LDPE)

Standards & Practices
of Plastics Molders

Material
Low Density Polyethylene
(LDPE)

Note: The *Commercial* values shown below represent common production tolerances at the most economical level. The *Fine* values represent closer tolerances that can be held but at a greater cost. Any addition of fillers will compromise physical properties and alter dimensional stability. Please consult the manufacturer.

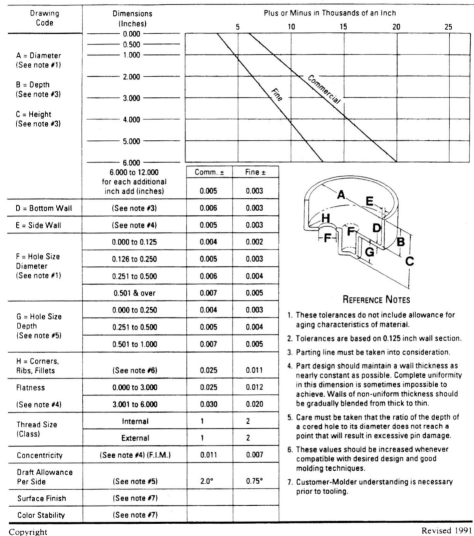

Drawing Code	Dimensions (Inches)	Comm. ±	Fine ±
A = Diameter (See note #1) B = Depth (See note #3) C = Height (See note #3)	0.000 — 0.500 — 1.000 — 2.000 — 3.000 — 4.000 — 5.000 — 6.000		
	6.000 to 12.000 for each additional inch add (inches)	0.005	0.003
D = Bottom Wall	(See note #3)	0.006	0.003
E = Side Wall	(See note #4)	0.005	0.003
F = Hole Size Diameter (See note #1)	0.000 to 0.125	0.004	0.002
	0.126 to 0.250	0.005	0.003
	0.251 to 0.500	0.006	0.004
	0.501 & over	0.007	0.005
G = Hole Size Depth (See note #5)	0.000 to 0.250	0.004	0.003
	0.251 to 0.500	0.005	0.004
	0.501 to 1.000	0.007	0.005
H = Corners, Ribs, Fillets	(See note #6)	0.025	0.011
Flatness	0.000 to 3.000	0.025	0.012
(See note #4)	3.001 to 6.000	0.030	0.020
Thread Size (Class)	Internal	1	2
	External	1	2
Concentricity	(See note #4) (F.I.M.)	0.011	0.007
Draft Allowance Per Side	(See note #5)	2.0°	0.75°
Surface Finish	(See note #7)		
Color Stability	(See note #7)		

Plus or Minus in Thousands of an Inch: 5, 10, 15, 20, 25

REFERENCE NOTES

1. These tolerances do not include allowance for aging characteristics of material.

2. Tolerances are based on 0.125 inch wall section.

3. Parting line must be taken into consideration.

4. Part design should maintain a wall thickness as nearly constant as possible. Complete uniformity in this dimension is sometimes impossible to achieve. Walls of non-uniform thickness should be gradually blended from thick to thin.

5. Care must be taken that the ratio of the depth of a cored hole to its diameter does not reach a point that will result in excessive pin damage.

6. These values should be increased whenever compatible with desired design and good molding techniques.

7. Customer-Molder understanding is necessary prior to tooling.

Copyright

Revised 1991

© The Society of the Plastics Industry, Inc.
1801 K Street, N.W., 600K
Washington, D.C. 20006-1301

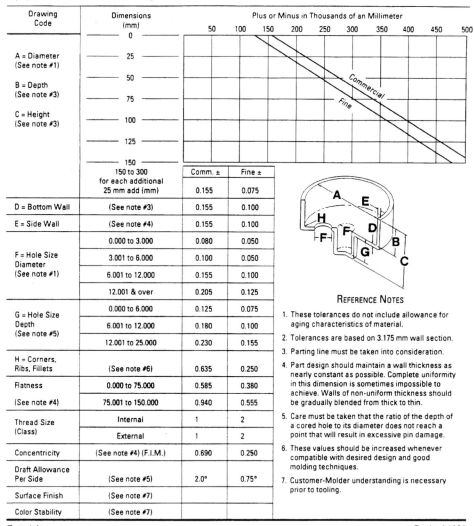

Standards & Practices of Plastics Molders	**Material** High Density Polyethylene (HDPE)

Note: The *Commercial* values shown below represent common production tolerances at the most economical level. The *Fine* values represent closer tolerances that can be held but at a greater cost. Any addition of fillers will compromise physical properties and alter dimensional stability. Please consult the manufacturer.

Drawing Code	Dimensions (mm)	Comm. ±	Fine ±
A = Diameter (See note #1) B = Depth (See note #3) C = Height (See note #3)			
	150 to 300 for each additional 25 mm add (mm)	0.155	0.075
D = Bottom Wall	(See note #3)	0.155	0.100
E = Side Wall	(See note #4)	0.155	0.100
F = Hole Size Diameter (See note #1)	0.000 to 3.000	0.080	0.050
	3.001 to 6.000	0.100	0.050
	6.001 to 12.000	0.155	0.100
	12.001 & over	0.205	0.125
G = Hole Size Depth (See note #5)	0.000 to 6.000	0.125	0.075
	6.001 to 12.000	0.180	0.100
	12.001 to 25.000	0.230	0.155
H = Corners, Ribs, Fillets	(See note #6)	0.635	0.250
Flatness (See note #4)	0.000 to 75.000	0.585	0.380
	75.001 to 150.000	0.940	0.555
Thread Size (Class)	Internal	1	2
	External	1	2
Concentricity	(See note #4) (F.I.M.)	0.690	0.250
Draft Allowance Per Side	(See note #5)	2.0°	0.75°
Surface Finish	(See note #7)		
Color Stability	(See note #7)		

REFERENCE NOTES

1. These tolerances do not include allowance for aging characteristics of material.

2. Tolerances are based on 3.175 mm wall section.

3. Parting line must be taken into consideration.

4. Part design should maintain a wall thickness as nearly constant as possible. Complete uniformity in this dimension is sometimes impossible to achieve. Walls of non-uniform thickness should be gradually blended from thick to thin.

5. Care must be taken that the ratio of the depth of a cored hole to its diameter does not reach a point that will result in excessive pin damage.

6. These values should be increased whenever compatible with desired design and good molding techniques.

7. Customer-Molder understanding is necessary prior to tooling.

© The Society of the Plastics Industry, Inc.
1801 K Street, N.W., 600K
Washington, D.C. 20006-1301

Figure 5-14 Polyethylene, high density (HDPE)

Standards & Practices of Plastics Molders

Material
High Density Polyethylene
(HDPE)

Note: The *Commercial* values shown below represent common production tolerances at the most economical level. The *Fine* values represent closer tolerances that can be held but at a greater cost. Any addition of fillers will compromise physical properties and alter dimensional stability. Please consult the manufacturer.

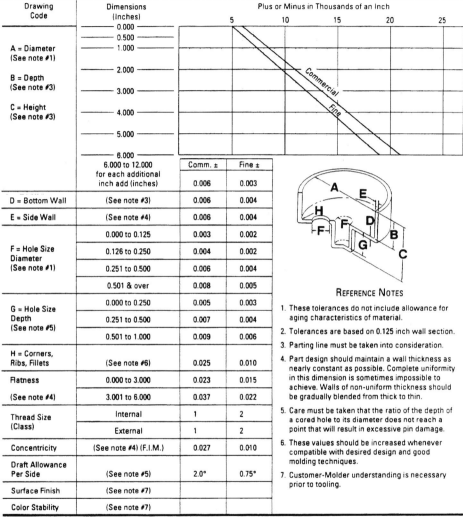

Drawing Code	Dimensions (Inches)	Comm. ±	Fine ±
A = Diameter (See note #1) B = Depth (See note #3) C = Height (See note #3)	6.000 to 12.000 for each additional inch add (inches)	0.006	0.003
D = Bottom Wall	(See note #3)	0.006	0.004
E = Side Wall	(See note #4)	0.006	0.004
F = Hole Size Diameter (See note #1)	0.000 to 0.125	0.003	0.002
	0.126 to 0.250	0.004	0.002
	0.251 to 0.500	0.006	0.004
	0.501 & over	0.008	0.005
G = Hole Size Depth (See note #5)	0.000 to 0.250	0.005	0.003
	0.251 to 0.500	0.007	0.004
	0.501 to 1.000	0.009	0.006
H = Corners, Ribs, Fillets	(See note #6)	0.025	0.010
Flatness	0.000 to 3.000	0.023	0.015
(See note #4)	3.001 to 6.000	0.037	0.022
Thread Size (Class)	Internal	1	2
	External	1	2
Concentricity	(See note #4) (F.I.M.)	0.027	0.010
Draft Allowance Per Side	(See note #5)	2.0°	0.75°
Surface Finish	(See note #7)		
Color Stability	(See note #7)		

REFERENCE NOTES

1. These tolerances do not include allowance for aging characteristics of material.

2. Tolerances are based on 0.125 inch wall section.

3. Parting line must be taken into consideration.

4. Part design should maintain a wall thickness as nearly constant as possible. Complete uniformity in this dimension is sometimes impossible to achieve. Walls of non-uniform thickness should be gradually blended from thick to thin.

5. Care must be taken that the ratio of the depth of a cored hole to its diameter does not reach a point that will result in excessive pin damage.

6. These values should be increased whenever compatible with desired design and good molding techniques.

7. Customer-Molder understanding is necessary prior to tooling.

Copyright   Revised 1991

Standards & Practices of Plastics Molders	Material Polyphenylene Oxide (PPO)

Note: The *Commercial* values shown below represent common production tolerances at the most economical level. The *Fine* values represent closer tolerances that can be held but at a greater cost. Any addition of fillers will compromise physical properties and alter dimensional stability. Please consult the manufacturer.

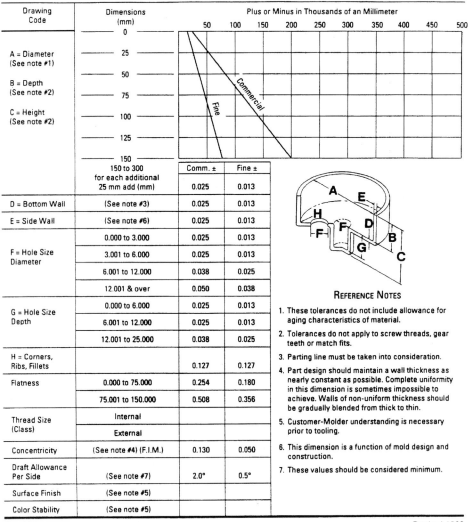

Drawing Code	Dimensions (mm)	Comm. ±	Fine ±
A = Diameter (See note #1) B = Depth (See note #2) C = Height (See note #2)			
150 to 300 for each additional 25 mm add (mm)		0.025	0.013
D = Bottom Wall	(See note #3)	0.025	0.013
E = Side Wall	(See note #6)	0.025	0.013
F = Hole Size Diameter	0.000 to 3.000	0.025	0.013
	3.001 to 6.000	0.025	0.013
	6.001 to 12.000	0.038	0.025
	12.001 & over	0.050	0.038
G = Hole Size Depth	0.000 to 6.000	0.025	0.013
	6.001 to 12.000	0.025	0.013
	12.001 to 25.000	0.038	0.025
H = Corners, Ribs, Fillets		0.127	0.127
Flatness	0.000 to 75.000	0.254	0.180
	75.001 to 150.000	0.508	0.356
Thread Size (Class)	Internal		
	External		
Concentricity	(See note #4) (F.I.M.)	0.130	0.050
Draft Allowance Per Side	(See note #7)	2.0°	0.5°
Surface Finish	(See note #5)		
Color Stability	(See note #5)		

REFERENCE NOTES

1. These tolerances do not include allowance for aging characteristics of material.
2. Tolerances do not apply to screw threads, gear teeth or match fits.
3. Parting line must be taken into consideration.
4. Part design should maintain a wall thickness as nearly constant as possible. Complete uniformity in this dimension is sometimes impossible to achieve. Walls of non-uniform thickness should be gradually blended from thick to thin.
5. Customer-Molder understanding is necessary prior to tooling.
6. This dimension is a function of mold design and construction.
7. These values should be considered minimum.

Copyright   Revised 1993

Figure 5-15   Polyphenylene oxide (PPO)

Standards & Practices of Plastics Molders

Material
Polyphenylene Oxide (PPO)

Note: The *Commercial* values shown below represent common production tolerances at the most economical level. The *Fine* values represent closer tolerances that can be held but at a greater cost. Any addition of fillers will compromise physical properties and alter dimensional stability. Please consult the manufacturer.

Drawing Code	Dimensions (Inches)		Plus or Minus in Thousands of an Inch				
			5	10	15	20	25

A = Diameter (See note #1)

B = Depth (See note #2)

C = Height (See note #2)

Dimension scale (Inches): 0.000, 0.500, 1.000, 2.000, 3.000, 4.000, 5.000, 6.000

(graph showing *Fine* and *Commercial* curves)

Dimensions (Inches)	Comm. ±	Fine ±
6.000 to 12.000 for each additional inch add (inches)	0.001	0.0005
D = Bottom Wall — (See note #3)	0.001	0.0005
E = Side Wall — (See note #6)	0.001	0.0005
F = Hole Size Diameter — 0.000 to 0.125	0.001	0.0005
0.126 to 0.250	0.001	0.0005
0.251 to 0.500	0.0015	0.001
0.501 & over	0.002	0.0015
G = Hole Size Depth — 0.000 to 0.250	0.001	0.0005
0.251 to 0.500	0.001	0.0005
0.501 to 1.000	0.0015	0.001
H = Corners, Ribs, Fillets	0.005	0.005
Flatness — 0.000 to 3.000	0.010	0.007
3.001 to 6.000	0.020	0.014
Thread Size (Class) — Internal		
External		
Concentricity — (See note #4) (F.I.M.)	0.005	0.002
Draft Allowance Per Side — (See note #7)	2.0°	0.5°
Surface Finish — (See note #5)		
Color Stability — (See note #5)		

REFERENCE NOTES

1. These tolerances do not include allowance for aging characteristics of material.

2. Tolerances do not apply to screw threads, gear teeth or match fits.

3. Parting line must be taken into consideration.

4. Part design should maintain a wall thickness as nearly constant as possible. Complete uniformity in this dimension is sometimes impossible to achieve. Walls of non-uniform thickness should be gradually blended from thick to thin.

5. Customer-Molder understanding is necessary prior to tooling.

6. This dimension is a function of mold design and construction.

7. These values should be considered minimum.

Revised 1993

© The Society of the Plastics Industry, Inc.
1801 K Street, N.W., 600K
Washington, D.C. 20006-1301

Standards & Practices of Plastics Molders	Material Polypropylene (PP)

Note: The *Commercial* values shown below represent common production tolerances at the most economical level. The *Fine* values represent closer tolerances that can be held but at a greater cost. Any addition of fillers will compromise physical properties and alter dimensional stability. Please consult the manufacturer.

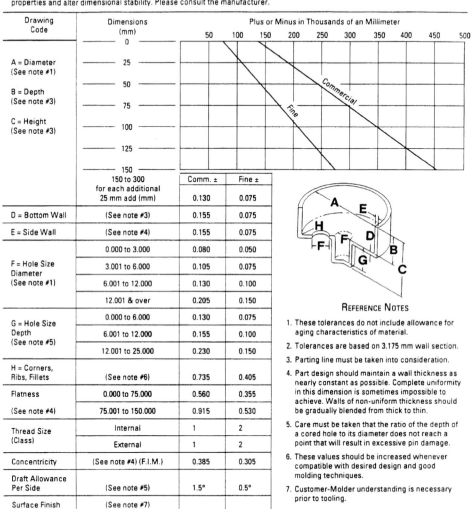

Drawing Code	Dimensions (mm)	Comm. ±	Fine ±
A = Diameter (See note #1) B = Depth (See note #3) C = Height (See note #3)	150 to 300 for each additional 25 mm add (mm)	0.130	0.075
D = Bottom Wall	(See note #3)	0.155	0.075
E = Side Wall	(See note #4)	0.155	0.075
F = Hole Size Diameter (See note #1)	0.000 to 3.000	0.080	0.050
	3.001 to 6.000	0.105	0.075
	6.001 to 12.000	0.130	0.100
	12.001 & over	0.205	0.150
G = Hole Size Depth (See note #5)	0.000 to 6.000	0.130	0.075
	6.001 to 12.000	0.155	0.100
	12.001 to 25.000	0.230	0.150
H = Corners, Ribs, Fillets	(See note #6)	0.735	0.405
Flatness	0.000 to 75.000	0.560	0.355
(See note #4)	75.001 to 150.000	0.915	0.530
Thread Size (Class)	Internal	1	2
	External	1	2
Concentricity	(See note #4) (F.I.M.)	0.385	0.305
Draft Allowance Per Side	(See note #5)	1.5°	0.5°
Surface Finish	(See note #7)		
Color Stability	(See note #7)		

REFERENCE NOTES

1. These tolerances do not include allowance for aging characteristics of material.
2. Tolerances are based on 3.175 mm wall section.
3. Parting line must be taken into consideration.
4. Part design should maintain a wall thickness as nearly constant as possible. Complete uniformity in this dimension is sometimes impossible to achieve. Walls of non-uniform thickness should be gradually blended from thick to thin.
5. Care must be taken that the ratio of the depth of a cored hole to its diameter does not reach a point that will result in excessive pin damage.
6. These values should be increased whenever compatible with desired design and good molding techniques.
7. Customer-Molder understanding is necessary prior to tooling.

© The Society of the Plastics Industry, Inc.
1801 K Street, N.W., 600K
Washington, D.C. 20006-1301

Figure 5-16   Polypropylene (PP)

Standards & Practices of Plastics Molders	Material Polypropylene (PP)

Note: The *Commercial* values shown below represent common production tolerances at the most economical level. The *Fine* values represent closer tolerances that can be held but at a greater cost. Any addition of fillers will compromise physical properties and alter dimensional stability. Please consult the manufacturer.

Drawing Code	Dimensions (Inches)	Plus or Minus in Thousands of an Inch
A = Diameter (See note #1)	0.000 / 0.500 / 1.000 / 2.000 / 3.000	
B = Depth (See note #3)		
C = Height (See note #3)	4.000 / 5.000 / 6.000	

	Dimensions (Inches)	Comm. ±	Fine ±
	6.000 to 12.000 for each additional inch add (inches)	0.005	0.003
D = Bottom Wall	(See note #3)	0.006	0.003
E = Side Wall	(See note #4)	0.006	0.003
F = Hole Size Diameter (See note #1)	0.000 to 0.125	0.003	0.002
	0.126 to 0.250	0.004	0.003
	0.251 to 0.500	0.005	0.004
	0.501 & over	0.008	0.006
G = Hole Size Depth (See note #5)	0.000 to 0.250	0.005	0.003
	0.251 to 0.500	0.006	0.004
	0.501 to 1.000	0.009	0.006
H = Corners, Ribs, Fillets	(See note #6)	0.029	0.016
Flatness	0.000 to 3.000	0.022	0.014
(See note #4)	3.001 to 6.000	0.036	0.021
Thread Size (Class)	Internal	1	2
	External	1	2
Concentricity	(See note #4) (F.I.M.)	0.015	0.012
Draft Allowance Per Side	(See note #5)	1.5°	0.5°
Surface Finish	(See note #7)		
Color Stability	(See note #7)		

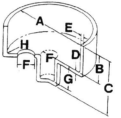

REFERENCE NOTES

1. These tolerances do not include allowance for aging characteristics of material.

2. Tolerances are based on 0.125 inch wall section.

3. Parting line must be taken into consideration.

4. Part design should maintain a wall thickness as nearly constant as possible. Complete uniformity in this dimension is sometimes impossible to achieve. Walls of non-uniform thickness should be gradually blended from thick to thin.

5. Care must be taken that the ratio of the depth of a cored hole to its diameter does not reach a point that will result in excessive pin damage.

6. These values should be increased whenever compatible with desired design and good molding techniques.

7. Customer-Molder understanding is necessary prior to tooling.

Copyright   Revised 1991

Standards & Practices of Plastics Molders	Material Polystyrene (PS)

Note: The *Commercial* values shown below represent common production tolerances at the most economical level. The *Fine* values represent closer tolerances that can be held but at a greater cost. Any addition of fillers will compromise physical properties and alter dimensional stability. Please consult the manufacturer.

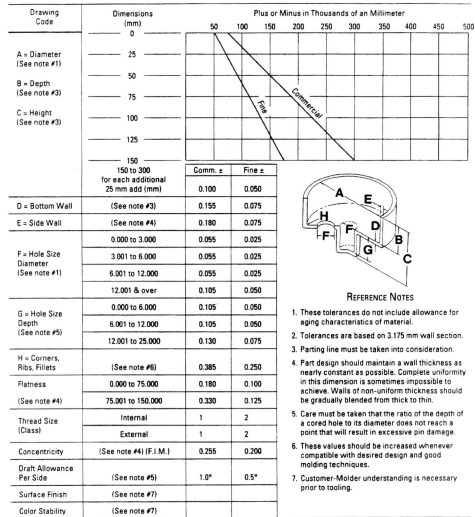

Drawing Code	Dimensions (mm)	Comm. ±	Fine ±
A = Diameter (See note #1) B = Depth (See note #3) C = Height (See note #3)	150 to 300 for each additional 25 mm add (mm)	0.100	0.050
D = Bottom Wall	(See note #3)	0.155	0.075
E = Side Wall	(See note #4)	0.180	0.075
F = Hole Size Diameter (See note #1)	0.000 to 3.000	0.055	0.025
	3.001 to 6.000	0.055	0.025
	6.001 to 12.000	0.055	0.025
	12.001 & over	0.105	0.050
G = Hole Size Depth (See note #5)	0.000 to 6.000	0.105	0.050
	6.001 to 12.000	0.105	0.050
	12.001 to 25.000	0.130	0.075
H = Corners, Ribs, Fillets	(See note #6)	0.385	0.250
Flatness	0.000 to 75.000	0.180	0.100
(See note #4)	75.001 to 150.000	0.330	0.125
Thread Size (Class)	Internal	1	2
	External	1	2
Concentricity	(See note #4) (F.I.M.)	0.255	0.200
Draft Allowance Per Side	(See note #5)	1.0°	0.5°
Surface Finish	(See note #7)		
Color Stability	(See note #7)		

REFERENCE NOTES

1. These tolerances do not include allowance for aging characteristics of material.

2. Tolerances are based on 3.175 mm wall section.

3. Parting line must be taken into consideration.

4. Part design should maintain a wall thickness as nearly constant as possible. Complete uniformity in this dimension is sometimes impossible to achieve. Walls of non-uniform thickness should be gradually blended from thick to thin.

5. Care must be taken that the ratio of the depth of a cored hole to its diameter does not reach a point that will result in excessive pin damage.

6. These values should be increased whenever compatible with desired design and good molding techniques.

7. Customer-Molder understanding is necessary prior to tooling.

Copyright   Revised 1991

© The Society of the Plastics Industry, Inc.
1801 K Street, N.W., 600K
Washington, D.C. 20006-1301

Figure 5-17   Polystyrene (PS)

Standards & Practices of Plastics Molders	Material Polystyrene (PS)

Note: The *Commercial* values shown below represent common production tolerances at the most economical level. The *Fine* values represent closer tolerances that can be held but at a greater cost. Any addition of fillers will compromise physical properties and alter dimensional stability. Please consult the manufacturer.

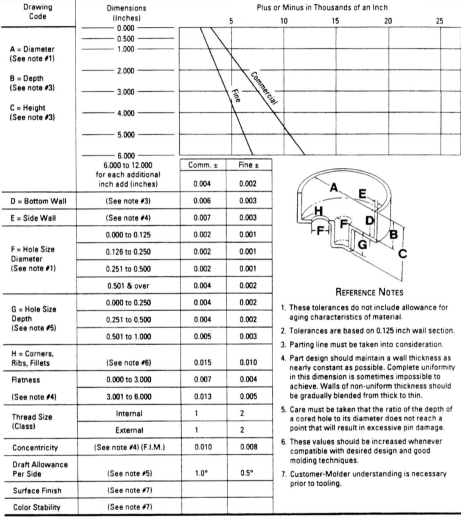

Drawing Code	Dimensions (Inches)	Plus or Minus in Thousands of an Inch
A = Diameter (See note #1) B = Depth (See note #3) C = Height (See note #3)	0.000 0.500 1.000 2.000 3.000 4.000 5.000 6.000	(graph: Commercial and Fine curves, scale 5, 10, 15, 20, 25)

Drawing Code	Dimensions (Inches)	Comm. ±	Fine ±
6.000 to 12.000 for each additional inch add (inches)		0.004	0.002
D = Bottom Wall	(See note #3)	0.006	0.003
E = Side Wall	(See note #4)	0.007	0.003
F = Hole Size Diameter (See note #1)	0.000 to 0.125	0.002	0.001
	0.126 to 0.250	0.002	0.001
	0.251 to 0.500	0.002	0.001
	0.501 & over	0.004	0.002
G = Hole Size Depth (See note #5)	0.000 to 0.250	0.004	0.002
	0.251 to 0.500	0.004	0.002
	0.501 to 1.000	0.005	0.003
H = Corners, Ribs, Fillets	(See note #6)	0.015	0.010
Flatness	0.000 to 3.000	0.007	0.004
(See note #4)	3.001 to 6.000	0.013	0.005
Thread Size (Class)	Internal	1	2
	External	1	2
Concentricity	(See note #4) (F.I.M.)	0.010	0.008
Draft Allowance Per Side	(See note #5)	1.0°	0.5°
Surface Finish	(See note #7)		
Color Stability	(See note #7)		

REFERENCE NOTES

1. These tolerances do not include allowance for aging characteristics of material.

2. Tolerances are based on 0.125 inch wall section.

3. Parting line must be taken into consideration.

4. Part design should maintain a wall thickness as nearly constant as possible. Complete uniformity in this dimension is sometimes impossible to achieve. Walls of non-uniform thickness should be gradually blended from thick to thin.

5. Care must be taken that the ratio of the depth of a cored hole to its diameter does not reach a point that will result in excessive pin damage.

6. These values should be increased whenever compatible with desired design and good molding techniques.

7. Customer-Molder understanding is necessary prior to tooling.

Copyright   Revised 1993

| **Standards & Practices of Plastics Molders** | **Material** Polyvinyl Chloride (PVC) (Vinyl) (Rigid) |

Note: The *Commercial* values shown below represent common production tolerances at the most economical level. The *Fine* values represent closer tolerances that can be held but at a greater cost. Any addition of fillers will compromise physical properties and alter dimensional stability. Please consult the manufacturer.

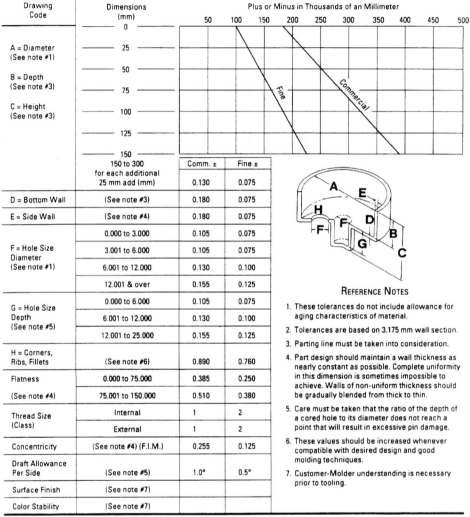

Drawing Code	Dimensions (mm)	Comm. ±	Fine ±
A = Diameter (See note #1) B = Depth (See note #3) C = Height (See note #3)	150 to 300 for each additional 25 mm add (mm)	0.130	0.075
D = Bottom Wall	(See note #3)	0.180	0.075
E = Side Wall	(See note #4)	0.180	0.075
F = Hole Size Diameter (See note #1)	0.000 to 3.000	0.105	0.075
	3.001 to 6.000	0.105	0.075
	6.001 to 12.000	0.130	0.100
	12.001 & over	0.155	0.125
G = Hole Size Depth (See note #5)	0.000 to 6.000	0.105	0.075
	6.001 to 12.000	0.130	0.100
	12.001 to 25.000	0.155	0.125
H = Corners, Ribs, Fillets	(See note #6)	0.890	0.760
Flatness	0.000 to 75.000	0.385	0.250
(See note #4)	75.001 to 150.000	0.510	0.380
Thread Size (Class)	Internal	1	2
	External	1	2
Concentricity	(See note #4) (F.I.M.)	0.255	0.125
Draft Allowance Per Side	(See note #5)	1.0°	0.5°
Surface Finish	(See note #7)		
Color Stability	(See note #7)		

REFERENCE NOTES

1. These tolerances do not include allowance for aging characteristics of material.
2. Tolerances are based on 3.175 mm wall section.
3. Parting line must be taken into consideration.
4. Part design should maintain a wall thickness as nearly constant as possible. Complete uniformity in this dimension is sometimes impossible to achieve. Walls of non-uniform thickness should be gradually blended from thick to thin.
5. Care must be taken that the ratio of the depth of a cored hole to its diameter does not reach a point that will result in excessive pin damage.
6. These values should be increased whenever compatible with desired design and good molding techniques.
7. Customer-Molder understanding is necessary prior to tooling.

Copyright   Revised 1991

© The Society of the Plastics Industry, Inc.
1801 K Street, N.W., 600K
Washington, D.C. 20006-1301

Figure 5-18 Polyvinyl chloride – Rigid (PVC)

Standards & Practices
of Plastics Molders

Material
Polyvinyl Chloride (PVC)
(Vinyl) (Rigid)

Note: The *Commercial* values shown below represent common production tolerances at the most economical level. The *Fine* values represent closer tolerances that can be held but at a greater cost. Any addition of fillers will compromise physical properties and alter dimensional stability. Please consult the manufacturer.

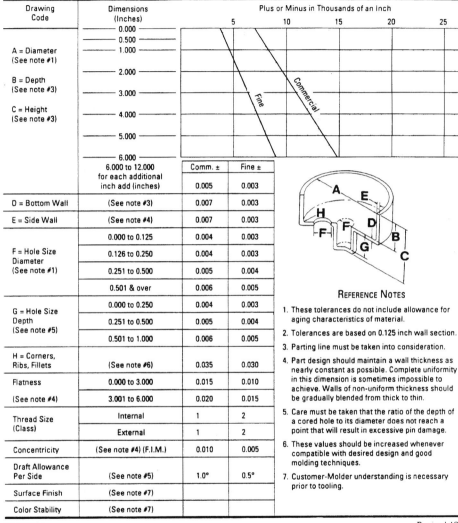

Drawing Code	Dimensions (Inches)		
A = Diameter (See note #1)	0.000		
	0.500		
	1.000		
	2.000		
B = Depth (See note #3)	3.000		
C = Height (See note #3)	4.000		
	5.000		
	6.000		
	6.000 to 12.000 for each additional inch add (inches)	Comm. ±	Fine ±
		0.005	0.003
D = Bottom Wall	(See note #3)	0.007	0.003
E = Side Wall	(See note #4)	0.007	0.003
F = Hole Size Diameter (See note #1)	0.000 to 0.125	0.004	0.003
	0.126 to 0.250	0.004	0.003
	0.251 to 0.500	0.005	0.004
	0.501 & over	0.006	0.005
G = Hole Size Depth (See note #5)	0.000 to 0.250	0.004	0.003
	0.251 to 0.500	0.005	0.004
	0.501 to 1.000	0.006	0.005
H = Corners, Ribs, Fillets	(See note #6)	0.035	0.030
Flatness	0.000 to 3.000	0.015	0.010
(See note #4)	3.001 to 6.000	0.020	0.015
Thread Size (Class)	Internal	1	2
	External	1	2
Concentricity	(See note #4) (F.I.M.)	0.010	0.005
Draft Allowance Per Side	(See note #5)	1.0°	0.5°
Surface Finish	(See note #7)		
Color Stability	(See note #7)		

Plus or Minus in Thousands of an Inch: 5, 10, 15, 20, 25

REFERENCE NOTES

1. These tolerances do not include allowance for aging characteristics of material.

2. Tolerances are based on 0.125 inch wall section.

3. Parting line must be taken into consideration.

4. Part design should maintain a wall thickness as nearly constant as possible. Complete uniformity in this dimension is sometimes impossible to achieve. Walls of non-uniform thickness should be gradually blended from thick to thin.

5. Care must be taken that the ratio of the depth of a cored hole to its diameter does not reach a point that will result in excessive pin damage.

6. These values should be increased whenever compatible with desired design and good molding techniques.

7. Customer-Molder understanding is necessary prior to tooling.

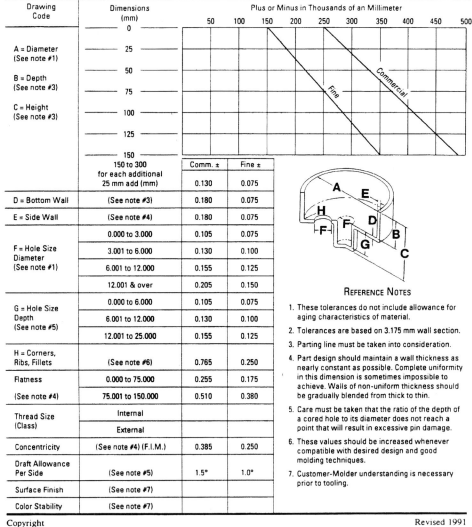

Standards & Practices of Plastics Molders

Material
Polyvinyl Chloride (PVC)
(Vinyl) (Flexible)

Note: The *Commercial* values shown below represent common production tolerances at the most economical level. The *Fine* values represent closer tolerances that can be held but at a greater cost. Any addition of fillers will compromise physical properties and alter dimensional stability. Please consult the manufacturer.

Drawing Code	Dimensions (mm)	Comm. ±	Fine ±
A = Diameter (See note #1) B = Depth (See note #3) C = Height (See note #3)	150 to 300 for each additional 25 mm add (mm)	0.130	0.075
D = Bottom Wall	(See note #3)	0.180	0.075
E = Side Wall	(See note #4)	0.180	0.075
F = Hole Size Diameter (See note #1)	0.000 to 3.000	0.105	0.075
	3.001 to 6.000	0.130	0.100
	6.001 to 12.000	0.155	0.125
	12.001 & over	0.205	0.150
G = Hole Size Depth (See note #5)	0.000 to 6.000	0.105	0.075
	6.001 to 12.000	0.130	0.100
	12.001 to 25.000	0.155	0.125
H = Corners, Ribs, Fillets	(See note #6)	0.765	0.250
Flatness	0.000 to 75.000	0.255	0.175
(See note #4)	75.001 to 150.000	0.510	0.380
Thread Size (Class)	Internal		
	External		
Concentricity	(See note #4) (F.I.M.)	0.385	0.250
Draft Allowance Per Side	(See note #5)	1.5°	1.0°
Surface Finish	(See note #7)		
Color Stability	(See note #7)		

Plus or Minus in Thousands of an Millimeter (graph columns: 50, 100, 150, 200, 250, 300, 350, 400, 450, 500; curves labeled *Fine* and *Commercial*)

REFERENCE NOTES

1. These tolerances do not include allowance for aging characteristics of material.

2. Tolerances are based on 3.175 mm wall section.

3. Parting line must be taken into consideration.

4. Part design should maintain a wall thickness as nearly constant as possible. Complete uniformity in this dimension is sometimes impossible to achieve. Walls of non-uniform thickness should be gradually blended from thick to thin.

5. Care must be taken that the ratio of the depth of a cored hole to its diameter does not reach a point that will result in excessive pin damage.

6. These values should be increased whenever compatible with desired design and good molding techniques.

7. Customer-Molder understanding is necessary prior to tooling.

Copyright   Revised 1991

Figure 5-19   Polyvinyl chloride – Flexible (PVC)

Standards & Practices of Plastics Molders	Material Polyvinyl Chloride (PVC) (Vinyl) (Flexible)

Note: The *Commercial* values shown below represent common production tolerances at the most economical level. The *Fine* values represent closer tolerances that can be held but at a greater cost. Any addition of fillers will compromise physical properties and alter dimensional stability. Please consult the manufacturer.

Drawing Code	Dimensions (Inches)	Plus or Minus in Thousands of an Inch				
		5	10	15	20	25

A = Diameter (See note #1)

B = Depth (See note #3)

C = Height (See note #3)

Dimensions (Inches): 0.000, 0.500, 1.000, 2.000, 3.000, 4.000, 5.000, 6.000

(graph with *Fine* and *Commercial* lines)

Drawing Code	Dimensions (Inches)	Comm. ±	Fine ±
	6.000 to 12.000 for each additional inch add (inches)	0.005	0.003
D = Bottom Wall	(See note #3)	0.007	0.003
E = Side Wall	(See note #4)	0.007	0.003
F = Hole Size Diameter (See note #1)	0.000 to 0.125	0.004	0.003
	0.126 to 0.250	0.005	0.004
	0.251 to 0.500	0.006	0.005
	0.501 & over	0.008	0.006
G = Hole Size Depth (See note #5)	0.000 to 0.250	0.004	0.003
	0.251 to 0.500	0.005	0.004
	0.501 to 1.000	0.006	0.005
H = Corners, Ribs, Fillets	(See note #6)	0.030	0.010
Flatness	0.000 to 3.000	0.010	0.007
(See note #4)	3.001 to 6.000	0.020	0.015
Thread Size (Class)	Internal		
	External		
Concentricity	(See note #4) (F.I.M.)	0.015	0.010
Draft Allowance Per Side	(See note #5)	1.5°	1.0°
Surface Finish	(See note #7)		
Color Stability	(See note #7)		

REFERENCE NOTES

1. These tolerances do not include allowance for aging characteristics of material.

2. Tolerances are based on 0.125 inch wall section.

3. Parting line must be taken into consideration.

4. Part design should maintain a wall thickness as nearly constant as possible. Complete uniformity in this dimension is sometimes impossible to achieve. Walls of non-uniform thickness should be gradually blended from thick to thin.

5. Care must be taken that the ratio of the depth of a cored hole to its diameter does not reach a point that will result in excessive pin damage.

6. These values should be increased whenever compatible with desired design and good molding techniques.

7. Customer-Molder understanding is necessary prior to tooling.

Copyright

Revised 1991

Standards & Practices of Plastics Molders		Material Styrene-Acrylonitrile (SAN)

Note: The *Commercial* values shown below represent common production tolerances at the most economical level. The *Fine* values represent closer tolerances that can be held but at a greater cost. Any addition of fillers will compromise physical properties and alter dimensional stability. Please consult the manufacturer.

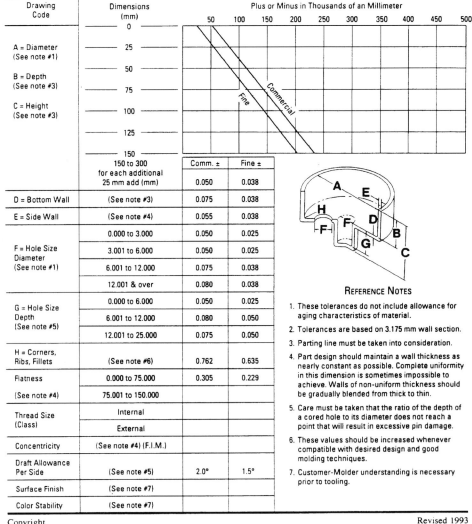

Drawing Code	Dimensions (mm)	Comm. ±	Fine ±
A = Diameter (See note #1) B = Depth (See note #3) C = Height (See note #3)	150 to 300 for each additional 25 mm add (mm)	0.050	0.038
D = Bottom Wall	(See note #3)	0.075	0.038
E = Side Wall	(See note #4)	0.055	0.038
F = Hole Size Diameter (See note #1)	0.000 to 3.000	0.050	0.025
	3.001 to 6.000	0.050	0.025
	6.001 to 12.000	0.075	0.038
	12.001 & over	0.080	0.038
G = Hole Size Depth (See note #5)	0.000 to 6.000	0.050	0.025
	6.001 to 12.000	0.080	0.050
	12.001 to 25.000	0.075	0.050
H = Corners, Ribs, Fillets	(See note #6)	0.762	0.635
Flatness	0.000 to 75.000	0.305	0.229
(See note #4)	75.001 to 150.000		
Thread Size (Class)	Internal		
	External		
Concentricity	(See note #4) (F.I.M.)		
Draft Allowance Per Side	(See note #5)	2.0°	1.5°
Surface Finish	(See note #7)		
Color Stability	(See note #7)		

REFERENCE NOTES

1. These tolerances do not include allowance for aging characteristics of material.

2. Tolerances are based on 3.175 mm wall section.

3. Parting line must be taken into consideration.

4. Part design should maintain a wall thickness as nearly constant as possible. Complete uniformity in this dimension is sometimes impossible to achieve. Walls of non-uniform thickness should be gradually blended from thick to thin.

5. Care must be taken that the ratio of the depth of a cored hole to its diameter does not reach a point that will result in excessive pin damage.

6. These values should be increased whenever compatible with desired design and good molding techniques.

7. Customer-Molder understanding is necessary prior to tooling.

Copyright   Revised 1993

© The Society of the Plastics Industry, Inc.
1801 K Street, N.W., 600K
Washington, D.C. 20006-1301

Figure 5-20   Styrene Acrylonitrile (SAN)

Standards & Practices of Plastics Molders

Material
Styrene-Acrylonitrile
(SAN)

Note: The *Commercial* values shown below represent common production tolerances at the most economical level. The *Fine* values represent closer tolerances that can be held but at a greater cost. Any addition of fillers will compromise physical properties and alter dimensional stability. Please consult the manufacturer.

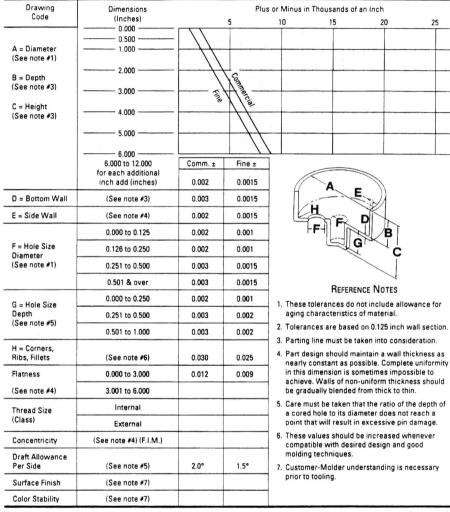

Drawing Code	Dimensions (Inches)	Plus or Minus in Thousands of an Inch				
		5	10	15	20	25
A = Diameter (See note #1)	0.000					
	0.500					
	1.000					
B = Depth (See note #3)	2.000					
	3.000					
C = Height (See note #3)	4.000					
	5.000					
	6.000					

		Comm. ±	Fine ±
	6.000 to 12.000 for each additional inch add (inches)	0.002	0.0015
D = Bottom Wall	(See note #3)	0.003	0.0015
E = Side Wall	(See note #4)	0.002	0.0015
F = Hole Size Diameter (See note #1)	0.000 to 0.125	0.002	0.001
	0.126 to 0.250	0.002	0.001
	0.251 to 0.500	0.003	0.0015
	0.501 & over	0.003	0.0015
G = Hole Size Depth (See note #5)	0.000 to 0.250	0.002	0.001
	0.251 to 0.500	0.003	0.002
	0.501 to 1.000	0.003	0.002
H = Corners, Ribs, Fillets	(See note #6)	0.030	0.025
Flatness	0.000 to 3.000	0.012	0.009
(See note #4)	3.001 to 6.000		
Thread Size (Class)	Internal		
	External		
Concentricity	(See note #4) (F.I.M.)		
Draft Allowance Per Side	(See note #5)	2.0°	1.5°
Surface Finish	(See note #7)		
Color Stability	(See note #7)		

REFERENCE NOTES

1. These tolerances do not include allowance for aging characteristics of material.

2. Tolerances are based on 0.125 inch wall section.

3. Parting line must be taken into consideration.

4. Part design should maintain a wall thickness as nearly constant as possible. Complete uniformity in this dimension is sometimes impossible to achieve. Walls of non-uniform thickness should be gradually blended from thick to thin.

5. Care must be taken that the ratio of the depth of a cored hole to its diameter does not reach a point that will result in excessive pin damage.

6. These values should be increased whenever compatible with desired design and good molding techniques.

7. Customer-Molder understanding is necessary prior to tooling.

Revised 1991

© The Society of the Plastics Industry, Inc.
1801 K Street, N.W., 600K
Washington, D.C. 20006-1301

6 Assembly Method Selection by Process

6.1 Introduction

Chapter 5 dealt with the selection of assembly based on the proposed plastic material. This chapter considers the selection of the appropriate assembly method based on the proposed processing method. Our brief overview presents the principal plastic processes, in alphabetical order, followed by a discussion of the assembly methods available for that process. Space limitations prevent a full discussion of each of these processes. Anyone wishing to pursue this topic in greater detail is referred to Irv Rubin's handbook *Handbook of Plastic Materials and Technology* (John Wiley & Sons, New York, 1990), which is in the opinion of this author, the single best reference for this type of information.

The processes are discussed in alphabetical order for easy reference. At the end of Section 6.23 "Process Selection", they are ranked in order of increasing production volume and a size chart (Table 6-1) is provided.

6.2 Blow Molding

6.2.1 The Process

Blow molding (Fig. 6-1), commonly known as the principal method of producing bottles in high volume, can be thought of as the plastic equivalent of glass blowing. In extrusion blow molding, the preform is extruded as a tube known as a parison. (Smaller complicated shapes can use an injection-molded parison. This is known as injection blow molding.) The two halves of the mold close on the tube and air is blown in, which causes the tube to take the shape of the mold. The mold opens, the part is ejected and the cycle begins anew.

Blow molding is confined to thermoplastic materials and, since it is a low pressure process, its tooling cost is fairly moderate. A blow molder's ideal part is about the shape of a soda bottle.

Large hollow industrial shapes can be produced by so-called industrial blow molding. From a product design standpoint, this is the variety of greatest interest. In addition, fairly flat dual wall panels (1 in. thick) have been successfully blow-molded. Such parts can have stretch problems in the corners, however. These parts can be strengthened with kiss-off (tack off) ribs and the cavity can also be filled with foam for further stiffening. Blow molding can make open parts by cutting a moldment into individual parts.

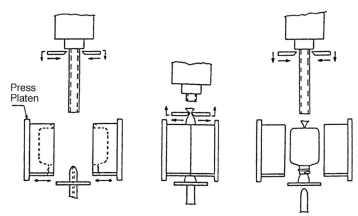

Figure 6-1 Schematic of an extrusion blow-molding process: left, extrusion of parison; center, inflation with compressed air inside blow mold; right, opening mold and ejection of product (Belofsky, Harold, *Plastics: Product Design and Process Engineering*, page 353)

6.2.2   Assembly Considerations

As the traditional bottle-making process, the creation of molded-in threads for the assembly of bottle caps is a highly refined practice. The bottle cap business itself is an extremely competitive one, and a wide variety of stock caps are available in both simple and highly sophisticated patented child-resistant, tamper-evident, and dispensing models. For custom caps, standard bottle finishes with established tolerancing exist which are readily accessed from cap and bottle suppliers. Neck finishes are the tightest tolerances held in blow molding. Industrial blow moldments are typically large hollow parts designed to combine several smaller components into one and eliminate a number of assembly operations. They are often held in place with straps or fasteners, the holes for the latter being cut in secondary machining operations, since it is not possible to mold in an open hole. A hole can be molded through a part, but this type of detail is usually reserved for larger openings such as bottle handles. If through holes are to be molded in place, their size, number, and location become an issue. Some fuel tanks are designed to stay in place because of gravity, an arrangement that helps to accommodate changes in volume through filling and thermal expansion. Crossed slots are used to permit the large tolerances necessary.

Assembly method selection decisions need to take into account the fact that tight tolerances are not economically feasible. Only the outside dimension is controlled by the mold, the inside size being determined by the outside dimension combined with the wall thickness. Since the wall thickness is determined by the stretching of the plastic into the contours of the mold, it is not uniform. Tight tolerances are best accomplished by post-molding machining operations.

Industrial blow molding applications requiring more than one mold are unusual; therefore mold-to-mold variation is not normally a problem. However, the outside dimensions are determined by the temperature of the melt, the pressure on the parison, and the cooling parameters. Thus, they cannot be held to the tight tolerances typically required by fine press and snap fits. Such fitments are sometimes accomplished on a crude scale by taking advantage of the flexibility provided by the thin walls attainable through this process, and therefore a limited amount of undercut is possible.

On occasion, press or snap fits are achieved by molding both parts as one piece. A segment between the two parts is then removed and they can then be fitted together. Since the two parts are molded as one piece, they mold oversize or undersize together. This practice, which avoids the problem of part-to-part variation, is often followed for rotationally molded parts as well.

As a thermoplastic process, blow molding can produce parts that can theoretically utilize most of the heat welding processes. In practice, however, the nature of the process does create some limitations. The precise detail required for the energy directors used in ultrasonic welding, for example, simply is not possible. For an interior wall, it would also be very difficult to provide the level flatness necessary for hot die or vibration welding.

Two-color blow molding can be achieved by means of coextruded parisons. Labels can also be molded into the part. Molded-in inserts are possible, with limitations, but postmolded inserts can also be emplaced with heat or ultrasonic welding. Adhesives could be used as well; however, since the bulk of blow molding is done with polyolefins, which accommodate adhesives only with difficulty, this is not often done. In the case of polyolefins, hot melt adhesives can be used.

Solvent welding falls into the same category. To determine whether a given material can be solvent-welded, refer to the listing for that material in Chapter 5, "Assembly Methods by Material."

6.3   Casting, Potting Encapsulation, and Embedment

6.3.1   The Processes

A variety of processes use both open and closed molds and are free of pressure or vacuum. The resin is composed of two components that react when mixed. They are poured into the mold and allowed to set by chemical reaction, as shown in Fig. 6-2.

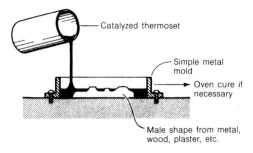

Catalyzed thermoset

Simple metal mold

Oven cure if necessary

Male shape from metal, wood, plaster, etc.

Figure 6-2 Casting of thermoset plastic (Rosato, D. V., *Rosato's Plastics Encyclopedia and Dictionary*, page 92)

Because of its tooling cost, this process is sometimes used for prototypes. Its principal materials are epoxy, polyester, silicone, urethane, and acrylic. Castable elastomers include room temperature vulcanizing rubbers, flexible urethanes, flexible silicones, and polysulfides.

6.3.2   Assembly Considerations

Casting is used as an assembly method in itself for applications that require an object or reinforcement to be emplaced within the moldment. Sometimes referred to as embedment, it is the only process by which objects can be completely embedded within a plastic product. It is also known as potting, impregnation, and encapsulation. Two-color casting is readily accomplished.

Casting is frequently used as a low pressure, closed-mold process; however mold costs are kept low because it is often possible to cast the mold directly from the model. Casting cycles are long and, consequently, piece part prices are high. The same methods are used for the assembly of castings and for machined thermosets. Although inserts can be easily molded in place, they are difficult to replace in the event of a reject moldment because the material is usually a thermoset and the inserts cannot be melted out. This aspect severely limits the heat weldability of the materials used in casting.

Since casting is a low volume process, the high volume assembly methods, which require expensive tooling, are not normally employed. The usual assembly methods are fasteners, typically in holes drilled after molding as casting tolerances must be very loose. Undercuts and side-cored holes for snap fits are not possible. It is difficult to cast threads and not easy to control wall thicknesses.

6.4   Coextrusion

See Section 6.8.2.

6.5   Co-Injection Molding

See Section 6.19.

6.6   Cold Press Molding

6.6.1   The Process

Cold press molding is a thermoset process that utilizes a closed mold. Reinforcements in the form of mats or preforms are placed in the mold. A predetermined amount of resin is sprayed or poured into the mold, and the press is closed. Pressures are low and cycles generally range from 3 to 6 min. The process is similar to compression molding (see below: Fig. 6-3) except there is no heat applied.

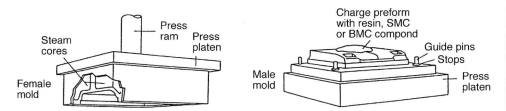

Figure 6-3   Compression or matched die composite molding (Belofsky, Harold, *Plastics: Product Design and Process Engineering*, page 446)

In the thermoset pecking order, this process falls between resin transfer molding (RTM) and compression molding for general applications when reaction injection molding (RIM) is omitted (owing to material limitations). Its tooling costs are roughly equivalent to those of RTM. However, it is not as well suited to intricate part design details such as undercuts and strategically located additional reinforcements.

6.6.2   Assembly Considerations

Cold press molded parts are often used as structural components of assemblies where another thermoset process (e.g., RTM or SMC) provides the class A outer section. Adhesives are the usual assembly method. Solvents are not generally selected because the materials used for cold molding are not readily solvent weldable.

Since cold press molding is a thermoset process, none of the heat-based assembly methods are usable. That eliminates hot gas welding, hot plate/fusion welding, induction/electromagnetic welding, ultrasonic welding, vibration welding, and staking. A version of staking in which a press-on fastener is positioned over a post is often used on thermoset parts. Such posts may be brittle, however, because they are "resin rich," – that is, they contain little glass fiber reinforcement.

Inserts cannot normally be molded in. However, under certain circumstances they can be molded on a limited basis. For example, labels can be molded in, but there can be no two-color molding. While postmolded inserts cannot be emplaced with heat or ultrasonics, press-in inserts can be and usually are installed with adhesives. Also, expansion, helical coil and self-tapping inserts can be used. Threads can be molded in on a very limited basis.

Translated, the foregoing limitations mean that press and snap fits are not often used. Also, molded-in tolerances are fairly difficult to control. Wall thicknesses require a tolerance of ±0.50 mm (±0.020 in.). Fasteners are the most commonly used means of assembly for applications that do not use adhesives.

6.7   Compression Molding

6.7.1   The Process

One of the oldest processes known to the industry, compression molding has spawned a number of variants. As shown in Fig. 6-3, the resin is placed in a heated mold and the mold is closed. Compressive pressure is applied and it, along with heat, causes the material to cross-link so it will not return to its original shape. The part is then removed from the mold and the cycle is repeated. Although nothing prevents compression molding from being used for thermoplastic materials, we think of this process and its kin as the principal means of processing thermosets.

Unreinforced or short fiber reinforced, the compression molding process finds it largest market in products like electrical components and such temperature-resistant housewares as plates and ashtrays.

For long fiber reinforced materials, the compression molding processes are usually referred to by the type of compound used.

6.7.2   BMC: Bulk Molding Compound

In BMC applications, the 0.25 to 0.5 in. long strands of reinforcement are placed in the material along with other additives. The mixture is then formed in the shape of a ball, slab, or log and placed in the mold. This is the least costly method of adding reinforcement and is used when parts of intermediate strength are called for.

6.7.3   SMC: Sheet Molding Compound

Reinforcement fibers range from the very smallest to those of indefinite length, although they seldom exceed 3 in. The fibers are spread into a resin paste to form a sheet, a predetermined amount of which goes into the mold. SMC is used for higher strength applications such as the truck tractor hoods and fenders referred to earlier.

A recent development in SMC is the availability of molding compounds that can be processed at compression pressures 20 to 40% of normal. The advantage of using such materials is an increase in the size of parts that can be molded in a press of a given size. A resin of this type is known as low pressure molding compound (LPMC) or low pressure SMC.

Compression molding is a process that offers economies of high volume. However, it requires a closed mold (composed of two halves) that must be strong enough to withstand high pressures. Therefore, the tooling is costly and usually has a long delivery time.

6.7.4   Assembly Considerations

Since compression molding is a thermoset process, none of the heat-based assembly methods (i.e., hot gas welding, hot plate/fusion welding, induction/electromagnetic welding, ultrasonic welding, vibration welding, and staking) are usable. A fastener approximation of staking in which a press-on fastener is positioned over a post is often used on thermoset parts. Such posts, however, may be "resin rich," or low in reinforcement content. When polyester is the material, the lack of reinforcement will make the parts brittle.

Undercuts are possible in SMC and BMC in a limited fashion. Therefore, undercuts are not used without a very special reason. Holes can be side-cored, permitting the use of snap fits in some cases. Tolerance control is good, thus press fits are possible.

Inserts can be readily molded in and often are; this is a time-consuming process, however, and the inserts in a reject part are not economically extracted. Thus molded-in inserts add a good deal of cost to the part. While postmolded inserts cannot be emplaced with heat or ultrasonics, press-in inserts can be installed with adhesives. Also, expansion, self-tapping, and helical-coil inserts can be used, and threads can be molded in to a limited degree. Labels can also be molded in. Two parts can be molded together if absolutely necessary, but joining is more efficiently accomplished by other processes.

Although both side-core holes and those in the compression direction can be molded in, it is usually better to machine them in after molding. Molding in a hole usually leaves the portion of the part beyond the hole without reinforcement, and greater precision can be achieved through machining. Fasteners are the most commonly used means of assembly for applications that do not use adhesives.

Adhesives are an important means of assembly for compression-molded parts. However, only those made of phenolic are readily solvent-welded, and then only to phenolic itself and to nylon.

6.8   Extrusion

6.8.1   The Process

Extrusion is a basic process in which resin from a hopper is fed into a heated chamber known as a barrel. A turning screw moves the material down the barrel, melting it as it goes. It then passes through a die, which creates it shape. Beyond the die is a cooling area, which causes the extrusion to keep its shape.

As illustrated in Fig. 6-4a, the equipment is composed of a heated extruder barrel with a hopper at one end and a die at the other. The material enters the barrel through the hopper. As the screw turns, the resin is mixed and plasticized as it moves toward the die.

The material moves through the screens and the breaker plate to the die. The die sets its shape, and the product now passes through a water bath to the water quench tank. Upon exiting the quench tank, the product has attained its final shape. Depending on the nature of the product (film, sheet, tubing, profile, etc.), it is cut to length and stacked or coiled. The full extrusion line is shown in Fig. 6-4b.

Profiles can be extruded with an integral hinge, thus eliminating the need for a separate hinge and its assembly. If the associated parts need to be more rigid, they can be molded

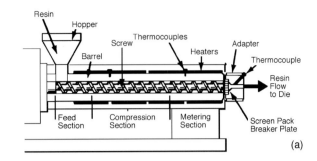

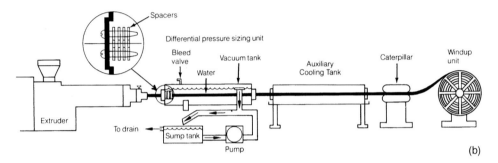

Figure 6-4   (a) Cross-section schematic of a single-screw extruder and (b) Example of calibration systems for pipe/tube extrusion line (Rosato, D. V., *Rosato's Plastics Encyclopedia and Dictionary*, pages 240 and 255)

of another material and force-fit through a tongue and groove. In some cases, it may be more desirable to join the parts to the hinge with solvents or adhesives.

To a limited degree, hollow parts beyond simple tubing can be extruded – thereby eliminating assembly operations. To use this process, product profile must be uniform. Extrusion blow molding or twin-sheet thermoforming processes should be considered if this condition cannot be met.

The extrusion process can create a variety of continuous shapes from soda straws to sheeting. Rigid materials must be cut to shape, but flexible materials can be rolled. This is essentially a thermoplastic process, although thermoset materials can be extruded with special equipment. This process is important as the first stage of the extrusion blow molding process. It is also the usual source of sheet goods for the thermoforming process.

6.8.2   Coextrusion

Laminates can be created when two compatible resins, coming from different extrusion barrels, are extruded through the same die. This process, known as coextrusion, uses heat to bond the layers together. It is capable of producing laminates of several layers.

Coextrusion basically uses two extruders with compatible materials or the same material in a different color. The resins are placed in a position that flows the two melts together in a way that results in one being welded to the other. Thus two parts that would otherwise have to be assembled are joined in the processing stage.

If one layer is placed over the other, it can form a barrier (known as a "cap sheet"), which will protect the base material from ultraviolet light, moisture vapor transmission, etc. Multiple layers are often used for sheet destined to be thermoformed for food and pharmaceutical packaging. In some cases, a rigid material is placed over a flexible material or vice versa.

The two materials can also be coextruded side by side – for example, in containers designed with one half in a color and the other in a clear material. Such a container is thermoformed to ensure that the colored section becomes the bottom part of the container and the clear section becomes the top. A hinge is formed between the sections on one side and, through the use of very thin gauge sheet or film, enough flexibility is attained to permit snap fits to be molded in. In this manner, the container is manufactured with no additional assembly operations except to close it when filled.

6.8.3   Assembly Considerations

From the assembly point of view, dimensional stability is good except in the cut-off direction. Extrusion can make shapes in the form of any letter of the alphabet in one stroke, a task that otherwise would require fabrication from several parts. Tubing can be coextruded into round, square, rectangular or free-form shapes. A semi-flexible material can be coextruded to make two shapes connected with a hinge. The only prerequisite is that the profile be continuous in one plane. Shapes in that plane and holes must be machined into the part. Side-cored holes and undercuts in the cross direction are impossible. Two-color extrusion is readily accomplished with coextrusion.

Since extrusion is a thermoplastic process, it can produce parts that can theoretically utilize most of the heat processes. In practice, however, the nature of the process does create some limitations. The kind of sharp detail required for the energy directors used in ultrasonic welding is very difficult, though perhaps not impossible, to make. Since one end is a cut edge, extrusion could theoretically be used for ultrasonic shear welding. However, that would require machining to maintain the necessary tolerance control. It would be an extraordinary application in which the extrusions were cut short enough for that edge to be used in this fashion. Vibration welding is also theoretically feasible, however it is difficult to imagine a configuration that would permit the driving surfaces necessary for this process. Spin welding could be accomplished either into a drilled hole or into the end of an extruded tube; however, a secondary machining operation would be required unless the spin welding fitment were in the mating piece.

A variety of hot plate/fusion welding is employed in the manufacture of bags used in packaging. The die seals and cuts thin-gauge extruded tubing in a process that can be highly automated. Heavier gauge, rigid tubes are often sealed with injection-molded caps, which are press-fit onto their ends. There is nothing to prevent the application of these concepts to other shapes as well.

Plastic can be extruded around an insert (e.g., wire). Extruded-in threaded inserts, on the other hand, are simply not possible. Post-extruded inserts could be employed, but the holes would need to be drilled in a secondary operation or placed in the ends of the extrusion. Unless the fastener was also to be attached to the end of the extrusion, holes for fasteners would likewise need to be drilled as a secondary operation. Inserts could be emplaced in drilled holes by heat or ultrasonic methods. Molded-in threads are impossible.

Welding could be accomplished with hot gas or solvents if the materials permit. Electromagnetic/induction welding, though theoretically possible, might not be economically feasible because the fitment would have to be machined in as a secondary operation.

6.9   Filament Winding

6.9.1   The Process

Filament winding is a specialized process traditionally used to create pressure vessels. Resin-impregnated or bath-dipped filaments (usually glass) are wound around a mandrel, as illustrated in Fig. 6-5. A variety of winding configurations, like those in Fig. 6-6, are

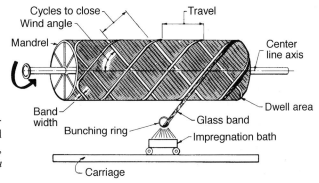

Figure 6-5 Schematic of filament winding using a helical winding pattern (Rosato, D. V., *Rosato's Plastics Encyclopedia and Dictionary*, page 289)

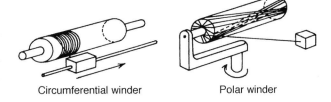

Figure 6-6 Schematic representation of circumferential and polar types of filament placement (Rosato, D. V., *Rosato's Plastics Encyclopedia and Dictionary*, page 289)

used, depending on the application. Other high strength applications beyond pressure vessels are rocket motor casings, rocket tubes, helicopter blades, automobile leaf springs, and aircraft parts. Small diameter pipe rated at 2000 psi is used in the chemical industry. Filament winding offers maximum strength with low investment.

6.9.2   Assembly Considerations

Assembly methods for filament winding are very limited. Although dimensions can be held reasonably well for fitments, none of the usual devices (undercuts, molded-in threads, side-cored holes) are possible. Since the materials are thermosets, none of the welding techniques are feasible either. There is no two-color filament winding.

Adhesives are the principal means of assembling filament-wound parts; however, metal inserts can be wound into the part in some cases. Also the outer contour can be shaped to provide a gripping shoulder.

6.10   Gas-Assisted Injection Molding

See Section 6.12.

6.11   Gas Counter Pressure Structural Foam Molding

See Section 6.19.

6.12   Injection Molding

6.12.1   The Process

Although a greater volume of material is processed by extrusion, the plastic process with far and away the widest number of design applications is injection molding. In certain respects, the two processes are similar in that the resin is fed through a hopper into a heated chamber in both cases. A rotating screw moves the resin toward the nozzle end. In injection molding, the entire screw moves forward pushing a "shot" of plastic into a mold under pressure. A backflow check valve prevents the material from pulling back into the chamber when the screw is retracted. An injection molding machine is illustrated in Chapter 15 (Fig. 15-17).

The mold capabilities, broad availability of materials, and extreme controllability are the keys to the success of this process. Moldmakers have succeeded in constructing incredibly complex molds with side cores coming in from virtually any angle conceivable. Collapsing cores are no longer a rarity and with special equipment, even thermosets can be injection molded. With combinations of heat and pressure, molders have been able to successfully fill these molds and create parts that combine many parts into one with the accompanying savings in assembly costs (although the tooling can be very costly). The widest possible material availability has created great opportunities for product designers.

In recent years, however, there have been a number of applications that attempt to do so much that they fail to attain their objectives. This can happen when molds are too costly to be written off in a reasonable length of time, are too costly to maintain at an acceptable level, or are too complex to permit a profitable number of cavities to be kept in operation.

Gas-assisted injection molding is a variation of this process in which a gas is injected into the center of the wall as the material enters the mold or immediately thereafter. This creates a hollow core in the center of the thicker sections, thus permitting a far wider variation in wall thickness than is obtainable with traditional injection molding, which in turn provides greater stiffness and easier material flow. Another benefit, assuming that wall thickness is not increased to accommodate the process, is reduction in weight and material usage. Gas-assisted injection molding has a slower cycle than standard injection molding, but it does offer an advantage over the foam processes. It might be used, for example, to change the "feel" or "heft" of a product component such as a steering wheel or a vacuum cleaner handle.

6.12.2   Assembly Considerations

These comments refer to thermoplastic injection molding. For a discussion of the assembly considerations for thermoset injection molding, please refer to Section 6.22: "Transfer Molding."

Virtually all the assembly methods can, in some way, be utilized for injection-molded products. Size is probably the major limitation. As the parts grow larger, the number of molding machines available to make them diminishes significantly. Garbage cans are injection-molded, but the number of custom molders who can quote on sizes in that range is quite limited. Even larger injection-molded parts have been manufactured, however those are usually for special applications, and the equipment that made them may have been specially built for a proprietary manufacturer.

Thermoplastic injection molding is capable of undercuts, side-cored holes, controlled wall thicknesses, extraordinary detail, and the tightest molded-in tolerances (machining can be tighter); thus press fits and snap fits are readily accomplished.

Injection molding is capable of molding two, or even three, parts in the same moldment. Although inserts can be molded in, they can significantly slow the molding cycle if too many are used in a single moldment. Shuttle presses, sometimes employed in these cases, permit the inserts to be loaded on one side of the mold while the part is being molded on the other. Inserts can be economically retrieved from reject thermoplastic parts. Postmolded inserts can be emplaced by either heat or ultrasonic vibrations. Expansion inserts can be used, but usually are avoided because the other methods are more efficient. The same can be said for adhesive emplacement of inserts.

Both self-tapping and thread-forming screws can be used, the type depending on the flexural modulus of the particular material. There are a number of specialty fasteners developed for certain applications. Male and female threads can be molded in. Injection molding is the principal method of manufacturing bottle caps.

Hot die/fusion, electromagnetic/induction, spin, vibration, and ultrasonic welding can all be used for injection-molded parts. Hot gas welding can be used, but is not often employed because the process is best suited to volumes lower than those typical of injection molding. The use of solvent welding is widespread for small parts, however its suitability is dependent on the material.

6.13   Lay-up and Spray-up

6.13.1   The Processes

Lay-up and spray-up are the processes by which the largest parts are made. An open mold is used, the limits of which have not yet been reached. The polyester laminate construction is composed of a surface or gel coat, a layer of reinforcement, and an interior coat of resin. In some cases, additional layers are provided for high strength requirements. Mold costs are low. Since, however, these processes are very slow, many molds would be required for high volumes.

The terms "lay-up" and "spray-up" refer to the method by which the reinforcement is applied. In lay-up, the reinforcement mats are simply laid in the mold over the gel coat. In spray-up, which is illustrated in Fig. 6-7a, glass rovings are sprayed over the gel coat. After the material has been applied, air bubbles are removed with a hand roller, as in Fig. 6-7b. Spray-up is much faster, but higher strengths can be achieved with lay-up. In either case, the process provides a part with a smooth surface on only one side.

Because the mold is open, it has only one side. Consequently, no internal detail can be molded in. Ribs and other internal structures must be emplaced with adhesives or screws. However, the sophisticated placement of reinforcement can often eliminate the need for some of the internal structure, thereby reducing the overall cost of the product.

6.13.2   Assembly Considerations

From an assembly standpoint, the principal advantage of the lay-up and spray-up processes is their ability to combine parts through size alone. Products that would have to be made of multiple components assembled in some fashion can be molded in one piece. For example, lay-up and spray-up have virtually eliminated the competition for large boat hulls. It is the rare vessel that is made up of individual sections in today's market. Even 150 ft long mine sweepers now have hulls made in this manner.

That is not to say that the goal of combining parts is easily attained. Delamination can occur in drilling or trimming. Tolerances need to be generous for these processes, and wall thicknesses are difficult to hold within ± 0.50 mm (± 0.020 in.). Handling of the finished goods can be a major expense, and the problem grows with increased size.

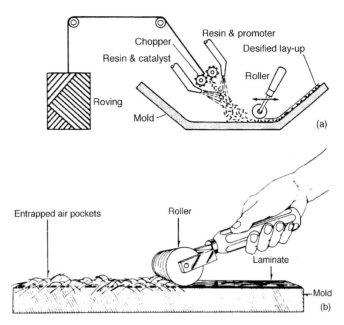

Figure 6-7   (a) Schematic of a two-pot airless sprayup system and (b) Use of roller to compact sprayup (Rosato, D. V., *Rosato's Plastics Encyclopedia and Dictionary*, page 707)

Parts made by these processes are usually quite large. Hence, press and snap fits are not normally applicable, although undercuts are feasible. Side-cored holes are not possible. Adhesives and fasteners are the customary assembly methods.

Threads cannot be molded in with these processes, however inserts can be molded in. The molding cycle is so long for these methods that the time required to place the inserts in the mold is relatively negligible. While postmolded inserts cannot be emplaced with heat or ultrasonics, they can be pressed in, in conjunction with adhesives. Also, expansion, helical coil, and self-tapping inserts can be used. Holes for fasteners cannot be readily molded in. They are better machined in after molding. It is not possible to mold in a second color, but labels can be molded in.

None of the heat-based assembly methods (hot gas welding, hot plate/fusion welding, induction/electromagnetic welding, spin welding, ultrasonic welding, vibration welding, and staking) are usable with lay-up and spray-up, since these are thermoset processes.

6.14   Machining

6.14.1   The Process

The majority of machining applications for plastics are as secondary operations to the primary plastics fabrication processes. That aspect is discussed separately in each of those sections. These comments mainly refer to machining when it is used as a primary process.

Often designers and engineers become so preoccupied with the various processes particular to plastics that some of the traditional methods are forgotten. Plastics can be machined just like other materials, and often with less scrap, since the stock is more readily formed to a size closer to the finished shape.

6.14.2   Thermoplastics

Most of the same machining processes applied to other materials can be used with thermoplastics; however, there are some limitations. Obviously, the heat generated by the machining operations can melt the plastic. Thus drills, grinding wheels, etc. will load with the melted plastic. This material must be kept to a minimum and removed frequently to maintain controlled dimensions. If the heat does not melt the plastic, it may still cause thermal expansion and create cutting difficulties. Also, many plastics are so flexible that they cannot be readily positioned for machining without deforming. Furthermore, they are subject to deflect away from the cutting tool when pressure is applied. Freezing the material prior to machining can sometimes be of help in resolving this problem. Some plastics are very brittle and must be machined with care.

The greatest limitation, however, is cost. Even the lowest priced plastics are expensive in comparison to most woods or metals. Plastics produce lower cost products because they can be worked with processes that use the minimum amount of material with little or no scrap production and because they offer considerable reductions in assembly and decorating operations. Utilizing the same processes as the less costly competing materials makes no economic sense unless the special properties of a given plastic are required.

With the ready availability of most machining equipment and the absence of tooling requirements, many plastic prototypes are machined. While the author has himself employed this technique on many occasions to resolve fitment, appearance, and ergonomic issues, the reader is warned that absolutely no strength or endurance testing should be done with parts made other than by the ultimate production process. The reason is that the internal structure and stresses will be significantly different from those resulting from machining operations. Thus, decisions based on the results of such testing will not be reliable. Most processes have developed prototyping methods that provide much more reliable results. However, even these need to be confirmed with production parts to avoid the development of problems further down the production road.

Probably the most common use of machining of plastics is as a necessary adjunct to normal plastics processes; such uses are discussed in connection with each process. Combining machining with other processes can also result in tooling benefits, however. For example, a drilled hole can eliminate a costly side action in a molded part. This approach can be useful for low volume applications that cannot justify expensive tooling or in reducing the "entry price" for a new product with a high risk factor.

6.14.3 Thermosets

Like thermoplastics, thermosets can be machined. In fact, most of the thermosets are somewhat easier to machine than the thermoplastics in that the melting problem is less of a factor. While localized heat at the machining surface can still be a problem, the temperatures are much higher, and charring or burning is the likely result.

A considerable amount of machining is done with thermosets because, with only a few exceptions, the mechanical removal of molding flash is necessary in thermoset processes. Drilling holes and cutting openings also are commonplace in some processes because it is difficult, bordering on impossible, to mold them in.

6.14.4 Assembly Considerations

The principal advantages of machining are as follows: it can provide the highest level of accuracy, and it requires little in the way of tooling expense. Machining can be used for both thermoplastics and thermosets. When the material is a thermoset, none of the heat-based assembly methods are usable. That eliminates hot gas welding, hot plate/fusion welding, induction/electromagnetic welding, spin welding, ultrasonic welding, vibration welding, and staking. Basically, that leaves adhesives and fasteners as the principal means of assembling thermoset machined parts. Holes may be tapped, or self-tapping screws may be used. Inserts can be self-tapping, helical coil, press in, of the expansion variety or emplaced with adhesives.

For all practical purposes, the foregoing statements could be applied to thermoplastic machined parts as well, but for a different reason – namely, because economic considerations are such that machining is usually a low volume process. Except for hot gas welding and spin welding, most of the heat processes do not lend themselves well to the typical volumes of the machining process.

Hot plate/fusion welding could be done in a crude manner without much investment, but production tooling would be volume sensitive. Ultrasonic energy directors would be far too costly to machine in on a production basis and very difficult to accomplish even on a prototype basis. Both heat and ultrasonics could readily be used to emplace threaded inserts. Although electromagnetic/induction and vibration welding are theoretically possible, it is difficult to imagine a circumstance in which these techniques would be economically feasible.

6.15 Pultrusion

6.15.1 The Process

By providing for the fiber reinforcement of profile shapes, pultrusion permits the use of such shapes in structural applications. Glass fiber strands are pulled off spools or racks through a resin bath, where they receive a coat of resin. They then pass into a heated die, which gives them their shape while curing them. Once cured, they are cut to the desired length. The complete pultrusion line is shown in Fig. 6-8.

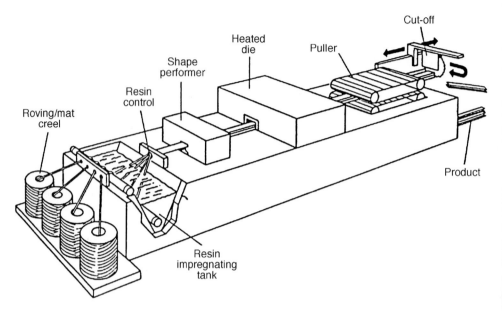

Figure 6-8   Schematic of a pultrusion system. Resins, such as polyester and epoxy, that cure without producing a condensation by-product are most suitable. Fiberglass cloth and surfacing veils may be added as required to the strands or rovings (Belofsky, Harold, *Plastics: Product Design and Process Engineering*, page 454)

Developed principally for thermosets, pultrusion has some thermoplastic applications. Industrial products made from this process include light poles, wind turbine blades and structural beams. Fishing rods, ski poles, and flag poles are some of its consumer products. The most notable attribute of this process is its ability to produce a part with extraordinary resilience. Pultruded products can absorb considerable deflection and return to their original shape.

Extrusion is usually the process of choice for profile applications unless the strength requirements are too great. In that case, the reinforcement available through pultrusion may be necessary. The processes compete when a high strength thermoplastic is called for, since at times extrusions are more costly than pultrusions.

6.15.2   Assembly Considerations

Since pultrusion is primarily a thermosetting process, none of the welding processes are available. Furthermore, pultruded shapes cannot be stress-relieved in notches and holes. In some cases, holes and reinforcements in the pultrusion direction can be incorporated. Pultrusion has no two-color capability. Dimensions can be held to fairly close tolerances. Fasteners and adhesives are the principal means of assembling pultrusions. Expansion, self-tapping, helical coil, and press-in inserts can be used. However, the latter should be reinforced with adhesives.

6.16   Reaction Injection Molding (RIM)

6.16.1   The Process

The RIM process is unique in that it is actually two processes in one. Not only is the part made, but the material is made as well. This is accomplished by mixing two or more liquid components that react with each other by high pressure impingement, as illustrated in Fig. 6-9. The resulting mixture is injected into the mold at pressures under 100 psi, where it completes the reaction. This low pressure permits the use of closed molds of lower strength than can be specified for compression or injection molding. Therefore, they are less expensive. The larger the mold, the greater this advantage becomes. Molding cycles are longer, however, resulting in higher piece part cost.

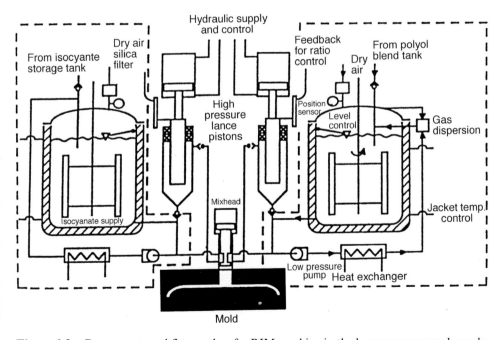

Figure 6-9   Components and flow paths of a RIM machine in the low pressure recycle mode. The low pressure recirculation paths are bounded by the dashed lines (Belofsky, Harold, *Plastics: Product Design and Process Engineering*, page 449)

The real drawback to RIM is material limitation. The principal material available is thermosetting polyurethane. Epoxy, nylon, polydicyclopentadiene (cross-linked polyethylene), and polyester are also used to a limited extent.

This process is best known for its applications in the automotive field, where a variety of parts such as fenders have been made. However, there have been a number of successful applications for large commercial parts. Reinforced reaction injection molding (RRIM) is a process variant that uses reinforcement in the form of a chopped glass preform.

6.16.2   Assembly Considerations

RIM is a thermoset process, and thus none of the heat-based assembly methods are usable. That eliminates hot gas welding, hot plate/fusion welding, induction/ electromagnetic welding, ultrasonic welding, vibration welding, and staking. A version of staking using a press-on fastener over a post is often used on thermoset parts.

Inserts can be molded in, however this time-consuming process increases the molding cycle, and the inserts in a reject part are difficult to remove. Postmolded inserts can be pressed in with adhesives but cannot be installed with heat or ultrasonics. Expansion, self-tapping, and helical coil inserts can also be used.

Holes for fasteners can be molded in, both in the direction of draw and as side cores. Threads can also be molded in. Fasteners and adhesives are the principal means of assembling RIM parts.

Undercuts are possible, but expensive. Therefore, they are not used without a very special reason. Also, tolerances can be held fairly well. Several of the RIM materials are sufficiently flexible to permit press and snap fits, but these assembly methods are not often used, since parts made from this process are usually too large to lend themselves well to these methods. Reaction injection molding does not have a two-color capability.

6.17   Resin Transfer Molding (RTM)

6.17.1   The Process

RTM is a closed-mold process that is basically a step up from lay-up and spray-up in terms of its ability to produce parts of greater complexity and wall thickness consistency. Reinforcement, which can be in the form of mats or preforms, is placed in the mold and the mold is closed. Polyester is then injected into the mold at low pressure. The process is illustrated in Fig. 6-10a. The equipment is shown in Fig. 6-10b.

This low pressure process permits the use of relatively inexpensive molds. Molding cycles, however, are quite long. Thus the advantage of low mold cost is lost as the

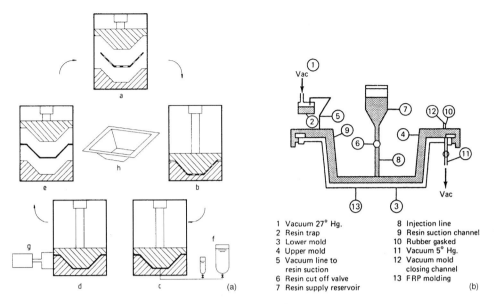

1 Vacuum 27° Hg.
2 Resin trap
3 Lower mold
4 Upper mold
5 Vacuum line to resin suction
6 Resin cut off valve
7 Resin supply reservoir
8 Injection line
9 Resin suction channel
10 Rubber gasked
11 Vacuum 5° Hg.
12 Vacuum mold closing channel
13 FRP molding

Figure 6-10 (a) Process cycle in RTM: a, placing reinforcement or preform; b, closing the mold; c, injection of liquid plastic mix; d, heat curing of plastic; e, demolding; f, material storage (view not shown); g, thermal control of mold; and h, molded part removed from RTM process and (b) Schematic of RTM with plastic liquid entering from top of mold (Rosato, D. V., *Rosato's Plastics Encyclopedia and Dictionary*, pages 644 and 645)

piece part price penalty makes itself felt. Therefore, RTM is an excellent process for low to moderate volumes that require a smooth surface both inside and out.

6.17.2 Assembly Considerations

Since RTM is a thermoset process, none of the heat-based assembly methods are usable. That eliminates hot gas welding, hot plate/fusion welding, induction/electromagnetic welding, ultrasonic welding, vibration welding, and staking. Dimensions can be held fairly well, and undercuts are possible with side cores or split molds. Generally, however, press and snap fits are not recommended. RTM does not have a two-color capability.

Inserts can be molded in reasonably well, but this is a time-consuming process, and the inserts in reject parts are not economically extracted. Thus, molded-in inserts add a good deal of cost to a part. Expansion inserts or self-tapping inserts can be used, and press-in inserts can be installed with adhesives. Helical coil inserts can also be used. In some cases, threaded holes for fasteners in the molding direction and side-cored holes can be molded in, but the practice is not recommended. Holes are usually better machined in after molding. Two-color molding is not possible, but labels can be molded in.

6.18   Rotational Molding

6.18.1   The Process

Rotational molding is a low pressure thermoplastic process used to make hollow parts. The principle of rotational molding dates back to the molding of hollow chocolates in the early part of the twentieth century. It has a long history in the plastics industry as the vinyl plastisol process. In theory, all thermoplastics can be rotationally molded. However, the principal commercial applications are for PVC, polyethylene, polypropylene, and polybutylene.

As shown in Fig. 6-11a, there are three stations to rotational molding. This process utilizes a closed female mold. At the first station the mold is opened and filled with a

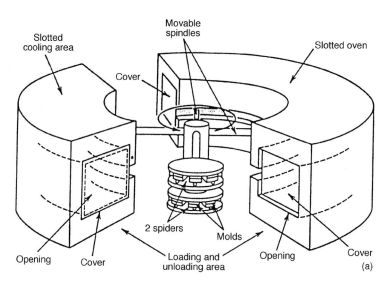

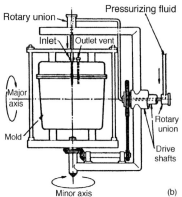

Figure 6-11 (a) Schematic rotational molding machine. Multiple spindles carry the molds through three stations to accomplish charge loading (stage 1), oven rotation (stage 2), demolding (stage 3) and recharging (back to stage 1) (Belofsky, Harold, *Plastics: Product Design and Process Engineering*, page 385) and (b) Feeding inlet to form hollow product inside a closed mold while the mold is rotated about two axes and heat is applied. This system permits molding multiplastic layers of different materials (corotational molding) (Rosato, D. V., *Rosato's Plastics Encyclopedia and Dictionary*, page 655)

predetermined weight of powder or liquid. It then swings to the heating station, where it is then rotated around its primary and secondary axes as in Fig. 6-11b. The oven heats the mold which, in turn, heats the powder. As the powder melts, it deposits uniformly on the interior walls of the mold. When all the material has been deposited on the walls, the mold then swings to the cooling station, where it is cooled by moving air, water fog, or water spray. When this stage is completed, the mold is then swung back to its original position, where the part is removed and a new charge of powder is placed in the mold.

The mold cost for this low pressure process is low. Unfortunately, there is a long molding cycle. Thus, the piece part price can be high for very small parts unless the part lends itself to a large number of cavities. This process is well suited to large parts.

Rotational molding is best known for large hollow parts, however it can also produce large double walled panels as thin as 5 times the nominal wall thickness.

Besides its hollow part capability, rotational molding can create open parts by molding two or more at once and cutting the moldment into its individual parts. The two parts might be the halves of a housing or the body and lid of a container. When entire units are molded at once, multiple assembly operations are eliminated.

Rotational molding can produce parts with outside radii as low as 0.13 in. with difficulty, however 0.375 in. is preferred. Inside radii can be as small as 0.06 in. No inside structures such as ribs can be produced, but hollow ribs that must show from the outside can be made. Strength can be further increased through kiss-off (tack off) ribs which meet at the center of the part. The hollow space can also be filled with foam for additional stiffness.

6.18.2   Assembly Considerations

The assembly aspects of rotational molding are similar to those for blow molding. Both processes make a hollow part, however blow molding is a high volume process, with the bulk of its applications of limited size, whereas rotational molding is typically used for larger parts of more modest volume. Like blow molding, rotational molding is often used for containers. There are stock fitments and caps available for these vessels, which are spin-welded in place, provided they are not made of cross-linked polyethylene. That material is a thermoset, and parts are normally assembled with fasteners.

Like industrial blow moldments, rotational molded parts are typically large parts designed to combine several smaller components into one and eliminate a number of assembly operations. They are often held in place with straps or fasteners, the holes for the latter being accomplished through secondary machining operations. It is possible to mold an opening through the part, but this type of detail is usually reserved for larger openings such as handles. If such openings are to be molded in place, their size, number, and location become an issue. Some fuel tanks are designed to stay in place because of gravity, an arrangement that helps to accommodate changes in volume through filling and thermal expansion. Crossed slots are used to permit the large tolerances necessary. Actual holes can be molded in, but only in a crude fashion. They are not recommended.

Tight tolerances are not economically feasible, the usual tolerance being on the order of 0.010 in./in. Only the outside dimension is controlled by the mold, the inside size being determined by the outside dimension combined with the wall thickness. Since the wall thickness is determined by the adhesion of the melted plastic into the contours of the mold, it is not uniform. Expect a variation of $\pm 20\%$. Tight tolerances are best accomplished by post molding machining operations, and this production detail needs to be considered during the planning stages.

Rotational molding applications requiring more than one cavity are not typical, therefore, cavity-to-cavity variation is not normally a problem. However, since the outside dimensions are difficult to control, they cannot be held to the tight tolerances typically required by fine press and snap fits. Undercuts, however, are possible, and these fits are sometimes accomplished on a crude scale by taking advantage of the flexibility and high shrinkage provided by the softer materials usable with this process.

On occasion, press or snap fits are achieved by molding both parts as one piece. A segment between the two parts is removed, and they can then be fitted together. Since the two parts are molded as one piece, they mold large or small together. In that way, they avoid the problem of part-to-part variation. This practice is often followed for blow-molded parts as well.

Rotational molding, being a thermoplastic process, can produce parts that can theoretically utilize most of the heat processes. In practice, however, the nature of the process does create some limitations. The detail required for the energy directors used in ultrasonic welding simply is not possible. Hot die welding is possible on a crude basis, although the kind of volume necessary to justify production equipment is not often evidenced. The same can be said for vibration welding. Threaded bosses and other fitments are frequently spin-welded in place.

Molded-in inserts are possible, and the cycle for this process is slow enough that the emplacement does not add to the time significantly unless the inserts are many or of awkward shape. However, postmolded inserts can also be installed with heat, induction, or ultrasonic welding. With these processes available, adhesive is seldom used to emplace the inserts. Threads and hollow bosses can also be molded in rotational parts, and two parts can be molded together.

Adhesives can be used for assembling rotational molded parts as can solvent welding. To determine whether a given material can be solvent welded, refer to Chapter 5, "Assembly Method Selection by Material."

6.19   Structural Foam Molding, Gas Counterpressure Structural Foam Molding, and Coinjection Molding

6.19.1   The Processes

Structural foam and coinjection molding processes are derived from injection molding, in which material is melted in the heating chamber and injected into a mold. In structural foam molding, a mixture of plastic and blowing agent is injected into the mold simultaneously. This creates a low pressure fill, which allows for a less substantial, and therefore less costly, mold than injection molding. The molding cycle, however, is much longer. Moreover, since the material does not completely fill the mold until the gas expansion takes place, the resulting part has a skin of relatively rigid material, with cells of increasing size toward the center of the wall. A reduction in part weight can be achieved with this process, but only in comparison to a solid part of the same wall thickness. If the wall thickness must be increased to utilize the process, this advantage is lost.

The principal advantage of structural foam molding is that its clamp tonnage requirement of a quarter-ton per square inch of part surface is one-twelfth that of injection molding. Thus, far larger parts can be made in the same size machine using structural foam.

One of the negatives associated with structural foam molding is the swirled surface that results from the broken cells. This can be avoided with two other techniques: gas counterpressure structural foam molding and coinjection molding. In the case of the former, gas is preinjected into a sealed mold to prevent foaming as the mixture enters it. This measure allows solid skins to be created. Then the gas pressure is released and the core is permitted to foam, yielding a surface similar to that obtained by means of conventional injection molding. However, the extra cost of sealing the mold reduces the savings in mold cost due to structural foam molding.

In coinjection molding, two different but compatible materials are used. The equipment is composed of two melting chambers joining together to inject with one nozzle inside the other, as illustrated in Fig. 6-12. The result is that the outer layer can be an unfoamed resin and the inner layer can be a foamed material. Thus, the surface imperfections associated with structural foam molding are essentially eliminated. A stronger, more costly material might be used for the outer skin and a weaker, less expensive material for the inner core.

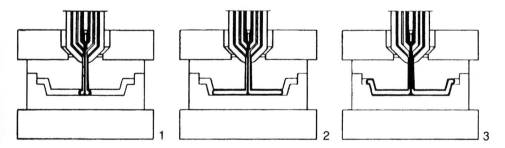

Figure 6-12   Coinjection three-channel system that simultaneously injects two different plastic molds (Rosato, D. V., *Rosato's Plastics Encyclopedia and Dictionary*, page 121)

Coinjection molding, like the structural foam processes discussed earlier in this section, uses closed molds and low pressures. The closed molds are more costly than the open molds used in thermoforming because there are two halves instead of one. However, the low pressure keeps the tool cost significantly lower than is possible with traditional injection molding. Yet the high pressure associated with injection molding is what permits its fast cycles. Thus, piece part prices are higher for these methods than they are for injection.

As in thermoforming, these processes are most competitive with injection molding for large parts. The larger the part, the greater the mold cost advantage over injection molding. Of course, if one designs parts of such intricacy that the amount of detail in the tool becomes the overriding cost factor, this advantage is negated.

Piece part costs are a different matter. As a broad statement vulnerable to all sorts of exceptions, parts from these processes will be less costly than those from thermoforming, but more costly than those from injection molding.

6.19.2   Assembly Considerations

The structural foam processes are capable of very thick wall sections. Thus, it is sometimes possible to eliminate an assembly operation by molding the inner and outer walls as one part. However, the core will be a foam structure, and that can complicate some assembly operations.

The use of structural foam processes makes it possible to change the "feel" or "heft" of a product or part. Since foam molding processes eliminate the need for assembly and the cost of the fasteners, a steering wheel or a vacuum cleaner handle could be made in one piece instead of by means of the older method, in which two injection-molded halves were assembled, usually with screws and nuts. The result is a lower product cost even though the injection molding process has a faster molding cycle. Theoretically, two parts can also be molded together as in injection molding, but this is not often done.

Since these are thermoplastic processes, all the assembly methods can be used with structural foam molded parts. These methods are capable of side-cored holes, undercuts, extraordinary detail, and fairly tight molded-in tolerances. However, the ease with which press fits and snap fits can be accomplished is limited by the stiffness the foam gives to the part.

Inserts can be molded in, however they slow the molding cycle if too many are used in a single moldment. If numerous inserts are required, shuttle presses can be employed which permit the inserts to be loaded on one side of the mold while the part is being molded on the other. Inserts can be economically retrieved from reject parts for recycling. Postmolded inserts can be emplaced by heat, electromagnetic/induction or ultrasonic vibration methods. Expansion inserts can be used, but usually are not because the other methods are more efficient. The same can be said for adhesive emplacement of inserts.

There are some additional factors to be considered from an assembly standpoint. The most significant of these is the basic fact that the center is not solid material. This means that the part will not have the strength that would have been available if it were a solid wall. Consequently, product strength computations based on the data supplied from the resin supplier will not be accurate because solid test specimens were used to create the data. While strength as high as 85% of the solid material can be attained, it is more prudent to use a much lower multiplier.

Obviously, any diminution in strength raises assembly concerns. It is still possible to use metal inserts, but the strength of adhesion is reduced. Inserts may be inserted ultrasonically, however the size of the hole is different. (*Note*: Check with the insert supplier for size and strength data beyond the brochure information before proceeding.) Other ultrasonic operations are also severely affected because of the absorption of vibrations by the structure. Use of this process is highly design dependent.

Both thread-cutting and thread-forming screws can be used. Since, however, the strength and stiffness levels approaching those of the solid material are found only near the relatively thin outer skin, the use of such screws is relatively limited. There are a number of specialty fasteners developed for certain applications. Both male and female threads can be molded-in structural foam molded parts.

Hot die/fusion welding, electromagnetic/induction welding, spin welding, and vibration welding can all be used for structural foam parts. Hot gas welding can be used but is not often employed because there is only a thin layer of solid material for most of the surface of a structural foam part. Solvent welding or adhesives can be used depending on the material.

6.20   Thermoforming

6.20.1   The Processes

Thermoforming is a thermoplastic process that uses heat and pressure to force sheet or film stock against the contours of a male or female mold. Vacuum, positive pressure and mechanical pressure in the form of a plug assist or matched die can be used to form the plastic. Thermoforming is inherently more expensive than some other processes because it must use sheet or film stock created by another process, usually extrusion.

Once known simply as "vacuum forming," this thermoplastic process has been expanded to encompass a variety of processes. In their most fundamental categorisation, they divide according to the thickness of the sheet used, namely, "thin-gauge" or "heavy-gauge" thermoforming. A further division is according to the die used, male or female.

In all the thermoforming processes, the sheet is first placed in clamps and then heated until it is soft enough to be stretched. Figure 6-13a illustrates a single-station machine in which the sheet is located immediately above the die. Rotary thermoforming machines, like the one in Fig. 6-13b, separate the functions of clamping, heating, and forming into separate stations.

In male thermoforming, sometimes referred to as "drape" thermoforming, the softened sheet is lowered over the male die. The softened sheet, drawn down over the die by a vacuum, takes the shape of the die. The part is then allowed to cool until it will hold its shape sufficiently to permit removal. Once removed, the part is trimmed to size. Female thermoforming is similar to male thermoforming except that a female die is used. Of the two processes, female thermoforming applications far exceed those that use male thermoforming.

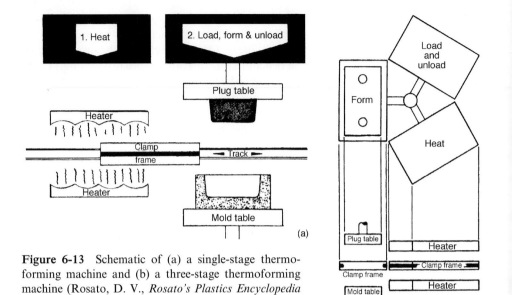

Figure 6-13 Schematic of (a) a single-stage thermo-forming machine and (b) a three-stage thermoforming machine (Rosato, D. V., *Rosato's Plastics Encyclopedia and Dictionary*, page 761)

6.20.2 Thin-Gauge Thermoforming

Thin-gauge thermoforming, which entails any thickness less than 0.060 in., is the variety of thermoforming generally associated with packaging. As such, it has tended toward thicknesses far below the upper limit due to market pressures. Sheet thicknesses below 0.015 in. are commonplace, and many applications can be found at half that level.

This area has developed independently of the rest of the thermoforming world because of its special requirements. Thin-gauge packaging sheet is often multilayered in barrier-layered sandwiches. At these thin gauges, it is usually made available in roll form. High speed, roll-fed machines predominate in these high volume applications.

Thin-gauge processes have two elements of interest from the assembly standpoint. The first is their ability to eliminate assembly operations by using coextruded sheet and integral features such as snap fits and hinges. The other is that they are incorporated into some of the most automated assembly lines in industry. Equipment exists in which roll stock is fed in line to where the package is first formed, then filled, sealed, die-cut, and finally placed in cartons, never touched by human hands. In some cases, there is sufficient volume to warrant an extruder dedicated to the application and placed directly on the line. There may also be an equally automated auxiliary line producing the product that feeds into the packaging line.

Thin-gauge thermoforming uses film wound on reels and multi-cavity molds for high volume applications. Generally, these are in packaging, however there are some product applications. Disposable cups and plates are two that come to mind.

6.20.3   Heavy-Gauge Thermoforming

Heavy gauge (thickness >0.060 in.) is the variety that normally concerns product designers. In its simplest form, it can produce relatively flat parts with rounded corners. However, its "pressure forming" version can produce detail that rivals some work done by injection molding.

Thermoforming is an "open mold" process; that is, the mold has only one side. Furthermore, it is a low pressure process. Consequently, the tooling is relatively inexpensive – among the lowest in cost of all the plastic processes. Thus, it is an excellent "entry level" process, opening the plastics world to many products that could not afford the higher tooling cost of a process like injection molding. This process can use most of the same materials as injection molding, although some of the high temperature materials can be handled only with difficulty. Unfortunately, it has a much longer processing cycle than injection molding, and the material must be processed into sheet first. Consequently, as the volume of a given product grows, the additional cost of injection molding tooling may reach a point that is more than offset by the reduction in piece part cost.

Since thermoforms are made from a sheet, the spare material must be removed. While traditional cutting tool methods are commonplace, high pressure water jets, hot wires, thermal laser, and robotic routers are often used. As thermoformers strive to be competitive, however, they have begun to introduce one or another of the automated devices. These methods are also used to cut any openings in the part, since openings cannot be formed in this process.

6.20.4   Pressure Thermoforming

Pressure thermoforming is a variety of thermoforming in which air pressure, sometimes with the addition of a plug assist, is used to force the material deeper into the mold cavity. This permits the use of a combination of flat and textured surfaces and the creation of very fine detail. Pressure thermoforming can produce inside and outside radii as low 0.015 in.; if cost is no object, 0.005 in. is possible. While the process may be capable of such inside radii, they would lead to very high internal stress, hence are not to be considered good practice.

The development of pressure forming has provided a notable expansion of the capabilities of the product designer as it extends the range of products that can use such detail to those with annual volumes far below levels that would be economic for injection molding. Typically, injection molding becomes a more serious contender as the volume of the application grows. Then the difference in piece part cost will justify the substantially greater cost of the injection molds. However, as the size of the object increases, the cost of the injection mold increases substantially, as does the difficulty with which the large part can be molded. Consequently, the payoff volume goes up significantly as the size of the part increases. Indeed, with thermoforming parts can be substantially larger than is possible with injection molding.

6.20.5   Other Forming Processes

For some plastics applications, the forming can be accomplished without the need to use heat at all. Two processes that do not use heat are known as "coining" and "cold forming."

6.20.6   Assembly Considerations

The ability to create deep undercuts with female thermoforms sometimes permits the combination of two housing parts into one. Also, two colors or materials can be thermoformed together through the use of coextruded sheet. However, the process tends to lose its competitive edge over other processes when the size falls below one square foot. Conversely, the very large part capability of thermoforming allows the combination of several parts into one in many cases, thus eliminating some assembly operations completely.

Thermoformed parts are made from sheet and share many of the assembly characteristics of sheet metal. Their holes generally have to be drilled or cut as a secondary operation and, unless cut away, a flange will be present around the outer perimeter. Thus, many of the fasteners originally developed for sheet metal (e.g., "J" or "U" nuts) are, in one form or another, used on the flanges. One technique applied to parts made by this process is the use of small blocks of material fixed to the thermoforming with solvents or adhesives. Typically, these blocks have inserts emplaced or holes for self-tapping screws. This method is used because the sheet stock is often too thin for self-tapping screws and too weak for sheet metal screws. When the gauge is sufficiently heavy, both self-tapping and thread-forming screws can be used, the variety depending on the flexural modulus of the particular material. There are a number of specialty fasteners developed for certain applications. Molded-in threads are not usable in most cases.

Thermoforming is capable of undercuts, molded-in hinges, press fits, and snap fits. However, their use is limited by the loose tolerance requirement of the process. Also, the ability to accommodate these devices generally diminishes as the wall thickness progresses from thin gauge to thick gauge. Because it is a low pressure, open-mold process, thermoforming cannot produce internal projections like ribs, pins, and bosses. External bosses can be formed. They are then opened by machining off the tip. Louvers are opened in the same manner.

In pressure thermoforming, inserts can be thermoformed in for light-duty applications with low tensile requirements. Inserts can be economically retrieved from reject thermoplastic parts for recycling. Postmolded inserts, emplaced by either heat, electromagnetic induction, or ultrasonic vibrations, are usually preferred because thermoforming cycles run several minutes even without molded-in inserts. Expansion, self-tapping, and helical coil inserts can be used but usually are not, since the other methods are more efficient. The same can be said for adhesive emplacement of press-in inserts.

Hot die/fusion welding, spin welding, vibration and hot gas welding can all be used for thermoformed parts. Ultrasonic energy directors can be pressure thermoformed and all thermoformed parts can be ultrasonically spot welded, however thermoforming cannot produce the tolerances required of shear joints. Likewise, electromagnetic/induction welding fitment details would be difficult to produce in thermoformed parts.

6.21   Twin-Sheet Thermoforming

6.21.1   The Process

A variety of thermoforming that can produce hollow parts is accomplished by means of a thermoforming machine with both upper and lower molds. Two sheets are formed and then pressed together around their perimeter while still soft, so they weld together where they contact. The result is a two-sided part with a cavity between the sides. Localized contacts, known as tack-offs, between the two parts are used to increase strength. The cavity can be filled with foam for further stiffness improvement.

This process is well suited to large flat panels. It can handle nearly all the thermoplastics, although some of the high temperature resins are difficult to process. Depending on the material used, both inside and outside radii can be as low as 0.060 in., however that is without taking into account internal stress considerations. All openings in the part must be machined and the outer flange must be cut from the original sheet, just as with single sheet thermoforming.

6.21.2   Assembly Considerations

Basically, all the assembly considerations for single-sheet thermoforming are valid for twin-sheet thermoforming. However, there are some additional features.

Although the tooling is closed, there is an uncontrolled internal space between the walls. Thus all ribs and bosses must be hollow. Additional structural support can be attained by designing the walls or bosses from each side to meet. These contact surfaces will weld together and are known as "tack-offs." Male and female threads can be molded in from the outside surfaces, as can threaded inserts for light-duty applications. Large insert plates can also be positioned out of sight between the walls of the part. Considerable strength and rigidity can be gained by placing foam between the walls.

6.22   Transfer Molding

6.22.1   The Process

When fine detail is required, a variant of compression molding known as transfer molding can be used. Here the resin is first placed in a transfer pot and then forced through a sprue and runner system into the cavity as shown in Fig. 6-14, much like injection molding. This process must compete with injection molding of thermosets for its applications. Transfer molding is confined to fairly small parts, whereas compression molding will handle some of the largest plastic parts.

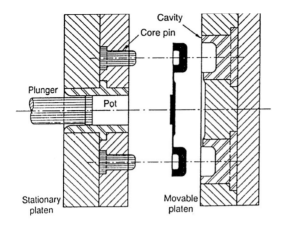

Figure 6-14 Plunger transfer with preheated preforms (Wright, Ralph E., *Injection/Transfer Molding of Thermosetting Plastics*, page 28)

6.22.2 Assembly Considerations

Transfer molding and injection molding of thermosets have common assembly characteristics. First, since they are thermoset processes, none of the heat-based assembly methods are usable. That eliminates hot gas welding, hot plate/fusion welding, electromagnetic/induction welding, ultrasonic welding, vibration welding, and staking. A version of staking in which a press-on fastener is emplaced over a post is often used on thermoset parts. Both transfer and thermoset injection molding can make these posts.

Inserts can be readily molded in, but they slow the cycle, which increases the molding cost. While postmolded inserts cannot be emplaced with heat or ultrasonics, they can be installed with adhesives. Also, expansion, self-tapping, and helical coil inserts can be used.

Both internal and external threads can be molded in. Holes for fasteners can also be molded in, both perpendicular and parallel to the parting lines. However, molding in a hole usually leaves the portion of the part beyond the hole with less reinforcement, and greater strength can be achieved through machining. Fasteners are the most commonly used means of assembly for applications that do not use adhesives.

External undercuts are possible but expensive to tool, since they require a side action or a split mold. Internal undercuts require difficult and expensive tooling. Tight tolerances are possible, which means that press and snap fits are feasible.

6.23 Process Selection

The key principle in process selection is to establish the most efficient means of manufacture. To do so, one must take into account the investment as well as the piece part cost. For example, injection molding might offer the lowest piece part price for a given project, but it might be hard to save enough to pay the difference in mold cost over structural foam if only a few thousand parts were to be made each year.

It would simplify the decision making tremendously if we could determine a given volume for each process beyond which the additional investment could be justified by the reduction in piece part cost. Actually, if all parts were identical in shape and size, that might be possible. Since they are not, each designer and engineer must learn through experience with the types of parts he or she creates. As a rough guide, however, one may presume that the larger and/or more complex the part, the greater the investment will be. As the investment grows, the volume must be greater in order to pay the difference with lower piece part costs within an acceptable period. That period also varies considerably between companies.

A further consideration lies in the size capability of each of the processes. Table 6-1 gives some indication of the size range each process is capable of. Note that on the high

Table 6-1   Approximate Part Size Ranges for the Principal Processes

Process	Approximate size ranges		
	Smallest known	Largest commercial	Largest known
Blow molding	$\frac{3}{8}$ in. $D \times 1\frac{1}{4}$ in. L	$9\frac{1}{2}$ ft. $L \times 1$ ft. $D \times 4$ in. thick 28 in. $D \times 44$ in. L	1320-gallon tank
Casting	No limit	Limited only by physical ability to handle molds and moldments	Limited only by physical ability to handle molds and moldments
Coinjection molding	$\frac{1}{4}$ in. $\times \frac{1}{4}$ in. $\times \frac{1}{4}$ in.	2 ft. $\times$ 5 ft. $\times$ 5 ft.	$2\frac{1}{2}$ ft. $\times$ 4 ft. $\times$ 10 ft.
Cold molding	$1\frac{1}{2}$ ft. $\times 1\frac{1}{2}$ ft.	10 ft. $\times$ 10 ft. $\times 1\frac{1}{2}$ ft.	14 ft. boat hull
Compression molding	$\frac{1}{4}$ in. $\times \frac{1}{4}$ in. $\times \frac{1}{16}$ in.	4 ft. $\times$ 5 ft. $\times$ 8 ft.	$1\frac{1}{2}$ ft. $\times 4\frac{1}{2}$ ft. $\times$ 14 ft.
Extrusion	No limit	12 in. $\times$ 12 in.	42 in. D
Filament winding	4 in. $D \times 8$ in. L	13 ft. $D \times 60$ ft. L	10 ft. $H \times 82\frac{1}{2}$ ft. D
Injection molding	0.008 in. $\times$ 0.020 in. $\times$ 0.020 in.	$2\frac{1}{2}$ ft. $D \times 3$ ft.	4 ft. $\times$ 4 ft. 6 in. $\times$ 7 ft.
Layup and sprayup	$\frac{1}{4}$ in. $\times$ 6 in. $\times$ 6 in.	150 ft. Minesweeper	Continuous roadway
Machining	No limit	10 ft. Wide or 15 in. D	Limited by size of stock available
Pultrusion	$\frac{1}{16}$ in. D	12 in. $\times$ 12 in.	15 in. $\times$ 100 in.
Reaction injection molding	4 in. $\times$ 12 in.	3 ft. $\times$ 4 ft. $\times$ 10 ft.	10 ft. $\times$ 10 ft.
Resin transfer molding	1 in. $\times$ 3 in. $\times$ 2 ft.	16 in. $\times$ 4 ft. $\times$ 8 ft.	4 ft. $\times$ 8 ft. $\times$ 28 in.
Rotational molding	$\frac{1}{2}$ in. Diameter sphere	6 ft. $D \times$ 18 ft.	12 ft. $D \times$ 30 ft. L
Structural foam molding	$\frac{1}{4}$ in. $\times \frac{1}{4}$ in. $\times \frac{1}{4}$ in.	$2\frac{1}{2}$ ft. $\times 6\frac{1}{2}$ ft. $\times 6\frac{1}{2}$ ft.	$2\frac{1}{2}$ ft. $\times$ 4 ft. $\times$ 10 ft.
Transfer molding	$\frac{1}{8}$ in. $\times \frac{1}{8}$ in. $\times \frac{1}{16}$ in.	2 ft. $\times$ 1 ft. $\times \frac{1}{2}$ ft.	$2\frac{1}{2}$ ft. $\times$ 1 ft. $\times$ 9 in.
Thermoforming			
Thin gauge	$\frac{1}{4}$ in. $\times$ 1 in. $\times$ 1 in.	3 ft. $\times$ 3 ft. $\times$ 3 ft.	3 ft. $\times$ 3 ft. $\times$ 3 ft.
Heavy gauge	6 in. $\times$ 6 in. $\times$ 6 in.	3 ft. $\times$ 10 ft. $\times$ 12 ft.	2 ft. $\times$ 6 ft. $\times$ 20 ft.
Twin-sheet thermoforming	6 in. $\times$ 6 in. $\times$ 6 in.	6 in. $\times$ 3 ft. $\times$ 6 ft.	6 in. $\times$ 3 ft. $\times$ 6 ft.

side we have indicated two sizes. The column titled "Largest known" lists the largest size the author has ever heard of. That size is not realistic for general use, since not all materials can be run at the limits for a given process, and the equipment that made a particular example may not be available for custom projects. Indeed, such equipment has typically been built for some company's proprietary needs. The column "Largest commercial" lists sizes at which a reasonable selection of competing bids might be obtained. When working near the limits of a given process, however, the product designer should be certain that the equipment is available for the project before proceeding with the design.

Beyond size, there is an approach to process selection based roughly on the relationship between investment, assembly, and moldment cost. What follows are brief discussions of the principal processes. The lists that begin Sections 6.23.1 to 6.23.5 indicate "pecking orders," which start with the very lowest volume processes for each category and proceed to those that have increasing investment commensurate with lower piece part cost.

6.23.1   Thermoplastic Open Shapes

1. Machining
2. Thermoforming and pressure thermoforming
3. Structural foam molding and coinjection molding
4. Gas-assisted injection molding
5. Injection molding

The thermoplastic process pecking order begins with machining. While this is a common means of producing prototypes, the volume does not have to climb very high before other processes become more cost-effective.

Thermoforming and pressure thermoforming become cost-effective at very low volumes, although parts under roughly one square foot in area are difficult to justify for these processes. Thus we need to develop several pecking orders, which are size dependent, to allow reliable costing for smaller parts.

For small parts, therefore, injection molding is the next logical process to consider. Fortunately, the injection molding industry has evolved relatively inexpensive methods, involving such tooling devices as insert and family molds, for producing small parts in low volume. There are, however, a limited number of molders interested in such work.

As the part increases in size beyond one square foot, other processing options beyond thermoforming, such as structural foam and coinjection molding, come into the pecking order. These processes can compete with injection molding to higher volumes as the part size increases because the cost differential with the injection mold grows disproportionately with size.

Actually, depending on the part configuration (such as a requirement for internal details), structural foam or coinjection molding may be the preferable starting process (after machining). Structural foam molds have traditionally cost significantly less than injection molds (although more than thermoforming dies) because the process is a low pressure one. While the structural foam cycle is much longer than injection molding, the mold cost differential can cover quite a volume before injection molding becomes more competitive. However, as product designers create structural foam designs with detail equivalent to that of injection-molded parts or require gas counterpressure molds, that advantage has diminished. When a structural foam mold costs nearly as much as an injection mold, the process will be selected for its unique attributes. Such features as lighter weight, increased heft, and the ability to make larger parts (owing to the low pressure) become the principal selection criteria.

For some parts, there might be one more stop on the pecking order before going to injection molding: gas-assisted injection molding. This process is of particular value in the elimination of assembly operations used to fabricate box structures. There have been successful applications in the elimination of parts in automobile door frames, where the gas has been used to create the box structures within the frame.

For most parts, the ultimate in production volume economies can be reached with injection molding. At a size of about 5 ft^2 of surface area, however, injection molding drops out of the competition because molding machines able to handle that size are very scarce and usually are reserved for proprietary applications. In that case, gas-assisted injection molding or structural foam molding may be the most efficient process.

Unfortunately, structural foam molding can handle a much more limited palette of materials than injection molding can. The materials that are readily molded with structural foam are ABS, ionomers, polycarbonate, polyethylene, and polypropylene. Most of the remainder can be handled only with difficulty and some, like styrene and acrylic, cannot be processed at all. The only high strength, high temperature material in the readily processed group, polycarbonate, may not be acceptable for cost or chemical compatibility reasons. That means that the next step up in volume economies for a part too large for injection molding and calling for higher temperature resistance than the structural foam materials can inexpensively offer may be a thermoset process like compression molding.

The hollow processes (blow molding, rotational molding and twin-sheet thermoforming) can also be used to make open parts. This can be done either by cutting open hollow parts or molding dual-walled panels. However, these are usually special applications which do not fit well into the pecking order. Examples of such applications are a container and its cover molded as one moldment and cut apart or a structural part like a dual-walled pallet.

6.23.2 Thermoset Open Shapes

1. Machining
2. Casting
3. Lay-up or spray-up (depending on shape and strength requirements)
4. Resin transfer molding
5. Cold press molding

6. Reaction injection molding
7. Compression molding, BMC, SMC, transfer molding, and injection molding of thermosets (depending on shape and strength requirements)

Thermoset materials like polyester often offer elevated physical properties at a much lower cost than the engineering thermoplastics. However, their very nature as thermosets means that the molding scrap cannot be reground and remolded. Consequently, some of that economic advantage is lost.

Since many of the thermoset processes are best suited to long fiber reinforced large parts, two pecking orders exist in this area as well. For short fiber reinforced parts up to around $0.25\,m^2$ ($2.7\,ft^2$), the next step after machining and casting, depending on part configuration, is compression molding, transfer molding, or injection molding. Of the three, the tooling cost for compression molding would be lowest and the piece part cost would be highest. Injection molding would offer the lowest part cost, but the tooling cost would likely be the most expensive. Transfer molding would be somewhere between them. Compression molding, however, cannot produce the same level of detail that the other two can.

For long fiber reinforced parts over approximately $0.25\,m^2$ ($2.7\,ft^2$) in area, the pecking order includes the full gamut of thermoset processes with the exception of transfer or injection molding.

6.23.3 Hollow Parts

The hollow part processes in this section are all for thermoplastics. Hollow shapes in thermosets can be made in one piece through filament winding. However, for applications that do not require extraordinary strength, it is usually less costly to fabricate the parts by assembling halves made by one of the other processes. When proceeding with the development of thermoplastic hollow shapes, the author recommends the following sequence.
1. Rotational molding or twin-sheet thermoforming (depending on shape)
2. Blow molding

Processes for hollow shapes have a much simpler pecking order. The ideal shape for rotational molding is a sphere. The ideal shape for twin-sheet thermoforming is a flat panel. The selection of a starting process normally is based on the shape of the particular part to be manufactured, although it should be noted that rotational molding can make parts with sections as thin as one inch. Additional factors include the size of the part (rotational can go larger) and the selected material, since each of these processes favors different materials.

Depending on the size and shape of the part, blow molding may become competitive as the volume grows. Since, however, the other two processes can make larger parts than are producible on the majority of blow-molding machines available, there are limits to the size to which this step can be taken.

6.23.4   Profiles

1. Extrusion
2. Pultrusion

To a certain degree, the profile processes can also create hollow shapes, but only if the shape is continuous in one plane. The capped tube is an obvious example. Beyond that, the bulk of extruded materials are open profile, much of it in the shape of sheet and film. Since extrusion is capable of enormous volumes, the profile processes are listed in order of increasing strength. Extrusion and pultrusion are specialized processes for which there is no real pecking order. True, there are some applications where extrusion and pultrusion may be competitive, but the size and capability relationships we have been considering with respect to the elimination of assembly operations would not apply. However, it should be noted that on some occasions the assembly of two parts made by another process, such as injection molding, may be competitive with an extruded or pultruded part that had to be capped, since the latter would require two operations whereas the former would require only one.

6.23.5   Ultra High Strength

1. Filament winding

Filament winding is a process that stands alone with no real pecking order. It is a high strength process that defines its own applications. Its principal competition is really metal fabrication, over which it can have significant assembly, cost, and weight advantages when it can be employed.

7 Adhesive and Solvent Joining

7.1 Advantages and Disadvantages

Adhesives and solvents are fundamentally different. It is logical to discuss them together, however, because they share a number of similarities in terms of surface preparation, joint design, etc. Adhesives join materials by adhering to the surface of the plastic, whereas solvents melt the surfaces of the plastics such that they intermingle, leaving a welded joint when the solvent evaporates. Solvent joining is sometimes referred to as "solvent sealing," "solvent welding," or "solvent cementing."

7.1.1 Advantages

The use of adhesives and solvents in the assembly of plastic parts is widespread. Solvent joining is one of the least expensive techniques available. As technology leads to new material combinations, applications for adhesives increase. In many cases, they are the only alternative to the use of fasteners. Recent developments in adhesive technology itself have led to many new applications. As in every method, there are advantages and disadvantages. Here are the advantages.

1. *Uniform stress distribution* One of the principal problems associated with the design of plastic assemblies is the control of the stress. The processing techniques used in the manufacture of plastic parts result in varying degrees of internal stress. Additional stresses can result from assembly methods. Fasteners, for example, concentrate stresses around the hole or boss in the part. Thus an injection molded part will have molding stress around the fastener holes as well. Spot welds can also cause stress concentrations.

Most of the joint designs associated with adhesives have much less molded-in stress than designs that include holes. Some, such as lap joints, have none other than the basic molding stresses related to the structure of the part. Furthermore, the load for an adhesive joint is spread uniformly over the surface on which the adhesive is spread. The improvement in the resistance of the assembly to shock, fatigue, and vibration thus achieved can sometimes permit a reduction in wall thickness.

2. *Bonding of dissimilar materials* A great number of adhesive applications are driven by the need to join dissimilar materials: either the adhesion of a plastic to another material or the joining of two plastics that cannot be solvent or heat welded to each other. Adhesives, along with fasteners, are the principal means of joining thermoset plastics. When two dissimilar materials are joined, their respective coefficients of linear thermal expansion may be so different that mechanical fasteners are out of the question. Other such materials, such as paper and glass, may be too brittle or not suitable for any other bonding methods. Adhesives have permitted the construction

of composite materials with unique combinations of physical properties. The use of solvents, however, is limited to compatible resins.

3. *Maintenance of bonded material integrity* Fasteners, the principal competitor to adhesives for many joining applications, require holes in the parts. The holes remove material and, if molded in, result in a weld line at some point around each hole. Both these features tend to reduce the strength of the structure.

4. *Maximum fatigue resistance* Many adhesives have the capability of absorbing the vibration that could lead to fatigue failure if transmitted to the structure. The ability of the adhesive to provide this benefit will vary according to the flexibility of the adhesive. Solvent welds do not exhibit this characteristic.

5. *Ability to seal as well as bond* When the adhesive or solvent is applied in a continuous bead, it will seal the interior of the assembly from the environment. While a hermetic seal is possible, care is necessary in the surface preparation of the joint and the application of the adhesive or solvent. These seals can break down in time.

6. *Weight reduction* Adhesive and solvents are low in specific weight, and they are spread very thin. In addition, the wall thickness of the plastic can sometimes be thinned through their use, resulting in a weight reduction. Thus adhesive and solvent joining may offer lower overall product weight than is possible with fasteners.

7. *Smooth, blemish-free surface* For applications in which protruding fastener or stake heads and spot weld blemishes would interfere with the cosmetic or functional requirements of the product, adhesive and solvent joining can provide a smooth and blemish-free surface. The use of these methods thus can provide greater mechanical and aesthetic design freedom.

8. *Thin or flexible substrate bonding* The use of adhesives, most often in the form of tapes, on very thin, flexible plastic substrates, is commonplace. Other than the heat bonding of films, there is virtually no other way to accomplish such joining. The thinner the substrate wall, the more likely it is that adhesives will be the optimum assembly method. Solvent welding is more limited for very thin films because some resins will become too soft to apply or may disintegrate completely.

9. *Lower total product cost* When measured simply by the price per pound of the adhesive, cost may seem prohibitive. However, when the total product cost is considered, adhesives or solvents may turn out to be the optimum method of assembly. To begin with, the original investment is usually quite low, so the amortization factor is virtually nonexistent. Second, depending on the number of fasteners required, the application of adhesives may require less labor time than assembling the fasteners. Next, some of the preparation work (e.g., drilling holes), is eliminated. Finally, the reduction in wall thickness, which may be possible through the use of adhesives, will reduce the material and processing cost of the parts. This, in itself, can provide a significant cost savings to more than justify the use of adhesives.

7.1.2   Disadvantages

While the advantages to adhesive joining are many, there are some disadvantages.

1. *Cost* While it is important to consider the total product cost in determining the optimum assembly method, the cost of the disposal of waste adhesive and of the adhesive itself are often insurmountable. This is particularly true for many of the thermoplastics that can be assembled by a variety of assembly methods that have no cost associated with an additional material. For press and snap fits, there is not even a power expense to be considered – although the additional part and tooling costs associated with snap fits are often overlooked in making the comparison. When similar amorphous thermoplastics are to be joined, solvents can be used that are much less costly than adhesives, but the ventilation issue remains.

2. *Recycling difficulties* The adhesive is a contaminant and must be cut or broken out of the material to be recycled even when the two parts are made of the same material and might not otherwise have needed to be disassembled. Pure solvent chemicals do not present a problem. Since the precise ingredients of commercially prepared varieties are proprietary, however, there is no guarantee that such products contain no contaminants.

3. *Joint strength uncertainty* While solvents evaporate in the atmosphere, adhesives are affected by environmental conditions such as heat and humidity just as the plastics themselves are. Furthermore, their properties change as they age. Also, adhesives are coatings and, as such, their adhesion to a substrate is highly dependent on the surface condition of the substrate. Surface treatment may be necessary. Therefore, it is difficult to predict the strength of the joint. Total confidence in the bond strength may be difficult to assure without costly 100% testing, although testing of seal integrity is more economically accomplished. For critical joints, additional methods, such as fasteners, may be required.

4. *Assembly rate limitations* Adhesive and solvent assemblies often require clamping, which is difficult to do at the high speeds necessary for some applications, particularly for odd-shaped parts. Heat may also be needed, but that is more easily accomplished. The time that a solvent joint is open is critical, and a long setup time may be required.

5. *Special handling* Most of the solvents that attack polymers and the solvents associated with adhesives are flammable. Consequently, their receipt and usage must be recorded, and they must be isolated and stored in a special facility. For this reason, there is a strong trend toward aqueous-based adhesives.

6. *Emission control* In the United States, some states have strict limitations on the amount of volatile organic compounds that can be emitted into the atmosphere. Ventilation and capture of toxic vapors may be required, and any additional equipment or labor needed can detract from the cost-effectiveness of the process.

7. *Crazing* Solvents act as stress relief agents and can result in tiny fissures in the surface known as "crazing." Gate locations must be carefully placed away from the joint because they are highly stressed areas that are readily attacked by both the solvents used in solvent welding and some of the solvents used in adhesives. Although solvents are widely used to join polystyrene parts, styrenics are known to be particularly prone to this problem, and care must be taken in the application of solvent welds with these materials.

8. *Solvent sensitivity* Solvents trapped inside a joint may lead to porosity or weakness. In addition, the finished assembly cannot be exposed to the solvent used for

solvent joining or for solvent-based adhesives. Solvents can also migrate in some plastics.

9. *Surface preparation* The surface of the substrate or adherend must be kept clean for proper adhesion. Furthermore, separate preparation procedures usually are necessary to attain the necessary level of bond strength.

Disadvantages notwithstanding, adhesives and solvents are an important weapon in the designers' arsenal. They must be seriously considered because they are the optimum mode of assembly for many applications and practically the only option for some applications.

7.2   Basic Theory and Terminology

First, a little terminology. The words "adhesion" and "cohesion" appear repeatedly in any discussion of adhesives. *Adhesion* refers to the adherence of two bodies to each other. In Fig. 7-1, there are three such bodies: adherend A, the adhesive, and adherend B. There are two adhesions, the one between adherend A and the adhesive and the one between adherend B and the adhesive. Adhesive failure would, therefore, refer to the failure of either of these adhesions. It is usually due to a poor adherend surface condition.

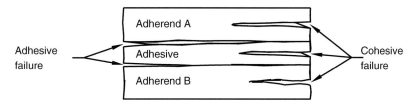

Figure 7-1   Adhesive vs. cohesive failure

Cohesion refers to the force within a body holding it together. Cohesive failure, therefore, refers to the failure within the body. Cohesive failures in each of the three bodies is illustrated on the right-hand side of Fig. 7-1. When testing to failure, **100% cohesive failure is ideal** and indicates that the adhesion attained was stronger than the surrounding material. Bear in mind that the adhesive is one of the surrounding materials. If it fails in a cohesive fashion, a stronger adhesive or more coverage may be needed. If the adherend fails in a cohesive manner below the required strength level, a stronger material may be called for.

Failures are often given as percentages. Thus, an 80% cohesive failure would mean that 80% of the failed surface was cohesive and 20% was adhesive. As long as there is some adhesion failure, there is the potential for improvement. There is, however, a time effect. Some joints would fail adhesively if tested shortly after assembly and cohesively if tested after some time had passed. This is particularly true of adhesives that gain strength slowly over a long period of time.

Solvent Welds. In the case of solvent welding, there is no adhesion. Therefore, all the failure would be cohesive failure. Here, one needs to determine what percentage of the joint surface melted and formed a bond.

Mechanical Adhesion. The theory of adhesion states that adhesive flows into micro-cavities in the surface of the substrate and hardens forming a mechanical bond. To validate this theory, the trapped air at the interface must be displaced and the adhesive must be observed to penetrate cavities on the surface of the adherend. Mechanical abrasion of the surface benefits this type of adhesion.

Adsorption Adhesion. This theory of adhesion, also known as *specific adhesion*, states that molecular contact between two surfaces and the consequent surface forces result in adhesion. (Taken together, the mechanical adhesion and the adsorption adhesion are sometimes referred to as the *effective adhesion*.) For this to take place, there must be good adhesion of the liquid adhesive to the solid adherend. This is a phenomenon known as "wetting," which is illustrated in Fig. 7-2.

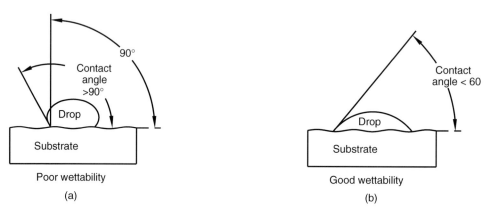

Figure 7-2   Wettability

Contact angle measurements are used to measure wettability. The contact angle is measured from the horizontal plane of the substrate to the tangent to the edge of the drop of liquid on the far side when equilibrium has been reached. Figure 7-2a illustrates poor wettability, where the contact angle is greater than 90°. Adhesion would be very weak under this condition. In good wettability, illustrated in Fig. 7-2b, the contact angle is less than 60°. If complete wetting were to be achieved, the contact angle would approach zero.

For an alphabetic listing of some of the most important terms used in the adhesive industry, see the glossary near the end of the Chapter (Section 7.10).

7.3   Methods for Measuring the Wettability of a Plastic Surface

7.3.1   Contact Angle Test

A contact angle meter (goniometer) is the most precise method used for measuring the contact angle of a drop of distilled water. The device shown in Fig. 7-3 is available from Tantec, Inc. (Refer to Section 7.11 "Sources").

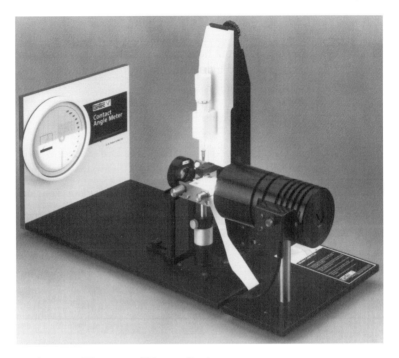

Figure 7-3   Contact angle meter (Courtesy of Tantec, Inc.)

7.3.2   Wetting Tension Test (ASTM D-2578-73, Wetting Tension of Polyethylene and Polypropylene Films)

Although designed for the polyolefins named, the ASTM wetting tension test has been used for other plastics as well. Solutions composed of percentages of Cellosolve and formamide are tested on the substrate surface for wetting. The percentage at which wetting takes place defines the surface energy of the substrate. As the percentage of formamide increases, the surface energy increases. For example, if the drop wetted at 100% methyl Cellosolve, the surface energy of the substrate would be 30 dyn/cm. If it wetted at a mixture of 48.5% formamide and 51.5% Cellosolve, the surface energy would be 37 dyn/cm. At 99% formamide and 1% Cellosolve, the surface tension would be 56 dyn/cm. This is the most widely accepted method of determining surface energy. This test is capable of accuracy to 2 dyn/cm, however an accuracy of 4 dyn/cm is more typical.

7.3.3   Adhesion Ratio Test (Tentative ASTM D-2141-63R)

This test measures the amount of force required to remove a given amount of the same tape before and after surface treatment to assess the effectiveness of the surface treatment.

7.3.4   Water Spreading Test

A drop of water is placed on a horizontal surface of the substrate and the effectiveness of the treatment is evaluated by measuring the drop's diameter before and after surface treatment.

7.3.5   Dye Stain Test

The substrate surface is dipped in a special dye and placed in a vertical position. If the surface is stained, it has been treated.

Table 7-1   Surface Energy of Solid Materials Versus Surface Tension of Liquid

Surface energy (dyn/cm)	Solids	Liquids	Surface tension (dyn/cm)
100⁺	Metals, glass, ceramics		
100			100
80			80
		Water	72
		Glycerol	63
60			60
		Formamide	58
50	Polyether sulfone		
48	Styrene butadiene		
47	Phenolic, polyphenylene oxide		
46	Polycarbonate		
42	Nylon		
41	Polyester, PET, polysulfone, polyarylsulfone		
40	Polyimide		40
39	PVC		
38	Polyphenylene sulfide		
37	PVA		
36	Acetal, epoxy, PEEK		
35	ABS		
33	Polystyrene, EVA, Acrylic		
32	PBT		
31	Polyethylene	Cellosolve	30
29	Polypropylene	Toluene	29
		N-Butanol	25
24	Silicone rubber	Alcohol	22
20	Fluorinated ethylene propylene (Teflon)		20
18			0

7.3.6   Ink Retention Test

Ink or paint is placed on the substrate surface and a 10×10 grid is cut in it with 11 blades set 0.031 in. apart (this tool is available commercially). When the assembly is dry, a specified pressure-sensitive tape is applied. The tape is then removed and the number of squares of dye remaining is counted to arrive at a percentage. This test can be used to compare surface treatments, but is more likely to be used to measure the adhesion of the coating itself as a quality assurance test on the production line.

The property of a substance that determines its wettability is its surface energy, also known as its surface tension in liquids. It is measured in dynes per centimeter. To ensure a proper bond between liquid and substrate, the substrate's surface energy must exceed that of the liquid by 10 dyn/cm. Table 7-1 compares the relative surface energies and surface tensions of several plastics and liquids.

Clearly, some of the plastics at the low end of the scale will require surface treatment to achieve adequate wettability. Additives such as antiblocking agents, slip agents, antistatic agents, and antioxidants will result in actual values lower than those published for the base resin. The shorter the time period between the assembly operation and the molding, the less time there is for these additives to bloom to the surface and affect the surface energy.

7.4   Surface Treatments

Regardless of the level of surface energy of the substrate material, the first step in any adhesive application procedure is to provide a clean surface. If there is dirt, machine oil, mold release agent, grease, moisture, or weak oxide layers on the surface of the substrate, the adhesive will attempt to bond to the contaminant rather than the substrate. To obtain good adhesion, therefore, a clean surface must be achieved. Regardless of the relative surface energies of the plastic and the adhesive, or whether any further steps will be needed, the first step in the preparation of a surface for bonding should be solvent cleaning. A clean surface is defined as one having no visible soil or foreign matter when inspected by the naked eye.

The *water-break-free test* is a common method for determining whether a surface is clean enough for adhesive bonding. The surface is rinsed in uncontaminated distilled water and, if no break in the film of water appears on the surface, it is regarded as clean enough for bonding. A break in the surface is an indication of contamination.

7.4.1   Solvent Cleaning

These surface treatments will not change the chemical or physical characteristics of the substrate. However, small amounts of material may be removed from the surface of the substrate, thereby altering it slightly. The quality of solvent cleaning is highly operator sensitive. The quality of the surface for the adhesive application operation is equally dependent on the careful handling of the parts after the solvent cleaning process.

Toxicity, flammability, hazardous incompatibility, and the use of hazardous equipment are all issues that must be dealt with when solvent cleaning methods are used.

The solvent cleaning methods can be applied to plastics by immersion, spray, or wipe methods, as described in the subsections that follow, according to Table 7-2.

Table 7-2   Plastic Cleaning Solvent Recommendations

Plastic	Solvent
ABS	Ketone
Acetal	Ketone
Butadiene styrene	Aromatic
Cellulose acetate	Ketone
Cellulose acetate butyrate	Ketone
Cellulose acetate propionate	Ketone
Diallyl phthalate	Ketone
Epoxy, phenolics	Ketone
Ethylene vinyl acetate	Alcohol
Furane, ionomer, melamine resins, SAN, polysulfone or rigid vinyl	Ketone
Fluorocarbons	Ketone or alcohol
Polyamide (nylon)	Chlorinated, aromatic or ketone
Polycarbonate	Alcohol
Polyester	Ketone
Polyethylene, polypropylene, or chlorinated polyether	Ketone
Polyimide or polymethylmethacrylate	Ketone alcohol or chlorinated
Polyphenylene oxide	Alcohol
Polystyrene	Alcohol
Polyurethane	Alcohol
Silicone	Ketone
Vinyl – flexible	Ketone

7.4.1.1   Solvent Immersion

Parts may be cleaned by immersion in a solvent bath, where they are soaked and scrubbed. Following this step, they are solvent-rinsed with a spray or flowing liquid. The cleanliness observed in this final step is important because the parts cannot be any cleaner than the solvent used in the last rinse. This process is recommended for light soil only.

7.4.1.2   Solvent Wiping

Solvent is placed on a clean cloth, which is used to wipe the parts. The cloth must be changed often because the dirt accumulates on it and may simply be transferred from one part to the next. Much of the contaminant would then be moved from one surface to another. The surface to be adhered to should be wiped until there is no residue on either the cloth or the surface, and the cloth must not contact the solvent container. After cleaning, the parts should be dried under a clean, dry hood with the vapors controlled. This process is adequate only for oil, grease, and light soil.

7.4.1.3 Solvent Spray

The solvent spray method benefits from the scrubbing effect produced by the impingement of the surface being treated by high speed particles. The loosened dirt is washed away by the spray. Efficiency is very high for this process.

7.4.1.4 Vapor Degreasing

Vapor degreasing can be highly effective for high volume applications involving highly soluble soils such as waxes, greases, and oils. More consistent than solvent wiping, this process is based on a thermally balanced distillation in which hot solvent vapors contact the cooler surface of the adherend forming a condensate that attacks and dissolves any organic soil. Baskets, wire hooks, and other devices are used to suspend the parts in a fluorinated or chlorinated solvent vapor bath. The solvent distillate is used until the accumulated soil has significantly lowered its boiling point.

7.4.1.5 Ultrasonic Vapor Degreasing

An ultrasonic cleaning device can be added to the clean solvent rinse tank. A mechanical scrubbing effect is created by the rapid cavitation and agitation of the solvent by the high frequency sound waves. This rapidly removes the soil from the parts.

7.4.1.6 Ultrasonic Cleaning with Liquid Rinse

When the ultrasonic cleaning device is used without the vapor degreasing procedure, liquid solvents produce a relatively high level of cleaning.

7.4.2   Abrasive Methods

Some applications do not have important appearance requirements and can tolerate abrasive treatment. These methods provide a higher level of surface treatment than the solvent methods listed in Section 7.4.1. While they produce no chemical change, they do remove small particles from the surface of the substrate, thus altering it physically. Consistency can be difficult to maintain, however, and complex shapes may have surfaces that cannot be accessed by abrasive methods. Abrasive cleaning can be applied to plastics according to Table 7-3.

7.4.2.1 Dry Abrasion

Dry abrasion is simply a light sanding of the surface with a 180 to 325 grit sand paper. A solvent cleaning is used to remove any surface soil prior to sanding, and another is used to clean it up afterward.

Table 7-3   Abrasive Cleaning Recommendations

Plastic	Method
ABS	Dry abrasion, wet or dry abrasive blast
Acetal copolymer	Dry abrasion, wet or dry abrasive blast
Acetal homopolymer	Dry abrasion
Butadiene styrene	Dry abrasion, dry abrasive blast
Cellulose acetate	Dry abrasion, wet or dry abrasive blast
Cellulose acetate butyrate	Dry abrasion, wet or dry abrasive blast
Cellulose acetate propionate	Dry abrasion, wet or dry abrasive blast
Diallyl phthalate	Dry abrasion, wet or dry abrasive blast
Epoxy, phenolics	Dry abrasive blast, wet abrasive scour, or detergent scrub
Furane, ionomer, melamine resins, SAN, polysulfone or rigid vinyl	Dry abrasion, wet or dry abrasive blast
Polyamide	Dry abrasion
Polycarbonate	Dry abrasion, wet or dry abrasive blast
Polyimide or polymethylmethacrylate	Dry abrasion, wet or dry abrasive blast, or abrasive scour
Polystyrene	Dry abrasion

7.4.2.2   *Dry Abrasive Blast*

The dry abrasive blast process replaces sandpaper with sand blasting. The particle size of the "sand" will vary according to the application and it may be composed of aluminum oxide, flintstone, glass beads, silica, or silicon carbide. The same solvent precleaning and postcleaning procedures are used as with dry abrasion.

7.4.2.3   *Wet Abrasive Blast*

The dry abrasive may be replaced by a 220–325 grit abrasive slurry composed of aluminum oxide or glass beads at a ratio of 1 part abrasive to 3 parts water. This is followed by a distilled water rinse. A water-break-free surface should be obtained.

7.4.2.4   *Wet Abrasive Scour*

Scouring powder and water are applied to the surface of the substrate with a metallic bristle brush or clean cloth. A rinse in clear running tap water is followed by a distilled water rinse. This technique should produce a water-break-free surface.

7.4.2.5   *Detergent Scrub*

The substrate surface is scrubbed with a nonmetallic brush using a 110–140 °F detergent solution composed of 1 to 5 oz of detergent per gallon of tap water. Scouring should be followed by tap water and then distilled water rinses. A water-break-free surface should be obtained.

7.4.3   Surface Energy Treatments and Process Selection Factors

Some plastics require significant improvement in surface energy. The bulk of these materials are the polyolefins (polyethylene, polypropylene, etc.), which have surface energies in the range of 30 dyn/cm. Four surface treatments are employed for this purpose: chemical, corona, plasma, and flame.

7.4.3.1   *Chemical Treatment*

When stronger surface treatments are required, chemicals can be applied. However, some polymers require very aggressive solutions that call for the level of ventilation associated with flammable or corrosive fumes. At best, they are expensive to use and dispose of. In addition, they are illegal in many places. They should be used only as a last resort, after a thorough search for an adhesive that does not require this extreme in surface treatment. Chemical treatments must be controlled in isolated storage, handled with care, and disposed of in a manner commensurate with their status as hazardous wastes.

Chemical treatment significantly improves the receptivity of the surface to adhesion by changing the chemical nature of the substrate surface. It is always preceded and is usually followed by solvent cleaning rinses, thereby performing that function as well. Such treatments are defined by their chemical, temperature, immersion, rinse, and drying parameters. Specific treatments can be found in most adhesive texts. They do not last long and should be performed just prior to the application of the adhesive.

7.4.3.2   *Corona Treatment*

Corona surface treatment is a term applied to several technologies utilizing high voltage electrical discharge in air. The gas is ionized when the free electrons always present in air accelerate in a high voltage field when it is applied across an air gap. Electron avalanching occurs when a very strong electric field causes collisions of high velocity electrons with molecules of gas resulting in no loss of momentum.

The electrons generated in this discharge impact the surface with energies two to three times that necessary to break the molecular bonds on the surface of most substrates, thus creating free radicals. The very reactive free radicals thus created can react quickly in the presence of oxygen to form various chemical functional groups on the surface of the substrate. Increased surface energy results from this oxidation reaction. This occurs without significantly affecting the balance of the part.

The process illustrated in Fig. 7-4a is the original corona treatment, the so-called conventional corona treatment. It is a process best known for the treatment of film or flat sheet with a maximum thickness of 0.25 in. Voltages of 10 to 12 kV are used in this process. The version shown in Fig. 7-4b is a more recent development. It is referred to by the manufacturer as a "three-dimensional electrical surface treatment" and uses voltages up to 50 kV to achieve three-dimensional surface treatment on parts as high as 1.5 in. Beyond the height limitation, the size can be unlimited.

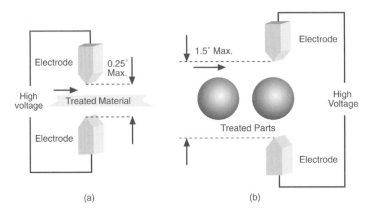

Figure 7-4   Corona surface treatments (Courtesy of Tantec, Inc.): (a) conventional surface treatment and (b) three-dimensional surface treatment

The apparatus produces surface energies from 50 to 72 dyn/cm with very high uniformity and can operate at speeds of up to 60 ft/min. Operating costs are low, but the initial investment can run from $12,000 to $200,000. The principal assets of this equipment are high speed and versatility; several other versions use gas plasma generated in a discharge to treat plastic substrates.

In one variety of the three-dimensional treatment, the "air-assisted arc treater," two wire electrodes are mounted across from each other in the treating area which also contains a fan mounted behind the electrodes. The electrodes receive a high voltage at a frequency of 60 Hz. The resultant arc is hot enough to melt the plastic substrate. However, the plasma generated is not, and it is blown onto the substrate surface in a plume that can reach an inch from the electrodes. Although this process is limited to small areas and treatment speeds up to 15 ft/min, it can handle slightly curved shapes as well as flat areas. There are no regulatory issues, and surface energies over 50 dyn/cm can be attained. The principal assets of this equipment are its low investment and operating costs plus its versatility, which lends itself well to production lines.

The effectiveness of this method is affected by the rate of speed at which the part passes the electrodes. For a polypropylene part, the surface energy can drop from 50 dyn/cm to 35 dyn/cm as the rate of speed increases from 10 ft/min to 30 ft/min. Distance is also a factor. For the same part, the surface energy can drop from 53 dyn/cm to 29 dyn/cm as the distance of the substrate surface is increased from 0.5 in. to 2.0 in.

Corona discharge systems are excellent for small parts, however they operate with high voltage and have a high initial investment. In addition, they require an ozone filter to cope with the ozone created by the process.

7.4.3.3   Plasma Treatment

Plasma treatment can improve bond shear strength by 200 to 300%, but the very high initial investment restricts use of this method to applications where a large number of very small parts can be fit into the treatment chamber. In addition, since plasma treatment is a batch process, it is not economical for use on large parts. The treatment speed and the part size and shape limitations depend on the size of the chamber, which in turn affects the investment. The process has been accomplished with an investment as low as $13,000 but initial costs can also run as high as $1 million. There are no regulatory issues, and a surface energy of 50 to 72 dyn/cm can be achieved.

In operation, a highly reactive ionized gas that contains ions, electrons, and other species is introduced to the treatment chamber. Then a high frequency (300 kHz to 15 mHz) electric field is applied, which excites the gas. The plasma fills the entire chamber and treats all sides of everything within. Plasma treatment is performed in a vacuum or specialized gas atmosphere and is best suited for intricate parts with adherend surfaces that are difficult to access by other methods.

7.4.3.4 Flame Treatment

Flame can also be used to increase the surface energy of the plastic, and it is often the technique of choice for three-dimensional plastic parts. Since it is inconvenient to turn the flame on and off frequently, however, a great deal of energy is needed to maintain a constant flame. Thin-walled sections cannot withstand this process without warpage, and the glossy substrate surface of some resins becomes marred by the flame treatment.

Furthermore, the flame can be difficult to control and consistency hard to sustain. There are four variables to control: air-to-gas ratio, burner output (in BTU), the distance from the substrate surface to the flame tip, and the length of time the substrate is under the flame. A very small window exists for optimum performance for each variable, and surface treatment levels drop if there is any deviation. However, a flame plasma analyzer can resolve problems of consistency in some cases.

Nonetheless, flame treatment is frequently the lowest cost method of providing a surface treatment, particularly when initial investment is taken into consideration. It is nontoxic, has no hazardous disposal problems, does not pin hole, and can attain higher surface energies than corona treatments, which last longer. It can also handle some, but not all, irregular shapes. Widely used for polyethylene and polypropylene, flame treatment has also been used for polyacetal, polyphenylene sulfide, and thermoplastic polyester, as well as other plastics.

7.4.3.5 Process Selection Factors

Ultimately, there are five decision factors involved in the selection of a surface treatment.
1. *Treatment level.* The cost of surface treatment is roughly commensurate with the level of treatment needed.
2. *Treatment speed.* The slower systems cost less to buy and operate.
3. *In-line or off-line surface treatment.* This aspect is related to treatment speed, since only the highest speed systems need be on-line.
4. *The shape and size of the parts to be treated.* Basically, flat is least costly.
5. *Costs.* Investment and operating costs are the operating factors in any business decision.

7.4.4   Shelf Life of Surface Treatments

The heightened level of surface energy created by the surface treatment is not permanent. In fact, it begins to drop off sharply immediately after the treatment. As
Fig. 7-5 illustrates, 10 dyn/cm is lost within 3 h; after a day, 18 dyn/cm is gone. The
loss is due to the migration to the surface of low molecular weight components such
as amides and antiblocking agents. The purity of the resin is a determining factor,
but the storage environment and type of surface treatment are significant as well.
Note, however, that solvent washes can restore much of the loss.

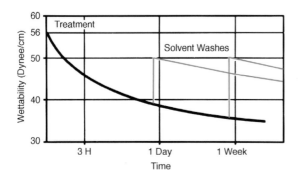

Figure 7-5   Effect of additive migration to the surface on wettability of
corona treated polymers (Courtesy of
Tantec, Inc.)

7.5   Design for Adhesion

To design for adhesion, it is first necessary to understand the forces the adhesive or
solvent joint must withstand. There are four: shear, tensile, cleavage, and peel. The
latter two describe the same type of loading, namely, a tensile pull from one end of
the joint. However, one refers to the type of stress that results when both adherends
are rigid and the other pertains to the type of stress created when one part is rigid
and the other is flexible. A fifth stress, compressive stress, is also applied to adhesive
and solvent joints, but failures of this type would be extraordinary. Fatigue failure
modes for adhesive joints include crazing, interfacial failure, debonding, void
growth, multidirectional cracking, plus crack initiation and growth.

7.5.1   Shear Stress

As depicted in Fig. 7-6, shear stress is the type that pulls across the adherends, trying to
cause them to slide against each other. Most adhesive joints are strongest in this direction, and most are designed to ensure that the bulk of the load is in shear. Note that the
load is applied to all the bonding surface. Tensile and shear strengths are usually almost
equal in flexible epoxies, whereas tensile strength is greater in rigid and toughened
epoxies.

7.5.2   Tensile Stress

Figure 7-7 shows tensile stress attempting to pull the two parts apart. The stress is spread uniformly across the entire surface so that all the bonded surface is under load. Adhesive and solvent joints are designed to put the load that is not in shear in tensile stress.

7.5.3   Cleavage

Cleavage is a load applied to one end of the joint as illustrated in Fig. 7-8. The joint is far more vulnerable to this type of stress than it is to shear or tensile stress because the adhesive at one end of the joint must withstand most of the load. As failure occurs, the load moves progressively across the bond surface. In practice, the relative stiffness of the adherends and the nature of the adhesive determine the distribution of the load. This effect is more pronounced in peel stress.

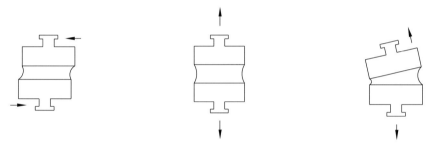

Figure 7-6   Shear stress   **Figure 7-7**   Tensile stress   **Figure 7-8**   Cleavage

7.5.4   Peel

Figure 7-9 represents two adherends, one rigid and one flexible. The flexible nature of the left one results in a greater concentration of the load on the adhesive at the edge of the joint. The extent of this loading is also affected by the nature of the adhesive. A solvent or rigid adhesive, such as that shown in Fig. 7-9a, constricts the stress-bearing area to a very narrow band. The flexible adhesive in Fig. 7-9b results in a much greater stress-bearing area. (Solvents do not, themselves, provide flexibility, since they evaporate and are no longer present.)

7.5.5   Adhesive Joint Designs

7.5.5.1   *Load-Bearing or Non-Load-Bearing Joints*

While solvent and adhesive joints are normally thought of as load bearing, many sealing applications call for joints that emphasize non-load-bearing functions. Principal

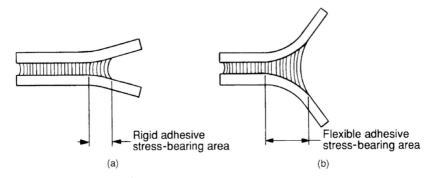

Figure 7-9   Peel: (a) rigid adhesive stress bearing area and (b) flexible adhesive stress bearing area

among these are joints designed for sealing or cushioning applications. The adhesives used for such purposes are frequently supplied in liquid or hot melt form.

Some adhesives perform both functions to a reasonable degree in that they fill in the crevices of rough surfaces or compensate for part distortion. That type of condition is illustrated in Fig. 7-10. Solvent joint surfaces need to be smooth for sealing to be accomplished.

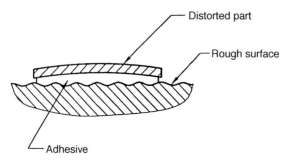

Figure 7-10   Sealing adhesive

In general, adhesives are stronger in shear and tension than they are in peel and cleavage. Consequently, adhesive joints are designed to place the loads in a manner, that utilizes their best qualities. The designer must also be cognizant of the effect of thermal expansion on the joint.

Shear stresses can increase dramatically when the two bonded materials differ significantly in coefficient of linear thermal expansion.

Obviously, the more severe the thermal operating conditions of the application, the more significant this factor becomes.

7.5.5.2   Lap Joints

Most joint discussions make the assumption that only two flat sheets are to be attached to each other. In reality, that is rarely the case. Most parts to be joined have configurations in which the bond surfaces tend to reinforce each other and create a much stronger structure. The author has taken this property into consideration when commenting on the relative desirability of various joints, a refinement that may result in apparent discrepancies between the comments the reader will find here and in other sources.

The "simple lap joint" provides an excellent example of the inadvisability of making the assumption mentioned above.. Most references state that this design is undesirable because it has poor resistance to cleavage and peel. For two rigid pieces of sheet stock, that is true. However, the simple lap joint is not often used that way. When used such that the adhesive or solvent is applied around the sides of a part, it is quite strong. In fact, that is probably the most common method of solvent joining plastic components.

In lap joint, one material overlaps the other. There are several varieties of lap joint. The one illustrated in Fig. 7-11a, the *simple lap joint*, works well when the two materials are reasonably flexible or the wall thicknesses are thin enough to flex. If one of the flat pieces is too stiff, the joint is vulnerable to peel stress, as illustrated in Fig. 7-9. If both are too stiff, cleavage like that depicted in Fig. 7-8 will result. Should the walls become too thick, the forces pulling on the part will be out of alignment, resulting in the deformation depicted in exaggerated fashion in Fig. 7-11b.

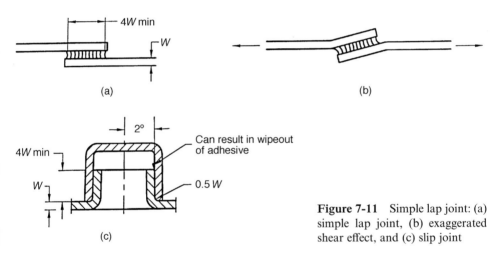

Figure 7-11 Simple lap joint: (a) simple lap joint, (b) exaggerated shear effect, and (c) slip joint

The previously mentioned version of a lap joint commonly used to assemble small thermoplastic parts, known as a *slip joint*, is illustrated in Fig. 7-11c. In this case, the part is dipped in solvent up to the level of the joint and the two parts are pressed together. The bond is on the complete structure, which thereby acts as a reinforcement, reducing the effect of cleavage stress and permitting the use of rigid materials. When used in this fashion, the design is sometimes referred to as an *overlap slip joint*. Care must be taken in the preparation of the solvent, for if it becomes too viscous, wipeout of the adhesive may occur during assembly.

The alignment of forces issue can be resolved by going to the *joggle lap joint* shown in Fig. 7-12a, where the shear load is now in line. A generous inside radius must be provided, lest a sharp corner stress concentration be created. In addition, Fig. 7-12b

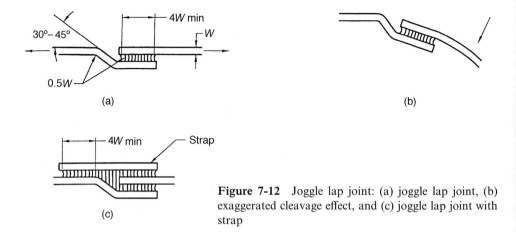

Figure 7-12 Joggle lap joint: (a) joggle lap joint, (b) exaggerated cleavage effect, and (c) joggle lap joint with strap

demonstrates that some of the bending moment that would otherwise result in cleavage stress can be accommodated by this design if the components are of moderate stiffness. Reinforcement of this type of joint can be accomplished with a strap, as illustrated in Fig. 7-12c.

The *strap joint* can be used as a joint in itself, as in Fig. 7-13a. However, it is much stronger when used in the double-strap configuration shown in Fig. 7-13b. The recessed version illustrated in Fig. 7-13c is designed to improve aesthetics by hiding the joint. This modification leaves a flush surface and makes it easier to control the oozing of excess adhesive out of the joint. Note that we have specified radii in inside corners to

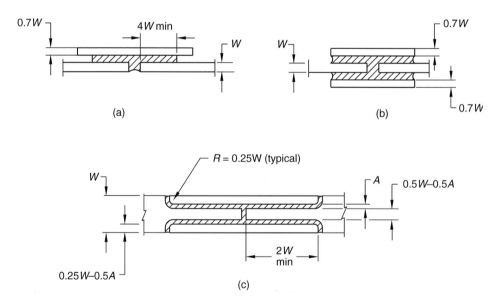

Figure 7-13 Strap joint: (a) single strap joint, (b) double strap joint, and (c) recessed double strap joint

reduce the stress concentrations so vulnerable to solvents. Since this is typically a machined joint, those radii would require custom-made cutting heads or blades.

> The improvement in part quality resulting from the elimination of sharp inside corners is well worth the extra cost of custom-made cutters.

A plastic part is, after all, no stronger than its weakest link. Other joint designs in this chapter will also illustrate radii on inside corners.

The joggle lap joints shown in Figs. 7-12a and 7-12c could be made by all the processes that make flat parts; they lend themselves best, however, to the thermoset processes and thermoforming. Straps can be put on anything, but they are most likely to be found on low volume machined applications. Machining is also the most likely means of producing the recessed strap shown in Fig. 7-13c. The only other processes that lend themselves well to this configuration – injection and transfer molding – would typically be designed to use a higher volume type of joint design.

The bending moment can be handled in a different fashion with the *tapered simple lap joint*, also a design for unstructured parts. Since many plastic parts become more flexible as their wall thickness is reduced, a feathering of the edge can be used, as illustrated in Fig. 7-14a. This, too, will permit bending and avoid cleavage with some resins. This joint could be produced by all the flat part processes, but it would most likely be machined.

The *double-butt lap joint* depicted in Fig. 7-15a takes a different approach. The butt joints located at the ends of the lap joint cause the stress at the ends to be either in compression or tension, depending on direction, as shown in Fig. 7-15b. The greater the wall section, the more robust this joint becomes. This is another very effective design for flat sheets. While having thick wall sections may be an acceptable condition

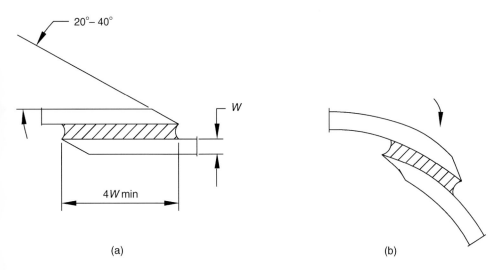

Figure 7-14   Tapered simple lap joint: (a) tapered simple lap joint and (b) exaggerated shear effect

if a small volume of machined units are to be made, it is unacceptable for products molded in large numbers. The thin wall configuration illustrated in Fig. 7-15c has draft figured in for a $3W$ length and would be more desirable from a molding stand-point, where it is more often referred to as a *semi dovetail joint*. Once again, the designer must be cognizant of the potential for inside sharp corner stress. This type of joint would have to be machined into a cold molded, RTM, or compression-molded part

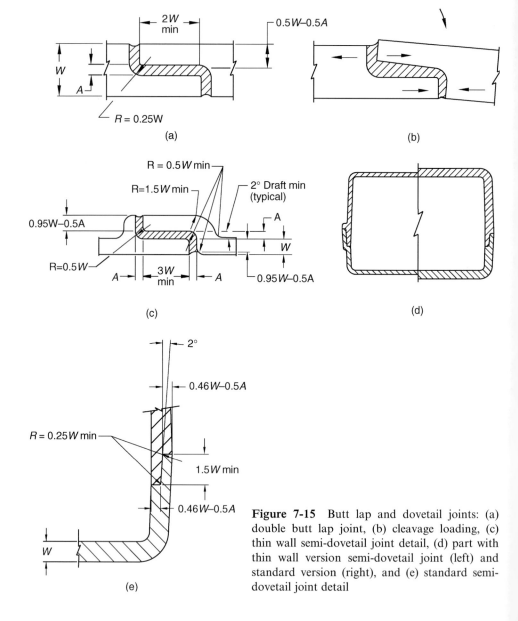

Figure 7-15 Butt lap and dovetail joints: (a) double butt lap joint, (b) cleavage loading, (c) thin wall semi-dovetail joint detail, (d) part with thin wall version semi-dovetail joint (left) and standard version (right), and (e) standard semi-dovetail joint detail

and is not really feasible for a thermoformed, blow-molded, or rotationally molded part.

The left side of Fig. 7-15d illustrates how the semi dovetail joint would appear as part of a thin-walled molded part with a standard version shown on the right. A standard version with suggested dimensions is illustrated in Fig. 7-15e. Note that these dimensions differ slightly from the ones shown in Fig. 7-15a (a joint that is intended to be machined), whereas the joint in Fig. 7-15e is intended to be molded and has a draft angle of 2° included for a 1.5W length. For another combination of draft angle and length of joint, the thickness of the end of each half of the dovetail would be determined by calculating the effect due to the draft with trigonometry and taking half from each side of the joint. Care must be taken not to thin the wall below minimum strength and moldability limits for the material.

The *scarf joint* represented in Fig. 7-16a provides an increase in the bonded surface for a given wall thickness and less opportunity for loads to result in cleavage stress. However it provides poor positioning and is time-consuming to machine.

The *double scarf joint* shown in Fig. 7-16b is more robust than the double butt lap for flat panels, but its inside corners are more severe, and it is even more difficult to tool or machine. The *Butt scarf lap joint* represented in Fig. 7-16c is a compromise between the two. This joint would suffer the same processing limitations as the double scarf joint. All the scarf joints are best used for assembling sheet stock as used in furniture, large housings, and so on.

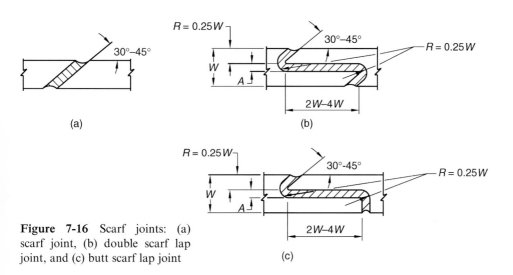

Figure 7-16 Scarf joints: (a) scarf joint, (b) double scarf lap joint, and (c) butt scarf lap joint

7.5.5.3 Butt Joints

In a butt joint, the adhesive is applied to the ends of the walls. This is another joint that is held in low esteem by those who think of it only as a method of joining flat panels.

In fact, it is probably the most common method of joining fastener blocks and supports to the insides of thermoformed, cold-molded, RTM, and compression-molded parts. Fixtures are used to locate the parts when no internal locating features have been provided.

The design illustrated in Fig. 7-17a is a *simple butt joint*. Although inexpensive to machine in that form, it is rarely used to join two pieces because it would be extremely vulnerable to cleavage stress. A more typical configuration is the version in Fig. 7-17b where the joints are part of a larger structure, which limits the amount of cleavage stress that can reach the joint. If the material is sufficiently flexible, it will absorb some of the load as well. When the butt joints are formed as angles, they are referred to as *angle butt joints*. This is a configuration commonly used as stiffeners for large panels or in a gusset form between two walls as a reinforcement for thermoformed, layup/sprayup, RTM, cold-molded, and compression molded parts.

A larger reinforcement might be formed as represented in Fig. 7-17c. This type of construction is known as a *corrugated stiffener*. The extruded T shape shown in Fig. 7-17d is also used as a stiffener and it is referred to as a *T joint*. Fixtures would need to be employed to locate these stiffeners.

In many cases, however, the design permits location to be formed or molded into the part. This will improve the efficiency of the assembly process. Such a design is illustrated in Fig. 7-17e. The design can also be used to provide a well for the adhesive

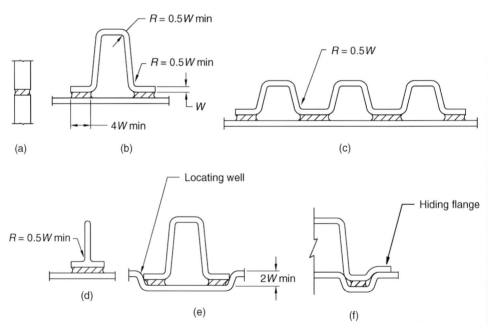

Figure 7-17   Butt joints: (a) simple butt joint, (b) angle butt joints, (c) corrugated stiffener, (d) T joint, (e) joint with integral location, and (f) integral located joint in adhesive well

or solvent, as in Fig. 7-17f, to help control the location of the adhesive so that it does not leak out of the joint area. The flange added to this example would also help hide any excess adhesive.

Taken to the next step, in Fig. 7-18a, the butt joint is placed deeper into the slot that provides side walls for support, thus becoming the well-known *tongue-and-groove joint*. Note that the radius at the tip of the dovetail can go to 0.2W. That would make it a full round, a configuration that minimizes stress in a plastic joint. If the sides are angled to a scarf configuration, the joint becomes a *landed scarf tongue-and-groove joint*, as shown in Fig. 7-18b. With the extra strength provided by the scarfs, the lands can be used to control the depth and, consequently, the adhesive line thickness. In a third form of this concept, the *scarf tongue-and-groove joint*, the lands come to a sharp point; this leaves very thin walls on each side, however, which is not a condition recommended for plastics. In addition to providing a controlled place for the adhesive, tongue-and-groove configurations are self-aligning during assembly. Finally, there is a version in which there are no lands and no butt either. That leaves a simple V configuration, which might be acceptable for some machined applications but would have little value for parts manufactured by the other processes.

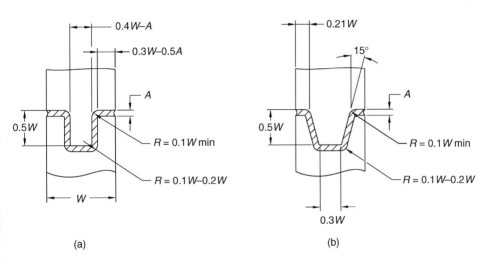

(a)   (b)

Figure 7-18   Tongue and groove joints: (a) tongue-and-groove butt joint and (b) landed scarf tongue-and-groove joint

These designs are well established in the furniture industry and can be used for the assembly of plastic sheet goods. However, they must be injection-molded, transfer-molded, cast, or machined into parts from one of the other processes. The semidovetail joint is better suited to the injection and transfer molding processes because these allow the available wall thickness to be split between two arms instead of three. Since location is provided by the joint on the other side of the part (long walls may need inside controlling ribs), the only reason to use a full dovetail is to provide a well for the adhesive or solvent. As for the other processes, this becomes a costly joint to create and should be used only for low volume applications as there are more cost-effective ways of handling higher volumes.

Some constructions call for corner joints. The *corner butt joint* depicted in Fig. 7-19a would be highly vulnerable to cleavage stress; normally, however, it is supported by other construction. If it does need further support, a reinforcement block or extruded angle can be added to the inside, as indicated by the dashed lines. Occasionally, support is added to the outside in the form of an angle – also represented by dashed lines.

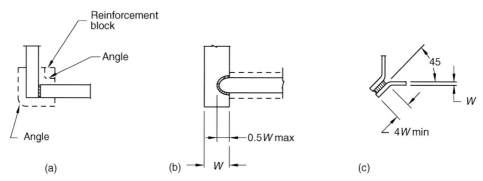

Figure 7-19   Corner joints: (a) corner butt joint, (b) rounded corner butt joint, and (c) flanged corner joint

The strongest corner joint for adhesives is represented by the *rounded corner butt joint* illustrated in Fig. 7-19b. That design converts all the stress to shear stress in which adhesives are at their strongest. However, it is more costly to machine and does not lend itself well to the other methods. Processes such as thermoforming, layup/sprayup, RTM, cold molding, and compression molding produce flanged parts quite readily. The *flanged corner joint* shown in Fig. 7-19c is well suited to these processes and can be readily clamped, although the 45° angle it calls for will result in some wastage.

7.5.5.4   Screw and Glue

Insurance against the possibility of an improperly prepared joint surface or poor joining technique is provided by the screw-and-glue method, which also can be used to provide clamping force when the joint is in a location that is difficult to reach with conventional clamps. (The reader is referred to Chapter 8, "Fasteners and Inserts," for a discussion of clamping force.) As illustrated in Fig. 7-20, an adhesive or solvent is placed on the bonding surface and the parts are joined in the normal fashion. Then a screw is emplaced in a secondary operation. Normally, this would be a self-tapping screw, since this is a permanent joint with no intent to reopen and consequently no need for a nut or threaded insert. This concept is not to be confused with the practice of placing a drop of adhesive in a hole that is about to receive a screw.

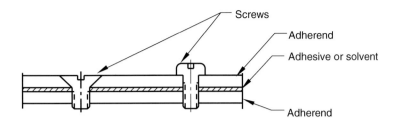

Figure 7-20
Screw and glue

In that case, the adhesive is merely intended to prevent the screw from vibrating loose, the screw is in fact the principal joining device.

7.6   Adhesives

Adhesives have benefited from considerable research and are available in a wide variety, some with strengths far beyond those that were previously obtainable. In some cases, commercial preparations consist of multiple adhesive formulations. One adhesive might provide a quick bite but have a lower ultimate strength, while the second might have a higher ultimate strength but require too much time to set up to permit immediate further assembly operations. By combining the two, the desired characteristics of both are obtained.

The adhesives are discussed in a generic manner and are listed in alphabetical order. However, some manufacturers attempt to obscure the true nature of the proprietary formulations in their catalogs. The author has attempted to identify manufacturers of the generic adhesives where possible. Individual recommendations for each plastic are listed in Tables 7-4b, c, and d (the key to the number codes is found in Table 7-4a). The adhesives listed there are provided as general direction for the designer to pursue.

Individual formulations must be investigated to be certain that there are no solvents or other chemicals present which would attack the plastic materials.

Stress cracking is a common result when this rule is ignored.

7.6.1   Acrylics

Acrylic adhesives are structural adhesives that provide good resistance to low temperatures and high temperature resistance up to the 300 to 350 °F range. Chemical resistances are good and humidity resistances are good to excellent. They provide an excellent bond to plastics and a good bond of plastics to metals. Strengths are on the order of 1000 to 3000 psi. Bonds to polyolefins are only fair, and those involving wood, paper, and cardboard are not recommended. Impact strength and ability to fill in crevices are good, and flexibility is good to excellent. Acrylics are more effective

Table 7-4a   Key to Adhesives Listed by Number in Tables 7-4b,c,d

Elastomeric
 1. Natural rubber
 2. Reclaim
 3. Neoprene
 4. Nitrile
 5. Urethane (also thermosetting)
 6. Styrene butadiene

Thermoplastic resin
11. Polyvinyl acetate
13. Acrylic
14. Cellulose nitrite
15. Polyamide
16. Hot melt copolymer blends (EVA, EEA, etc.)
25. Cyanoacrylate

Thermosetting resin
22. Resorcinol phenolic
23. Epoxy
24. Urea formaldehyde
26. Reactive acrylate monomer systems
31. Butyral phenolic
36. Polyester (also thermoplastic)
37. Anaerobic

Miscellaneous
41. Rubber lattices (natural or synthetic-water based)
42. Resin emulsions (water-based)

than epoxies for bonding plastics to metals because they can be bonded to slightly oily metal surfaces, whereas epoxies require clean surfaces.

Acrylics have a medium viscosity and are available as a two-part polymerization with a catalyst cure at room temperature or as a one-part adhesive with a heat or ultraviolet light cure. UV cures are very fast, with much less energy usage than ovens, however deep cures can require an additional cure mechanism. Acrylics have a medium viscosity and are available as liquids and gels. Acrylics can be applied with tubes, bottles, brushes and meter-mix equipment.

Typical Applications Potting, circuit boards
Sources Devcon, Dymax, Fuller, Hardman, Humi-Seal, Loctite, Lord, Permabond, Tech Spray

7.6.2   Anaerobics

The anaerobic adhesives have the unusual property of curing only in the absence of oxygen. They have good sealant qualities and environmental resistance. On an overall

Table 7-4b Adhesives for Bonding Plastics to Plastics[a]

Surfaces	ABS	Acetal	Cellulosic[b]	Diallyl phthalate	Epoxy	Ethyl cellulose
ABS	4, 5, 23, 25, 26	4, 16	4, 16, 36	4, 31	4, 25	5, 16
Acetal	4, 16	4, 5, 16, 23, 31	4, 5, 16	4, 23, 31	4, 23, 31	4, 5, 16
Cellulosic[b]	4, 16, 36	4, 5, 16	4, 5, 14, 16, 36	4, 36	4, 36	4, 5, 14, 16, 36
Ethyl cellulose	5, 16	5, 16	5, 14, 16	4, 5, 36	4, 36	4, 5, 14, 15, 16, 36
Nylon	5, 23, 25, 26	5, 23	5, 36	3, 5, 23, 36	3, 23, 36	5, 36
Phenylene oxide-based resin	5, 23, 25, 26, 36	4, 5, 16, 23	4, 36	3, 5, 23, 31, 36	4, 23, 25, 31, 37	4, 5, 36
Polycarbonate	5, 25, 26	4, 16	4, 16	4	4	5, 16
Polyether sulfone	5, 25, 26, 36	5, 16, 23	5, 36	5, 31, 36	25, 31, 36	5, 36
Polyethylene[c]	5, 16	5, 23, 31	5, 36	23, 31, 36	23, 31, 36	5, 15, 16, 36
Polyethylene terephthalate	5, 25, 26, 36	4, 5	4, 5, 36	4, 5, 36	4, 36	4, 5, 36
Polymethyl methacrylate	25, 26	5, 31	5, 36	5, 31, 36	31, 36	5, 36
Polyphenylene sulfide	23, 25	4, 5, 16, 23	4, 14, 16	23, 31, 36	4, 23, 36	4, 16
Polypropylene[c]	5, 16	5, 23, 31	5, 36	23, 31, 36	23, 31, 36	5, 36
Polystyrene	16, 25	5, 16, 23, 31	5, 16, 36	23, 31, 36	23, 31, 36	5, 16, 36
Polyurethane	4, 25, 36	4, 5, 23	4, 5, 36	4, 23, 36	4, 23, 36	4, 5, 36
Polyvinyl chloride	4, 25, 26	4, 5	4, 5, 36	4, 36	4, 36	4, 5, 36
Tetrafluoro-ethylene	15, 23	23	5	5, 23	23	5, 15
Diallyl phthalate	4, 31	4, 23, 31	4, 36	3, 4, 5, 23, 31, 36	3, 4, 23, 31, 36	4, 5, 36
Epoxy	4, 25	4, 23, 31	4, 36	3, 4, 23, 31, 36	3, 4, 23, 31, 36	4, 36
Melamine	4, 16	4, 16, 23, 31	4, 16, 36	3, 4, 23, 31, 36	3, 4, 23, 31, 36	4, 16, 36
Phenolic	5, 16, 25	4, 5, 16, 23, 31	4, 5, 16, 36	3, 4, 23, 31, 36	3, 4, 23, 31, 36	4, 5, 16, 36
Polyester	23, 26, 36	4, 23, 31	4, 36	4, 23, 31, 36	4, 23, 31, 36	4, 36
Urea	4, 16	4, 16, 23, 31	4, 16	3, 4, 23, 31	3, 4, 23, 31	4, 16

Table 7-4b Adhesives for Bonding Plastics to Plastics[a] *Continued*

Surfaces	Melamine	Nylon	Phenolic	Phenylene oxide-based resin	Polycarbonate	Polyester
ABS	4, 16	5, 23, 25, 26	5, 16, 25	5, 23, 25, 26, 36	5, 25, 26	23, 26, 36
Acetal	4, 16, 23, 31	5, 23	4, 5, 16, 23, 31	4, 5, 16, 23	4, 16	4, 23, 31
Cellulosic[b]	4, 16, 36	5, 36	4, 5, 16, 36	4, 36	4, 16	4, 36
Ethyl cellulose	4, 16, 36	5, 36	4, 5, 16, 36	4, 5, 36	5, 16	4, 36
Nylon	3, 23, 36	3, 5, 22, 23, 25, 26, 36	3, 5, 23, 25, 36	3, 5, 23, 26, 37	25, 26	23, 26, 36
Phenylene oxide-based resin	4, 23, 31	3, 5, 23, 26, 37	3, 23, 31, 37	5, 13, 23, 25, 26, 36	4, 16, 25	4, 23, 26, 36, 37
Polycarbonate	4, 16	25, 26	4, 16, 25	4, 16, 25	4, 15, 16, 25, 26	4, 26
Polyether sulfone	23, 31, 36	5, 25, 26, 36	5, 25, 31	23, 25, 36	25, 26	26, 31, 36
Polyethylene[c]	16, 23, 31, 36	5, 23, 36	5, 16, 23, 31, 36	23, 31, 36	15, 16	23, 31, 36
Polyethylene terephthalate	4, 36	5, 25, 26, 36	4, 5, 25, 36	4, 5, 36	4, 25, 26	4, 26, 36
Polymethyl methacrylate	31, 36	5, 25, 26, 36	5, 25, 31, 36	5, 31, 36	25, 26	23, 31, 36
Polyphenylene sulfide	4, 23, 36	5, 22, 23, 25	4, 16, 25, 26	23, 25, 26	4, 15, 16, 25	4, 26, 36
Polypropylene[c]	23, 31, 36	5, 23, 36	5, 23, 31, 36	23, 31, 36	15, 16	23, 31, 36
Polystyrene	16, 23, 31, 36	5, 23, 25, 36	5, 16, 23, 25, 31, 36	23, 31, 36	16, 25	23, 31, 36
Polyurethane	4, 23, 36	5, 23, 36	4, 5, 23, 36	4, 23, 36	4	4, 23, 36
Polyvinyl chloride	4, 36	5, 26, 36	4, 5, 36	4, 36	4, 26	4, 26, 36
Tetrafluoro-ethylene	23	5, 22, 23	5, 23	5, 23	15	23
Diallyl phthalate	3, 4, 23, 31, 36	3, 5, 23, 36	3, 4, 23, 31, 36	3, 5, 23, 31, 36	4	4, 23, 31, 36
Epoxy	3, 4, 23, 31, 36	3, 23, 36	3, 4, 23, 31, 36	4, 23, 25, 31, 37	4	4, 23, 31, 36
Melamine	3, 4, 16, 23, 31, 36	3, 23, 36	3, 4, 16, 23, 31, 36	4, 23, 31	4, 16	4, 23, 31, 36
Phenolic	3, 4, 16, 23, 31, 36	3, 5, 23, 25, 36	3, 4, 5, 16, 23, 25, 31, 36	3, 23, 31, 37	4, 16, 25	4, 23, 31, 36
Polyester	4, 23, 31, 36	23, 26, 36	4, 23, 31, 36	4, 23, 26, 36, 37	4, 26	4, 23, 26, 31, 36
Urea	3, 4, 16, 23, 31	3, 23	3, 4, 16, 23, 31	4, 23, 31	4, 16	4, 23, 31

Table 7-4b Adhesives for Bonding Plastics to Plastics[a] *Continued*

Surfaces	Polyether sulfone	Polyethylene[c]	Polyethylene terephthalate	Polymethyl-methacrylate	Polyphenylene sulfide	Polypropylene
ABS	5, 25, 26, 36	5, 16	5, 25, 26, 36	25, 26	23, 25	5, 16
Acetal	5, 16, 23	5, 23, 31	4, 5	5, 31	4, 5, 16, 23	5, 23, 31
Cellulosic[b]	5, 36	5, 36	4, 5, 36	5, 36	4, 14, 16	5, 36
Ethyl cellulose	5, 36	5, 15, 16, 36	4, 5, 36	5, 36	4, 16	5, 36
Nylon	5, 25, 26, 36	5, 23, 36	5, 25, 26, 36	5, 25, 26, 36	5, 22, 23, 25	5, 23, 36
Phenylene oxide-based resin	23, 25, 36	23, 31, 36	4, 5, 36	5, 31, 36	23, 25, 26	23, 31, 36
Polycarbonate	25, 26	15, 16	4, 25, 26	25, 26	4, 15, 16, 25	15, 16
Polyether sulfone	25, 26	5, 31, 36	5, 13, 25, 26	5, 13, 25, 31	25, 26	5, 31, 36
Polyethylene[c]	5, 31, 36	5, 15, 16, 23, 31, 36, 41	5, 36	5, 31, 36	5, 23	5, 23, 31, 36, 41
Polyethylene terephthalate	5, 13, 25, 26	5, 36	4, 5, 13, 25, 26, 36	5, 13, 25, 26, 36	5, 25, 26	5, 36
Polymethyl methacrylate	5, 13, 25, 31	5, 31, 36	5, 13, 25, 26, 36	2, 5, 6, 13, 25, 26, 31, 36	5, 25, 26, 36	5, 26, 31
Polyphenylene sulfide	23, 25	5, 23	5, 25, 26	5, 25, 26, 36	23, 25, 26	5, 23, 26
Polypropylene[c]	5, 31, 36	5, 23, 31, 36, 41	5, 36	5, 26, 31	5, 23, 26	5, 15, 16, 23, 31, 36, 41
Polystyrene	5, 13, 25, 31	5, 23, 31, 36	5, 13, 25, 26	2, 5, 6, 13, 25, 31, 36	5, 23, 25, 36	5, 23, 31, 36
Polyurethane	5, 25, 36	5, 23, 36	4, 5, 36	5, 36	5, 23, 36	5, 23, 36
Polyvinyl chloride	5, 13, 25, 26	5, 36	4, 5, 13, 26, 36	5, 13, 26, 36	5, 25, 26, 36	5, 36
Tetrafluoro-ethylene	5	5, 15, 23	5	5	5, 22, 23	5, 15, 23
Diallyl phthalate	5, 31, 36	23, 31, 36	4, 5, 36	5, 31, 36	25, 31, 36	23, 31, 36
Epoxy	25, 31, 36	23, 31, 36	4, 36	31, 36	4, 23, 36	23, 31, 36
Melamine	23, 31, 36	16, 23, 31, 36	4, 36	31, 36	4, 23, 36	23, 31, 36
Phenolic	5, 25, 31	5, 16, 23, 31, 36	4, 5, 25, 36	5, 25, 31, 36	4, 16, 25, 26	5, 23, 31, 36
Polyester	26, 31, 36	23, 31, 36	4, 26, 36	26, 31, 36	4, 26, 36	23, 31, 36
Urea	31	16, 23, 31	4	31	3, 23	23, 31

Table 7-4b Adhesives for Bonding Plastics to Plastics[a] *Continued*

Surfaces	Polystyrene	Polyurethane	Polyvinyl chloride	Tetrafluoroethylene[c]	Urea
ABS	16, 25	4, 25, 36	4, 25, 26	15, 23	4, 16
Acetal	5, 16, 23, 31	4, 5, 23	4, 5	23	4, 16, 23, 31
Cellulosic[b]	5, 16, 36	4, 5, 36	4, 5, 36	5	4, 16
Ethyl cellulose	5, 16, 36	4, 5, 36	4, 5, 36	5, 15	4, 16
Nylon	5, 23, 25, 36	5, 23, 36	5, 23, 36	5, 22, 23	3, 23
Phenylene oxide-based resin	23, 31, 36	4, 23, 36	4, 36	5, 23	4, 23, 31
Polycarbonate	16, 25	4	4, 26	15	4, 16
Polyether sulfone	5, 13, 25, 31	5, 25, 36	5, 13, 25, 26	5	31
Polyethylene[c]	5, 23, 31, 36	5, 23, 36	5, 36	5, 15, 23	16, 23, 31
Polyethylene terephthalate	5, 13, 25, 36	4, 5, 36	4, 5, 13, 26, 36	5	4
Polymethyl methacrylate	2, 5, 6, 13, 25, 31, 36	5, 36	5, 13, 26, 36	5	31
Polyphenylene sulfide	5, 23, 25, 36	5, 23, 36	5, 25, 26, 36	5, 22, 23	3, 23
Polypropylene[c]	5, 23, 31, 36	5, 23, 36	5, 36	5, 15, 23	23, 31
Polystyrene	2, 5, 6, 13, 16, 23, 25, 36	5, 23, 36	5, 13, 36	5, 23	16, 23, 31
Polyurethane	5, 23, 36	4, 5, 23, 36	4, 5, 36	5, 23	4, 23
Polyvinyl chloride	5, 13, 36	4, 5, 36	4, 5, 11, 13, 26, 36, 42	5	4
Tetrafluoroethylene	5, 23	5, 23	5	5, 15, 22, 23	23
Diallyl phthalate	23, 31, 36	4, 23, 36	4, 36	5, 23	3, 4, 23, 31
Epoxy	23, 31, 36	4, 23, 36	4, 36	23	3, 4, 23, 31
Melamine	16, 23, 31, 36	4, 23, 36	4, 36	23	3, 4, 16, 23, 31
Phenolic	5, 16, 23, 25, 31, 36	4, 5, 23, 36	4, 5, 36	5, 23	3, 4, 16, 23, 31
Polyester	23, 31, 36	4, 23, 36	4, 26, 36	23	4, 23, 31
Urea	16, 23, 31	4, 23	4	23	3, 4, 16, 23, 31

[a] Adhesive number codes shown in Table 7-4a.
[b] Cellulose acetate, cellulose acetate butyrate, cellulose nitrate.
[c] Will require surface treatment.
Source: Goodman, S. H. and Schwartz, S. S., *Plastics Materials and Processes*, p. 791.

Table 7-4c Adhesive Types for Bonding Nonplastics to Thermoplastics

Thermoplastics	Nonplastics						
	Ceramic	Fabric	Leather	Metal	Paper	Rubber	Wood
ABS	6, 16	3, 5	16, 23	23, 25, 26	4, 42	5, 16, 21, 25	4, 26, 42
Acetal	16, 23	4, 23	4, 23	4, 23	4, 16, 23	4, 16	16, 23
Cellulosic[b]	4, 16	4, 5, 42	4, 5, 42	3, 4	16, 42	1–5, 16	4, 16
Ethyl cellulose	14, 16	14	14	14	14, 16	14, 16	14, 16
Nylon	4, 23, 26	3, 4	3, 4, 26	3, 23, 25, 26	4, 41	2, 3, 25	3, 4, 26
Phenylene oxide-based resin	5, 24	4	16, 31, 36	23, 25, 31, 37	4, 42	3, 4	16, 26, 42
Polycarbonate	16, 23, 26, 36	23, 26	16, 23, 26	23, 25, 26	16, 36	5, 16, 25, 36	16, 23, 26, 36
Polyethersolfone	16, 23, 25	3, 4	4, 31, 36	23, 25	5, 41	2, 6, 16, 25	16, 26, 36
Polyethylene	3, 16, 41	3, 41	3, 41	3, 31, 41	16, 41	3, 16, 41	3, 16, 41
Polyethylene terephthalate	26, 36	5, 36	5, 26, 36	26, 36	5, 36	13, 36	26, 36
Polymethylmethacrylate	3, 4, 26	4	3, 4, 16, 42	3, 4, 25	42	1–5	3, 4, 26, 42
Polyphenylene sulfide	16, 23	3, 5, 42	16, 31	4, 23, 25, 26	4, 16, 41	5, 16, 25	16, 26
Polypropylene[c]	1, 16, 41	1, 41	1, 41	1, 2	1, 2, 16, 41	1, 2, 16, 41	1, 2, 16, 41
Polystyrene	16, 41, 42	3, 5	5, 31, 36	25, 31	5, 16, 31, 36	2, 16, 25	16, 31, 36
Polyurethane	4	5, 36	4, 5	4, 5	5, 36	5, 36	36
Polyvinyl chloride	4, 5, 26	4, 5, 42	4, 5, 16, 41, 42	3, 4, 15, 26, 36	42	4, 5, 15	4, 26, 36, 42
Tetrafluoroethylene[c]	23	22	22	22, 23	22, 23	23	23

[a] Adhesive number code is shown in Key.
[b] Cellulose acetate, cellulose acetate butyrate, cellulose nitrate.
[c] Special surface treatment recommended.
Table developed by Bostik Div., USM Corp., and updated by the present authors.

Table 7-4d Adhesive Types for Bonding Nonplastics to Thermosets[a]

Thermosets	Nonplastics						
	Ceramic	Fabric	Leather	Metal	Paper	Rubber	Wood
Diallyl phthalate	5, 24	36	31, 36	31	31, 36	31	31, 36
Epoxy	23, 31	4	4	23, 31	4	4	23, 31
Melamine	3, 16	4	3, 4	4	16, 41, 42	2, 3, 4, 16	3, 16
Phenolic	3, 16	4	3, 4	3, 25	16, 42	3, 4, 16, 25	3, 16, 42
Polyester	3, 26	4	5, 26	5, 26	41	1–5	3, 26
Urea	4, 16	4, 42	3, 4	3, 4	16, 42	1–5, 16	3, 16, 42

[a] Adhesive number code is shown in Key.
Table developed by Bostik Div., USM Corp., and updated by the present authors.

scale, their bonding to metals is fair; they bond poorly to plastics, and not at all to polyolefins or natural cellulosics. Care must be taken not to use anaerobic adhesives with highly notch-sensitive plastics. Thread locking compounds have experienced problems with stress cracking of threads exposed to uncured adhesive and high thread stresses.

Anaerobics have a low viscosity and a medium window and cure time. They are usually dispensed a drop at a time from a bottle or a dispenser and are toxic.

Typical Applications Principal applications are locking fasteners to prevent them from loosening with vibration

Sources Loctite, Permabond

7.6.3   Cyanoacrylates

Also known as CAs, cyanoacrylates are one-part adhesives widely known for their instant bite and small operating window. Cyanoacrylates have reasonably high strength on the order of 1000 to 2000 psi. High temperature resistance is 150 to 250 °F. Chemical and cold temperature resistances are fair. Cyanoacrylates do not fill in gaps well and have poor to fair resistance to shock and moisture. They bond very well to metals and polyolefins and extremely well to most of the other plastics, however their bond to natural cellulosics is poor. Cyanoacrylates require a clean surface and are known as the adhesive that can bond skin.

Cyanoacrylates have low viscosity, are available as liquids or gels, and are dispensed from bottles and pressure dispensers. To cure, they polymerize in the presence of a slightly basic surface, and a small amount of moisture. They cure very slowly in extremely low humidity. Cyanoacrylates require a thin bond line to cure, or an activator in thicker bonds. CAs may also be used as a fixturing adhesive for slower curing epoxies, etc. They are the most expensive adhesives when calculated on a price per pound of solid basis and are best used very sparingly.

Typical Applications Automobile gaskets, radio buttons

Sources Devcon, Henkel, Loctite, Lord, Permabond, 3M

7.6.4   Epoxies

Epoxies are tough, rigid structural adhesives that combine toughness with the highest strength potentials of the adhesives at levels of 2000 to 6000 psi. They cure at room temperatures without excessive shrinkage and are available in a wide range of properties. Their high temperature resistances can reach 350 °F and their moisture and chemical resistances are good to excellent. Epoxies fill gaps well but have only fair flexibility and resistance to cold temperatures. They can be used as coatings, and there are potting compounds available as well as those for electrical and thermal conductivity. Epoxies bond very well to metals and woods, less well to most of the plastics, and poorly to the polyolefins.

Epoxies are more difficult to use than some adhesives and are available in two-part with catalyst or one-part heat cure configurations. They are applied with dual-barrel syringes, flow guns, meter-mix ovens, and radiation cure units and are supplied in the form of liquids, pastes, and film. Irregular production rates are hard to maintain because of limited application windows. Epoxies have medium to thick viscosities. Very slow cure times make automated assembly operations difficult, however cures can be accelerated by a variety of methods. While far less costly than cyanoacrylates, epoxies are the second most costly structural adhesive system on a price per pound of solids basis.

Typical applications Automotive, aerospace, and electronic small parts

Sources Armstrong, Devcon, Emerson and Cuming, Fuller, Hardman, Loctite, Lord, Permabond, Techform, 3M

7.6.5   Hot Melts

Hot melts are basically thermoplastics, usually polypropylene, ethylene vinyl acetate, polyesters, or polyamides. Polypropylenes offer good adhesion to polyolefins and moderate performance, with temperature resistance of 170 °F. Ethylene vinyl acetates have lower performance, with heat resistance in the 120 °F range, but they are less expensive. Polyesters provide moderate to high performance with temperature resistance at 200 °F. Polyamides offer high performance with temperature resistance of 300 °F.

These adhesives are heated to their molten state and applied to the substrate; they form a bond when they cool. They are often used to join dissimilar materials like plastic to wood. Hot melts are not particularly strong because of viscosity and temperature limitations, their strengths falling in the range of 100 to 500 psi. In addition, these adhesives can result in residual stresses at the joint interface during bonding between plastics and metals which can significantly effect joint strength and durability. Strengths for these adhesives are much higher with wood than they are with plastics and metals. High temperature resistance ranges are from 130 to 300 °F. Low temperature resistance is fair. Humidity and moisture resistance range from poor to good, however hot melts can be used as moisture vapor barriers. Chemical resistances are poor to fair, but gaps are filled in very nicely. Hot melts are sometimes used as a fixturing device holding the parts in position while another type of joining method is applied.

These adhesives require special equipment and are applied with sprays, bulk units, and handheld guns. They are 100% solids and available as chips, ropes, and sticks. They have good storage stability because they do not freeze, however they can degrade with continuous heating. Hot melts have a quick bite and a very short application window. They set up quickly and simply by cooling and crystallizing. No additional equipment is required. Preheating of adherends may be necessary, particularly if metal is involved because the heat is conducted away very quickly. Coating weight is difficult to control, however hot melts are not flammable. Polyesters and polyamides cost about twice as much as EVAs and polypropylenes.

Typical Applications Costume jewelry, luggage, toys, carpet, foam cushions, point-of-purchase displays
Sources Bostik, Fuller, 3M

7.6.6   Phenolics

Phenolics are structural adhesives that can provide strengths from 2000 to 3500 psi and high temperature resistances from 100 to 300 °F. Chemical resistance is good and moisture resistance is good to excellent.

Phenolic adhesives are available in liquids and powders and are dispensed with flow guns, meter-mix ovens, and radiation cure units. They are cured with heat or catalysts.

7.6.7   Polyurethanes

Polyurethanes can seal as well as perform as structural adhesives with strengths in the range of 1000 to 2500 psi. Shear and peel strengths are good. Their moisture and chemical resistances are fair, however cold temperature and gap filling capabilities are good. High temperature limits range to 250 °F. Flexibility, sealing, and durability qualities are good. Polyurethanes bond extremely well to most plastics and wood; bonding is good to polyolefins and limited to metals. Not recommended for paper, cardboard, and wood; polyurethanes can be used with glass only with primers.

Polyurethanes are available as liquids or pastes and are applied with flow guns and meter-mix heated guns. They are available as two component room temperature cures or single-component moisture cures. Single-component adhesives are dispensed from cartridges or heat guns. They have a medium working window, medium viscosity, and a variety of cure times. They are significantly less expensive than epoxies on a cost per pound of solids basis.

Typical Applications Used for bonding plastics like polycarbonate, nylon, ABS, FRP, and Metton
Sources Ciba-Geigy, Emerson & Cuming, Fuller, Hardman, Humi-Seal, Lord, Tech Spray, 3M

7.6.8   Polysulfides

Polysulfides are sealants with excellent gap filling capabilities and chemical resistance. Flexibility, humidity resistance, and resistance to high and low temperatures are good. These sealants have fair adherence to most plastics and good adherence to metals. They are not recommended for polyolefins or natural cellulosics. Polysulfides are very viscous and have a medium application window and cure time.

Typical Applications Windows and aerospace
Sources Products Research, Thiokol

7.6.9   Pressure-Sensitive Adhesives

These adhesives are notable for their ability to be removed without leaving a residue. Usually rubber based, their properties are quite low: strengths under 10 psi, heat resistance from 100 to 140 °F, poor chemical resistance, and poor to fair moisture resistance. However, they have the advantage of being flexible.

Pressure-sensitive adhesives (PSAs) are available in liquids or films. They cure by solvent evaporation and are applied by hand, spray, and brush, with bonding effected by light pressure.

Sources Ashland, National Starch, 3M

7.6.10   Silicones

Silicones are noted for their temperature range and as a sealant. Low temperature resistance can reach -100 °F and high temperature resistance can attain 600 °F. Silicones are used to fill voids and have excellent flexibility and chemical and humidity resistance. Adhesion strengths, however, are modest to metals and plastics; woods are not recommended. Compatibility of reactive components with specific plastic resins must be checked before usage.

Silicones are quite viscous and are available in tubes and handheld guns. They have a large application window and a medium cure time.

Typical Applications Sealing applications of all types
Sources Dow Corning, General Electric, Rhone-Poulenc

7.6.11   Solvent-Based Adhesives

Solvent-based adhesives (i.e., one-part solvent evaporation adhesives) offer good moisture resistance, flexibility, and hot and cold temperature resistance, with moderate strength and gap-filling capability. They have the advantage of being able to wet some difficult materials but have only fair chemical resistance. They bond well to metals and woods, but bonding to plastic substrates is only fair. Solvent-based adhesives are reasonably priced and are available as contact adhesives and dip coatings. They have a good initial tack, low to medium viscosity, and a medium working window; however they are a fire hazard and care must be taken in their storage and use. In addition, they are toxic and must be used only in well-ventilated areas.

Typical Applications Appliqués, plastic-to-wood laminates
Sources Camie-Campbell, Fuller, Goodyear, PDI, 3M

7.6.12   Water-Based Adhesives

Currently, there is a strong trend toward the adoption of water-based adhesives (i.e., one-part water evaporation adhesives), which have received a great deal of research in recent years as companies seek to replace toxic and flammable solvent-based adhesives. Early water-based adhesives were principally used for wood, cardboard, and paper as in boxes and envelopes. These adhesives still adhere best to those materials, and adherence to plastics and metals, some of which they can corrode, is only fair to poor. Flexibility, gap-filling capability, and chemical and water resistance are also poor. However, high and low temperature resistances are fair. Water-based adhesives are available in a wide range of solids contents, and a high concentration of high molecular weight material can be used.

Available in a wide range of viscosities, these adhesives have a medium to slow cure time and a medium working window. They can freeze if stored at low temperatures, after which most of them are still relatively useful. Water-based adhesives are less expensive than the equivalent solvent-based adhesives.

Typical Applications Packaging, plastic-to-wood laminates
Sources Fuller, National Starch, Swift, Techform, 3M

7.7   Solvents

Solvents function by separating the polymer chains of the plastics to be joined and causing their bond surfaces to soften. They are then pressed against each other so their polymer chains can intermingle and held in that position until the solvent has evaporated. Only amorphous thermoplastics can be solvent-welded at room temperature. Thermosets cannot be solvent-welded, and semicrystalline thermoplastics can be solvent-welded only at elevated temperatures.

> The specific solvent recommendations are listed for each plastic in Chapter 5, "Assembly Method Selection by Material."

Solvent formulations are often composed of more than one solvent. Most resins have several solvent options. Solvents evaporate at different rates; those with lower boiling points doing so more rapidly than the ones with higher boiling points. Solvents can evaporate so quickly that the lowering of the surface temperature results in the formation of condensate at the joint, or "blush" as it is often referred to. The addition of 10 to 20% of a slow drying solvent, such as Methyl Cellosolve, can delay the drying of a low boiling point solvent such as acetone enough to prevent such precipitation of water vapor. These solvents are referred to as "retarders," and some manufacturers only use them on days when the humidity is high. Microwave-assisted solvent bonding has been used to substantially improve joint strength and reduce the time required to remove the solvent.

Frequently, a small amount of resin is added to the solution to increase the viscosity or "give it body." Solutions such as these typically contain 1 to 7% resin, however they can sometimes contain up to 50% plastic. This improves the gap-filling capability of the formulation which has the added benefit of permitting greater tolerances and less surface preparation because wider space between the bonding surfaces can be tolerated. In most circumstances, it also extends the drying time of the solution

Table 7-5   Recommended Solvent–Plastics Combinations

Plastic	Acetone	Cyclohexanone	N,N-Dimethyl formamide	Dioxane	Ethyl acetate	Ethylene dichloride	Formic acid	Glacial acetic acid	Methyl ethyl ketone	Methyl cellosolve	Methylene chloride	N-Methyl pyrrolidone	O-Dichlorobenzol	Perchlorethylene	Tetrahydrofuran	Toluene	80 Toluol–20 Ethanol	Trichlorethylene	Xylene
ABS					▨			▨						▨					
Acrylic	▨				▨				▨										
Cellulose acetate	▨								▨										
Cellulose acetate butyrate		▨																▨	
Ethyl cellulose										▨									
Nylon							▨												
Polyaryl ether	▨											▨							
Polyaryl sulfone												▨							
Polycarbonate											▨								
Polystyrene		▨		▨				▨	▨		▨								
Polysulfone		▨									▨	▨							
Polyvinyl chloride	▨								▨										
PPO base (Ndryl)																			
Styrene-acrylonitrile	▨								▨										
Vinylidene chloride	▨	▨																	
Vinyl chloride–vinyl acetate	▨	▨							▨										

Source: Goodman, S. H. and Schwartz, S. S. *Plastics Materials and Processes*, p. 787.

which provides a longer application window. Bodied solvents are also known as "doped solvents".

Plastics that are dissimilar can be solvent-welded provided they are soluble in the same solvent. If the parts are not free to expand, it will be necessary to take into account the difference in their respective coefficients of linear thermal expansion. This can be more critical with solvent bonds than with some adhesive joints because many adhesives have the flexibility to deal with CLTE differentials. If there is no common solvent for two resins, solutions can be created composed of two different miscible solvents.

Generally , solvents evaporate quite readily, and there is no permanent residue other than what might have been added to thicken the solution. However, there are joint designs that could cause solvents to become trapped, which can soften the plastic in that region. Solvents also migrate, thus affecting an area beyond the joint surface itself. In addition, they act as stress relief agents, causing crazing to occur where there are stresses. Therefore, it is important to locate stresses, such as injection molding gate stress, well away from a solvent weldment. Finally, to be readily solvent-welded, a plastic must be vulnerable to attack by a solvent. That means that it is also susceptible to attack by that and similar solvents at any time. The designer must, therefore, be careful to investigate the potential for solvent attack in the products' manufacturing, shipping, and functional environments.

Specific plastic recommendations are provided in Table 7-5.

Safety Note: Most solvents are both toxic and flammable. Processing areas must be properly vented and storage areas secured accordingly.

7.8   Adhesive and Solvent Assembly Techniques

The inherent nature of adhesives and solvents makes their handling somewhat unique among plastics assembly methods. For one thing, there is the time factor. Some of these compounds have a limited shelf life. Once out of their container, most of these formulations have a relatively short period of time during which they can be applied to the joint surface. Once beyond their application window, these materials must be discarded. In some cases, this means that the production line cannot be easily shut down without replacing all the open solvent. Worse, there is the insecurity of not being able to check whether the work has been done properly without 100% testing. Thus increased safety factors must be used to allow for a greater margin of security.

Fixturing and clamping are quite important in achieving a good assembly, as proper strength and location of the finished article are highly dependent on these elements. For some application methods, technique is critical – particularly if a high level of appearance is necessary. In that case, these processes become extremely operator sensitive.

7.8.1   Fixturing

Fixturing can be simplified a great deal by incorporating location features into the design itself. This can often be done in a manner that provides a natural well into which the adhesive or solvent can be placed.

When that method is not feasible, external fixturing will be necessary. Figure 7-21a illustrates a simple variety of fixturing in which blocks of wood have been used to locate a plastic part by trapping it between them. In this simple version, they are held in place with adhesive tape. While this is an acceptable practice for very low volumes, it would clearly be too slow for higher production rates. The toggle clamp illustrated in Fig. 7-21b would be much more efficient. For very high production, more sophisticated fixtures would be made which could be attached to a conveyor. The conveyor would run under hot air or into an oven to speed up the drying process.

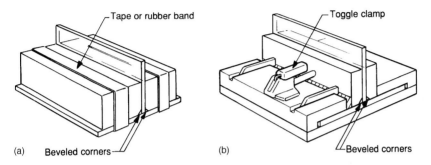

Figure 7-21   Fixturing: (a) simple fixture and (b) toggle clamp fixture

7.8.2   Clamping

The simplest means of applying pressure to the joint is illustrated in Fig. 7-22a. A weighted bag (containing shot, screws, washers, etc.) is placed over the joint. Another means of achieving the same objective is the use of wedges, as shown in Fig. 7-22b. A spring-loaded block, such as the one depicted in Fig. 7-22c, is more efficient for higher volume applications. There are a variety of clamps like the alligator clamp in Fig. 7-22d that can be used for small parts. These are readily available from suppliers who normally service cabinet makers.

High production rates can be achieved with a clamping conveyor such as that illustrated in Fig. 7-23. The parts are joined, typically by the dip method in conjunction with a slip joint, and placed in nests built into a conveyor. They then pass under a second overhead conveyor, which gradually applies pressure until a predetermined loading is achieved. The rate of speed and length of the conveyor are calculated to apply pressure for the desired time period.

It is possible to design a clamp into the part in the form of a press or snap fit, the former being the more desirable because of its lower tooling expense. While more costly in terms of investment and part cost, it can result in a significant labor savings in assembly. Again, this would best be utilized in conjunction with the dip method. Such a design is shown in Fig. 7-24.

Also, the "screw and glue" technique discussed previously for use with adhesives can be used for solvents as well.

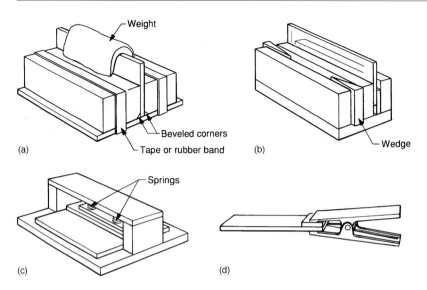

Figure 7-22 Clamping: (a) simple clamp, (b) wedge clamp, (c) spring clamp, and (d) alligator clamp

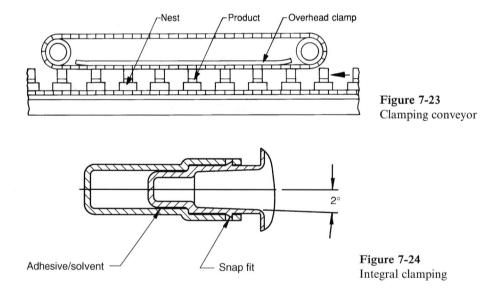

Figure 7-23
Clamping conveyor

Figure 7-24
Integral clamping

7.8.3   Application Methods

Just about every means of placing a coating on a material has been used for the application of adhesives and solvents. Brushes, sprays, rollers, coaters, mix-metering machines, bottles with needle adapters, single- and double-barreled syringes, handheld guns, and applicators are commonplace. The selection of the precise variety of means is largely dependent on the viscosity of the adhesive, the length of time during which it can be applied, the production rate required, and whether it is a two-part adhesive that requires mixing. The adhesive manufacturer will know which application method is best suited to a specific adhesive. Masking is often used to keep adhesives and solvents from finished surfaces. Hot melts have the added complication of the heat factor. There are, however, two techniques unique to adhesives and solvent welding which should be discussed further.

7.8.3.1 Capillary Method

The capillary method is principally a solvent method, since most adhesives are too viscous for it to be effectively employed. The parts are positioned to each other and the solvent is applied along the edge with a cannula tipped bottle, as illustrated in Fig. 7-25. It then flows into the joint through capillary action. The edges do not have to be absolutely smooth as this method can work with sawcut edges, however the smoother the surface is, the better it will look. The clear, transparent joints sometimes found on acrylic furniture are achieved by sanding or filing the surface after cutting it to remove the saw marks and then flaming the file marks off by running a torch across the surface. The flaming also improves the wettability of the surface. A hypodermic needle is sometimes used in place of the bottle. An eyedropper can also be used, however the vertical piece would need to be tipped slightly, about 1° from vertical, for it to work effectively.

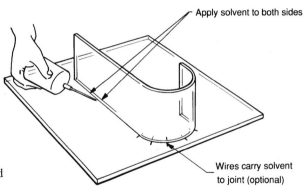

Apply solvent to both sides

Wires carry solvent to joint (optional)

Figure 7-25 Capillary method of solvent application

Note the small wires located around the curved section of the workpiece in Fig. 7-25. They are used as shims in some cases. Although the time would vary depending on the drying rate of the solvent in use, they are usually removed in about three minutes.

7.8.3.2 Dip or Soak Method

Dipping and soaking really comprise one joining technique that is also mainly for solvent welding. Whether this is a dip method or a soak method depends on the length of time the part is left in the solvent. For some plastic and solvent combinations, the parts can be soaked in the solvent as illustrated in Fig. 7-26, then removed and assembled. Other combinations might call for the two parts to be fixtured together and then placed in the solvent tray. The soak method can result in solvent reaching appearance surfaces. This can be controlled through the use of commercial masking compounds or ordinary adhesive tape, which is preferable to masking tape.

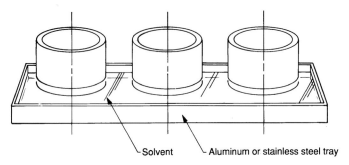

Solvent   Aluminum or stainless steel tray

Figure 7-26   Dip or soak method of solvent application

The way to achieve high speed production, however, is to use the dip method. To do this manually, a part is picked from a bin with each hand, dipped in a solvent tray, dabbed in a piece of absorbent cloth, and then joined with a slip fit. This process can be automated with certain materials, however there can be difficulties in maintaining quality without a human observer monitoring solvent viscosity or watching for solvent drips or spatter. The key is really in creating the optimum solvent combination. Depending on the circumstances, it may be desirable to utilize the clamping conveyor illustrated in Fig. 7-23.

Regardless of whether the dip or soak process is used, the problem with these inexpensive techniques for assembling amorphous thermoplastics is the ventilation of toxic fumes. Both methods use open trays that are constantly evaporating toxic fumes, which must be controlled and ventilated. Good ventilation and exhaust hoods are the minimum requirement. Individuals tolerance to these vapors vary, and more sophisticated methods may be necessary, some of which are quite cumbersome and add cost to the process. This evaporation process also results in the continuous thickening of the solvent mix. That means that it must be replenished from time to time, a procedure which complicates the use of an enclosed chamber.

7.9   Adhesive and Solvent System Selection

The decision to join with an adhesive or solvent is normally based on one or more of the advantages discussed at the beginning of this chapter. That fundamental reason for selecting this approach is usually the driving force in the decisions that follow. For example, if the need to join dissimilar plastics was the principal factor, the adhesive or solvent selection will be limited to those that will join these two resins. They, in turn, will dictate which application methods are available, and the volume requirement will make the final determination. Presumably, that volume requirement will have been one of the key factors in the process selection for the two parts, and the processes will determine which joint configurations are available. The amount and type of stress applied to the joint would drive the final selection. Translating this to a procedure comes out as follows:

1. Identify the process, resin, and configuration that best meet the needs of the overall product design requirements.
2. Establish the principal attribute or attributes for which adhesive or solvent bonding was selected as a joining method.
3. Combine the two to determine which joint design, adhesive, or solvent and application options are available.
4. Proceed with the lowest cost option unless it becomes obvious that not all product requirements can be met. If such a point is reached, proceed with the next lowest cost option until a successful conclusion is attained.

In the execution of this procedure, be certain to include the following decision factors:

1. *Differential in coefficients of linear thermal expansion* This is a common problem, particularly with the use of adhesives, because they are so often used to join very different materials (e.g., plastic to metal). The difference in CLTEs will be considerable for some of these combinations and, if the product is subjected to a wide range of thermal conditions, the resultant stresses can cause failure. Some adhesives are more elastic than others and can better accommodate these stresses. A thicker coating of adhesive can also resolve this problem if the adhesive is one that can handle flexure well.

2. *Effects of temperature on adhesive properties* Like the plastic substrates themselves, the properties of adhesives vary with the temperature. This can have disastrous results, since adhesives that have sufficient strength at one end of the operating range may be too brittle at the other. Properties of the proposed adhesives need to be determined for the extremes of the products' thermal exposure to insure that they are adequate.

3. *Bond surface area* The strength of a particular joint is determined by the area of bonding surface as well as the strength of the adhesive or solvent. The cost of a higher strength adhesive can often be avoided by simply increasing the joint area. The width of the bond surface can be as important as its length. Adhesives and solvents with gap-filling properties can have a greater bond area even with the same size of joint.

4. *Determination of joint stress* Stresses on adhesive and solvent joints can be difficult to ascertain. This must be done carefully because adhesives have different capabilities with each type of stress, and joints are affected by the direction from which it is applied.

5. *Changes in stress loadings* As loads increase, plastics tend to deform in varying degrees according to the particular resin. This deformation can cause the stress on an adhesive joint to move from shear to cleavage, resulting in a change in the stress loading that could lead to joint failure.

6. *Differential adhesion* A given adhesive is likely to show different adhesion to the two adherends. This difference can be particularly acute when bonding a plastic to a metal. There are primers available that can be used to improve the adhesion to the metal.

7. *Solvent attack* The solvents used in solvent cementing and those used in solvent-based adhesives can act as a stress relief agent causing fine cracks to appear in highly stressed areas of the part. The stress concentration need not be in the bond area, as solvents can migrate in plastics. Solvents trapped inside a joint may lead to porosity or weakness. Finally, the finished product cannot be exposed to harmful solvents in its shipping, storage, or operating environments.

8. *Heat resistance* The heat used to cure adhesives must be below that which will affect the plastic substrate. Be alert to the temperatures the substrate may reach as a result of an exothermic reaction.

9. *Moisture resistance* The ability of different adhesives to withstand moisture varies considerably. If a great deal of moisture is present in the operating environment, the long-term effects of this environment should be tested. Be particularly alert to the effects of thermocycling and of high heat and humidity in combination.

10. *Effects of aging* The adhesive, the substrates, and their surfaces will all be affected by long-term aging. If exposed to sunlight, there will be an additional effect due to ultraviolet light. Since many projects do not have enough time to permit extended aging tests, this factor must be taken into consideration when one is establishing a safety factor for the joint strength. The aging information available from the adhesive manufacturer may not apply precisely to the application at hand.

11. *Cost evaluation* It is not enough to simply calculate the amount of adhesive that will be used and add in the application labor to determine the cost of an adhesive or solvent joining method. There will also be the cost of cleaning the bonding surfaces and possibly costs of surface treatment, ventilation, curing, and the special handling required of hazardous chemicals.

With the appropriate attention to the special requirements of adhesive and solvent joining, long-lasting bonds of excellent strength can be achieved. This is the economic means of putting many parts together and the only way to assemble some parts.

7.10 Glossary

Like most industries, the adhesives industry has it own jargon, and persons using adhesives need to become familiar with the following terms.

Adhesion   The adherence of two bodies to each other (Fig. 7-1).

Adsorption adhesion   Adhesion produced by the surface forces that result when molecular contact occurs between two surfaces.

Angel hair   Strands of adhesive stretched between the applicator and the bond surface. Although this characteristic is not shared by all adhesives, it can slow assembly rates when present.

Bite   This term refers to how quickly an adhesive or solvent takes hold of the adherend. The rubber-based adhesives used in adhesive tape, for example, usually have a very quick "bite." An adhesive can lose its bite in time. Other terms for this characteristic are "tack," "quickstick," and "grab."

Bleed   Adhesive that has oozed out from under the edges of the bond area. Soft adhesives are more likely to exhibit this characteristic.

Blush   The formation of condensate in a solvent bond resulting from the cooling of the surface due to the rapid evaporation of a low boiling point solvent.

Caliper   The thickness of the adhesive, usually measured in "mils" or thousandths of an inch. A mil is sometimes called a "point."

Cohesion   The force within a body holding it together.

Conformability   The ability of an adhesive to conform to a rough substrate surface; usually refers to pressure-sensitive adhesives.

Cure, accelerated   Cures can be accelerated to permit handling strength and optimum product strength to be reached in a shorter period of time.

Cure, full   The time required to reach 100% of the ultimate strength of a product. This is also the time that should elapse before test and evaluation are attempted. If testing occurs prematurely, results will show less strength than would have been recorded after full cure had been accomplished.

Dopes   Solvent formulations consisting of the solvent plus some of the substrate resin. These are used to increase the viscosity of the solution so that it will better fill in voids.

Dwell   The length of time a pressure-sensitive adhesive is adhered to a surface before it is tested.

Effective adhesion plus adsorption adhesion.

Flow out   The ability of an adhesive to spread out and fill in the crevices in the surface of a substrate.

Grab   See Bite.

Handling strength   The time needed to reach 50 psi in overlap shear (OLS). Parts small enough not to overly stress the bondline can be moved or handled for further processing at this time.

Mechanical adhesion   The force that produces a mechanical bond when adhesive flows into surface microcavities and hardens.

Migration   The movement of chemical components of the solvent or adhesive into the substrate or vice versa. The plasticizer used in some plastics can migrate into the adhesive, causing it to soften and degrade.

Mil   0.001 inch

Nonstructural adhesive   An adhesive for non-load-bearing applications such as gasketing, cushioning, sealing, and adhering light materials such as decorative trim and weather stripping.

Permanent adhesive   An adhesive whose ultimate strength is sufficiently high that the bond cannot be readily broken.

Pressure-sensitive   Describing an adhesive that is tacky at room temperature and will adhere to the substrate with a pressure light enough to be applied by hand.

Quickstick   See Bite. Can also refer to a measure of this property immediately after application.

Removable adhesive   An adhesive that has an ultimate strength low enough to permit ready removal without leaving a residue. For this purpose, it must have a relatively high cohesive strength and a low adhesive strength. This type of adhesive can be used to provide fixturing or light clamping during assembly operations.

Retarder   A slowly drying solvent used to retard the evaporation rate of a rapidly drying solvent to prevent the formation of condensate on bond surface due to too rapid cooling.

Service temperature   The temperature an adhesive will withstand 72 h after application.

Set time   The time necessary to obtain a hard, tack-free surface; normally longer than work life and shorter than handling strength.

Shelf life   The length of time an adhesive can be stored under specified conditions and still be usable.

Skip   A break in the bond area where no adhesive has been applied.

Specific adhesion   The characteristic of adhesives to have different adhesion levels to different substrate materials such that a given adhesive can be removable from some materials and permanently bonded to others.

Structural adhesive   Adhesives used for load-bearing joints. They can often be as strong or stronger than the adherends.

Tack   See Bite.

Ultimate adhesion   The maximum strength of an adhesive. Most adhesives reach their ultimate strength within 96 h of application.

Wettability   The ability of a liquid to spread over a solid. In the case of adhesive bonding, it refers to the extent to which the adhesive is able to contact the substrate.

Window   The period of time during which an adhesive or solvent can be applied to a joint before becoming too viscous to be effective.

Work life   The time during which an adhesive will adequately wet out on a substrate. At the end of an adhesive's work life it will not set up. Shorter worklife products can be accelerated from their room temperature cure schedule, but not to the degree possible for longer work life products. This is because the heat given off during the reaction of room temperature curing adhesives (called exotherm) causes an internal acceleration of the cure. The additional heat has minimal effect on the rate of cure.

Work time   See Work life.

7.11   Sources

Adhesives Research, Inc., 400 Seaks Run Rd., Glen Rock, PA 17327, (800) 774-5996, (717) 235-7979, fax (717) 235-8320, www.adhesivesresearch.com

Adhesive Technologies, Inc., 3 Merrill Industrial Dr., Hampton, NH 03842, (800) 544-1021,ext. 123, (603) 926-1616, fax (603) 926-1780, www.adhesivetech.com

Armstrong Products Co., Resin Technology Group LLC, 28 Norfolk Ave., S. Easton, MA 02375, (508) 230-8070; fax (508) 230-2318, www.resintechgroup.com

Ashland Chemical, Inc., Composite Polymers Div., 5200 Blazer Parkway, TW-1, Dublin, OH 43017, (800) 545-8779, (614) 790-3333, fax (614) 790-4119

Bostik Findley, Inc., 11320 Watertown Plank Rd., Wauwatosa, WI 53226, (414) 774-2250, fax (414) 479-0645, www.bostikfindley-us.com

Camie-Campbell, Inc., 9225 Watson Industrial Park, St. Louis, MO 63126, (800) 325-9572, (314) 968-3222, fax (314) 968-0741, www.camie.com

Ciba Specialty Chemicals Corp., Formulated Systems, 4917 Dawn Ave., East Lansing, MI 48823, (517) 351-5900, fax (517) 351-6255

ITW Devcon, 30 Endicott St., Danvers, MA 01923, (800) 933-8266, (978) 777-1100, fax (978) 774-0516,www.devcon.com

Dow Corning Corp., 2200 Salzburg, P.O. Box 994, Midland, MI 48686-0994, (989) 496-7881, fax (989) 496-6731, www.dowcorning.com

Dymax Corp., 51 Greenwoods Rd., Torrington, CT 06790, (860) 482-1010, fax (860) 496-0608, www.dymax.com

Eastman Chemical Co., 100 N. Eastman Rd., P.O. Box 511, Kingsport, TN 37662, (800) 327-8626, (423) 229-2000, fax (423) 229-1194, www.eastmanchemical.com

Ellsworth Adhesive Systems, W129 N10825 Washington Dr., P.O. Box 1002, Germantown, WI 53022-8202, (800) 888-0698; fax (262) 253-8619, www.ellsworthadhesive. com

Emerson & Cuming, Inc., 869 Washington St., Canton, MA, 02021, (781) 828-3300, fax (781) 828-3104,www.emersoncuming.com

H. B. Fuller Co., 1200 Willow Lake Blvd., P.O. Box 64683, St. Paul, MN 55164-0683, (888) 423-8553, (651)236-5900, fax (651) 236-5898, www.hbfuller.com

GE Silicones, 260 Hudson River Road, Waterford, NY 12188, (800) 255-8886, (518) 237-3330, fax (518) 233-2642, www.gesilicones.com

Goodyear Tire & Rubber, Chemical Div., 1144 E. Market St., Akron, OH 44316-0001, (330) 796-2120, fax (330) 796-2222, www.goodyearchemical.com

Grace Specialty Polymers, W.R. Grace & Co., 55 Hayden Ave., Lexington, MA 02173, (617) 861-6600, fax (617) 863-1436, www.medizin.li

Hardman Div., Harcros Chemicals Inc., 600 Cortlandt St., Belleville, NJ 07109, (973) 751-3000, fax (973) 751-8407

Henkel Loctite Corp., 1001 Trout Brook Crossing, Rocky Hill, CT 06067-3910, (800) 562-8483, (860) 571-5100, fax (860) 571-5465, www.loctite.com

Hernon Manufacturing Inc., 121 Tech Dr., Sanford, FL 32771, (800) 527-0004, (407) 322-4000, fax (407) 321-9700, www.hernonmfg.com

Humi-Seal Div., Chase Corp., 2660 Brooklyn Queens Expressway, P.O. Box 770445, Woodside, NY 11377, (888) 854-5693, (718) 932-0800), fax (718) 932-4345, www. humiseal.com

Lord Corporation, Industrial Adhesives Div., P.O. Box 10038, Erie, PA 16514, (800) 458-0434, (919) 468-5980, fax (919) 469-5923, www.lordadhesivews.com

Master Bond Inc., 154 Hobart St., Hackensack, NJ 07601, (201) 343-8983, fax (201) 343-2132, www.masterbond.com

National Starch - See Permabond

Pacer Technology, 9420 Santa Anita Ave., Rancho Cucamonga, CA 91730, (800) 538-3091, (909) 987-0550 , fax (909) 987-0490, www.pacertech.com

PAM Fastening Technology, Inc., 2120 Gateway Blvd., P.O. Box 669063, Charlotte, NC 28208, (800) 699-2674, (704) 394-3141, fax (704) 394-9339, www.pamfast.com

Plasti Dip International, 3920 Pheasant Ridge Dr., Blaine, MN 55449 (800) 969-5432; fax (763) 785-2058, www.plastidip.com

Permabond Engineering Adhesives, Div. National Starch & Chemical Co., 10 Finderne Ave., Bridgewater, NJ 08807-3300, (800) 653-6523, (908) 575-7200, fax (908) 575-7203, www.permabond.com

Plexus Adhesive Systems (see ITW Devcon)

Products Research & Chemical Corp., 21800 Burbank Blvd., P.O. Box 4226, Woodland Hills, CA 91364, (818) 702-8900, fax (818) 702-7499

Quantum Chemical Co., 11500 Northlake Dr., P.O. Box 429550, Cincinnati, OH 45249-1694, (800) 543-5900, (513) 530-6500, fax (513) 530-6560

Rhodia, Inc., 259 Prospect Plains Road, P.O. Box CN-7500, Cranbury, NJ 08512, (888) 776-7337, (609) 860-4000, fax (609) 860-0138, www.us.rhodia.com

SAF-T-LOK International Corp., 300 Eisenhower Lane N., Lombard, IL 60148, (800) 222-2087, (630) 495-2001, fax (630) 495-8813, www.saftlok.com

Shell Resolution Performance Products, 1600 Smith St., Houston, TX 77002, (713) 241-2996, www.shellchemicals.com

Stik-II Products, 41 O'Neill St., Easthampton, MA 01027, (800) 356-3572, (413) 527-7120, fax (413) 527-7249, www.stik-2.com

M. Swift & Sons, Inc., 10 Love Lane, Hartford, CT 06112, (800) 628-0380, (860) 522-1181, fax (860) 249-5934, www.mswiftandsons.com

Tantec, Inc., 630 Estes Ave., Schaumburg, IL 60193, (888)382-6832, (847) 524-5506, fax (847) 524-6956, www.tantecusa.com

Techform, Inc., 1553 Carter St., P.O. Box 270, Mt. Airy, NC 27030, (336) 789-2115, fax (336) 789-2118, www.techforminc.com

Tech Spray, P.O. Box 949, Amarillo, TX 79105, (800) 858-4043, (806) 372-8523, fax (806) 372-8750, www.techspray.com

3M Engineered Adhesives Div., 3M Center Bldg. 220-7E-01, St. Paul, MN 55144, (800) 362-3550, (651) 733-1771, fax (651) 736-4776, www.3m.com

Uniroyal Technologies Corp., Adhesives and Sealants Div., 2001 W. Washington St., South Bend, IN 46628, (800) 999-GLUE, (574) 246-5000, fax (574) 246-5425, www.uniroyaluas.com

ZENECA Specialties, Luxtrak Adhesives Business, 1800 Concord Pike, Wilmington, DE 19850, (302) 886-7979, fax (302) 886-2378, www.chemscope.com

8   Fasteners and Inserts

8.1   Advantages and Disadvantages

The need to reopen an assembly for maintenance limits the number of assembly methods that can be utilized. The traditional means of disassembling a product has been the use of fasteners. However, the elimination of fasteners is one of the major trends in assembly today. They cost money to buy and to emplace. In addition, they can be expensive to remove when the product must be disassembled for recycling. Nonetheless, the appliance industry alone used over 4.37 trillion fasteners in 1995. Clearly, there are many applications for which fasteners remain the optimum solution.

8.1.1   Advantages of Using Fasteners

1. *Reopenability* The principal reason for most fastener applications lies in their ability to be reopened. This makes them ideal for applications where replacement or servicing of mechanical components and batteries may be required.

2. *Low technology requirement* This is a long-established technology, the fundamentals of which most people learn as children. The cost of tooling, even torque-limiting applicators, is comparatively modest.

3. *Low volume capability* Most fasteners do not require any special mold features that would prevent their use at the very lowest production levels.

4. *High volume capability* As the volumes increase, features can be molded into the product which speed up the process. Ultimately, automated emplacement can be achieved for some applications.

5. *Availability* The small size and widespread use of fasteners mean that they are nearly always in stock and available at once with no wait for special equipment, tooling, or training.

6. *Ability to join thermosets and dissimilar materials* Most of the other processes used to join plastics require some level of compatibility. Fasteners require none and can be used to join thermosets, dissimilar plastics, and plastics to other materials.

8.1.2   Disadvantages of Using Fasteners

1. *Concentration of stress* Fasteners concentrate their loads at the points where they are located. Other assembly methods such as adhesives (their principal competitor for the joining of thermosets) or some of the welding techniques, distribute them across a wide area. In plastics, this can lead to failure due to creep, stress relaxation, crazing, or notch sensitivity unless these phenomena are properly addressed.

2. *Loosening of fasteners* Creep, moisture, and stress relaxation can cause fasteners to loosen in plastic parts, particularly when in the presence of vibration. These, too, are challenges the engineer must deal with when designing assemblies using fasteners.

3. *Notch sensitivity* Many plastics exhibit high levels of notch sensitivity, and screw threads are nothing if not a series of notches. Unfortunately, these notches are at a point where very high stress loadings occur and one that is not visible from the exterior of the part. Thus, these combinations of stresses can result in sudden failure, without a warning.

4. *Crazing* The tiny cracks known as crazing can be the beginning of a more serious crack leading to total failure of the part. Many plastic materials are vulnerable to crazing, and screws, in particular, tend to produce high levels of localized stress.

5. *Reclosure limitations* Since the emplacement of screws into metal is the norm, it is the customary practice among many to tighten screws to the maximum. However, the strengths of the various plastics vary enormously and it is difficult to discern the strength of the resin at hand. Once out of the hands of the manufacturer, there is no way to control the amount of torque that will be applied to the screw. Thus, threads in plastic materials, which cannot withstand the same levels of torque as metals, are often stripped and cannot be relied upon for repeated reclosures.

6. *Differentials in the coefficients of linear thermal expansion* Since parts of different materials are commonly assembled, differences in CLTEs can lead to substantial stresses on the joint when temperature extremes are encountered. Ideally, the assembly is designed so that one of the parts is free to float relative to the other(s). Unfortunately, this is not often feasible.

7. *Loss of properties due to moisture* Some plastics are sufficiently hygroscopic to allow moisture to affect their strength and stiffness. This can result in the loosening of a fastener. It is sometimes necessary to add a desiccant to the product storage or shipping container.

8. *Cost of fasteners* Fasteners vary considerably in cost, and many are very inexpensive. However, even the lowest cost fasteners have a cost, and there is the additional cost of emplacement of the fastener. With assembly methods available which eliminate this cost, their advocates claim the savings of several thousands of dollars per year per fastener.

9. *Cost of disassembly for recycling* Since most fasteners are metal, they are a contaminant in the recycling waste stream and must be removed. Even simple unscrewing, which will do for most fasteners that are reopenable, has a significant expense associated with it. However, permanent fasteners may need to be melted or broken out at a greater cost.

10. *Unpleasant appearance* Industrial designers have long objected to the indiscriminate placement of nut and bolt heads on appearance surfaces for nonindustrial applications. While flat-head screws will help in some cases, it is often necessary to make other design concessions to accommodate aesthetic requirements. In some cases, the substitution of cap nuts or other decorative fasteners may suffice.

11. *Need to access both sides of the part* This is a disadvantage essentially confined to nuts and bolts. When it becomes a problem, an alternative solution such as threaded inserts may be required. The need to access both sides of the part can be partially

alleviated by molding a recess in the shape of a nut in the outside of one of the parts. While it must still be held in place during assembly, it does not need to be prevented from turning.

8.2   Basic Design Considerations for Fasteners

Despite the advantages of fasteners, product failures related to them are a major problem for manufacturers. Nearly 90% of these failures are the result of improper joint design or assembly. In one industry's investigation of product recalls, 20% were related to fasteners alone. Warranty costs associated with fasteners came to 70% of the total.

Many of the failures are due to the use of fasteners with plastic parts. Often, there is a casual disregard for the special considerations required for materials that are to be assembled. Many engineers simply proceed in the same manner they are accustomed to using for metal parts. Not only are there special fasteners for use with plastics, but there are also significant differences in the way in which standard fasteners are to be used.

The principal issue lies in the characteristics of the various plastics themselves. There are several that significantly affect the use of fasteners. Besides the fundamental strength of the plastic, there are creep, relaxation, craze resistance, notch sensitivity, and stiffness. Furthermore, these properties will be affected by changes in temperature as well as the presence of moisture and fillers. In addition to the physical disadvantages associated with the use of fasteners, there are cost negatives as well.

8.2.1   Creep Effects

The term "creep" refers to the continuous deformation of a plastic under a constant load. Creep can be induced by tensile, compressive, flexural, and shear stress. Creep rates for a given material will vary with temperature. The term is often used interchangeably with the term "cold flow," however the latter normally refers to this property within the normal range of human living conditions. Creep rates vary considerably with the plastics, and they can have a pronounced effect on fastener performance. In some cases, the clamping pressure supplied by a screw can drop 25% in the first 24 h.

Creep data usually are plotted as a function of tensile strain versus time, as illustrated in Fig. 8-1. There are three stages to creep strain. The first stage is the primary creep, which occurs immediately after loading. It is largely elastic and diminishes rapidly. The second stage is secondary creep, which is more or less constant. Depending on the material, temperature, loading, and environmental conditions, this stage can go on for extended periods of time ranging into years. Secondary creep is not recoverable. The third, or tertiary creep stage, occurs at a very high rate of strain just prior to fracture. The ultimate strength varies with the same properties that affect the creep rate. Creep fracture will take place more quickly as the amount of stress is increased.

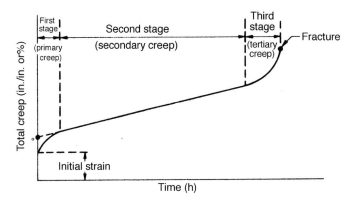

Figure 8-1   Creep curve (From Rosato, *Plastic Encyclopedia and Dictionary*, p. 158)

Creep is measured by the creep modulus E_{cr}.

$$E_{cr} = \sigma/e_{cr} \tag{8-1}$$

where σ = constant stress and e_{cr} = creep strain. Both the creep modulus and creep strain are time and temperature dependent.

8.2.2   Stress Relaxation Effects

Stress relaxation is similar to creep except that this case applies to the condition where the strain is held constant. Consequently, when a part is constrained in a bending, shear, tension, or compression state, stress relaxation takes place in much the same fashion that creep occurs when the stress is held constant. The net result is that the holding force, known as the "clamp load," of a fastener is reduced over a period of time. If the fastener is a screw, the drop in clamp load will permit it to loosen and turn out of the hole.

Stress relaxation is measured by the relaxation modulus E_r,

$$E_r = \sigma/e \tag{8-2}$$

where σ = stress and e = strain. The relaxation modulus and the stress are both time and temperature dependent.

The relaxation modulus is often similar to the creep modulus for a given material for low levels of stress and short time periods. However, as the stress level and time period increases, the values differ. When stress relaxation data are not available, creep data can provide a crude, if unreliable, guide.

8.2.3   Notch Sensitivity

The notch sensitivity of many plastics makes plastic threads vulnerable to this phenomenon. As illustrated in Fig. 8-2a, thread-forming screws have less notch-sensitive

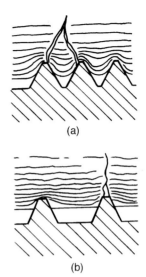

Figure 8-2 Notch sensitivity: (a) thread-forming screw and (b) thread-cutting screw (From *Assembly Engineering*, September 1982)

threads than thread-cutting screws because the former have slightly rounded thread forms; nonetheless, stress cracks can form as shown. These screws can be used in ductile plastics that permit the threads to be formed. If the hole is molded instead of drilled, they do not break the skin of the plastic either, thus preserving the "skin effect," which reduces notch sensitivity.

Stiffer materials require thread cutting screws. The thread-cutting screws shown in Fig. 8-2b must use their sharp-edged teeth to cut through the skin, and the resins they are intended for tend to be brittle and more notch sensitive. Both these factors result in a greater level of notch sensitivity for this type of screw.

A gauge of a material's notch sensitivity is derived from a comparison of the notched Izod and unnotched Izod impact test results. This property can vary with temperature.

8.2.4   Craze Resistance

Stress in plastic parts often makes its presence known by tiny voids, cracks, or fissures, which appear in the surface of the part. This stress may be residual, resulting from the molding process, or it may be induced by external loadings such as fasteners. Screws, in particular, tend to produce high levels of localized stress. The material's ability to withstand this stress is diminished by chemical attack. While most parts will start to exhibit crazing at stress levels far below those that would result in a catastrophic failure, crazing is the beginning of a crack that could ultimately lead to such a fracture. Consequently, it is wise to avoid specifying screws for a part made of a material that is prone to crazing and must perform its function in a chemically aggressive environment. When that is not possible, the stress resulting from the fastener should be spread as widely as possible and the mold and molding conditions designed to minimize residual stress.

8.2.5   Stiffness Considerations

Incorrect fastener selection is often related to the failure to consider the stiffness of the plastic. This applies to self-tapping and thread-forming screws in particular. Stiffness is measured in terms of a material's flexural modulus. General recommendations for the selection of such screws will be found in the section of this chapter devoted to these screws. Specific recommendations for the correct screw for each material will be found in Chapter 5, "Assembly Method Selection by Material."

8.2.6   Differentials in the Coefficients of Linear Thermal Expansion

Since parts of different materials are commonly assembled, differences in CLTEs can lead to substantial stresses on the joint when temperature extremes are encountered. Methods for determining these stresses are found in Chapter 2, "Designing for Efficient Assembly." Ideally, the assembly is designed so that one of the parts is free to float relative to the other(s). Unfortunately, this is not often possible. Therefore, methods such as those illustrated in Fig. 8-3 are employed to alleviate these stresses.

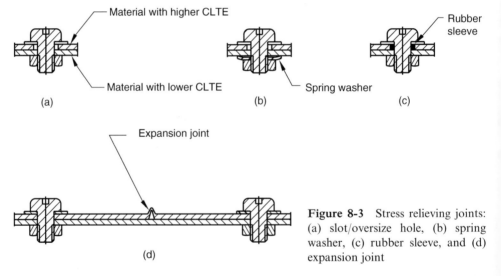

Figure 8-3   Stress relieving joints: (a) slot/oversize hole, (b) spring washer, (c) rubber sleeve, and (d) expansion joint

A slot or oversize hole beneath the screw, as in Fig. 8-3a, eliminates the constraint of the upper hole against the side of the screw and permits a degree of relative motion, depending on the clamping force. Unfortunately, it also results in a loss of control of lateral positioning. The hole is preferred to the slot unless the direction of the stress can be assured. In most cases, a washer under the screw head will be required.

Another approach is shown in Fig. 8-3b. In this case a spring washer is employed to accommodate the load resulting from thermal expansion.

Figure 8-3c depicts a slightly oversize rubber sleeve placed around the screw to receive the principal clamp load as well as that due to thermal expansion.

If the material with the higher coefficient of thermal expansion is fairly flexible, an expansion joint such as the one illustrated in Fig. 8-3d can be designed into that part.

8.2.7   Loss of Properties Due to Moisture

Some plastics are sufficiently hygroscopic to allow moisture to affect their strength and stiffness. This can result in the loosening of a fastener. It is sometimes necessary to add a desiccant to the product storage or shipping container.

8.2.8   Clamp Load

The primary design stress to be concerned with is the compressive stress, which is created by the screw. Also known as the prestress, this is the stress (σ) created when the screw is turned beyond the point where the parts are just held together. There are two methods of determining the clamp load: by the strain created when the screw is turned beyond the seating point, and by the applied torque.

8.2.8.1   Strain Method

The stress σ is determined by multiplying the modulus of elasticity in compression E_c of the plastic at 72 °F by the applied strain e.

$$\sigma = E_c e \tag{8-3}$$

The steel bolt will be in tension; however, since it is so much stronger than the plastic part, it is presumed that it will not stretch. The applied strain e is determined by dividing the number of turns N_t multiplied by the pitch p of the screw by the distance L over which the load is being applied.

$$e = \frac{N_t p}{L} \tag{8-4}$$

Referring to Fig. 8-4, we have a $\frac{1}{4}$-20 screw applying its load over a 1 in. length. Assuming these parts were made of acetal (Celcon) with a compressive modulus of 450,000,

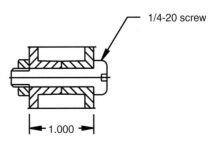

Figure 8-4   Example for Equation 8-4

and were preloaded $\frac{1}{4}$ turn, the applied strain would be 0.0125 in. and the clamp load would be 5625 psi.

$$e = \frac{\frac{1}{4} \times \frac{1}{20}}{1.00} = 0.0125 \, \text{in.}/\text{in.}$$

Therefore,

$$\sigma = 450{,}000 \times 0.0125 \, \text{psi} = 5625 \, \text{psi}$$

We note that 5625 psi exceeds the compressive strength of the material, which is 4500 psi. We, therefore, must decide whether to change the material, decrease the amount of turn, increase the length of engagement, or use a screw with a shorter pitch. The easiest approach is to decrease the amount of turn, however that is difficult to control on the production line and, therefore, probably not a good idea. Thus, it remains to determine which is the least costly to change: the material, the size of the screw, or its length of engagement. Changing the size of either the screw or material will lead to complications that would need to be examined. Therefore, in most cases, increasing the length of engagement would seem the easiest approach.

A word about the compressive modulus. This is a property that is not always available from the manufacturer. For most unfilled materials, the tensile modulus can be substituted. However, the tensile modulus is often higher than the compressive modulus for filled materials.

Clamp load can be controlled with the use of shoulder screws. However, in most cases, it is not necessary to pay the additional cost of shoulder screws.

8.2.8.2   Torque Method

The equations for the torque method make liberal use of approximations to shorten them for convenient application. Therefore, the results must be regarded as approximate as well. With this method, we first determine the compressive force F generated by a given torque T over the outside diameter of the screw D_0.

$$F = \frac{5T}{D_0} \tag{8-5}$$

For a torque of 8 in.-lb and a screw with an outside diameter of 0.250 in., we have:

$$F = \frac{5 \times 8}{0.250} = 160 \, \text{lb}$$

With this information, we can determine σ, the stress under the head of the screw, by dividing the force F by the area A under the head of the screw. For a $\frac{1}{4}$-20 pan-head machine screw, the head diameter would be 0.482 in. That would make the radius (r_{hd}) 0.241 in. The area of the hole must be subtracted from the total area of the head to determine the area under the head. If a hole clearance of 0.010 in. is allowed, the diameter of the hole would be 0.260 in., providing a hole radius r_h of 0.130. The area under the head is therefore:

$$A = \pi(r_{hd}^2 - r_h^2) = \pi(0.058 - 0.017) = 0.129 \, \text{in.} \tag{8-6}$$

Therefore, the stress under the head of the screw would be:

$$\sigma = \frac{F}{A} = \frac{160}{0.129} = 1240.3 \, \text{psi} \tag{8-7}$$

For this example, the clamp load would, therefore, be 1240.3 psi. However, it would not remain at that level for very long. Stress relaxation will promptly cause the clamp load to lessen, while an increase in temperature will cause it to increase. That is because the differential in the coefficient of linear thermal expansion between the two materials will make itself felt. The CLTE for the plastic (acetal) is 45×10^{-6} in./in./°F, while that of the steel bolt is 9×10^{-6} in./in./°F. The difference between the two is 36×10^{-6}; therefore the plastic will expand that much more than the metal for each Fahrenheit degree over 72 °F, thereby increasing the clamp load accordingly. Conversely, the plastic will decrease in size much faster than the metal and the prestress will drop accordingly.

It will be necessary to determine that the increase in temperature will not combine with the prestress to cause failure within the design limits of the product. In addition, the combined effects of stress relaxation and thermal contraction of the plastic parts must not result in loss of clamp force to the point where vibration could cause the fastener to loosen. The effects of thermal expansion and contraction can be reduced through the use of plastic screws in many cases. That option is discussed in further detail in Section 8.21, "Machine Screws."

8.2.9   Vibration Resistance

The thread of a threaded fastener is basically an inclined plane. Therefore, there is a constant force attempting to move down that plane and thereby unscrew the fastener. It is the resistance of the force due to friction which prevents it from doing that. It is generally accepted that vibration causes a momentary relaxation of that force, which permits the screw to loosen. Lockwashers, locknuts, and screws with protrusions under the head are used to dig into the plastic and provide added resistance to vibration. For permanent assemblies, a drop of adhesive can also be placed in the hole before the screw is emplaced to prevent loosening. Screws and nuts can also be purchased with glue in place on their threads which is activated by the heat generated in assembly.

8.3   Methods of Using Fasteners with Plastics

There are a variety of ways to use fasteners to join plastics. Among them are press-in fasteners, self-tapping screws, machine screws with inserts, boss caps or nuts, press-on fasteners, and panel fasteners. There are a wide number of variations for each of the types of fastener. Besides the variations on the threads themselves, there are a number of head styles available. Indeed, even the socket-head screws come in a variety of different socket shapes, each with claimed advantages for specific applications. With so many options available, the precise selection of fastener is usually done experimentally once the field has been narrowed to a limited group. The following sections discuss the features of each of these fastener types to assist in narrowing the field.

8.3.1   Press-in Fasteners

The fasteners generally referred to by the term "press-in" have some sort of device protruding from the shaft that creates an interference fitment with the hole in the plastic. (Two examples of this type are illustrated in Figs. 8-5a and 8-5b.) In time, creep occurs and the plastic that has been displaced by the protrusions upon assembly flows back in around them. This helps to keep the fastener in place when the unit is exposed to thermal expansion and vibration. As a result, these fasteners cannot be reassembled once removed, owing to the damage created on removal. Also, since they lack the inclined-plane loading provided by screw threads, they provide little in the way of clamping force.

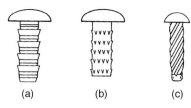

(a)   (b)   (c)   **Figure 8-5**   Press-in fasteners: (a) and (b) interference-type fitments; (c) U-type screw

This type of fastener assembles very quickly and lends itself well to automated assembly. The selection of the exact fastener is application specific; however the greater the interference, the greater the application force and, therefore, the removal force. These fasteners can be used with plastics that have a flexural modulus of 200,000 psi or lower. However, heat can be used to assist assembly either by enlarging the hole to create greater interference on cooling or to soften a borderline material. In some cases, that heat can be provided without additional cost if emplacement can be performed while the part is still hot from molding.

Threaded fasteners such as the type U screw illustrated in Fig. 8-5c can also be pressed into place. While they are capable of providing some clamping force, it is an order of magnitude lower than that provided by a normal thread form. Generally, these fasteners cannot be replaced after removal as reliably as the driven screws, since their threads that are formed on first assembly are easily damaged on reassembly.

The press-in fasteners are rarely used for anything besides injection-molded thermoplastics. The bulk of the materials that can accept them are thermoplastics, and the other thermoplastic processes cannot provide the type of close tolerance holes required for their use.

8.3.2   Self-Tapping Screws

Self-tapping screws comprise the class of reopenable fasteners that uses the fewest parts, since no female-threaded part is required. They have wide application in plastics assembly because they create their own threads and eliminate the need to tap the hole. In addition, it is not often cost-effective to mold female threads for fasteners into plastic parts.

Self-tapping screws are the fastest screws to assemble and the least expensive in terms of total cost. (Machine screws, without nuts, inserts, or tapping cost, are individually less costly.) If dimensions are not held closely, chipping and cracking of the molded part can result. There is a noticeable loss of holding power upon reassembly after the initial installation. Reassembly can damage the threads, and these screws are not recommended for applications requiring reassembly exceeding six times. The holding power of the plastic threads is lower than the threads present in a metal nut or insert.

8.3.2.1   Strength of Plastic Threads

Regardless of how the plastic threads are created, their strength in shear will be a determining factor in their application. Stripped threads will be the result of a shear failure. The following equations will enable the engineer to determine the approximate level at which failure will occur. Other factors, such as thread form, spacing, hole diameter, molding conditions, and notch sensitivity, will affect the actual stress at which failure will occur. In addition, these equations have been simplified and, therefore, the results must be regarded as approximate.

The shear stress τ at which the threads will strip equals the force F divided by the area A

$$\tau = \frac{F}{A} \tag{8-8}$$

where the area of stress A is equal to 0.5π times the effective length of thread engagement L times the outside diameter of the screw D_0, and the force F is equal to three-quarters the torque T divided by the dynamic coefficient of friction C_f. (As a practical rule, the length of engagement should be at least two to three times the screw diameter for a standard screw.)

$$A = 0.5\pi L D_0 \tag{8-9}$$

$$F = \frac{0.75T}{C_f} \tag{8-10}$$

The area equation (A) is correct for a standard thread like that found on a type C or T (23) self-tapping screw and illustrated in Fig. 8-6a. The decimal represents that portion of the length of thread engagement L that is not occupied by the metal thread L_p; therefore it would vary for screw threads with a longer lead proportionately. As shown in Fig. 8-6b, the thread form remains the same for these screws, however the lead changes. Thus for a #14 type C thread-forming screw, which has 20 threads per inch, 50% of the length would be metal and 50% plastic – therefore the multiplier is 0.5. The metal

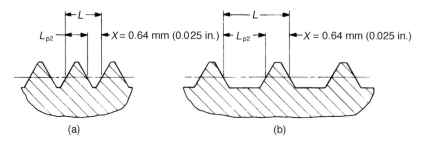

Figure 8-6 Plastic portion of thread engagement: (a) type C or T (23) $-L_{p2} = 0.5L$ and (b) type AB, B, BF, or BT (25) $-L_{p2} = 0.65L$

thickness at the root of each thread (x) would be 0.025 in. If the screw were changed to a type AB, the root thickness would still be 0.025 in., but there would only be 14 threads per inch. Thus, the metal would occupy 0.35 of the length (14×0.025) and the multiplier would become 0.65 ($1.00 - 0.35$). For different sizes of type AB, B, BF, or BT (25) screws, the equation would need to be similarly altered. For screws of other types, the reader is obliged to modify the equation accordingly.

The effectiveness of force equation (8-10) suffers from the problem of establishing a reliable value for the dynamic coefficient of friction C_f. To begin with, C_f will be different for every combination of materials. Furthermore, actual test values for a given pair of materials can vary enormously according to the load, rate of speed, surface finish, etc. In practice, a small amount of lubricant, hand lotion, mold release, or other contaminant can substantially alter the actual loading. Hence, the use of a substantial safety factor is recommended.

Most of the published figures for the dynamic coefficient of friction between two materials are based on test data conducted with the combination of low load and high speed. The emplacement of screws is, however, performed with the combination of high load and low speed. Since this equation is of value only as an approximation, and a substantial safety factor is presumed, a value of 0.15 is recommended for the dynamic coefficient of friction C_f between a metal screw and a plastic material.

The outward radial force component is a function of the thread form. For the standard unified thread form, which has a 60° included angle, as depicted in Fig. 8-7a, the approximate value for that component is 58% of the force F, a number that is determined by calculating the tangent of half the included angle. It is spread over the thread engagement cylinder, resulting in an internal pressure in the boss. The approximate value for this pressure P is determined by the following equation:

$$P = 0.58 \frac{F}{\pi L D_0} \qquad (8\text{-}11)$$

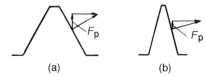

(a)   (b)

Figure 8-7 Outward radial force component: thread forms with (a) 60° and (b) 30° included angle

The narrower thread form in Fig. 8-7b has an included angle of 30°. That provides an outward force component F_0 that is only 0.27 of the axial force F component. The equation then becomes:

$$P = 0.27\frac{F}{\pi L D_0} \tag{8-12}$$

The pressure thus calculated can then be used to obtain an approximate value for the hoop stress σ based on simple thin-walled hoop stress equation:

$$\sigma = \frac{P(D_0 + 2t)}{2t} \tag{8-13}$$

The danger is that the boss will crack as a result of excessive hoop stress. Therefore, results of this computation must be evaluated against the tensile stress capability of the intended material. The approximations inherent in producing these equations require that a considerable safety margin be maintained.

In addition to these approximations, there is the matter of the weld or knit line. This is the weakest part of the boss and, under the best of conditions, will not surpass 85% of the strength of the surrounding material. For a further discussion of bosses, the reader is referred to Section 8.25: "Boss Designs."

8.3.2.2 Thread-Forming and Thread-Cutting Screws

There are two basic varieties of self-tapping screws: thread-forming and thread-cutting. The standard varieties of these screws, originally designed long ago for sheet metal, are shown in Fig. 8-8.

Thread-forming screws create threads by using the screw as a forming tool to push the walls of the plastic hole into the shape of a thread as illustrated in Fig. 8-2a. The threads at the tip of the screw are smaller in diameter, so that the pressure on the wall of the boss is applied gradually. As the full depth of the threads is reached, the compression pattern indicated in Fig. 8-2a takes place. The type C screw uses a standard machine screw thread, which creates a very high level of stress between the threads. Given the notch-sensitive nature of many plastics, this can lead to stress cracks as depicted in Fig. 8-2a. Fine threads such as this are used when the plastic is sufficiently ductile to withstand such forces and a high level of vibration and strip-out resistance is required. Type C screws can be replaced by ordinary machine screws, should they become lost on removal. For reasons of the previously mentioned weaknesses plus the high driving torques required for type C screws, however, they are not often used.

By increasing the distance between the metal threads, the compression pattern is relieved. This is the type of thread form used on the standard type AB and B screws. Thus, these screws have half as many threads per inch as standard machine screws, which results in twice as much plastic material between each metal thread. These are very similar screws, the difference being that the AB screw has a gimlet point that provides a lead into the hole but also requires a deeper hole for the same length of thread engagement.

Type	ANSI Standard	Manufacturer
	AB	AB
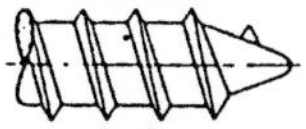 Not recommended - use type AB**	A	A
	B	B
	BP	BP
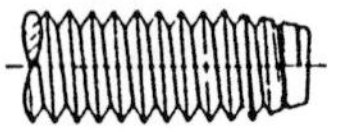	C	C
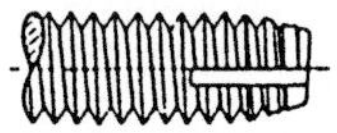	D	1
	F	F
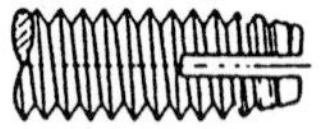	G	G
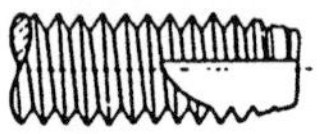	T	23
	BF	BF
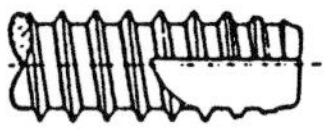	BT	25

Figure 8-8   Standard thread-forming screws (ASME B18.6.4-1981, courtesy of the American Society of Mechanical Engineers, New York)

Type AB screws have a thread with a 60° included angle, which results in a considerable outward force component. Consequently, the hole diameter must be selected carefully. Referring to Fig. 8-9, we note that the amount of material displaced from the side of the hole must not exceed 70 to 90% of the thread depth of the screw, the latter being for materials with the lower flexural moduli, with the percentage decreasing as the stiffness of the material increases. The proper hole size can be determined accordingly. The length of thread engagement should be determined by the amount of pullout strength required, but twice the screw diameter should be regarded as a minimum for most plastic materials. Thread engagement beyond three times the diameter is rarely required.

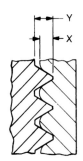

Figure 8-9   Type AB thread-forming screw thread ratio to hole diameter, $X/Y = 0.7$ to 0.9 for plastics with a flexural modulus less than 400,000 psi

The standard thread-forming screws can safely be used with plastics that have a flexural modulus of 200,000 psi and below. In most cases, they can be used for plastics with a flexural modulus of 400,000 psi and below; however the hole would need to be on the higher side. Because of the additional space between their threads, type AB and B screws can sometimes be used with some glass-reinforced materials. It should be noted, however, that some resin manufacturers do not recommend the use of thread-forming screws with their materials at any time owing to the high hoop stresses created in the boss.

For most materials with flexural moduli over 400,000 psi, and for some applications with flexural moduli over 200,000 psi, thread-cutting screws are the most desirable option.

The type T or 23 thread-cutting screw has a thread form that matches those found on standard machine screws. Like the type C thread-forming screw, the self-tapping screw can be replaced with a standard machine screw after its initial use. However, because the metal thread is so much harder than the plastic it is cutting, there is a danger of damaging the thread when the thread-cutting screw is reassembled multiple times. One way to eliminate that possibility is to remove the thread-cutting screw after it has been used to cut the thread and replace it immediately with a standard screw. One might, therefore, ask why not simply tap the hole and use a machine screw in the first place. The answer to that question is that installing a self-tapping screw is usually much faster than tapping a hole, since the screw is self-aligning and can be installed on the production line with no more than a hand drill, while tapping a hole requires the part to be clamped in place and the use of a drill press.

Like the type C screw, the type T uses a fine thread that creates a very high level of stress between the threads. This can lead to the same type of stress crack depicted in Fig. 8-2b. This type of screw is used when a high level of vibration and strip-out resistance is required or when repeated reassembling is anticipated.

8.3.3   Special Screws for Plastics

The limitations inherent in the use of standard sheet metal screws have resulted in the development of a number of screws specially designed for plastics. In general, these screws fall into one of four basic categories: screws with narrower than normal thread forms, screws with threads of alternating height, screws with asymmetrical thread forms, and those with the special thread configurations.

8.3.3.1 Narrow Thread Forms

Narrow thread screws are screws that have a thread form less than the standard 60° thread form. The narrower thread results in a lower outward force component. The reduction in hoop stress can lessen the likelihood of boss cracking or permit the use of bosses with one-third less wall section material. They also have extended leads to provide more plastic material between each metal thread. Therefore, there is more polymer to distribute the shear load over. The reduced profile also serves to lower installation torque. These screws are particularly useful for thermoplastics.

8.3.3.2 Alternating Thread Heights

The Hi-Lo® screw with alternating thread heights is probably the oldest screw designed specially for plastics. The high thread has a 30° included angle, while the low thread has a 60° included angle. This screw offers much the same advantages as the narrow thread form, however there are fewer high threads to carry the load. The low threads are claimed to reduce wobble during driving. The SMC Automotive Alliance recommends this screw for SMC compression applications.

8.3.3.3 Asymmetrical Thread Forms

A thread form with a very shallow angle on one side of the thread is an asymmetrical thread form. When the shallow angle is on the pressure flank, or thrust side, of the thread, it increases resistance to pullout and serves to further reduce the outward force component, permitting bosses with thinner walls. That is a desirable feature for thermoplastics because it reduces their tendency to sink at the base of the boss. Also, the greater angle on the entry side of the screw improves material displacement and flow. When it is on the nonthrust side of the screw, it serves to provide more plastic material between the threads.

8.4   Selection of Self-Tapping Screws

There are so many different self-tapping screws available that it is impossible to offer specific recommendations. However, it is possible to provide direction and establish criteria for making the final selection. Once the field has been narrowed to a limited number of candidates, final selection can be made experimentally.

8.4.1   Cost Criteria

The standard screws are the least expensive, since the manufacturers do not have to pay a royalty to sell them. However, royalties increase the cost of the screws by only 10 to 20% and, since screws are not usually an expensive product to purchase, this added amount may be of little consequence. If the specialty screw turns out to be less expensive to install, it might be more economical in the overall cost scheme. Consider also the number of fasteners required. Lack of suitable locations for bosses or the need to maintain flatness may alter the number and size of fasteners from the ideal, thereby affecting the cost of assembly. Obviously, one uses the fewest fasteners possible consistent with sound engineering principles. However it is not always wise to use the smallest units possible because very small fasteners can be slower to handle and assemble than those of medium size and larger.

8.4.2   Fail/Drive Ratio and Differential

The ratio between the torque at which a screw will fail (strip its threads) and the torque necessary to drive it into its hole is called the fail/drive ratio. While it is possible for the tempered and hardened metal screw thread to fail, it is sufficiently unlikely that, for practical purposes, it is presumed that the plastic threads will fail. Figure 8-10 illustrates the difference between the two for a given application. The fail/drive ratio is the fail torque divided by the torque required to drive the screw in. Although a ratio of 3:1 is regarded as a minimum for metal applications, plastic applications should be at 4:1 because the values are lower. However, lower ratios can be used under tightly controlled circumstances. It may be more significant to consider the differential between the two because of the difficulty in controlling the application torque under actual production conditions.

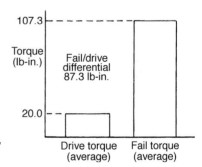

Figure 8-10   Fail/drive ratio and differential (*Assembly Engineering*, September 1982)

The fastener manufacturer may be able to supply these data for the combination of materials under consideration. However, it would be useful for comparison purposes only, as the actual values will be affected by other factors such as screw finish, hole size and finish, length of thread engagement, drive speed, and temperature.

8.4.3   Strength Criteria

The equations for clamp force, thread, and boss strength provided earlier are useful in providing direction only, since they are approximations and the data they rely on usually are based on tests conducted with samples and under conditions significantly

Table 8-1 Relationship Between Boss Cracking and Low Stripping Torque

Low stripping torque problem	Feature	Boss cracking problem
Decrease	Hole size	Increase
Increase	Length of thread engagement	Decrease
Increase	Screw diameter	Decrease
Increase	Flexural modulus of plastic	Decrease
Add	Nibs under head[a]	
	Flow ribs to boss	Add
Add	Support sleeve	Add
Increase	Thread angle	Decrease[b]
Decrease	Thread spacing	Increase
	Boss diameter	Increase[c]
Go to thread-forming screw from thread-cutting screw	Screw type	Go to special screw for plastics
Change to plating with higher coefficient of friction	Screw plating	

[a] Or other type of underhead friction.
[b] Less than 60° thread angle desirable. However beware of stress concentrations created by sharp corners.
[c] Increase in outside diameter of boss without changing inside diameter increases likelihood of "boss sink" on exterior of part.

varied from those of any given application. To keep well below the stripping torque, the clamp force should not exceed half the difference between the maximum drive torque where clamping begins and the stripping torque.

Comparison of data with existing applications is desirable when possible. Furthermore, comparison data sometimes are available from the screw manufacturer. "Fail tension" is the term used to describe the maximum clamp load attained before the screw thread fails. "Prevailing off-torque" is the torque required to back the screw out of the hole. It is a measure of the screw's resistance to loosening due to vibration.

Selection of screw type is not the only strength determinant. Hole size, length of thread engagement, screw diameter, and material selection can be adjusted to compensate for strength problems. The problems of low stripping torque and boss cracking are interrelated. Measures taken to improve one condition often lead to difficulties with the other. These effects are illustrated in Table 8-1.

There are some measures that benefit one phenomenon without affecting the other. The stripping torque can be increased by barbs or ribs under the head of the screw or on a boss cap, although the improvement is modest. Boss cracking can benefit from a boss cap or a support sleeve, the latter providing more support than the former.

8.4.4   Thread Cutting or Thread Forming

Historically, the rule we used was, "thread cutting when the flexural modulus of the material is over 400,000 psi, thread forming when it is under 200,000 psi, and trial and error when it is between these levels." For standard sheet metal screws, that is probably still a good rule to follow except in the case of higher modulus materials that are foam molded. However, there has been a profusion of materials, many of which have combinations of fillers etc., such that special thread forming screws designed for plastic have been successfully employed for applications with flexural moduli well above 400,000 psi. This despite the admonitions of most manufacturers of materials in this range to avoid using thread forming screws with their materials.

These applications have provided the higher stripping torque, pullout strengths, and superior reusability available from thread-forming screws despite their higher boss stresses and driving torque requirements.

Most screw manufacturers offer a variety of different screws and can also be helpful in recommending which may best apply to a given project. Some maintain a fastener laboratory that can provide assistance in selecting the best hole size for the selected screw.

For critical applications that involve expensive injection molding tooling, it may be cost-effective to construct an inexpensive prototype mold of the boss configuration. This prototype can be tested to ensure that the boss is adequate for the selected combination of screw, design, and material. Extra core pins can be made to determine the best hole size and to establish the effects of parts made to the extremes of the hole tolerance. If that is not done, the tests necessary to make the final determination will need to be performed when a part is available made of the production material process.

Do not test parts made of another material or by another process.

The results will not be comparable.

8.4.5   Tapped or Molded-in Threads

Internal threads can be tapped or molded into plastic parts. While the holding power of integral threads is less than that of molded-in or postmolded inserts, it is greater than that of self-tapping screws. There is very low stress on the thread, and the assembly of the screw is eased. Although the screws themselves are less expensive, neither tapped or molded threads are very desirable from a cost standpoint when there are more than a few holes to be threaded.

Tapping of drilled holes can be facilitated by molding the proper size hole instead of drilling it. In the case of glass fiber reinforced holes in bosses, however, the glass fibers may not flow into the boss sufficiently to maintain strength. Tapping itself is not often done because most of the other means of creating threaded holes are usually more cost-effective for the larger volumes that warrant the cost of molds. That is because tapping is a slower, and therefore more costly, means of creating threads. In some cases, the process has been speeded by using self-tapping screws to create the threads, then removing those screws and replacing them with standard machine screws.

Molding threads in the size range of screws requires an expensive unscrewing mold because they are usually too small for standard collapsing cores or to be stripped out of the mold. Internal threads are created by spinning cores in the mold, which are operated by a rack and pinion. Racks require a great deal of space in the mold. Therefore, to keep the number of racks required to a minimum, one attempts to place the holes in a line. This works well with rows of bottle cap molds where there is one core to be unscrewed per cavity and the cavities can be aligned practically at will. However, this effectively limits the number of holes that can be conveniently unscrewed in a typical plastic part. In general, molded-in threaded holes should be no smaller than a 0.250 in. diameter and no deeper than 0.500 in., with a minimum pitch of 24 threads per inch. For a further discussion of molded-in threads, the reader is referred to Chapter 18, entitled "Threads: Tapped and Molded-in."

8.5   Threaded Inserts: Advantages

There are occasions when plastic threads will not do. While improvements in self-tapping screws have permitted their use for a large number of plastic applications requiring repeated reassembly, in many cases metal threads are required. Threaded inserts are one of the principal means of providing them. Since their purchase, installation, and removal for recycling, brings extra costs, there have to be sound reasons for their use.

1. *Strength* In some cases, the plastic threads available from self-tapping screws will simply not provide the strength necessary for the application, and metal threads are required.

2. *Reusability* The metal threads of inserts are completely reusable. Self-tapping screws of both types, but thread-cutting screws in particular, can have threads damaged on reassembly.

3. *Standard screws* Standard machine screws are used with metal inserts, which make them easily replaced in the field should one be lost.

4. *Fine threads* The use of fine threads to provide resistance to loosening from vibration is limited with self-tapped threads. Finer threads and stronger fine threads are available through the use of metal inserts.

5. *Location and electrical contact* Inserts with flanges at a point part way down the length can be used for location in conjunction with holes in the mating part. For electrical applications, the flanges can be used for electrical contacts. There are standard expansion inserts made in this configuration.

6. *Resistance to abuse* Once beyond the control of the manufacturer, the product is susceptible to abuse in the form of forced plastic threads, overtorqued screws, application on an angle, or even the replacement of a lost screw with an inappropriate substitute. The metal insert limits the ability of unskilled individuals to abuse the fastening system.

The basic varieties of inserts are illustrated in Fig. 8.11. There are quite a variety made, as the widespread use of this technique has led to many specialized devices.

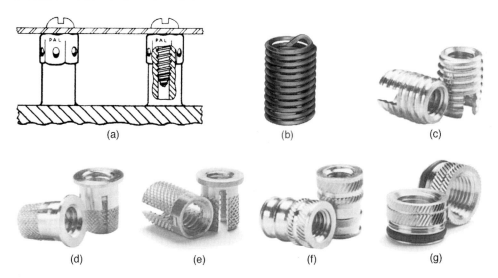

Figure 8-11 Threaded inserts: (a) boss cap (Courtesy of the Palnut Company), (b) helical coil (Courtesy of Helicoil/Emhart), (c) self-tapping (Dodge/Emhart), (d) spreader plate (Dodge/Emhart), (e) expansion (Dodge/Emhart), (f) Ultrasonic (Dodge/Emhart), (g) ultraseal © (Dodge/Emhart).

8.6 Boss Cap

The simplest variety of insert is the boss cap illustrated in Fig. 8-11a, which actually fits over the boss. Stamped from sheet metal, it is pressed over the top of the boss with a simple press to provide additional support against the radial and hoop stresses created upon installation of a self-tapping screw. Some versions have barbs on the inside to help hold the boss cap in place and to prevent rotation during installation. Since nearly a full turn of thread is provided by the cap, it can be used with sheet metal screws for very light loads without tapping. Boss caps provide fair holding power, are economical to purchase and quick to install, and can be used for both thermoplastics and thermosets. They can also be used to reinforce bosses that require additional strengthening.

8.7 Helical Coil Inserts

There is a variety of insert, shown in Fig. 8-11b, that provides a standard machine screw thread in metal to reinforce the plastic thread. It is screwed into a pretapped hole by means of a wire tang which, when twisted with a special tool, reduces in diameter to coil into the threads. When the insert has been fully installed, the tool is removed and the torsion is relaxed, allowing the coil to take its full diameter. This outward loading creates friction on the tapped threads, preventing the coil from unscrewing when the screw is unscrewed. The tang is then broken off with a special tool at a prenotched point at the top of the thread. This type of insert provides the highest quality, strength, and precision of all the postmolding inserts. Helical coil inserts provide even load distribution and very high tensile and torsional strengths, and they avoid most of the radial and hoop stresses associated with other types of insert. However, they are the most expensive of the postmolding inserts, and even with a premolded hole, the tapping and insertion operations are relatively slow and costly.

8.8   Self-Tapping Inserts

An insert that will provide metal threads is the self-tapping insert illustrated in Fig. 8-11c. This type of insert works in the same way that self-tapping screws do, except that it provides a metal thread for reassembly. An oversize hole is sized to fit the outside of the insert. The insert is then driven in place, usually via slots or grooves that permit a driving device to grip the insert. The inserts stay in place because the threads keep them from pulling out and friction prevents them from turning out. These inserts are available in both thread-forming and thread-cutting varieties, with the thread-forming variety being used for the softer plastics and the thread-cutting for the harder ones; however slotted inserts are suitable for most plastics applications. Inserts used with very hard or abrasive-filled plastics are sometimes made of case-hardened steel. Basically, the same parameters are used as for self-tapping screws. Extremes in hole or insert tolerances can result in high driving torque or poor insert retention.

Self-tapping inserts are easy to use, can be installed with a screwdriver, are suitable for both thermoplastics and thermosets, and do provide a metal thread. They are stronger than expansion inserts, with good resistance to pullout. They also prevent the destruction of threads during reassembly and permit the use of standard screws. Unfortunately, they cannot provide fine threads, provide only limited strength improvement over plastics threads themselves, and take longer to install than most of the other types of inserts. Self-tapping inserts are also more expensive than expansion inserts. These inserts are generally restricted to the size range from #8 to $\frac{1}{2}$ in.

8.9   Press-in Inserts

The press-in type of insert permits the user to install inserts without the use of heat or ultrasonics. They can be installed quickly with a simple press, or installation can be fully automated. The knurls, barbs, or flanges on the outside diameter of the insert push the plastic aside on entry. It then cold-flows back into the voids, thus creating resistance to torque or tensile forces. Obviously, installation is best done at the molding machine while the part is still hot, because the part will be both softer due to the heat and larger because it has not yet finished shrinking. These factors are particularly useful because the hole for a press-in insert should have an absolute minimum draft on the order of $\frac{1}{4}°$/side, or less if the material will tolerate it. No draft would be ideal from a mechanical standpoint because it distributes the stress in the most uniform fashion, however a slight draft is preferable to the use of mold release, which cannot be tolerated, or to distortion resulting from lack of draft. Normal draft in the hole can be used below the level of the insert. If the boss is undersized, it can be cracked by the insert.

While this design simplifies the assembly, it also results in an installation with less strength than the versions that involve melting the plastic and should not be considered for structural applications. In addition, pullout strengths can be uncertain owing to their reliance on tight tolerances. Also, high stresses are induced on installation. Consequently, the interference fit should be designed to limit the strain to 0.5%. Clearly, these inserts are best suited to plastics with a lower flexural modulus. Press-in inserts are available for both thermoplastic and thermoset materials.

8.10   Glue-in Inserts

Fasteners can be glued in place in addition to the previously described methods. This method is typically used for thermoset parts manufactured by the compression molding, resin transfer molding, and cold press molding methods. Barbed or knurled press-in inserts are the type normally glued into bosses; however they have open ends and sometimes need to have the surplus adhesive chased from their threads after installation. Consequently, closed-end inserts intended to be molded in are sometimes used in place of press-in inserts. If the volumes are large enough to warrant a special run, custom press-in inserts can be made with closed ends. These can be nearly as strong as molded-in inserts and are sometimes used to salvage an expensive moldment when a molded-in insert has been omitted. Care must be taken to avoid using adhesives that contain solvents that will result in crazing or cracking of the plastic in this highly stressed area. Bond strength can be uncertain owing to the requirement for clean surfaces.

8.11   Expansion Inserts

A stronger variety of press-in insert is the type that uses radial pressure to force the insert knurls, barbs or flanges into the surrounding plastic. The Dodge-Helicoil® version illustrated in Fig. 8-11d uses an internal spreader plate to force the wall of the insert into the plastic. It requires a blind hole to prevent the spreader plate from being pushed out the end. The expansion variety in Fig. 8-11e uses the pressure from the screw itself to apply pressure to the plastic. The advantage of the former is that it cannot fall out when the screw is loosened or removed; the latter has fewer parts and is inherently more economic. There is also a variety of expansion insert in which a cone on a stem is inserted with the insert. The cone is pulled upward, expanding the insert from the bottom, after which the stem is broken off and discarded.

Expansion inserts provide rotational and pullout strengths that are higher than press-fit inserts but lower than strengths obtainable from self-tapping or molded-in inserts. However, they can still be strong enough to cause the screw to shear before the insert is pulled out. They do not have the high installation stress associated with press-in inserts, but they also should be designed not to exceed an initial assembled strain of 0.5%. Tests conducted by DuPont indicate that an expansion insert should perform adequately for 1 to 10 years if it has a designed strength 140% of that required.

Expansion inserts are quick to install, with rates of 60 per minute being attained with automatic installation. They are widely used for thermoset parts, where they compete with molded-in inserts for applications. Advantages over those inserts include the elimination of rework to remove flash in threads and the elimination of floating inserts, which break core pins and damage the mold. Also, expansion inserts permit faster molding cycles by eliminating the need to load inserts, a significant advantage for applications that require a large number of inserts.

Expansion inserts can be used for thermoplastics as well; particularly with the larger sizes, however, there may be failure due to hoop stress. Besides, more cost-effective varieties of inserts are available for these materials.

8.12   Molded-in Inserts

Inserts can be molded directly into the part. In this case, the inserts are individually placed on pins in the mold. When the plastic fills the cavity, the melt runs into recesses in the insert, locking it into the part when the polymer cools. Molded-in inserts provide the highest levels of torsional and tensile stress resistance of any of the inserts with the possible exception of helical wire inserts for some applications. In addition, the strength and reusability of metal threads is provided. The limitations of molded-in inserts are the residual stress created when the plastic shrinks more than the metal insert and the additional molding costs. Molding costs increase because the molding cycle must be extended to permit the loading of the inserts, missing inserts result in reject parts, floating inserts cause mold damage, insert threads may need to be cleared of flash, and large inserts may require preheating to pre-expand them and improve flow and cure-in. Molded-in inserts are discussed in detail in Chapter 13, "Insert and Multipart Molding."

8.13   Ultrasonic Inserts

The style of insert depicted in Fig. 8-11f is one of a variety of inserts primarily designed to be installed with ultrasonic vibrations, which cause a narrow area of the plastic around the insert to be melted through friction. As the polymer is melted, it flows in and around the undercuts and crevices of the insert. On cooling, the plastic solidifies around the insert, thus providing resistance to extraction. However, a microscopic gap remains; therefore the compressive load that molded-in inserts receive is not present. This tends to reduce the likelihood of cracking the boss. Thus, the installation forces associated with ultrasonic installation of inserts are very low. The difference in cooling rate between the plastic and the insert leads to the creation of a stress relief area where the insert meets the plastic. Hermetic seals are possible only with special inserts (see Section 8.16: "Hermetic Seals").

Ultrasonic insertion is widely used to emplace inserts and has largely replaced the practice of molding-in inserts in thermoplastics because of the inherent disadvantages of that method (see Section 8.12: "Molded-in Inserts"). Ultrasonic inserts are restricted to thermoplastic materials only. Besides the reduced stress surrounding the insert and the low installation forces associated with this mode of insertion, ultrasonic insertion can be accomplished in a very short period of time: often less than a second; however it can range up to 3.5 s for large inserts. Thus, the process lends itself well to automation, and some inserts are designed with leads at both ends to eliminate the need for orientation upon entry to the hole. If the leads are located sufficiently close to each other, multiple inserts can be welded at one time. Compared to threaded inserts, ultrasonic inserts can use holes with more liberal tolerances. Compared to molded-in inserts, all the mold and molding problems are eliminated and smaller diameter bosses can be used. Insert cost is, however, slightly higher than molded-in or expansion types.

Ultrasonic insertion can be accomplished either by driving the plastic onto the insert or by forcing the insert into the plastic as shown in Fig. 8-12. The contact with the metal insert produces wear on the horn, requiring it to be made of hardened steel. Horn tips designed for ease of replacement reduce the cost of that operation. High stress applications may require horns of titanium even though they are less abrasion resistant than those of hardened steel. Conversely, driving the plastic onto the insert results in the horn being in contact with the plastic, which improves horn wear and is less noisy.

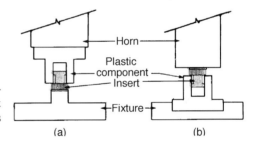

Figure 8-12 Types of ultrasonic insert installation: (a) plastic onto insert and (b) insert into plastic (Courtesy of Branson Ultrasonics Corporation, Danbury, CT)

The inserts are manufactured by a number of companies that offer patented versions with knurls, barbs, helical teeth, sharp recesses, etc. Inserts composed principally of undercuts are designed to emphasize resistance to pullout forces. Those with their surface largely devoted to helical slots are designed to resist torsion stresses. Most designs have some combination of these features. Manufacturers can supply relative tensile and torsional strength data for the different styles of inserts in most of the more commonly used plastics and should be consulted for the appropriateness of the application, the hole size, and its tolerances. The values for foamed materials will be much lower than those for solid resins and the recommended hole sizes will be slightly smaller.

The hole for ultrasonic insertion shown in Fig. 8-13 is generally 0.38 to 0.50 mm (0.015 to 0.020 in.) smaller than the diameter of the insert. Insufficient interference will result in a reduction in holding power; however excessive interference can cause cracking of the boss, excessive installation cycle time, or melted plastic to fill the hole or flow over the top during installation. If the interference is correct, but the

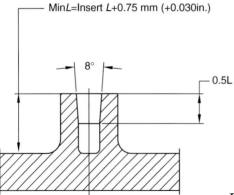

MinL=Insert L+0.75 mm (+0.030in.)

8°

0.5L

Figure 8-13   Boss design for ultrasonic inserts

material still flows over the top, an insert with a flange can be used. An 8° included angle lead-in is provided to guide the insert into place, reduce the time required to emplace it, and keep it in place while it is handled prior to welding. The lead-in can be designed into the insert instead of into the hole, however it should extend half the length of the insert in either case. The wall under the insert should be able to support the pressure of the weld, since insufficient support under the weld can result in an extended cycle time, part failure, or an inability to weld.

The hole must be 0.75 mm (0.030 in.) deeper than the length of the insert to provide space for the melt-out. Since most inserts used for ultrasonic insertion are open-ended, if the hole is too shallow for the insert, some of this melt can flow back into the threaded section of the insert, reducing the amount of usable thread and possibly rendering the insert useless because the screw cannot be inserted to the desired depth. It may not be obvious that this has occurred until the screw presses against the plastic in the bottom and jacks the insert loose so that it pulls out readily. This failure may also occur if too short an insert or too long a screw is used.

Ultrasonic insertion is performed with a low amplitude, medium pressure, and slow downstroke on the insert. When multiple inserts are to be installed or the insert diameter is 9.5 mm (0.375 in.) or larger, greater pressures are used and the ultrasonic vibrations are started before contact is made with the insert, a practice known as pre-triggering. The insertion of multiple inserts is more complicated because inconsistencies such as nonuniform melting and insert height variations can result from incorrect equipment settings.

Such settings can cause problems or extended cycle times for single insert installations as well. The boss can crack if the insert is pushed into the hole before the plastic has melted or if the insertion pressure is too great. This may also be the result of too thin a boss wall thickness. An incorrect stop setting may not drive the insert deep enough into the hole, or it might cause the insert to be driven too deep into the hole, resulting in plastic flowing over the top or filling the threaded hole. An insert set below the surface of the part will also be vulnerable to being jacked out of the hole by the screw. The horn should be two to four times the diameter of the insert, and its travel should be limited so the top of the insert winds up flush to the surface of the part or slightly above it. This provides the greatest torque resistance and pullout strength. The use of internal lubricants can cause dramatic reductions of holding power.

In addition to inserts, a variety of other fasteners such as studs, screws, pins, terminals, hinges, ferrules, and metal meshes can be inserted with ultrasonics. For a further explanation of ultrasonic assembly equipment, the reader is referred to Chapter 19, "Ultrasonic Welding."

8.14   Heat-Installed Inserts

Heat from a thermal press or inexpensive handheld inserter can be used to install inserts in place of ultrasonic vibrations. The insert is placed on a special pin that is used to transfer heat to the inside minor diameter of the insert thread. The heat then melts the plastic surrounding the insert until it is soft enough for the insert to be pressed in place. The same hole design used for ultrasonics is used for the installation of inserts with heat, and the same general rules apply, except for those relating to the particulars of ultrasonics. A thermal installer is illustrated in Fig. 8-14.

Figure 8-14 Thermal installation press (Courtesy of Sonitek, Milford, CT)

This method has advantages over ultrasonics for certain applications. Foremost are the cost of the equipment, which is about half that of an ultrasonic welder, and the reduction in expense for tool replacement due to wear. The method also handles filled materials better and is less sensitive to voids, openings, and contours in the plastic part. It can emplace several large inserts at different levels at the same time, works well on structural foam, and requires less force for insertion. Its adherents maintain that it is easier on the boss in that it permits greater tolerances, more ovality, and thinner walls without splitting (because there is less hoop stress). Also, its melt reportedly fills in the knurls and undercuts of the insert better, creating a 20% stronger bond. Finally, there are no operational requirements for ear protection or a sound enclosure because the method is much quieter and the handheld inserter is easier to use in difficult-to-access locations (it is also very good for prototypes and short production runs).

There are, however, some disadvantages as well. First, thermal insertion is slower than ultrasonic insertion because it takes longer to melt the plastic with heat (the entire insert must be heated) and because more plastic must be melted to ensure that enough of the hole is melted to emplace the insert. Also, the zone of melted plastic surrounding the insert is larger, and this can lead to lateral movement of the insert during installation. This makes it more difficult to hold tight location tolerances with heat insertion than with ultrasonic installation. Finally, the process is operator sensitive in that the application of excessive heat can result in heat degradation of the material surrounding the hole, with inherent loss of physical properties.

Inserts designed for heat or ultrasonic insertion can be used in unfilled thermoplastic materials and in thermoplastics with up to 30% fill of glass fiber or mineral fiber.

8.15   Induction Inserts

Induction heating can be used to heat the insert to be installed in the plastic part. (For a further discussion of induction heating, the reader is referred to Chapter 12, "Induction/Electromagnetic Welding.") In this case, the insert is mounted to a stainless steel push rod as depicted in Fig. 8-15. The push rod is attached to an air cylinder, which provides vertical motion. The insert to be installed is placed on the push rod and positioned in the coil from the induction heater where it is heated. Then the hot insert is plunged into the plastic part positioned below it, the molten plastic filling into the knurls and undercuts in the insert.

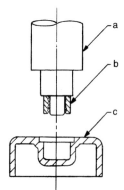

Figure 8-15   Insert installation with induction heating: (a) air cylinder, (b) insert, and (c) plastic part

The time required to heat the insert ranges from under 1 s to 6 s or more. Differences are due to the size of the insert and the material it is made from. The shortest melt times are achieved by small steel inserts and the longest times by larger aluminum and brass inserts. Steel inserts are harder than brass or aluminum; however they can corrode if their plating fails. Consequently, brass is the preferred insert material for most applications. However, brass begins to anneal at 450 °F, so brass inserts must be heated and inserted quickly to prevent annealing of the thread. Most inserts designed to be molded in, pressed in, self-tapped, expanded or ultrasonically installed can be used with induction heating, but those with deep knurls and undercuts work best.

The installation of inserts with induction heat is claimed to be able to hold the tightest tolerances of all the insert techniques including molding them in place. That is because the inserts can be installed with precision tooling, which accurately locates the insert in the same place each time even if the molded-in hole varies. Inserts can be located with an x-y positioning table or a robotic arm as well as a variety of simpler location devices. Multiple inserts can be simultaneously heated through the use of coils with more than one position. There is one example of ten 14.3 mm (0.563 in.) diameter brass inserts being installed in a glass-filled nylon engine manifold. The heating portion of the cycle is 4 seconds.

8.16   Hermetic Seals

As a result of the significant difference in coefficient of linear thermal expansion between the plastic hole and the metal insert, the plastic part will expand more than the insert. Thus, none of the inserts previously discussed can provide a hermetic seal. There is, however, a variety of ultrasonic insert that incorporates a rubber O-ring to accommodate this differential in expansion. It is the Dodge® "Ultraseal" illustrated in Fig. 8-11g. These inserts are designed to be installed with heat or ultrasonic vibrations and are provided with standard pipe threads instead of machine screw threads.

8.17   Studs

While most inserts used in plastics contain female threads, there is no reason why they cannot contain male threads. Several manufacturers produce studs, and these items can be custom-made if volume warrants. Studs can be pressed in, molded in or inserted by heat or ultrasonics. There also is a kind of stud that has a self-tapping screw on the end that fits into the plastic boss, thus offering a double-ended stud with a machine screw thread available for further assembly.

8.18   Insert Design Considerations

Inserts are customarily designed to withstand tensile pullout, torque rotation, and jackout. Inserts are designed to withstand pullout and torque rotation, with different designs

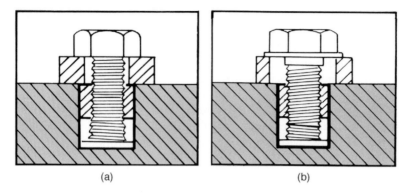

Figure 8-16 The problem of jack-out: (a) correct assembly (mating part prevents jack-out) and (b) incorrect assembly (potential for jack-out) (Courtesy of Dodge)

to emphasize one or the other. Jack-out is the condition illustrated in Fig. 8-16a in which the insert can be pulled out of its hole by the action of the screw. This problem can be eliminated by carefully designing the hole in the mating part to be smaller than the outside diameter of the insert, as shown in Fig. 8-16b. This prevents the insert from being pulled out of its hole.

8.19   U- or J-Clips

Female threads can be provided on flanges with inexpensive clips such as those shown in Figs. 8-17a, b, and c. The simplest variety has one turn of thread stamped in the hole; however some versions have a threaded tube with more turns of thread available. These clips are available for wall thicknesses ranging from 0.025 to 0.200 in.; however the thread sizes for each range of wall thickness are more limited. They can be very useful for parts from processes that must have their holes drilled or for which flanges are commonplace such as thermoforming, compression molding, resin transfer molding, and cold press molding. In some cases, a molded-in recess is used to locate the clip, as shown in Fig. 8-17d.

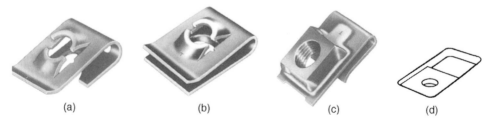

Figure 8-17 (a) J-clip, (b) U-clip, (c) multiple thread U-clip, and (d) J- or U-clip recess (Courtesy of Eaton Corporation)

8.20   Tee Nuts

Another device with a threaded tube that can be used for flanges or panels is the tee nut. One version uses a curved washer that digs into the back of the flange; the other uses prongs.

8.21   Machine Screws

Nuts and bolts are among the easiest of plastics assembly methods known since, for some applications, they require little preparation other than the drilling of mating holes in the parts to be assembled. This method lends itself well to low volume applications where the cost of molded-in assembly details would be hard to justify.

Machine screws offer several advantages over self-tapping screws and other fasteners. For one thing they offer metal-to-metal thread engagement. Since metals are much stronger than plastics, that means that a smaller screw can be used to achieve the same strength. Looked at in another way, if there is only enough space for very small screws, machine screws in combination with nuts or inserts can provide much greater strength.

A second benefit is the ability of machine screws to be readily disassembled for recycling without the part destruction associated with welded joints or metal inserts. In some cases, even the fasteners themselves can be recycled.

A third consideration is sink on the surface of the parts due to the wall thickness of structural bosses. While threaded inserts and self-tapping screws do require heavy structural bosses, nuts and bolts can do with much thinner wall bosses. The reduced wall thickness lessens the tendency of sink to appear on the outside of the part.

In some cases, the wide part-to-part variation inherent in many plastics processes calls for adjustment of alignment on assembly. It is common practice to assemble machine screws to a "finger-tight" level for handling during assembly, with final tightening taking place after the parts have been properly positioned.

Machine screws are also useful for applications where wide part-to-part dimensional variations are inherent to the nature of the process. The parts to be assembled can then be positioned to each other and a hole drilled through both parts, as illustrated in Fig. 8-18. The disadvantage to this approach is the loss of interchangeability between parts. If one of the pieces should require replacement, the user must drill the mating holes or the entire assembly must be returned to the factory for that purpose.

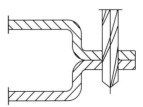

Figure 8-18   Drilling in place

This approach is well suited to the flanged parts often produced by the thermoforming, RTM, cold press molding, layup/sprayup, and compression molding processes. However, it can truly be used by all the plastics processes.

When the parts are assembled on their flanges as in Fig. 8-18, they cannot be distorted when they are clamped together. However, when there is a gap between the walls, as in Fig. 8-19a, the walls can deflect inward when the bolt is tightened, as shown by the dashed lines.

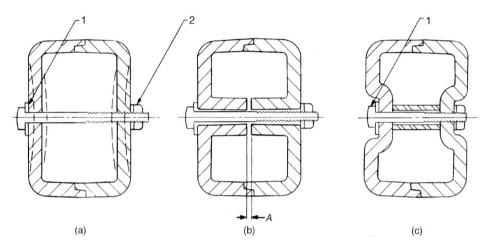

Figure 8-19 Design to avoid deflection: (a) unsupported walls (1, washer; 2, exposed nut and bolt), (b) boss supports, and (c) sleeve support (1, recessed screw head)

However, when bosses are designed into the parts, as in Fig. 8-19b, compression is limited to the amount preset by gap *A*. The amount of gap required is dependent on the shrinkage of the particular resin. To close on the bosses while there is still space at the outer walls is undesirable because it leaves a gap around the outer perimeter. Therefore, the dimensions are set such that there is contact without compression around the exterior when the bosses are at their maximum toleranced height. That means there will be a gap when they are at their minimum height. How much gap can be permitted will vary according to the location of the bosses relative to the outside wall, the overall size of the parts, and its stiffness as determined by its flexural modulus and wall thickness. For plastics with low to moderate shrink rates, a height tolerance of ±0.010 in. is reasonable. A tolerance of ±0.005 in. would be regarded as extremely tight and might result in an increase in the piece part cost.

In some cases it may be desirable to use a piece of cut metal or plastic tubing as a sleeve to provide support for the housing, as illustrated in Fig. 8-19c. The metal tubing option, in particular, will protect the assembly from overtightening, which may occur once the product is out of the manufacturer's control.

Note that examples (Fig. 8-19) show the heads of both the nut and bolt completely exposed. While not attractive, this provides easy access for application wrenches. These illustrations also have washers under both the screw head and the nuts. This helps to spread the stress resulting from the clamp load.

In Fig. 8-19c, both the head of the bolt and the nut have been recessed into the surface of the housing. There is a space around the nut to allow for a socket wrench. However, that is also a site for dirt and grease to accumulate. The shape of the recess could be molded to that of a hex to exactly match the nut. That would eliminate the dirt recess and the need for a wrench and access to the back side; however it would also subject the housing to the stress of tightening the bolt. For this type of design, the corners of the hex must be radiused to the maximum permissible by the nut to reduce the stress from that source.

Similar logic applies to the head of the bolt. Here, the example shows the walls of the recess just clearing the screw head. That would work if a slotted or socket-head screw were used; however the recess would need to be sufficient to clear the socket wrench if a hex-head bolt were specified. That would also provide space for a washer, which would help distribute the stress.

8.22   Tapping and Stud Plates

Tapping plates, such as the one shown in Fig. 8-20, are a useful device for parts made of low strength polymers or for high load applications. Basically, they are nothing more than tapped holes in a metal plate. However, they provide the strength of metal threads with the forces distributed over a wide area. Light-duty tapping plates can be made of sheet metal with the thread formed in the part. More demanding applications require heavier gauge stock.

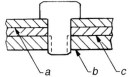

Figure 8-20   Tapping plate: *a*, top plastic part; *b*, tapping plate; *c*, bottom plastic part

Tapping plates can be swaged, riveted, glued, and staked to the plastic part in addition to the sandwich construction in the illustration. However, that is the strongest configuration. These plates can also be made with studs attached to the plate in place of tapped holes. In that form these are known as "stud plates."

Stud and tapping plates are commonly used for parts made from processes in which it is difficult to incorporate fastener details such as thermoforming, resin transfer molding, cold press molding, SMC compression molding, and sprayup/layup.

8.23   Plastic Screws

The problem of the differential in coefficient of linear thermal expansion between the metal screws and plastic parts can be substantially resolved through the use of plastic screws. In addition, the corrosion problem is effectively eliminated. Although these fasteners are limited by the reduced strength of material, there are many applications where they can be of use. However, the thread strength needs to be verified for both male and female threads when plastic screws are used. Plastic screws are available in the unified coarse, unified fine, and metric systems. Materials available are nylon 6/6, polycarbonate and, with varying degrees of availability, a number of other plastics. There are also some combination screws with metal cores and plastic threads. Finally, an intermediate step in CLTE between plastic and steel would be the use of aluminum as a screw material.

8.24   Screw Heads and Washers

Pan and truss heads, which are larger in diameter than ordinary round heads, are more desirable for plastics applications because they spread the load over a greater area. Washers are also extremely useful in distributing the clamp load on plastic assemblies and should be used with plastics. Screws are available with washers, combined with the head as a single unit, or loose but permanently attached to the screw just under the head. Screw heads with barbs or nibs under the heads are also manufactured. Lock-washers can be used to resist vibration just as with metal applications; however with plastics they also provide a significant benefit in resisting the effects of creep, cold flow, and stress relaxation.

The flat-head screw of the type shown in Fig. 8-21a is one type of screw head that must be avoided. This is because its angled contour on the bottom side of the head creates a lateral force component that places the material in tension due to hoop stress. If a flush surface is required, it is best to use a pan-head screw in a recess as in Fig. 8-21b.

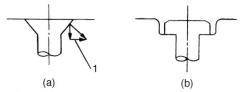

(a)   (b)

Figure 8-21 Flush surfaces: (a) flat head screw (1, lateral force component) and (b) pan-head screw in recess

8.25   Boss Designs

8.25.1   Design Criteria

As illustrated in Fig. 8-22, there are two fundamental criteria for boss design: sink and structural. The sink criteria shown in Fig. 8-22a is the principal one for bosses that either are nonstructural, such as locating bosses, or do not have radial or hoop stress, such as those used for machine screws and onserts. To eliminate sink, the fundamental requirement is that the thickest section not exceed 1.25 times the wall thickness. Structural bosses as in Fig. 8-22b are the type used for self-tapping screws or threaded

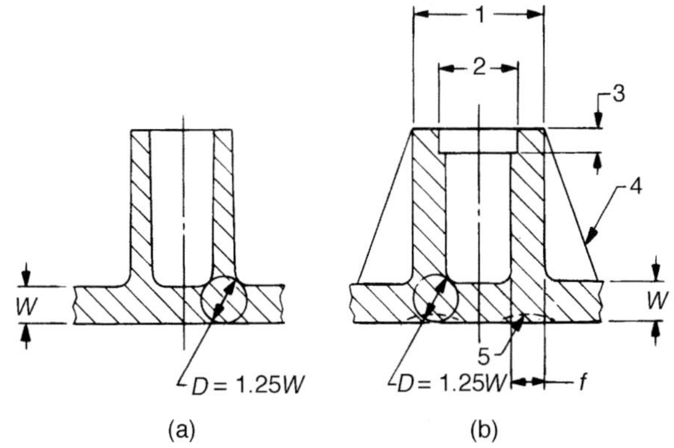

Figure 8-22 Boss design criteria: (a) boss design to avoid sink and (b) boss design for strength (1, boss diameter = 2 to 3 times insert diameter; 2, clearance diameter; 3, clearance = 1 full turn of thread minimum; 4, support gusset; 5, sink)

inserts. They have to be much stronger, and that requirement takes precedence over the sink criteria.

The hole for a self-tapping screw should have a clearance recess to a minimum depth of one full turn of thread. The outside diameter of a boss for a self-tapping screw or threaded insert is recommended to be two to three times that of the insert. Clearly, the wall will exceed the maximum ($1.25W$) to avoid sink. The actual minimum outside diameter will vary with the material and style of screw or insert, and the suppliers should be contacted for directions. However, an increase in that diameter ratio from 2:1 to 3.5:1 will improve the pullout strengths of press-in inserts by 40%, expansion inserts by 100%, and self-tapping inserts by 300%. Table 8-2 provides suggested minimum wall thicknesses for bosses to be used with inserts.

Supporting gussets should be used for screws of 6 mm (0.234 in.) and larger. The use of supporting gussets, which increase the resistance to bending and twisting while improving the flow of material to the boss, may permit the diameters of the bosses to be kept toward the low side for some applications.

8.25.2 Boss Sinks

There are several methods of dealing with boss sink.

8.25.2.1 Coring

How the hole is cored has an effect on the sink mark. Basically, sink is the external symptom of a high level of molded-in stress created by nonuniform cooling of the melt. (A void would be the internal symptom.) This condition becomes noticeable when the difference in wall thickness exceeds 25%, as illustrated in Fig. 8-22b. The sink criteria for the boss design in Fig. 8-22a deal with this issue by reducing the wall thickness accordingly. However, that approach results in a boss outside diameter that is too small to handle the stresses created by inserts and self-tapping screws. The accumulation of wall thickness resulting from the strength criteria can be reduced by increasing the depth of the core, as depicted in Fig. 8-23a.

Table 8-2 Minimum Wall Thickness (in.) for Insert Bosses (From *Plastics Engineering Handbook* of the Society of Plastics Industry, Inc., p. 765)

Material	Diameter of inserts (in.)									
	$\frac{1}{8}$	$\frac{1}{4}$	$\frac{3}{8}$	$\frac{1}{2}$	$\frac{3}{4}$	1	$1\frac{1}{4}$	$1\frac{1}{2}$	$1\frac{3}{4}$	2
Phenolics										
General purpose	$\frac{3}{32}$	$\frac{5}{32}$	$\frac{3}{16}$	$\frac{7}{32}$	$\frac{5}{16}$	$\frac{11}{32}$	$\frac{3}{8}$	$\frac{13}{32}$	$\frac{7}{16}$	$\frac{15}{32}$
Medium impact	$\frac{5}{64}$	$\frac{9}{64}$	$\frac{5}{32}$	$\frac{11}{64}$	$\frac{9}{32}$	$\frac{5}{16}$	$\frac{11}{32}$	$\frac{3}{8}$	$\frac{13}{32}$	$\frac{7}{16}$
High impact (rag)	$\frac{1}{16}$	$\frac{1}{8}$	$\frac{9}{64}$	$\frac{3}{16}$	$\frac{1}{4}$	$\frac{9}{32}$	$\frac{5}{16}$	$\frac{11}{32}$	$\frac{3}{8}$	$\frac{13}{32}$
High impact (sisal)	$\frac{5}{64}$	$\frac{9}{64}$	$\frac{5}{32}$	$\frac{3}{16}$	$\frac{1}{4}$	$\frac{9}{32}$	$\frac{5}{16}$	$\frac{11}{32}$	$\frac{3}{8}$	$\frac{13}{32}$
High impact (glass)	$\frac{1}{16}$	$\frac{3}{32}$	$\frac{1}{8}$	$\frac{1}{8}$	$\frac{3}{16}$	$\frac{3}{16}$	$\frac{1}{4}$	$\frac{1}{4}$	$\frac{5}{16}$	$\frac{5}{16}$
High heat resistant, general purpose type	$\frac{1}{8}$	$\frac{3}{16}$	$\frac{7}{32}$	$\frac{1}{4}$	$\frac{11}{32}$	$\frac{3}{8}$	$\frac{13}{32}$	$\frac{3}{16}$	$\frac{15}{32}$	$\frac{1}{2}$
High heat resistant, impact type	$\frac{5}{64}$	$\frac{9}{64}$	$\frac{5}{32}$	$\frac{13}{64}$	$\frac{9}{32}$	$\frac{5}{16}$	$\frac{11}{32}$	$\frac{3}{8}$	$\frac{13}{32}$	$\frac{7}{16}$
Low loss	$\frac{5}{32}$	$\frac{7}{32}$	$\frac{1}{4}$	$\frac{9}{32}$	$\frac{3}{8}$	$\frac{13}{32}$	$\frac{7}{16}$	$\frac{15}{32}$	$\frac{1}{2}$	$\frac{17}{32}$
Special for large inserts	$\frac{3}{64}$	$\frac{7}{64}$	$\frac{1}{8}$	$\frac{5}{32}$	$\frac{7}{32}$	$\frac{1}{4}$	$\frac{9}{32}$	$\frac{5}{16}$	$\frac{11}{32}$	$\frac{3}{8}$
Polyester, colors	$\frac{3}{32}$	$\frac{5}{32}$	$\frac{3}{16}$	$\frac{7}{32}$	$\frac{5}{16}$	$\frac{11}{32}$	$\frac{3}{8}$	$\frac{13}{32}$	$\frac{7}{16}$	$\frac{15}{32}$
Polyester, sisal-filled	$\frac{5}{64}$	$\frac{9}{64}$	$\frac{5}{32}$	$\frac{3}{16}$	$\frac{1}{4}$	$\frac{9}{32}$	$\frac{5}{16}$	$\frac{11}{32}$	$\frac{3}{8}$	$\frac{13}{32}$
Polyester, glass-filled	$\frac{1}{16}$	$\frac{1}{8}$	$\frac{9}{64}$	$\frac{3}{16}$	$\frac{1}{4}$	$\frac{9}{32}$	$\frac{5}{16}$	$\frac{11}{32}$	$\frac{3}{8}$	$\frac{13}{32}$
Diallyl phthalate										
"Orlon"-filled	$\frac{1}{8}$	$\frac{3}{16}$	$\frac{7}{32}$	$\frac{5}{16}$	$\frac{11}{32}$	$\frac{3}{8}$	$\frac{13}{32}$	$\frac{7}{16}$	$\frac{15}{32}$	$\frac{1}{2}$
Mineral-filled	$\frac{3}{32}$	$\frac{5}{32}$	$\frac{3}{16}$	$\frac{7}{32}$	$\frac{5}{16}$	$\frac{11}{32}$	$\frac{3}{8}$	$\frac{13}{32}$	$\frac{7}{16}$	$\frac{15}{32}$
Glass-filled	$\frac{5}{64}$	$\frac{9}{64}$	$\frac{5}{32}$	$\frac{3}{16}$	$\frac{1}{4}$	$\frac{9}{32}$	$\frac{5}{16}$	$\frac{11}{32}$	$\frac{3}{8}$	$\frac{13}{32}$
Cellulose acetate	$\frac{1}{8}$	$\frac{1}{4}$	$\frac{3}{8}$	$\frac{1}{2}$	$\frac{3}{4}$	1	$1\frac{1}{4}$	$1\frac{1}{2}$	$1\frac{3}{4}$	2
Cellulose acetate butyrate	$\frac{1}{8}$	$\frac{1}{4}$	$\frac{3}{8}$	$\frac{1}{2}$	$\frac{3}{4}$	1	$1\frac{1}{4}$	$1\frac{1}{2}$	$1\frac{3}{4}$	2
Ethyl cellulose	$\frac{1}{16}$	$\frac{3}{32}$	$\frac{1}{8}$	$\frac{5}{32}$	$\frac{3}{16}$	$\frac{7}{32}$	$\frac{1}{4}$	$\frac{9}{32}$	$\frac{5}{16}$	$\frac{11}{32}$
Urea formaldehyde	$\frac{3}{32}$	$\frac{5}{32}$	$\frac{3}{16}$	$\frac{7}{32}$	$\frac{5}{16}$	$\frac{11}{32}$	$\frac{3}{8}$	$\frac{13}{32}$	$\frac{7}{16}$	$\frac{15}{32}$
Melamine formaldehyde[a]										
MF	$\frac{3}{32}$	$\frac{5}{32}$	$\frac{3}{16}$	$\frac{7}{32}$	$\frac{5}{16}$	$\frac{11}{32}$	$\frac{3}{8}$	$\frac{13}{32}$	$\frac{7}{16}$	$\frac{15}{32}$
CF	$\frac{1}{8}$	$\frac{3}{16}$	$\frac{7}{32}$	$\frac{5}{16}$	$\frac{11}{32}$	$\frac{3}{8}$	$\frac{13}{32}$	$\frac{7}{16}$	$\frac{15}{32}$	$\frac{1}{2}$
Vinylidene chloride resin	$\frac{3}{32}$	$\frac{1}{8}$	$\frac{3}{16}$	$\frac{1}{4}$	$\frac{3}{8}$	$\frac{1}{2}$	$\frac{3}{4}$	$\frac{9}{32}$	$\frac{5}{16}$	$\frac{11}{32}$
Methyl methacrylate resin	$\frac{3}{32}$	$\frac{1}{8}$	$\frac{3}{16}$	$\frac{3}{16}$	$\frac{7}{32}$	$\frac{1}{4}$	$\frac{5}{8}$	$\frac{3}{4}$	$\frac{7}{8}$	1
Polystyrene	$\frac{3}{16}$	$\frac{3}{8}$	$\frac{9}{16}$	$\frac{3}{4}$	$1\frac{1}{8}$	$1\frac{1}{2}$	$1\frac{7}{8}$	$2\frac{1}{4}$	$2\frac{5}{8}$	3
Polyethylene	$\frac{1}{16}$	$\frac{3}{32}$	$\frac{1}{8}$	$\frac{5}{32}$	$\frac{3}{16}$	$\frac{7}{32}$	$\frac{1}{4}$	$\frac{9}{32}$	$\frac{5}{16}$	$\frac{11}{32}$
Nylon										
"Zytel" 101 or equiv.	$\frac{1}{16}$	$\frac{3}{32}$	$\frac{1}{8}$	$\frac{5}{32}$	$\frac{3}{16}$	$\frac{7}{32}$	$\frac{1}{4}$	$\frac{9}{32}$	$\frac{5}{16}$	$\frac{11}{32}$
"Zytel" 31 or equiv.	$\frac{3}{32}$	$\frac{1}{8}$	$\frac{5}{32}$	$\frac{7}{32}$	$\frac{1}{4}$	$\frac{5}{16}$	$\frac{11}{32}$	$\frac{13}{32}$	$\frac{7}{16}$	$\frac{15}{32}$
"Zytel" 63 or equiv.	$\frac{3}{32}$	$\frac{5}{32}$	$\frac{3}{16}$	$\frac{1}{4}$	$\frac{5}{16}$	$\frac{11}{32}$	$\frac{13}{32}$	$\frac{7}{16}$	$\frac{1}{2}$	$\frac{9}{16}$
"Zytel" 69 or equiv.	$\frac{1}{8}$	$\frac{7}{32}$	$\frac{9}{32}$	$\frac{11}{32}$	$\frac{7}{16}$	$\frac{1}{2}$	$\frac{19}{32}$	$\frac{21}{32}$	$\frac{23}{32}$	$\frac{13}{16}$
"Zytel" 105 or equiv.	$\frac{1}{16}$	$\frac{3}{32}$	$\frac{1}{8}$	$\frac{5}{32}$	$\frac{3}{16}$	$\frac{7}{32}$	$\frac{1}{4}$	$\frac{9}{32}$	$\frac{5}{16}$	$\frac{11}{32}$
"Zytel" 211 or equiv.	$\frac{3}{32}$	$\frac{1}{8}$	$\frac{5}{32}$	$\frac{7}{32}$	$\frac{1}{4}$	$\frac{5}{16}$	$\frac{11}{32}$	$\frac{13}{32}$	$\frac{7}{16}$	$\frac{15}{32}$
"Zytel" 42 or equiv.	$\frac{1}{16}$	$\frac{3}{32}$	$\frac{1}{8}$	$\frac{5}{32}$	$\frac{3}{16}$	$\frac{7}{32}$	$\frac{1}{4}$	$\frac{9}{32}$	$\frac{5}{16}$	$\frac{11}{32}$
Vinyl chloride–acetate resin	$\frac{3}{32}$	$\frac{1}{8}$	$\frac{3}{16}$	$\frac{1}{4}$	$\frac{3}{8}$	$\frac{1}{2}$	$\frac{5}{8}$	$\frac{3}{4}$	$\frac{7}{8}$	1

[a] MF, mineral-filled melamine ignition material; CF, cellulose-filled melamine, electrical grade.

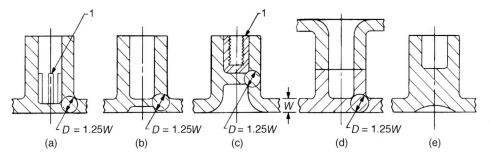

(a) $D = 1.25W$ (b) $D = 1.25W$ (c) $D = 1.25W$ (d) $D = 1.25W$ (e)

Figure 8-23 Methods of dealing with boss sink: (a) recessed core (1, insert height control ribs), (b) external recess, (c) deep recess (1, insert) (d) extension sleeve, and (e) poor practice

8.25.2.2 Location

When feasible, locate the bosses where their sinks will be the least noticeable. Locations under a hidden surface, a label or a recess, such as that as shown in Fig. 8-23b, are good examples of this concept. This concept works well for postmolded inserts and self-tapping screws where the depth of the hole can be extended beyond that which is needed for the screw or insert; however, as shown in Fig. 8-23c, it requires a deep hole for some molded-in inserts. A better approach is to minimize the height of the boss for the molded-in insert and extend the opposing boss to meet it, as illustrated in Fig. 8-23d, although this does require a longer screw. Under no circumstance is it wise to leave a large pocket of solid material, as depicted in Fig. 8-23e. Not only would such a design lead to sinks and high levels of molded-in stress, but the entire molding cycle would be delayed by the need to wait for the thick section to cool sufficiently to permit ejection.

8.25.2.3 Support

The base of the boss is not the only place that sink can occur. The perimeter (in the plan view) is also vulnerable to this problem, particularly when reinforcing ribs are used. Figure 8-24 shows several boss designs, some good and some very bad.

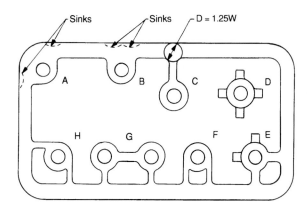

Figure 8-24 Boss configurations

First, let us consider the bad options. The configurations at A and B will result in sink for unfilled thermoplastics and, to a lesser degree, for some thermosets and filled thermoplastics. They can work for foamed thermoplastics, most thermosets, and some filled thermoplastics. In any case, they will still increase the molding cycle if they are the thickest section on the part.

Ribs and gussets should be used in place of integrating the boss into the wall. Again, the bosses in Fig. 8-24 represent examples of both good and bad practice. The corner boss at position A will sink (dashed lines) at the points where the material is the thickest. The corner boss at position H is better, but it still is likely to form sinks. However, these will be inside the part where the ribs are tangent to the boss. The corner boss design at position E is the strongest.

The midwall boss at position B will create sinks like the one at position A. To avoid sink on the outer wall completely, the boss must be moved in enough for a thin wall rib to be formed like the ones on the bosses at positions C, F, and G. However, the one at F can exhibit the same internal tangential boss sinks as the one at H. These designs shown all strive to reduce the wall section of the rib sufficiently to prevent the pocket of material in the circle from exceeding the nominal wall section by more than 25%. Depending on the radius at the intersection, that usually results in a rib thickness half that of the nominal wall.

The problem can be completely avoided by using freestanding bosses such as the one at position D in Fig. 8-24. While these are generally not as strong as those connected to the outer wall, their strength can be enhanced through the use of gussets such as on the one illustrated. Gussets are recommended for screws 6 mm (0.236 in.) and greater.

8.25.2.4 Material

High shrinking thermoplastics are the most vulnerable to sinks. Thermosets, filled thermoplastics, and foamed thermoplastics are far less likely to show noticeable sink. However, foamed materials will have considerably less strength than the same resin in a solid form. Also, the hole must be molded, not drilled, because the drill will remove the skin where most of the strength lies. One benefit of foamed holes is that the foam absorbs much of the radial and hoop stress, thereby compensating, somewhat, for the lowered strength of the moldment.

8.25.2.5 Surface Treatment

Gloss surfaces make sink marks more noticeable; however the sink area can be masked by an etched or matte surface in the mold. Hot stamping or heat transfers will cover minor sink marks, but can leave a gap in the mark if the sink is too deep.

8.25.3   Weld Lines

In addition to the sink problem, there is the matter of the weld or knit line. This is the weakest part of the boss. It is not formed as an extruded tube. As illustrated on the

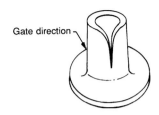

Gate direction

Figure 8-25   Boss weld/knit lines – partially formed boss

partially formed boss in Fig. 8-25, the melt runs up the side of the cavity closest to the gate and then fills around the core. The leading edges of the melt will be somewhat cooler than the remainder of the material. When these edges meet, there will not be complete fusion at the seam. Under the worst of molding conditions, that seam can actually be open. Under the best of conditions, the weld line might have a strength as high as 85% of the surrounding material. With rare exception, boss failures occur at their weld lines.

The material, the size of the boss, and its distance and flow path from the gate are also relevant to the strength of the boss. A small self-tapping screw, like a #4-40, could call for a boss with a wall thickness of 0.114 in. However, if it were to be made of a slow-flowing, glass-loaded material and located far from the gate at a site the melt could reach only by traveling an arduous path, there might not be enough pressure left to form a good boss. That could force the molder to resort to such remedies as an increase in the molding temperature, which could lead to degradation problems.

One solution is to provide a channel to run hot material over to the boss, a sort of internal runner within the part if you will. That concept is illustrated in Fig. 8-26.

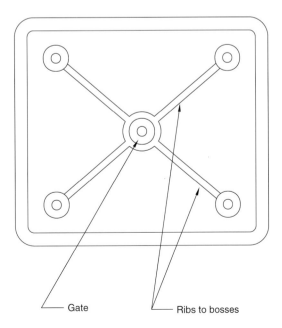

Figure 8-26   Internal runner system   Gate   Ribs to bosses

8.26   Self-Threading Nuts

Self-threading nuts, illustrated in Fig. 8-27, were originally created for the automotive industry. A self-threading nut spins onto a molded stud, creating its own thread as it does so. These nuts often have integral washers that spread the clamp force and offer excellent clamp strength. They are designed to resist loosening due to vibration and can be reassembled to the same stud if overtightening has not previously stripped the stud. The capped version shown in Fig. 8-27b is used when the stud needs to be covered. Depending on the nominal wall thickness, a sink mark can form under the stud on unfilled thermoplastics. Self-threading nuts can also be used on thermosets.

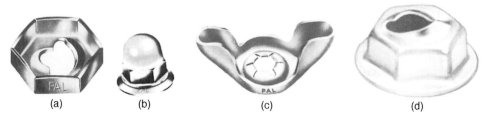

| (a) | (b) | (c) | (d) |

Figure 8-27   Self-threading nuts: (a) regular, (b) capped washer, (c) wing nut, and (d) washer (Courtesy of The Palnut Company)

8.27   Twist Nuts

For lower strength applications, there is a similar appearing, lighter duty variety that creates less torque and radial stress. Twist nuts can be installed partially threaded with a $\frac{1}{4}$ to $\frac{1}{3}$ turn. These fasteners can also be pushed on, removed, and reused; however, care must be taken not to distort them on assembly. They are shown in Fig. 8-28.

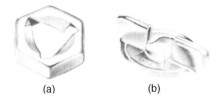

| (a) | (b) |

Figure 8-28   Twist nuts: (a) regular and (b) zip twist (Courtesy of The Palnut Company)

8.28   Press-on Nuts

Except that they are simply pressed onto the studs, press-on nuts are similar to the self-threading nuts. Consequently, they are assembled very quickly. Arched types provide spring pressure, which creates clamping force. Arched types are pressed to nearly flat at the base of the stake to provide that tension and depend on the teeth to maintain position. In order to do that, the bearing surface must be hard enough to permit the fastener to slide. Since they have no threads, they are highly resistant to vibration;

Figure 8-29 Flat, round press on nut (Courtesy of The Palnut Company)

Figure 8-30 Spring clip (Courtesy of The Palnut Company)

Figure 8-31 Push-in fastener (Courtesy of Eaton Corporation)

however they are limited to fairly light duty applications. The parts must have sufficient stiffness and strength for this type of fastener. Basically, these inexpensive fasteners must be regarded as nonreusable. They can be used with both thermoplastics and thermosets. A press-on nut is illustrated in Fig. 8-29.

8.29   Spring Clips

The spring clip is a removable device, shown in Fig. 8-30, with an integral spring arm that must be depressed for installation. When the spring is released, its teeth grip on the stud. Removal can be accomplished by depressing the spring arm with pliers and pulling the clip off.

8.30   Push-in Fasteners

Push-in fasteners, often referred to as panel fasteners, such as the one illustrated in Fig. 8-31, were originally designed for electrical applications involving the securing of panel components or, when extended with a spacer instead of capped, to separate printed circuit boards (standoffs). These plastic fasteners are removable (from the inside), reusable, and have now found many applications assembling thin-gauge materials. They have generally low strength relative to threaded fasteners and, in some cases, are used simply to plug a hole.

8.31   Rivets

There are four varieties of rivets used in plastic assembly: those with tubular shanks, solid shanks, semi solid shanks, and blind rivets. These rivets are illustrated in Fig. 8-32. For plastics, the most common variety of rivets selected are hollow and made of aluminum.

To withstand the high installation forces, rivets used for plastics should have large heads and be formed against a washer or metal surface, as illustrated in Fig. 8-33. While most fasteners, in general, are stronger in shear across their shanks than in tensile against their heads, this is particularly true of rivets. When possible, plastic applications

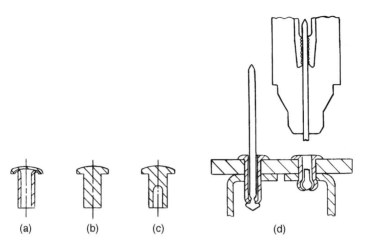

Figure 8-32   Rivets: (a) tubular, (b) solid, (c) semisolid, and (d) blind

using rivets should be designed with the primary load in shear across the rivet shanks, as opposed to having this load in tensile on their heads. However, depending on the resin used, the shear strength of the metal rivet may turn out to be stronger than the compressive strength of the plastic it is holding. In that case, the hole in the plastic will become elongated. The strength of both the rivet and the assembly materials need to examined. When necessary, resistance to shear deformation can be increased by increasing the diameter of the rivet.

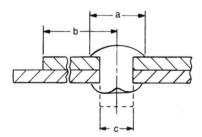

Figure 8-33   Rivet installation: a, head diameter; b, edge distance = 3 to 4 times rivet diameter; c, rivet diameter, hole diameter = 1.1 × shank diameter, clinch allowance = 0.5 to 0.75 times shank diameter

With respect to the rivet heads, the formed head or clinch is the most vulnerable of the two. The term to describe the amount of the rivet protruding beyond the inside surface and available to form the clinch is known as the clinch allowance. The optimum clinch allowance is 75% of the diameter of the rivet; the minimum being 50%. The hole for a rivet should be 10% greater than the diameter of the rivet, with a distance from the edge of the part of three to four times the diameter of the rivet. If sliding action is required between the two parts, the hole size and slot width should be greater. A shoulder rivet is desirable for sliding action; however it can also be useful in reducing the compressive loading on the plastic during head formation.

Rivets are well suited to products manufactured by processes that produce flanged parts such as thermoforming, layup/sprayup, compression molding, resin transfer molding, and cold press molding.

Since rivets are a permanent fastening method, there is no need for inter-changeability, and the drilling and riveting of two or more parts in place is perfectly feasible. Again, there is a need to be alert to differences in thermal expansion between the parts.

The topic in this chapter is too extensive to thoroughly cover in the space available in a handbook of this nature. The author would be remiss if he did not refer the reader to Lincoln, Gomes, and Braden's excellent coverage of this subject in their book, *Mechanical Fastening of Plastics* (Marcel Dekker, Inc./New York).

8.32 Sources

8.32.1 Fasteners and Inserts

Alpha Bolt Co., 1524 East 14 Mile Rd., Madison Heights, MI 48071, (810) 585-6050, fax (810) 583-1382

ATF , Inc., 3550 W. Pratt Ave., Lincolnwood, IL 60712, (847) 677-1300, fax (847) 677-9335, www.atf-inc.com

Camcar / Textron, 600 Eighteenth Ave., Rockford, IL 61104-5161, (800) 544-6117, (815) 961-5000, fax (815) 961-5345, www.camcar.textron.com

DFCI Solutions Inc., 425 Union Blvd., West Islip, NY 11795, (631) 669-0494, fax (631) 669-0785, www.dfcis.com

Dodge, Div. Emhart Fastening Teknologies, 50 Shelton Technology Center, P.O. Box 859, Shelton, CT 06484, (203) 924-9341, fax (203) 925-4481, www.emhart.com

Engineered Fastener Co., 1271 Hamilton Pkwy, Itasca, IL 60143, (630) 285-0001, Fax (630) 285-0000

Entegra Fastener Corp., 321 Foster Ave., Wood Dale, IL 60191, (630) 595-6250, fax (630)595-0336, www.entegrafastener.com

Halo-Krome Co., 31 Brook St., West Hartford, CT 06110, (860) 523-5235, fax (800) 243-3149

Hartwell Corp., 900 S. Richfield Rd., Placentia, CA 92870-6780, (800) 441-0327, (714) 993-4200, fax (714) 777-4031

Huck Manufacturing, Inc., P.O. Box 27207, 3724 E. Columbia St., Tucson, AZ 85726-7207, (520) 519-7400, fax (520) 362-7440, www.huck.com

Industrial Fasteners Corp., 57 Denton Ave., New Hyde Park, NY 11040, (516) 352-6220

ITW Anchor Fasteners, 26101 Fargo Ave., Bedford Heights, OH 44146, (216) 292-7161, fax (216) 292-6999

ITW Fastex Distributor Business, 226 Gerry Dr., Wood Dale, IL 60191, (708) 350-8200, fax (708) 595-2141

ITW Medalist, Inc., 2700 York Rd., Elk Grove Village, IL 60007, (847) 766-9000, fax (847) 766-3645, www.itwmedalist.com

ITW Shakeproof Nut Fastening Systems, 1201 St. Charles Rd., Elgin, IL 60120, (847) 741-7900, fax (847) 741-7914, www.itwshakeproof.com

Kamax L.P., 500 W. Long Lake Rd., Troy, MI 48098, (248) 879-0200, fax (248) 879-5850, www.kamax.com

Lang Fastener, Div. of L.E. Borden Mfg. Co., 15289 12 Mile Rd., Roseville, MI 48066, (586) 772-8180, fax (586) 772-0720, www.langfastener.com

Long-Lok Fasteners Corp., 10630 Chester Rd., Cincinnati, OH 45215 (800) 566-4565, (513) 772-1888, fax (513) 772-1888, www.medizin.li

Micro Plastics, Inc., #1 Industry Lane, Highway 178 North, P.O. Box 149, Flippin, AR 72634, (870) 453-2261, fax (870) 453-8676, www.microplastics.com

Multech Inc., 2108 Shawnee Rd., Baroda, MI 49101-0286, (888) 422-2353, (616) 422-1122, fax (616) 422-1371, www.multech.com

Nylok Fastener Corp., 15260 Hallmark Dr., Macomb, MI 48042-4007, (800) 826-5161, (586) 786-0100, fax (586) 786-0598, www.nylok.com

Orbitform Inc., 1600 Executive Dr., P.O. Box 1469, Jackson, MI 49203, (517) 787-9447, fax (517) 787-6609, www.orbitform.com

Panduit, 17301 Ridgeland Ave., Tinley Park, IL 60477, (888) 506-5400, (800) 777-3300, (708) 532-1800, fax (708) 532-1811, www.panduit.com

Penn Engineering, 5190 Old Easton Road, Danboro, PA 18916, (800) 237-4736, (215) 766-8853, fax (215) 766-3633, www.penn-eng.com

Reliant Fastener, Div. of Reliant Industries, 201 E. 2nd St., Rock Falls, IL 61071, (815) 625-4000, fax (815) 625-4021

Richco Inc., 8145 River Dr., Morton Grove, IL 60053, (800) 466-8301, (773) 539-4060, fax (773) 539-6770, www.richco-inc.com

Rivnut Engineered Products, Inc., 2750 Marion Drive, Kendallville, IN 46755, (260) 347-3903, fax (216) 666-8066, www.bollhoff-rivnut.com

Rockford International Group, 2501 Ninth St., Rockford, IL 61104-7197, (815) 397-6000, fax (815) 229-4326, www.rockfordinternational.com

SPS Technologies, Inc., Unbrako Engineered Fasteners, 4444 Lee Rd., Cleveland, OH 44128, (800) 255-5777 (216) 581-3000, www.spstechnologies.com

Textron Fastening Systems, 516 Eighteenth Ave., Rockford, IL 61104, (800) 544-6117, (815) 961-5300, fax (815) 961-5345, www.textronfasteningsystems.com

Tinnerman Palnut Engineered Products, 1060 W. 130th Street, Brunswick, OH 44212, (800) 221-2344, (330) 220-5100, fax (330) 220-5797, www.palnut.com

Trans Technology Corp., 150 Allen Rd., Liberty Corner, NJ 07938, (800) 221-2344, (908) 903-1600, fax (908) 903-1616, www.transtechnology.com

Vlier Products, 40 Guest St., Brighton, MA 02135-9105, (800) 821-1090, fax (800) 457-2020, www.vlier.com

Whitesell Mfg. Inc., 327 N. Pine St., Florence, AL 35630, (256) 248-8500, fax (256) 248-8585

8.32.2 Threaded-Insert

Emhart Fastening Teknologies Industrial Division, 50 Shelton Technology Center, P.O. Box 859, Shelton, CT 06484, (203) 924-9341, fax (203) 925-3109, www.emhart. com

Engineered Inserts & Systems, Inc., 22 Calender Road, Watertown, CT 06795, (860) 274-3628, fax (203) 274-7939

E-Z Lok, 240 E. Rosecrans Ave., Gardena, CA 90248, (800) 234-5613, (310) 323-5613, fax (310) 353-4444, www.ezlok.com

Groov-Pin Corp., 1125 Hendricks Causeway, Ridgefield, NJ 07657, (201) 945-6780, fax (201) 945-8998, www.groov-pin.com

Precision Fasteners, Inc., 3 Laurel Drive - Unit #16, Flanders, NJ 07836, (800) 447-2077, (973) 927-2077, fax (973) 927-3379, www.precisionfastenersinc.com

P.S.M. Fastener Corp., 12223 C R Koon Hwy. 76, Newberry, SC 29108 (803) 321-1300, fax (803) 364-7377, www.psminternational.com

Standard Insert Co., 5190 Old Easton Rd., Danboro, PA 18916, (800) DIAL-PEM, (215) 766-8853, fax (215) 766-3633, www.standardinsert.com

Thermosonics, P.O. Box 1287, 88 Spring Lane, Farmington, CT 06032, (203) 678-1280, fax (203) 678-0762

Tri Star Industries, Inc., 101 Massirio Dr., Berlin, CT 06037, (800) 882-8980, (860) 828-7570, fax (860) 828-7475, www.tristar-inserts.com

VMP Incorporated, 24380 Avenue Tibbitts, Valencia, CA 91355, (888) 486-7462; (661) 294-9934, fax (661) 294-0542, www.vmpinc.com

Yardley Products Corp., 10 W. College Ave., P.O. Box 357, Yardley, PA 19067-8357, (800) 457-0154, (215) 493-2723, fax (215) 493-6796, www.yardleyproducts.com

8.32.3   Thermal Insertion Equipment

Sonitek & Thermal Technologies, Inc., 84 Research Dr., Milford, CT 06460, (203) 878-9321, fax (203) 878-6786, www.sonitek.com

VMP Incorporated, 24380 Avenue Tibbitts, Valencia, CA 91355, (888) 486-7462; (661) 294-9934, fax (661) 294-0542, www.vmpinc.com

Yardley Products Corp., 10 W. College Ave., P.O. Box 357, Yardley, PA 19067-8357, (800) 457-0154, (215) 493-2723, fax (215) 493-6796, www.yardleyproducts.com

8.32.4   Induction Insertion Equipment

Ameritherm Inc., 39 Main Street, Scottsville, NY 14546, (800) 456-HEAT, (585) 889-9000, fax (585) 889-4030, www.ameritherm.com

8.33.5   Ultrasonic Insertion Equipment

can be found at the end of the chapter entitled "Ultrasonic Welding."

9 Hinges

9.1 Advantages and Disadvantages

9.1.1 Advantages

1. *Reopenability* When assemblies must be repeatedly reopened, hinges are the obvious solution. Hinge requirements vary from the very light duty variety common to packaging applications to the heavy-duty demands of large-scale housings.

2. *Part reduction* An increasing number of hinge applications are the result of other perceived benefits such as part reduction, the permanent attachment of the two parts, and lower total mold, molding and assembly costs.

3. *Size reduction* Integral hinges can often be designed within, or nearly within, the housing, resulting in a reduction in the overall size of the part.

9.1.2 Disadvantages

1. *Precision molds* Molds for living hinges, in particular, must be built to great precision with careful placement of cooling lines.

2. *Extended product development time* Except for companies with considerable living hinge experience, and sometimes even for them, development time will be extended while a single-cavity production mold is built to validate the design of the part and the mold.

3. *Complicated molds* Depending on the type of hinge used, side-action, split-cavity, or shutoff molds may be required.

9.2 One-Piece Integral Hinges

For discussion purposes, we have divided the hinges used for plastic parts into three types: one-piece integral, two-piece integral, and multipart hinges. Two of them could be regarded as integral hinges in that the hinge components are molded into the parts and require no additional pieces. However, the term usually refers to one-piece hinges where the two parts are molded as one complete unit. Two-piece hinges have all the hinge components molded into the plastic parts, but are manufactured as separate parts and assembled later. Multipart hinges require the use of additional parts such as rods or metal hinge components.

There are several methods of creating a one-piece integral hinge. Crystalline thermoplastics can be designed with either living, Mira, standard, or tab hinges. Hinges of these types can be produced by the injection molding process. Living hinges can also be produced by machining, extrusion, or cold forming, although these methods do not produce as durable a hinge. The latter methods lend themselves well to the thermoforming process, which works from extruded sheet. Polypropylene is the material of choice for most one-piece integral hinges, with polyallomer and polyethylene being the next most frequent selections. Thermosets or amorphous thermoplastics cannot be used for one-piece integral hinges.

9.2.1   The Living Hinge

The living hinge is a thin, flexible web of plastic that connects two parts so that they become one part. This discussion will focus on the use of polypropylene for a living hinge. In the author's opinion, this is the only material that should be selected by those with no prior experience in the design and development of living hinges.

This hinge design has enjoyed widespread exposure as the hinge that will last for one million flexures. In actual testing, a 0.43 mm (0.017 in.) polypropylene version of this hinge was flexed at the rate of 100 flexes per minute through a 135° angle at no stress for 3.1 million flexes. With a 6.895 MPa (1000 psi) applied stress, a flex life of 1.7 million flexes was achieved. One manufacturer of polypropylene resin states that to the company's knowledge, "no test has ever worn out a properly designed and molded hinge" of their material. In the author's personal experience, he has certainly seen "living hinges" that failed long before reaching thousands of cycles; therefore, the key words are "properly designed and molded." Let us examine the design of these hinges more closely.

The figures in the test just cited were achieved with parts molded under ideal conditions and initially flexed while still hot to achieve stretching ratios of 2 or 3 to 1. This was done to accomplish maximum and uniform molecular orientation in the hinge area. A great increase in hinge tensile strength is created by the stretching of those molecules on the outer surface. The mechanics for this flexure are shown in Fig. 9-1.

Flexure of the hinge immediately after molding is critical to achieving long life. However it is also dependent on the optimum molding conditions, inasmuch as flex life

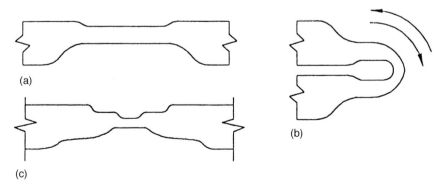

(a)

(b)

(c)

Figure 9-1   Flexure mechanics of the living hinge: (a) as molded, (b) flexing action, and (c) stretched hinge

varies with injection speed, pressure, and temperature. In the author's experience, the vast majority of living hinge applications are flexed only a modest number of times, perhaps 50 or fewer. Often the living hinge is used simply because it is a very compact style of hinge. Contrary to proper engineering practice, the initial hot flex of the hinge is sometimes omitted for short-life applications or for hinges that are simply being used to join two parts with no requirement for reopenability.

The selection of the correct polypropylene resin is also important in obtaining long life from the hinge. To the molder's dismay (high melt flow resins are easier to mold), lower melt flow resins with higher molecular weight are best. It is also important to have a wide molecular weight distribution, since longer molecular chains cannot relax as easily as short chains. Controlled rheology grades are not as good as reactor grades. Nucleation will improve hinge quality by helping freeze orientation. This is particularly true for thicker [0.38 mm (0.015 in.)], hinges, provided there is no melt flow hesitation, which might result in lower quality due to nucleation.

Homopolymer and random copolymer polypropylenes make better living hinges because they have higher starting tensile strengths and low blushing characteristics. However, these resins become brittle at low temperatures; therefore heterophasic copolymers are recommended for such applications. Special stabilization is required for long hinge life for hinges exposed to hot or wet environments and ultraviolet light, since the thin web is more vulnerable to stabilizer extraction and UV-induced degradation.

Hinge quality may be lost if fillers or reinforcements are used. High aspect ratio fillers like talc, mica, and glass fibers will not make good hinges. If the hinge is properly designed, low aspect ratio fillers, such as treated calcium carbonate, may be used for a limited number of flexes.

When flexures are anticipated to be absent or few in number, higher melt flow index polypropylenes may be used.

9.2.1.1 Living Hinge Design

The basic configuration of the standard living hinge is indicated in Fig. 9-2a. The thickness of the hinge is shown in the range of 0.15 to 0.38 mm (0.006–0.015 in.). It is the author's practice to stay "steel safe" (the dimension can be altered in the anticipated direction by the removal of steel from the mold, a much less expensive alternative to adding metal through welding or rebuilding the mold section) by initially specifying that dimension as 0.23 mm ± 0.02 (0.009 in. ± 0.001). Above the upper limit for a flexible living hinge, namely, 0.38 mm (0.015 in.), hinge life will be reduced as a result of less polymer orientation unless the angular motion of the hinge is small.

If the application is known to be one of short anticipated life, where the extra operation of flexure is likely to be omitted, the initial specification is reduced to 0.18 mm ± 0.02 (0.007 in. ± 0.001). This reduction holds even though the pressure drop, which occurs as the melt attempts to pass through the hinge restriction in the mold, will become severe enough to result in excessive shear heat buildup. Other possible effects are short or underpacked shots. For these applications, the pressure drop can be eased by modifying the design to provide a more gradual approach to the hinge point or by reducing the land length.

The land length is usually 1.5 mm (0.060 in.), since this has proven to be the optimum dimension. A greater land can cause an increased pressure drop, with the usual negative aspects. A shorter land results in too little back pressure. That can reduce the life of a living hinge by increasing the stress on bending due to nonuniform melt flow through the hinge.

Note the generous radii provided in the design illustrated in Fig. 9-2a. They are intended to ease the flow of the melt through the joint area. They are also designed to reduce the stress concentration that would occur in tight inside corners. The large 0.75 mm (0.060 in.) radius through the center of the joint has additional functions in that it controls the gap the melt must shoot and assures that the hinge line will be straight.

The 0.25 mm (0.010 in.) recess provides space for the 0.13–0.25 mm (0.005–0.010 in.) radius that is formed in the closed position (Fig. 9-2b). If it were not there, the surfaces of the two sides of the part would not lie flat against each other. Two flat plastic surfaces, however, do not seal well against each other. Therefore, if a seal is required,

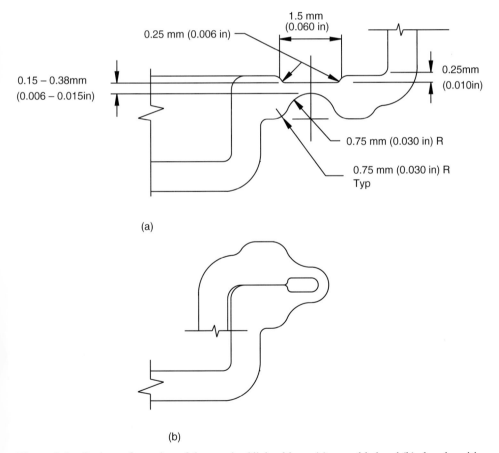

(a)

(b)

Figure 9-2 Basic configuration of the standard living hinge: (a) as molded and (b) closed position

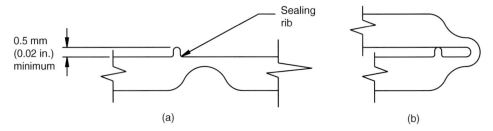

Figure 9-3   Living hinge with sealing rib: (a) as molded and (b) closed position

it will best be provided by a sealing rib such as that illustrated in Fig. 9-3. In this case, the height of the rib set at 0.50 mm (0.020 in.) minimum will accommodate the hinge radius, and no recess is required.

Living hinges will not lie flat open after molding once the hinge has been used. Therefore, if there is a requirement to offset the hinge, a double living hinge like the one depicted in Fig. 9-4 will be required.

Figure 9-4   Double living hinge

In some cases, such as two integral lids molded directly into a bottle closure, it is desirable to mold the parts at 90° to each other instead of the customary flat open (180°). The design is then adjusted as shown in Fig. 9-5.

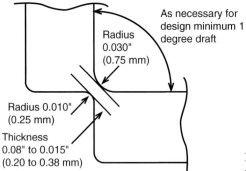

Figure 9-5   A 90° hinge design (Reprinted by permission of Montell USA Inc.)

Hinge Tear Resistance   The hinge joint may be subject to flexure, which can result in tearing the thin web of material beginning at the end of the hinge. There are several ways to combat this problem. The first consideration is to be certain that there are no inside sharp corners, which can create stress concentrations that give tears a place to begin. These details are shown in Fig. 9-6. In addition, it may also be effective to

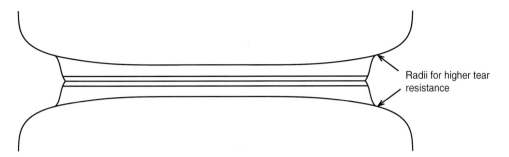

Figure 9-6   Hinge end design (Reprinted by permission of Montell USA Inc.)

increase the hinge thickness at each end by 0.25 to 0.50 mm (0.010–0.020 in.) for a distance of 0.50 to 1.00 mm (0.020–0.040 in.) from the end.

Beyond the elimination of sharp corners, two types of very thin web can be used. The version in Fig. 9-7a is in the plane perpendicular to that of the hinge. It will become highly oriented, after it is first flexed, and will prevent crack initiation. The type illustrated in Fig. 9-7b, which is in the plane of the hinge, performs the same function. A 0.13 mm (0.005 in.) film of this type is reported to increase the torque-to-failure tenfold.

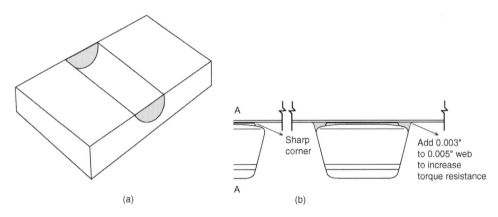

Figure 9-7   Hinge end webs: (a) webs perpendicular to the plane of the hinge increase tear resistance and (b) webs in the plane of the hinge (Reprinted by permission of Montell USA Inc.)

9.2.1.2   Living Hinge Molding Considerations

In molding, the molten plastic flow across one side of the part toward the hinge encounters the restriction and accumulates in the pocket just before the hinge segment. When sufficient pressure has built up, the melt bursts across the hinge and fills the other side. Gate location is, therefore, critical to the function of the part if the proper amount of molecular orientation is to be attained and a weld line, which would result in failure of the hinge or the entrapment of air in the hinge area, is to be avoided.

The gate for a living hinge is traditionally located on the far edge of the simplest part, as shown in Fig. 9-8. Either the large flash gate in the center or two gates in the corners are acceptable. The melt then rushes across that part, crosses the hinge, and proceeds to fill the other side. There is less pressure available on the opposite side of the hinge, making it hard to fill difficult details, bosses, etc. on that side. That is why the recommended procedure is to start with the minimum feasible hinge thickness and determine whether (and if so, how much) to increase it through experimentation with the different grades of polypropylene available.

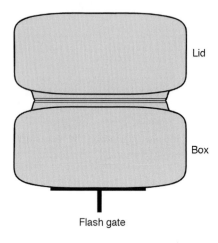

Lid

Box

Flash gate

Figure 9-8   Traditional gate locations for flat living hinge parts (Reprinted by permission of Montell USA Inc.)

The difficulty of filling the cavity on the opposite side of the hinge from the gate sometimes tempts the unwary designer to move the gate closer to the hinge, as illustrated in Fig. 9-9a. The melt seeks the path of least resistance. When it reaches the obstruction of the hinge, it halts in that direction and proceeds to fill the balance of the cavity on the gate side of the hinge. When that side is filled, the melt bursts across the hinge and proceeds to fill the other side. However, the material that first reached the hinge area at the center and stopped had begun to set up. Therefore, it has a cold, solidified edge and cannot proceed to fill the other side. Consequently, that part fills from its sides, resulting in a weld line that runs through the center of the part to the hinge and along the hinge, where it meets the material that has frozen at that point. In addition, there may be some air entrapment along the hinge, as illustrated in Fig. 9-9b.

Confusion over the acceptability of gating a living hinge part at this location may arise over the observation of deep boxes gated as shown in Fig. 9-10. In the case of a deep box, however, the material must first flow up the walls of the box. In doing so, it balances out its flow such that all sides reach the plane of the hinge at approximately the same time; consequently, the melt flow reaches the full length of the hinge simultaneously. That is the key point. It is difficult to calculate the melt path manually, but computer programs are available for this purpose.

As the depth of the box decreases from that shown in Fig. 9-10, the optimum location of the gate will tend to move toward the side away from the hinge. When it

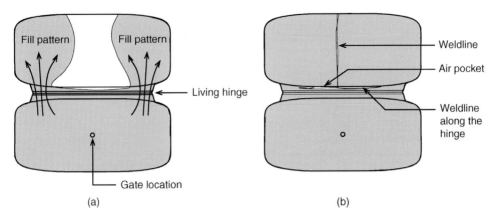

Figure 9-9   Flow patterns: (a) improper gate location and flow path for flat parts and (b) resulting air pockets and weld lines (Reprinted by permission of Montell USA Inc.)

reaches the general configuration of a flat box, the ideal location will be on the far edge, as depicted in Fig. 9-8.

For a double hinged part, the ideal gating location will be at two points on the centerline of the segment between the hinges, as shown in Fig. 9-11. Flow uniformity across the hinge will be improved by increasing the wall thickness in this section 20–30%.

Sometimes wall thicknesses are thinner than 1.0 mm (0.040 in.), sides are very complex (with numerous holes, bosses, ribs, etc.) or sides are very large and have tight tolerances. In such cases, it may be necessary to gate the part on both sides of the

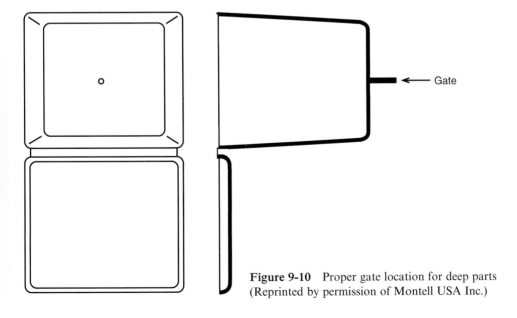

Figure 9-10   Proper gate location for deep parts (Reprinted by permission of Montell USA Inc.)

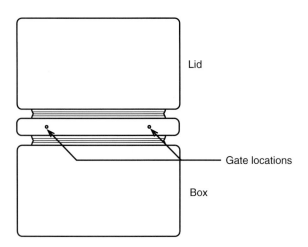

Lid

Gate locations

Box

Figure 9-11   Gating of a multiple-hinge part (Reprinted by permission of Montell USA Inc.)

hinge because the pressure drop encountered at the hinge would result in too little melt pressure on the ungated side of a mold gated on one side. Not long ago, such applications were dismissed as unsuited to a living hinge because gating on both sides of the hinge was not recommended based on the likelihood that the weld line would appear in the hinge area. This would result in a short-lived hinge, and there would promptly be two separate parts. However, the advent of better equipment control and melt flow programs has made such gating a viable alternative. Nonetheless, there are still situations that call for a thicker hinge, and the hinge is cold-worked, or coined, as a secondary operation. There are some recommended practices for gating to keep the weld out of the hinge area.

The gates for both sides are placed on the bottom of each side, as shown in Fig. 9-12. The larger cavity receives two gates and the smaller, one. The gate on the smaller side is set further from the hinge so that the melt from the larger side crosses the hinge and the weld line forms beyond it. There are other ways to ensure this result. One would be to vary the size of the gates. Another would be to use valve gates with sequential timing. If a hot runner system is used, the temperatures of the gates can be varied.

Regardless of which device is used, the proper development of a living hinge calls for a certain amount of trial and error. Even when both part and mold are designed perfectly, there still must be some experimentation with the process and the precise formulation of the material. Therefore, the construction of a single-cavity mold with a full complement of cooling and ejection systems just as anticipated in the production mold is strongly recommended. (It can be one cavity of the production mold.) This tool should be unhardened to facilitate modifications. It is unwise to risk the investment of the full production mold until all the issues discussed above have been resolved.

Cooling, the one issue that has not been discussed thus far, is nevertheless critical with respect to living hinges. The mold must be cool to preserve the initial molecular orientation from the mold filling stage by reducing molecular relaxation, which

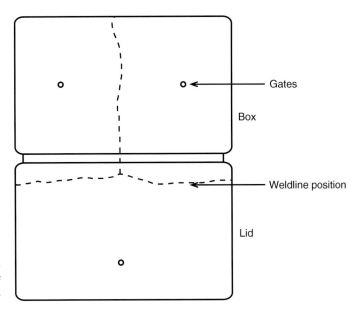

Figure 9-12 Gating on both sides of a hinge (Reprinted by permission of Montell USA Inc.)

occurs while the part cools. Ideally, the cooling channels should be as close to the part walls as possible. Although it is recognized that the water lines must evade the ejectors, core pins, and gates, it is recommended that their location be given a higher priority than usual for living hinge molds. (A further discussion on mold cooling can be found in Chapter 15, "Snap Fits.")

9.2.1.3   Living Hinges by Other Processes

a. Blow Molding   Living hinges can be manufactured through the blow-molding process by creating the hinge with a pinch-off. The process is difficult to control, and the final mold closing speed needs to be adjusted very carefully. Allow a 0.25 mm (0.010 in.) gap at the hinge pinch-off.

b. Coining, Cold Working, and Stamping   The complexity of some parts makes the use of injection-molded hinges impossible because the pressure drop incurred in the passage of the melt through the hinge area does not leave enough pressure to properly fill out the part. Cold working, usually referred to as "coining," the hinge after molding permits the use of thicker hinge sections, which in turn allows such complex sections to be filled. Furthermore, the direction of melt flow is not relevant to coined hinges. Coining can also be performed as part of the thermoforming process thereby creating hinged thermo-formed parts (Fig. 9-13).

In Fig. 9-13a the die is shown poised over the part to be coined. The process can be performed cold, but a shortened coining cycle and a reduced pressure requirement can be achieved by heating the part to the 175–230 °F range (for polypropylene) or by

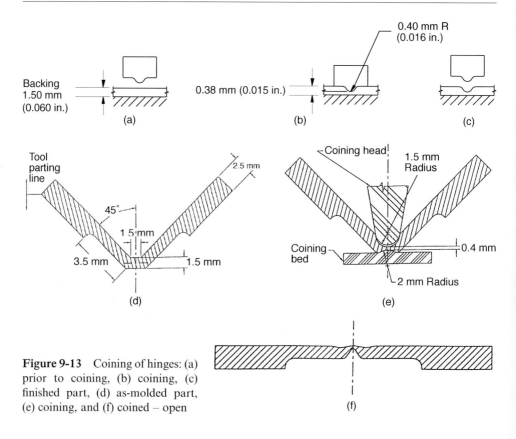

Figure 9-13 Coining of hinges: (a) prior to coining, (b) coining, (c) finished part, (d) as-molded part, (e) coining, and (f) coined – open

performing this operation while the part is still hot from molding. In general, the temperature should be below the glass transition temperature of the material. For the hinge to retain its postcoined shape, the material must be compressed beyond its yield strain. It cannot be compressed beyond its yield strain limit, however, or the hinge will have a very brief life. The actual coining is illustrated in Fig. 9-13b. The time and temperature will vary according to the material. For polypropylene that has not been preheated, however, it would be in the general range of 275 °F for approximately 10 seconds. The coined part is shown in Fig. 9-13c. Steel or another hard material is customarily used for backing, however a softer backing such as hard rubber produces a thinner hinge by stretching the hinge material.

Figures 9-13d through 9-13f illustrate a coined hinge design for nylon. The as-molded shape is shown in Fig. 9-13d. In this case, there are two coining tools, one from each side, as in Fig. 9-13e. The final illustration shows what the hinge would look like in the fully open position.

Extruded sheet is usually used for stamping living hinges as well. The hinge thickness is increased to 0.50 mm (0.020 in.) and will, consequently, have a shorter life. However, tear resistance will be improved.

Heat and pressure can be used to form the hinge in two ways. When formed from a heated sheet, usually by thermoforming, a cold die is used. Otherwise, the die is heated to form a cold sheet. This approach is more energy efficient than heating a sheet that does not have to be heated for another reason. Roller dies are not recommended because they produce unfavorable molecular orientation. Speed must be controlled, since too great a speed will not give the polypropylene enough time to flow and orient its molecular chains.

c. Extrusion It is possible to extrude living hinges. However, the life of the hinge will be shortened because the molecular orientation runs along the hinge, as opposed to across it, and the hinge cannot be flexed while it is still hot because it must proceed through the cooling section directly out of the die. In addition, the nonuniform wall thickness will cause distortion due to nonuniform cooling, which will result in differential shrinkage. To minimize warpage, the nominal wall thickness beyond the hinge should not exceed two to three times the thickness of the hinge.

Another approach is to form the hinge with a secondary forming die. The hinge section is extruded in the form shown in Fig. 9-14a. It is then passed through a forming die (Fig. 9-14b) to create the shape shown in Fig. 9-14c. This type of hinge has been used successfully for some packaging applications, particularly with very thin gauges. A hinge that is formed while the material is still hot will have a life similar to that of a coined hinge. Coextrusion, when used in conjunction with thermoforming, permits the creation of assemblies with two different colors or with combinations of clear and opaque resins.

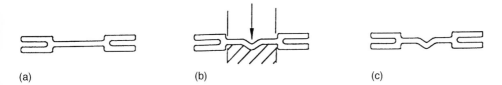

(a)   (b)   (c)

Figure 9-14   Extruded hinge: (a) as extruded, (b) forming die, and (c) finished hinge

d. Machining While this process is rarely used for anything but prototypes or extremely low volume production runs, living hinges can be machined from extruded polypropylene sheet stock. The same design used for an injection-molded hinge can be used, but care must be taken to cut the hinge in the direction perpendicular to the extrusion for proper molecular orientation.

For machining polypropylene, the resin manufacturer recommends the highest possible cutting speed without overheating the tool. (The low thermal conductivity of the polypropylene will result in the cutting tool receiving nearly all the heat due to friction.) The polymer will be softened by heat buildup and will tend to adhere to the cutting tool. A water-soluble oil or kerosene is recommended to reduce heat and friction. Metal cutting tools can be used.

9.2.2   The Mira Spring Hinge

A polypropylene leaf spring hinge element called a Mira spring is typically used in combination with a standard hinge to keep it open or snap it shut. Figure 9-15a shows the hinge in its open position. Parts A and B are joined by both types of hinge. The Mira hinge joins segment A to B in one section and the living hinge joins A to B in the adjoining section. The A side of the Mira hinge is the leaf spring and the B side is the tension arm. The leaf spring supplies the pressure to cause the hinge to open or close. When it is deflected, it exerts a force that resists the deflection.

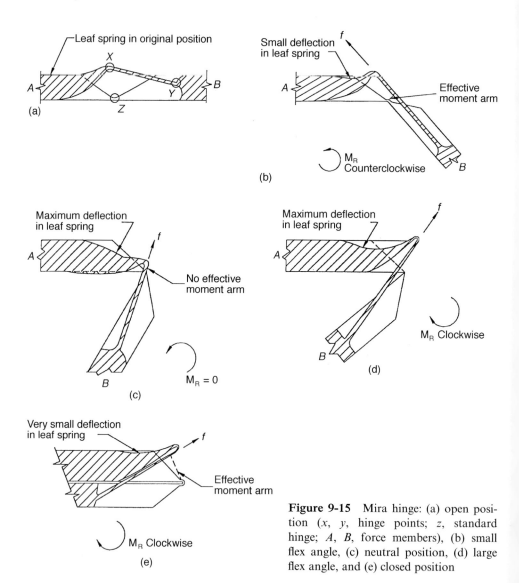

Figure 9-15   Mira hinge: (a) open position (x, y, hinge points; z, standard hinge; A, B, force members), (b) small flex angle, (c) neutral position, (d) large flex angle, and (e) closed position

In Fig. 9-15b the hinge has been deflected 45°. As this occurs, the tension arm exerts a force (F) on the leaf spring, causing it to deflect. If the pressure deflecting the spring were released, the spring would snap back to its original position.

When the hinge has been deflected to a 105° position (Fig. 9-15c), the leaf spring is deflected to its maximum position. This is the neutral position; therefore there is no moment arm in either direction. Changes in the spring geometry will vary this position. Figure 9-15d illustrates the point beyond the neutral position, where the leaf spring returns toward its original position. This motion creates a pull on the tension arm, causing a counter clockwise motion that snaps the hinge closed (Fig. 9-15e). Suggested dimensions for a Mira hinge are shown in Fig. 9-16.

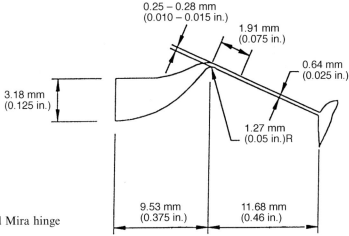

Figure 9-16   Suggested Mira hinge dimensions

The dimensions proposed in Fig. 9-16 necessarily violate the rule of avoiding wall thickness variation greater than 25% to prevent high levels of molded-in stress. However, the transitions must be made as gradual as possible to keep such stress to a minimum.

The snap force increases in direct proportion to the increase in the maximum deflection of the leaf spring, which, in turn, can be increased by increasing the hinge's rotation angle. This can also be accomplished by extending the force member further beyond the plane of the standard hinge section. Since this results in a loss of rigidity and a reduction in snap force, compensation is required to arrive at the desired level.

It is also possible to alter the snap force through changes in the material stiffness or geometry. To add to that force, the rigidity of the leaf spring needs to be increased. That can be achieved by shortening the leaf spring, by making it deeper, or by adding ribs lengthwise along its top or bottom. The spring will be stiffened by any increase in its moment of inertia. The reverse is also true, and a decrease in the rigidity of the leaf spring will result in a reduction in snap force.

The width of the Mira hinge should be in the range of 6.4 mm to 12.7 mm (0.25–0.50 in.). A typical packaging application calls for a standard hinge section flanked by a pair of Mira hinges. For other types of products, the Mira hinges may themselves be flanked by another pair of standard hinges.

9.2.3   Standard Hinges

There are a number of different configurations that fall in the category of standard hinges. Two such designs are illustrated in Fig. 9-17.

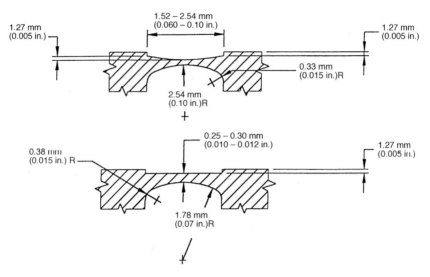

Figure 9-17   Standard hinge designs

9.2.4   Tab Hinges

Parts made from softer materials, such as low density polyethylene or flexible polyvinyl chloride, can use a tab hinge. Figure 9-18 depicts such a design.

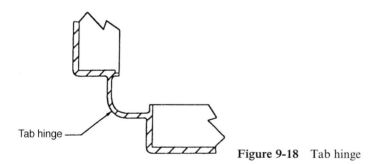

Figure 9-18   Tab hinge

9.3   Two-Piece Plastic Hinges

While one-piece hinges can be successfully used with the more flexible semicrystalline thermoplastics, the stiffer amorphous thermoplastics require two- or three-piece hinges. Basically, this type of hinge requires the use of snap fits, the mechanics of which are discussed Chapter 15. Because of the tight tolerances required, injection molding is the principal processing technique employed for these joints. They can be used with some thermosets but thermoplastics generally provide the most suitable combinations of properties for their application.

9.3.1   Ball-and-Socket Hinges

The type of hinge depicted in Fig. 9-19 can be executed as a full ball-and-socket unit (Fig. 9-19a) or as a half ball and socket (Fig. 9-19b). The ball is the easier part to mold because it can be formed with half the ball on each side of the mold. The socket is what is known as an undercut in the mold. To remove the socket from the mold, the ears in which it is formed must deflect outward sufficiently to clear the core. The cavity that formed the outer portion must be removed before that action can take place. This must be done while the part is still hot enough to flex, to ensure that the socket is properly formed and the ear not distorted or broken. This becomes a matter of experimentation best done with a prototype mold until the best combination of material and geometry has been established, particularly since the assembly must also be joined and must work effectively once that has been accomplished. (This topic is further discussed in Chapter 15, "Snap Fits.")

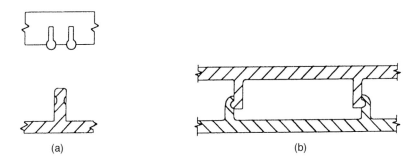

(a)   (b)

Figure 9-19   (a) Full ball-and-socket hinge; (b) half-ball-and-socket hinge

On assembly, the ball is pressed into the socket. This can be done when the ears are cold if they are sufficiently flexible. Otherwise, it must be done while the ears are still hot from molding, or they must be reheated for assembly. The full ball and socket offers a firmer grip because there is support on both sides of the joint. However, that style is generally more fragile and difficult to assemble than the half ball and socket, which provides the opportunity for much greater leverage in assembly. Thus, the ears could be physically flexed out more without the use of heat. The disadvantage of this type of hinge is that it must protrude well beyond the edges of the part. The protrusion could be avoided for an opening arc less than 180°, however that design would lead to a much more costly mold incorporating the use of lifters.

9.3.2   Two-Piece Lug-and-Pin Joint

The lug-and-pin joint can be made in either two or three pieces. The two-piece version is shown in Fig. 9-20. The snap-off lug version in Fig. 9-20a is similar to the half-ball-and-socket joint, however the use of the lug eliminates the need for tight tolerance control in the lateral direction. These details would be molded in much the same fashion as the ball-and-socket joints.

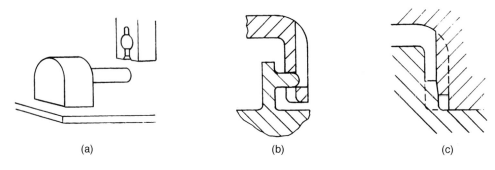

(a)   (b)   (c)

Figure 9-20   Two-piece lug-and-pin hinges: (a) with snap-off lug, (b) with shutoff molded lug; (c) mold detail for shutoff molded lug

Figure 9-20b shows a version in which the pin fits into a hole in the side of the mating part. That hole could be formed by a costly side action in the mold, of which two are required. However, in the form illustrated, it is created by a mold device known as a shutoff. The mold detail for that concept is shown in Fig. 9-20c. The hole is created by a piece of steel on the core which overlaps a piece in the cavity. Where they meet, a hole is created in the part. While this eliminates the side actions, it does leave a visible recess on the outside of the part for the piece of steel coming from the cavity side.

9.3.3   Hook-and-Eye Joint

Another version of a two-piece molded hinge can be molded with no side actions or undercuts. The two parts of the hook-and-eye design shown in Fig. 9-21 can be removed when it is opened to the 180° position.

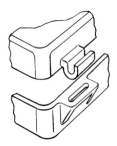

Figure 9-21   Hook-and-eye hinge

9.4   Three-Piece Hinges

Three-piece hinge designs are used for a wide range of applications for which one- or two-piece hinges are not well suited, including those with heavy loadings, low production volumes, rigid thermoplastics, thermosets, and all the processes beyond injection molding and, to a degree, extrusion and thermoforming. The third piece in this category refers to anything from a pin to a complete hinge assembly.

9.4.1   Three-Piece Lug and Pin

The three-piece lug-and-pin design substitutes a metal pin for the molded one found in Fig. 9-20. Solid pins can be pressed into the hole in one lug and slip-fit into the lug in the other. In some cases, other devices such as the spring pin or rivet are more suitable. For a long-lasting, smooth-operating hinge, a rivet, or eyelet for thin sections, can be placed in the hole of the moving part. The shoulder screw is another effective means of providing a smooth pin.

The version in Fig. 9-22a shows how eyes can be rotomolded or blow-molded and the holes drilled after molding. Similar hinge details can be thermoformed. Holes can also be molded directly into the walls of mating pieces as in Figs. 9-22b and 9-22c. Holes for hinges are generally drilled in place after molding for all the thermoset processes; those that cannot mold in the lugs can have them applied with adhesives.

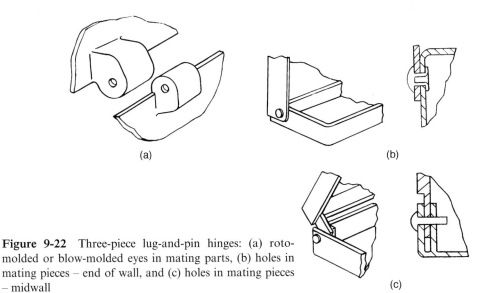

(a)

(b)

Figure 9-22 Three-piece lug-and-pin hinges: (a) roto-molded or blow-molded eyes in mating parts, (b) holes in mating pieces – end of wall, and (c) holes in mating pieces – midwall

(c)

9.4.2   Piano Hinge

The so-called piano hinge is a high strength, long-lasting, and smooth-operating hinge. However, the cost of the hinge, the fasteners required to secure it, and the time required to install it make this a costly alternative.

There are lighter duty, less costly hinges that would be installed in the same manner as piano hinges. The reader is referred to Chapter 8 for a further discussion of the methods of installing these hinges.

9.5   Latches

9.5.1   Snaps

While a variety of commercial snaps are available, molded-in latches are more commonly employed. This is particularly true of the one- or two-piece latches where the hinge details are also being molded into the product. In most cases, a simple snap fit is employed, the mechanics of which are provided in Chapter 15.

9.5.2   Rathbun Spring

The Rathbun spring is illustrated in Fig. 9-23. It is an internal device that provides closing pressure that eliminates the need for a latch in light-duty applications. It works by way of a curved spring clip that catches on both parts and permits the spring to open approximately 85°.

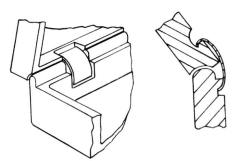

Figure 9-23   Rathbun hinge

9.6   Number of Hinges and Location

Although each application has its own special requirements for number and location of hinges, there are some general guidelines to follow. First, however, one establishes the minimum number of hinges that will provide adequate support. This can be determined using the basic strength equations.

The majority of light-duty hinge applications call for two hinges. Each one can be located between the end of the part and a point in from the end one-quarter the length of the hinged side. The location would depend on the length of the hinged side and is an engineering judgment. For longer sides, a third hinge might be added at the center, in which case the outer hinges would be relocated near the edges of the part. If additional hinge segments were required, they would be located accordingly.

10 Hot Plate/Hot Die/Fusion and Hot Wire/Resistance Welding

10.1 Advantages and Disadvantages

10.1.1 Description

Hot plate welding, also known as "fusion welding" and "hot die welding," is a technique that can be performed with literally nothing more than a hot plate, or it can be highly automated with sophisticated equipment. In its simple form, this technique is often used to fabricate prototypes or products with a very low production volume. Hot plate welding is the process of choice for welding polyolefin pipe, and special equipment has been developed for this purpose. Automatic equipment is used to manufacture a variety of parts ranging from those that are small and flimsy like whiffle balls to those that are large and require a hermetic seal, such as a 1372 mm (54 in.) automotive air duct.

This process is particularly useful for contoured or irregular joints such as automobile tail light lenses, items that require a hermetic seal, and parts made of materials like polyolefins, which are difficult to join by other means. If platen space permits and the parts are of the same material and reasonably similar, more than one part can be welded at the same time. The minimum wall thickness for this process is 1.60 mm (0.063 in.) in most cases, however for certain applications thinner walls are acceptable. Cycle times can range from 15 seconds to several minutes depending on the size of the joining surface.

Hot plate or fusion welding can also be considered a competitor to vibration welding because the joint strengths and material compatibilities of these processes are approximately equivalent. However, while hot plate welding has a greater range in size and shape, it has a much longer cycle time and higher equipment, fixture, energy and maintenance costs than vibration welding.

Hot wire welding, also known as "resistance welding," is slightly different from hot plate welding in that the hot-wire heated surface remains in the finished part. It is discussed later in this chapter.

10.1.2 Advantages

1. *No additional materials* Hot plate welding uses no additional materials such as fasteners, inserts, electromagnetic preforms, adhesives, or solvents. Therefore, it is inherently lower in cost than methods that have such requirements, and products are less expensive to disassemble for recycling.

2. *Ease of assembly* Hot plate welding requires only the placement of the two parts in the fixture of the welder.

3. *Entrapment of other parts* Additional parts can be captured between the two parts to be hot-plate-welded provided they are located such that they do not interfere with the welding.

4. *Permanence* Hot plate welding creates permanent assemblies that cannot be reopened without damaging the parts. The effects of creep, cold flow, stress relaxation, and other environmental limitations are absent in the joint area hot plate welds, although they may occur elsewhere if other parts of the design are under load. Owing to material limitations, differences in thermal expansion and moisture absorption are rarely of concern once the parts have been welded. Joint strengths equivalent to those of the surrounding walls are possible under ideal conditions.

5. *Materials* Hot plate welding handles most thermoplastic materials well, including the crystallines. In some cases dissimilar polymers can be hot-plate-welded by using two different hot plates at different temperatures (e.g., PC to PEI, PC to PBT). Thermoset composites can be hot-plate-welded through the process of embedding them with thermoplastic.

6. *Shape freedom* Hot plate welds can be made with contoured parts of practically any shape and with fairly thin wall thicknesses of 1.60 mm (0.063 in.) as long as the joining surfaces can be created within the limits of the hot plate.

7. *Hermetic seals* Hot plate welding creates excellent hermetic seals, which are usually a requirement for its applications.

8. *Clean atmosphere* Unlike adhesive and solvent joining systems, no ventilation equipment is necessary for the removal of toxic fumes with most low temperature welding materials. (ABS, acrylic, nylon, and polypropylene produce smoke that should be vented. PVC fumes are hazardous and must be vented.)

10. *High production rates* High production rates are possible with smaller parts, particularly if multiple parts are welded with each cycle.

11. *Process freedom* Parts made from virtually all the thermoplastic processes can be hot-plate-welded. However, fillers and additives can affect weldability.

12. *Large part capability* Equipment is available that can weld parts up to 1372 mm (54 in.) long. Part size is limited only by heat platen size. The largest current production model can weld a part 508 mm × 1828.8 mm (20 in. × 72 in.).

13. *Noise freedom* The noise associated with ultrasonic, vibration, and spin welding is not present.

14. *Internal walls* Hot plate welding can be applied to internal walls.

10.1.3   Disadvantages

1. *Energy cost* Hot plate welders have a high energy cost relative to the other thermoplastic welding processes, a consideration that tends to limit their use to applications too large or too poorly configured for ultrasonic or spin welding or for which the cost of vibration welding equipment is prohibitive.

2. *Slow start-up* Hot plate welders must wait for the plate to heat up on start-up, a period that may range from 15 to 30 minutes, during which heat is used without production.

3. *Stresses* The process can cause stresses in the bond area that result in stress cracks. (However, stresses can be more apparent with ultrasonic and spin welding.)

4. *Cleaning* Platen surfaces accumulate melted plastic and must be cleaned for high temperature welding materials.

5. *Equipment cost* While production hot plate welders are less costly than vibration welders of equivalent size, they are more expensive than ultrasonic and spin welders.

10.2   Materials

Hot plate welding is capable of producing welds with tensile strengths approaching those of the surrounding material. Table 10-1 provides the approximate levels that can be achieved under ideal conditions.

Hot plate welding can be used for parts made from most thermoplastics, including foamed materials and those manufactured by all of the thermoplastic processes.

Table 10-1   Tensile Strengths Achievable in Hot Plate Welding Under Ideal Conditions

Plastic Material	% of Surrounding Material Strength
ABS	80
Acetal	90
Acrylic	80
Cellulose acetate	80
Cellulose acetate butyrate	80
Cellulose acetate propionate	80
Ethylene vinyl acetate	80
Nylon	90
Polycarbonate	80
Low density polyethylene	100
Ultra high molecular weight polyethylene	90
High density polyethylene	100
Polypropylene	100
Polystyrene, GP	80
Polystyrene, high impact	90
Polysulfone	60
PVC (rigid or flexible)	90
Thermoplastic rubber	90

However, it is somewhat difficult for such materials as styrene butadiene, acrylic styrene acrylonitrile, polyethylene terephthalate, polybutylene terephthalate, and cellulose acetate, depending on the application. Nylons must be thoroughly dried to prevent frothing, and polycarbonates can leave bubbles at the joint. Thermosets cannot be hot-plate-welded.

Fillers reduce the strength of the bond because they do not weld, and they replace resin, which does. The greater the level of filler, the lower the strength of the joint. However, filled thermoplastics can be hot-plate-welded provided the percentage of filler does not exceed 40%. At that level, there is not enough of the base polymer remaining to create an adequate bond. Some difficulty in making hermetic seals can be encountered at levels of 35%. Rotating wire brushes may also be needed to remove any filler materials that adhere to the heat platens. Talc actually pulls or is drawn to the heat source or platens.

10.3   The Process

The basic principle of hot plate welding is clearly demonstrated in Fig. 10-1 which shows the simple version of the process. The edges of the pieces to be joined are heated on a thermostatically controlled hot plate until their surfaces are melted. This has to be done with light pressure because the softened surfaces are easily malformed by excess pressure. The pieces are then removed from the heat and the softened surfaces

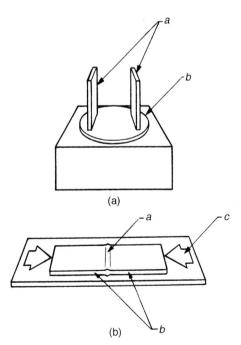

Figure 10-1 Simple hot plate sealing: (a) Resin heat-up (*a*, plastic pieces to be joined; *b*, hot plate), and (b) joining (*a*, weld; *b*, plastic pieces; *c*, contact pressure)

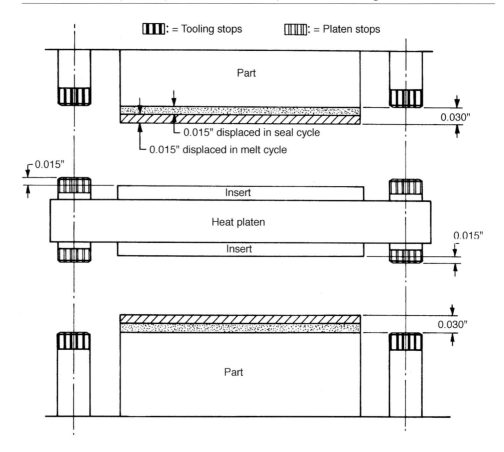

Figure 10-2   Production hot plate welding equipment with positive stops (Diagram courtesy of Forward Technology Industries, Inc.)

are pressed together. Again, excess displacement or travel will displace molten material, causing a cold seal. The parts are held in position until they solidify.

In contrast, the production equipment illustrated in Fig. 10-2 is far more sophisticated. There are now holding fixtures into which the parts to be joined are loaded. These fixtures are mounted to jaws that are part of the equipment. The parts are held in the fixtures with collets, vacuum cups, gripping fingers, and other mechanical devices that must grip firmly to avoid warpage. The hot plate has been replaced by a movable heating platen (also known as the core), which reciprocates in and out between the holding fixtures. It is heated by cartridge heaters, and inserts bolted to each side of it are machined to exactly match the weld bead area, to guarantee that the part surface is melted only where needed. For contoured parts, the contours will be machined on the inserts.

Note that there are tooling stops on the holding fixtures and melt stops on the heating platen. They create positive location, which provides consistency by controlling the melt dimension, the seal dimension, and adjustment for part-to-part dimensional variations. The tooling stops are designed to displace 0.38 mm (0.015 in.) from each part on the melt cycle and an additional 0.38 mm* from each part half on the seal cycle. They also eliminate cold seals, which result from the displacement of molten material during the seal cycle, leaving only cold material to make weakened welds. The stops remove the major causes of rejects for the process.

The process is illustrated in Fig. 10-3. The initial position (Fig. 10-3a) shows the holding fixtures with a part in each, the heating platen, and the tooling and melt stops. The platen is moved to one side so the parts can be placed in the holding fixtures.

In Fig. 10-3b, the platen has moved between the parts. The holding fixtures then close, and the parts are pressed against the platen, where they are heated until their surfaces are melted. This is the first of the three phases of the actual welding (Fig. 10-3c). Melting continues for a predetermined distance until the holding fixture's tooling stops contact the heating platen's melt stops. The parts are held in this position long enough for the joining surfaces to be melted to the ideal depth for the application, usually around 0.38 mm (0.015 in.). The total contact with the heating platen will be 1–6 seconds for high temperature materials and 8–15 seconds for low temperature materials. Small sinks and other imperfections on the melt surface will be removed by this process, leaving a smooth melted surface. The correct platen temperature is critical to the quality of the weld.

As shown in Fig. 10-3d, the next step is the opening of the holding fixtures and the withdrawal of the platen. This is the second or "changeover phase," and it is performed quickly to minimize cooling. In Fig. 10-3e, the fixtures have been quickly closed again and the plastic surfaces are compressed, usually around 0.38 mm (0.015 in.) on each part half, until they cool. This is the third and final or "joining" phase. This brings the total compression between the heating and joining phases to around 0.75 mm (0.030 in.) per part half, normally the starting point in setting up the cycle. Adjustments are then made until the optimum settings have been determined by trial and error. The open time between the end of the melt stage and the reclosing of the parts – usually less than 2 seconds – is a critical factor in establishing quality welds, since too long an open time will result in a cooled skin of plastic forming on the weld surface, which reduces weld strength.

Cooling time normally falls in the range of 6–15 seconds. In Fig. 10-3f, the holding fixtures have been opened and one side has released, leaving the completed assembly in the other side waiting for removal. If the volume warrants, removal can be done automatically. The total cycle time, not including loading and unloading of the parts, which varies with size and design, is 35–40 seconds for low temperature polymers and 15–20 seconds for high temperature materials. Hot plate welding can overcome a limited degree of distortion on the parts, however hermetic seals require an even weld surface.

This process is known as "welding by distance." Its basic parameters are melt depth and time, seal depth, and time and platen temperature.

There is another technique referred to as "welding by pressure." It is similar to "welding by distance" except that the parts are brought into contact with the hot plate at a very high pressure to completely match the parts to the tool surfaces. The pressure is reduced as the molten polymer begins to flow outward to allow it to thicken. The part is separated from the tool when the desired thickness of molten plastic has

---

* The distance can vary somewhat according to the application.

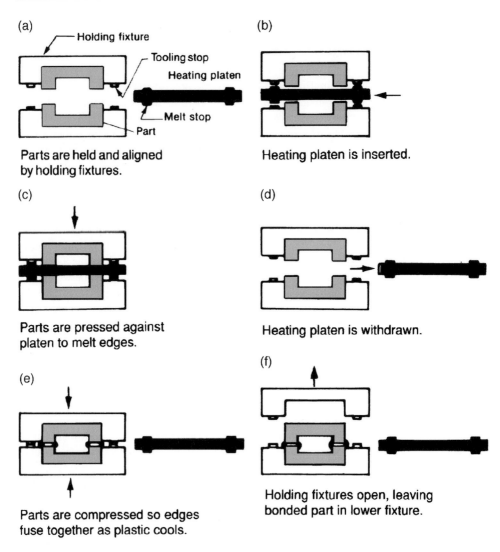

(a)

Holding fixture

Tooling stop

Heating platen

Melt stop

Part

Parts are held and aligned by holding fixtures.

(b)

Heating platen is inserted.

(c)

Parts are pressed against platen to melt edges.

(d)

Heating platen is withdrawn.

(e)

Parts are compressed so edges fuse together as plastic cools.

(f)

Holding fixtures open, leaving bonded part in lower fixture.

Figure 10-3 The hot plate welding process (Diagram courtesy of Forward Technology Industries, Inc.)

been reached. This is followed by the changeover and joining phases, except the latter is controlled by pressure. Distance is not controlled in "welding by pressure."

Figure 10-4 illustrates the relationship between the platen temperature and the time of melt. Clearly, the length of time required to melt the surfaces goes up sharply as the temperature goes down. It has reached 45 seconds on some occasions.

Most applications involve the assembly of like materials. However, by using platen inserts heated to different temperatures, compatible, yet dissimilar materials can be joined. For example, an acrylic lens could be welded to a polycarbonate or ABS housing, but not to one made of polyethylene or polypropylene. Filled materials can

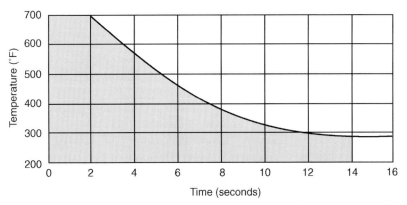

Figure 10-4   Platen temperature versus time of melt (Graph courtesy of Forward Technology Industries, Inc.)

also be welded, but it becomes more difficult as the percentage of filler increases because that leaves less polymer available for welding.

10.4   Types of Hot Plate Welding

The three types of hot plate welding are low temperature, high temperature, and noncontact.

10.4.1   Low Temperature Hot Plate Welding

Materials like polyethylene, polystyrene and polycarbonate can be welded at temperatures at or below 550 °F. These polymers can adhere to the heated surface, creating a stringy residue; therefore a Teflon coating can be applied to the heated inserts up to a temperature of about 520 °F. This practice raises maintenance considerations, however, since the coating has a limited life. Typically, three sets of inserts are made so that there are at least two sets available when one is being recoated. The precise type of Teflon coating will vary according to the application. It can be PTFE film, sinter-deposited PTFE, or the longer lasting PTFE-impregnated electroless nickel plating.

Low temperature welding has the advantage of providing a stronger weld strength than high temperature welding. It also degrades the material less and can weld a wider range of polymers. However, it requires more cycle time and cannot weld glass-filled materials; in addition, there is the matter of the extra inserts and their maintenance.

10.4.2   High Temperature Hot Plate Welding

Higher welding temperatures shorten the welding cycle and reduce the tooling costs. Contact time might drop from the 20 seconds typical of low temperature welding to the 4–6 second range. However, the higher temperatures not only limit the number of materials that are suitable but result in a discolored weld area and cannot produce as strong a weld.

There are two welding temperature ranges for this type of welding: 550 to 700 °F and 700 to 850 °F. Polypropylene, ABS, and polymethyl methacrylate are examples of materials that weld well in the lower high temperature range because they melt quickly and separate cleanly from the platen inserts, leaving only a light residue. That residue does not affect strength but can leave a discoloration.

For some materials, a higher temperature range (700–850 °F) can be used to oxidize the residue when buildup is unacceptable. The ash that may remain on the platen inserts when ABS is the material can be removed by rotating wire brushes between cycles. Such devices can also be used to remove the glass fiber buildup on the heat platen inserts that results from glass-filled nylon 6/6. Charring of the nylon is prevented by the glass fibers, which hold the heat. The heat platens and inserts will be made of bronze up to temperatures of 700 °F and of P-20 steel above that for better heat transfer, machinability, and resistance to oxidation. Extra safety and ventilation measures may be required for this process.

10.4.3   Noncontact Hot Plate Welding

Materials that are very difficult to weld can be noncontact hot-plate-welded provided they have flat joint lines and can utilize a horizontal heat platen or blade. This is necessary because the weld surfaces must be heated by radiant heat from a high temperature Nichrome blade, which precludes the use of contoured heat platen inserts. Also, the tooling cost is higher for this method and it may be difficult to achieve a uniform weld. Furthermore, noncontact hot plate welding has a high energy consumption, and the heat level annoys the operators.

Typical materials that warrant the use of this process are polycarbonate, which has a flash problem, acetal, which gives off a gas, and unfilled nylon, which chars when it contacts a heat platen and has poor release from it. Noncontact hot plate welding is well suited to materials that have a sharp melt transition range. An example of a noncontact hot plate weldment is a 6 in. by 9 in. nylon aircraft battery.

10.5   Hot Plate Welding Joint Designs

Hot-plate-welded joint designs are all butt joints, however some have designed-in devices to conceal the bead or flash. Figure 10-5a illustrates a straight butt joint. It is the simplest variety of joint in which the walls of the part are designed to be pressed together without special joining provisions. There will be a bead of plastic at the joint, which can be removed with a flash pinch-off operation during the welding process. It can also be machined off, as shown in Fig. 10-5b. Note that the total height of the assembly will equal the sum of the two individual parts less the total of the melt and seal displacements.

Figure 10-5c depicts a flanged butt joint. Whereas the straight butt joint's strength will be limited to that which is obtainable from the area of the wall thickness, the flange

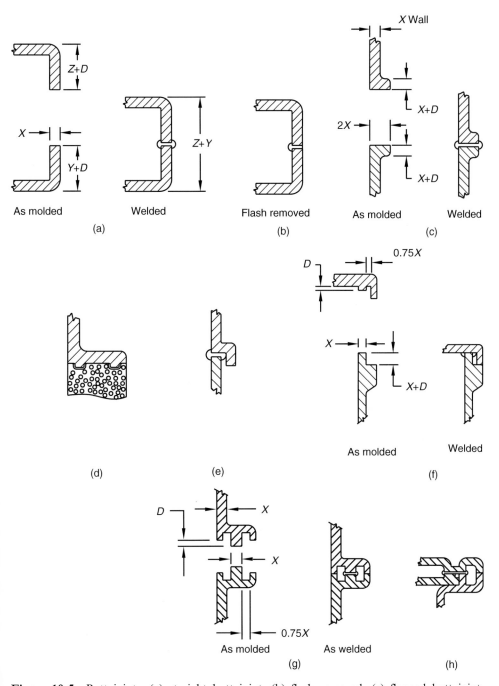

Figure 10-5 Butt joints: (a) straight butt joint, (b) flash removed, (c) flanged butt joint, (d) flange welded to plastic foam, (e) skirted butt joint, (f) recessed skirted butt joint, (g) double skirted butt joint, and (h) three-part assembly

allows the area to be increased as needed, thus creating a stronger joint. It is also the best type of joint for thermoformed parts, which inherently create a flange. It is a design that lends itself to rotational and blow-molded parts as well. In Fig. 10-5d, the flange has been welded to plastic foam.

The problem with the designs just mentioned is that they leave the flash common to the process visible unless it is somehow removed. For many products, this is undesirable because it suggests poor fit and finish. The skirted joint shown in Fig. 10-5e is designed to hide that flash from the outside. The recessed version in Fig. 10-5f is a joint with a more finished appearance. The butt joint with flash trap in Fig. 10-5g hides the flash from both sides. A joint design for the assembly of three parts is illustrated in Fig. 10-5h.

To obtain good welds, the surfaces must be clean and free of contaminants such as grease and mold release. These have the opposite effect on heat platens, causing the plastic to stick to them. This results in stringing, also known as "angel hair," and burning of the plastic. The charred plastic is degraded material that can weaken and discolor the joint.

Moisture in the plastic can also result in weakened joints. Parts made of hygroscopic resins need to be hot-plate-welded immediately after molding. Otherwise, to ensure good welds, they must be dried to the same level required for molding.

10.6   Equipment

Hot plate welding machines are available in both horizontal and vertical formats. Horizontal machines, like the one in Fig. 10-6a, are required when there are other parts to be placed into the assembly before sealing. They are also useful when there

(a)   (b)

Figure 10-6 Hot plate welding equipment: (a) horizontal platens, and (b) vertical platens (Photos courtesy of Forward Technology Industries, Inc.)

are other operations to be performed such as drilling, punching, or the ultrasonic welding of inserts or other components. Automatic loading and unloading equipment is available for horizontal hot plate welding machines.

Vertical hot plate welding machines, such as the one pictured in Fig. 10-6b, offer advantages for many applications. The heat and fog damage to parts resulting from the heat rise to the upper holding fixture is eliminated. Unless the parts have very small internal walls, which could deform, such damage is not usually a significant problem.

Computer-controlled hot plate welders that permit the changes in pressure or displacement during the weld cycle are now available, permitting the optimization of the hot plate welding process through better control of the cycle.

10.7   Hot Wire/Resistance Welding

Hot wire welding, also known as resistance welding, is similar to hot plate welding in that the surfaces of the two parts to be joined are melted and then welded together. However, in the case of hot wire welding, the heat is provided by conduction from a heated wire. The wire is heated by resistance when electrical current is passed through it. This technique eliminates the need for platens, inserts, and expensive equipment, making it ideal for prototypes, short runs, and very large parts that are difficult to join by other means. However, it is too slow and operator sensitive for volume production of smaller parts.

As illustrated in Fig. 10-7, the wire is placed on one of the surfaces to be joined. The parts are then placed together and clamped. The pressure is maintained while current is applied for an experimentally determined period of time to melt the surfaces. The wire remains in place and is thereby trapped in the part. The portions of the wire protruding beyond the assembly are clipped off after the joint has solidified, leaving the remainder in the part. Grooves or other locating devices may have to be placed in the part to keep the wire in its proper location.

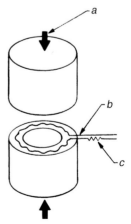

Figure 10-7   Hot wire/resistance welding: *a*, pressure; *b*, high resistance wire, and *c*, wire heat control

In another variation of this concept known as "Thermoband welding," a metallic tape is used in place of the hot wire. It is adhered to one surface to be welded and the other surface is placed in contact with it. A low voltage current is then passed through the tape to raise the temperature of the joint surfaces to molten level through electrical resistance.

10.8   Sources

Bielomatik, Inc., 55397 Lyon Industrial Dr., New Hudson, MI 48165, (248) 446-9910, fax (248) 446-6244, www.bielomatikinc.com

Branson Plastic Joining, Inc., 1001 Lehigh Station Rd., Henrietta, NY 14467, (716) 359-3100, fax (716) 359-1132

Branson Ultrasonics Corp., 41 Eagle Rd., P.O. Box 1961, Danbury, CT 06813-1961, (203) 796-0400, fax (203) 796-9838, www.bransonultrasonics.com

Dukane, 2900 Dukane Drive, St. Charles, IL 60174, (630) 584-2300, fax (630) 584-3162, www.dukane.com/us

Forward Technology, 260 Jenks Ave., Cokato, MN 55321, (320) 286-2578, fax (320) 286-2467, www.forwardtech.com

Manufacturing Technology Solutions, 14150 Simone Dr., Shelby Township, MI 48315, (586) 802-0033, fax (586) 802-0034, www.mts-telsonic.com

Service Tectonics Inc., 2827 Treat St., Adrian, MI 49221, (517) 263-0758, fax (517) 263-4145, www.padprinting.net

Sonics & Materials Inc., 53 Church Hill Rd. Newtown, CT 06470, (800) 745-1105, (203) 270-4600, fax (203) 270-4610, www.sonics.biz

T.P. Welders, 1514 E. 11 Mile Rd., Royal Oak, MI 48067, (248) 543-0240, fax (248) 543-0172

Ultra Sonic Seal Co., 200 Turner Industrial Way, Aston, PA 19014, (610) 497.5150, fax (610) 497-5195, www.ultrasonicseal.com

Young Technology, 48012 Fremont Blvd., CA 94538, (510) 490-6373, fax (510) 490-3214

11   Hot Gas Welding

11.1   Advantages and Disadvantages

11.1.1   Advantages

1. *Low investment* Relative to other assembly techniques, the equipment required for hot gas welding is very inexpensive.

2. *Shape freedom* Hot gas welding can be used for virtually any shape, its only limitation being the ability to access the joint site with the welding torch.

3. *Portability* The portable versions of the equipment used in hot gas welding are available in a carrying case that can be moved to the worksite when the work is too large or expensive to bring to the factory. This makes it well suited to field repairs. Hot gas welding emits no measurable quantities of particulates and very low levels of volatile organic components (VOCs).

4. *Size of work* Since the equipment is portable, there is virtually no limit to the size of the pieces this process can weld. Consequently, it is often used in the construction industry.

5. *Difficult-to-join materials* Polyolefins, like polyethylene and polypropylene, which cannot be joined with solvents, are readily hot gas welded.

6. *Trained welders* Because of the similarity in techniques, most technicians trained in brazing and gas welding of metals can be readily retrained to weld plastics. New trainees can manage the basics of the technique in a few hours and can achieve minimum production skills in a few days. (High levels of skill and safety-related applications require additional training.)

11.1.2   Disadvantages

1. *Speed* Relative to the other thermoplastic joining processes, hot gas welding has, historically, been the slowest technique, in most cases. This has limited its use to applications that are unsuited to other methods because of size or shape. However, the development of newer, high speed welding equipment has lessened this disadvantage somewhat.

2. *Additional materials* Hot gas welding employs the use of a welding rod, which is an extra cost item.

3. *Operator sensitivity* This manually operated process is highly operator sensitive, which makes it difficult to maintain consistency, particularly with multiple operators.

4. *Operator cost* Welding is regarded as a trade and welders must be paid at a tradesman's scale in some areas, whereas operators for the competing plastics techniques are normally employed at unskilled or semiskilled labor rates.

11.2   The Process

Three varieties of welding use hot gas: tack welding, permanent hot gas welding, and extrusion welding. All of them work on fundamentally the same principle. It is very similar to acetylene welding of metal except that a stream of hot gas is used in place of the flame. In its most basic form, the required equipment consists of a welding gun, welding tips, a compressed air supply, a regulator with gauges, line filters, a welding rod, and cutting tools. Most of the time, a nitrogen gas supply is also required, as the nitrogen is used to prevent material contamination and oxidation. In the absence of N_2 the air supply must be clean and free of oil, or moisture. In addition, the surfaces must be clean and free of contaminants like grease, oil or mold release. Solvent cleaning is recommended for high quality welds, as is roughing the surface with a light sanding. The solvents and other techniques used for surface preparation for adhesives discussed in Chapter 7 ("Adhesive and Solvent Joining") can be applied to welding as well. One must, however, exercise care in the selection of solvents, to avoid the choice of a substance that attacks the plastic. The surface must be dry before commencing welding.

11.2.1   Tack Welding

Tack welding (Fig. 11-1) is used to hold pieces together while permanent welding or other work is taking place. This is a temporary weld of the base materials performed

Figure 11-1   Tack welding (Illustration courtesy of Laramy Products Co., Inc.)

without a welding rod. Continuous welding is required for larger pieces or for rough handling; however, spot tacking is adequate for many applications. A special tacker tip is required for tack welding.

Apart from the lack of welding rod and the use of a tacker tip, there is little difference between tacking and permanent welding. The tacking tip is smaller and has a pointed orifice, which concentrates the hot gas to ensure fusion of the edges. The welder is held at an angle of approximately 80° from the horizontal, and the tip is slowly passed along the edges to be welded. The pieces can be held in position by fixtures or by hand depending on the degree of accuracy required. Once the parts have been tack-welded together, they can be handled sufficiently to remove the need for clamps and jigs for the permanent weld. However, tack welds are not strong and should not be considered permanent.

11.2.2 Permanent Hot Gas Welding

Permanent welding calls for the use of a welding rod made of the same material as the pieces to be welded. First a 45° chamfer is cut off the corners of the two edges to be butt-welded together, as in Fig. 11-2a. (Were this an inside corner weld, this step could be eliminated.) A plastic welding rod of the same material is heated to fill in the V-shaped groove created when the two chamfers are pressed to each other. While unfilled round rod is most common, the rod may be triangular or V-shaped. The hot gas, which is compressed air or nitrogen, passes through a heating element and is focused on the weld area by the welding tip. The weld area consists of the intersection of the V-groove and the welding rod. This method is also known as "hand welding."

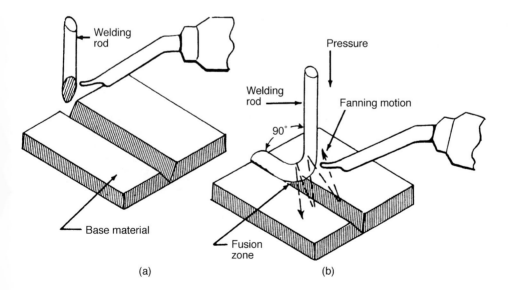

Figure 11-2   Permanent hot gas welding: (a) starting weld and (b) welding (Diagrams courtesy of Laramy Products Co., Inc.)

The tip of the welder is positioned 6.35 to 19.05 mm (0.25–0.75 in.) above the joint. The start of the weld is important because a poor start will result in a failed weld. When multiple layers of rod are used, it is wise to start them at different points along the weld. The joint and the rod should be heated and gently touched every few seconds until the rod will stick to the base. The weld rod diameter should be about the same as the thickness of the piece to be welded. For most materials, the rod is held down under a pressure of about 3 pounds and at an angle of 90° as shown in the illustration. For polyethylene and fluorocarbons, it is held at an angle of 45°.

The welding tip is fanned (to avoid discoloration) rapidly (2 fans per second) back and forth over the weld area until the V-groove and the welding rod are softened enough to weld when they are pressed together, often with a heated roller. Since the pieces to be welded have much more mass than the rod, the heat should be concentrated on them. About 12.7 mm (0.50 in.) of the welding rod and 9.5 mm (0.375 in.) of the material should be heated in front of the joint interface. The rod will become rubbery if it is overheated. The welding speed should average 102 to 152 mm (4–6 in.) per minute.

At the end of the weld, all forward motion is halted and the heat is briefly concentrated at the joint interface. Then it is removed and the downward pressure on the rod is continued for several seconds to set the end of the weld. Then the downward pressure is discontinued and the rod is twisted until it snaps off. Figure 11-3 is a photograph of hand welding.

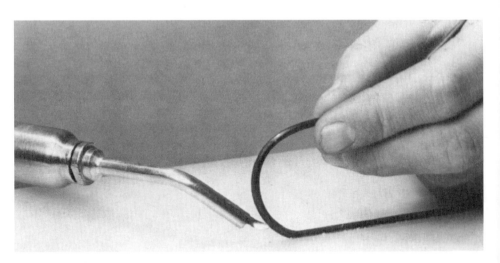

Figure 11-3   Hand welding (Illustration courtesy of Laramy Products Co., Inc.)

11.2.3   High Speed Welding

The welding process can be speeded up through the use of a "speed gun" such as the one in Fig. 11-4, which gives speeds of 610 to 762 mm (24–30 in.) per minute with materials such as polyethylene or polypropylene, a rate of 1016 mm (40 in.) can be

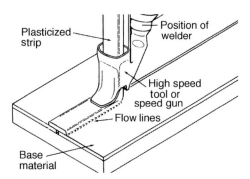

Figure 11-4   High speed welding (Diagram courtesy of Kamweld Products Co., Inc.)

attained with polyvinyl chloride. The speed gun used for "high speed welding" is designed to facilitate the technique by feeding the welding rod through the gun, where it is heated as it passes through. Beneath the rod there is an opening through which the hot gas is directed onto the weld area. The use of a speed gun eliminates the need to hold the rod, leaving one hand free to hold the work or feed in fresh rod.

The rod is inserted through the rod tube and slowly extended to the joint surface in the position shown in Fig. 11-5a. Pressure is exerted on the welder until the rod bends and reaches the position shown in Fig. 11-5b. The rod is then pulled in the direction of weld. The rod is fed with a slight pressure initially, however it feeds automatically after a few inches of weld. Flow lines should appear alongside the weld if welding is proceeding properly. If the weld is progressing too fast, there will be no such lines (the weld is too cold), and a poor weld will result. If welding is too slow, the weld will discolor, char, or burn if the material is polyvinyl chloride. Polyethylene or polypropylene will flatten out to an oversized bead if it is welded too slowly. When the correct rate of speed is reached, the weld should remain at the speed in a fairly constant manner.

To stop welding, the welder is tipped beyond the vertical position, as illustrated in Fig. 11-5c. The rod will be cut off by the end of the welding tip. Alternatively, the rod can be permitted to run through the end of the tip, whereupon it is cut or broken off.

For further information on hand and high speed welding techniques, the author recommends *Making Even Better Plastic Welds* by Donald W. Thomas and revised

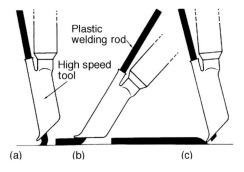

Figure 11-5   High speed welding positions (Diagrams courtesy of Kamweld Products Co., Inc.)

(a)   (b)   (c)

by J. Pierre Pottier. It is available from Laramy Products Co., Inc. (See "Sources" at the end of this chapter.)

11.2.4 Extrusion Welding

A further automated development of the hot gas welding technique uses an extruded welding rod that is pressed into the heated groove with either a roller or a welding shoe. The chief benefit of this process, known as extrusion welding, is that it eliminates the multiple passes required of many heavy-duty welding applications. That saves a great deal of time and reduces the stresses on the joint.

Extrusion welding is illustrated in Fig. 11-6. The filler resin is supplied either as pellets delivered through a hopper or as welding rod from a reel. The plastic is then extruded through the melting chamber with a screw driven by an electric motor. (For a further description of the extrusion process, the reader is referred to Chapter 6, "Assembly Method Selection by Process.") The heat is supplied by either an electric band heater or by hot air. The molten plastic is extruded out the shoe, which is shaped and sized to the desired bead. The joint interface is heated by the hot gas preheater, which is attached to the extruder. The rate of speed is controlled by the rate of extrusion.

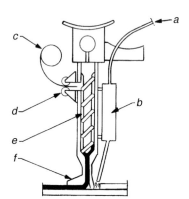

Figure 11-6   Extrusion welding: *a*, air supply; *b*, hot gas device; *c*, wire reel (shown) or granulate funnel; *d*, wire forward feed; *e*, small extruder; *f*, welding shoe

11.3   Joint Designs

A variety of welds are used according to the application, however they are similar to those used in the welding of metals. The types illustrated in Fig. 11-7 are some of the more common ones. The fillet and lap fillet welds shown here are easiest to use because they require no preparation beyond cleaning the joint surfaces. Typically, tack welds are used to position the pieces for welding, however fixtures or, in the case of the lap weld, C-clamps can perform this function.

Butt welds, edge welds, and some corner welds and fillet welds require that the joint edge be beveled. The bevel should not be cut to a sharp edge, but to a flat of about 0.80 mm (0.031 in.). The designs in Fig. 11-8 show a root gap to be used when a tacking

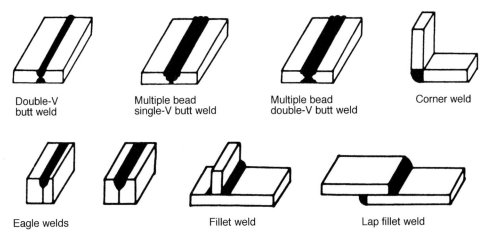

Figure 11-7   Weld types (Diagrams courtesy of Laramy Products Co., Inc.)

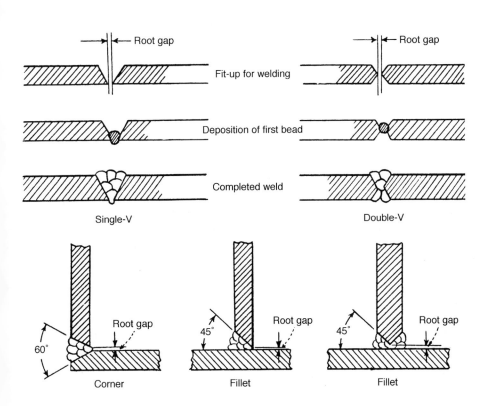

Figure 11-8   Weld beveling and preparation (Diagrams courtesy of Laramy Products Co., Inc.)

tip is not used. The root gap should be in the range of 0.40 to 0.80 mm (0.0156–0.031 in.), and the included angle of the bevel should fall between 60° and 90°, except for the fillet welds at 45°. The wall thicknesses of pieces typically gas welded range from 1.60 to 9.53 mm (0.063–0.375 in.). Although the material may be ductile, the welded joint will not be.

Hot gas welds are sometimes used as a sort of spot weld placed intermittently along lapped pieces. These are referred to as "rosette welds" and are illustrated in Fig. 11-9. One advantage is that access from only one side is needed, making this a useful technique in situations of limited access to the opposite side. Moreover, little preparation is required beyond drilling a hole and normal surface cleaning.

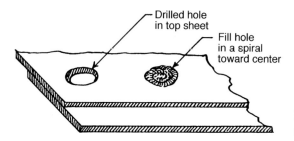

Figure 11-9   Rosette weld (Diagram courtesy of Laramy Products Co., Inc.)

11.4   Welding Practice

Hot gas welding is highly operator sensitive because it is done by hand. However, if the heat is properly applied, there should be no charring, discoloration, or warping. The strength of the weld can vary from none at all, if done improperly, to a maximum of 90% of the surrounding material for some materials (e.g., polyethylene, polypropylene, polyvinyl chloride, ABS). There are four basic essentials to high quality welds: proper temperature, pressure, angle, and speed. Cross sections of good welds are illustrated in Fig. 11-10.

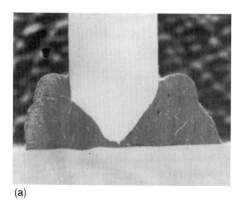

(a)

(b)

Figure 11-10   Cross sections of completed fillet welds: (a) fillet weld and (b) double-V weld (Diagrams courtesy of Laramy Products Co., Inc.)

Failure to execute the weld correctly can lead to a variety of problems, some of which are illustrated in Fig. 11-11.

11.4.1 Appearance Problems

The uneven appearance of the weld in Fig. 11-11a can be due to uneven pressure, excessive pressure or uneven heating. Corrective action would consist of using a slow, uniform, fanning motion, making certain to heat both the rod and the material, and ensuring that the correct rod angle is employed. Practice with the rod will also help.

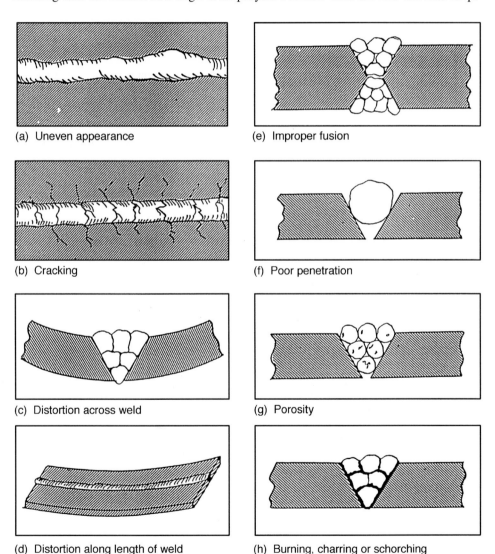

(a) Uneven appearance

(b) Cracking

(c) Distortion across weld

(d) Distortion along length of weld

(e) Improper fusion

(f) Poor penetration

(g) Porosity

(h) Burning, charring or schorching

Figure 11-11 Common welding problems (Diagrams courtesy of Laramy Products Co., Inc.)

11.4.2   Cracking Problems

Stress cracking (Fig. 11-11b) can be the result of a number of factors: incorrect welding temperature, excessive stress on the weld, improper matching of rod and part materials, and attack by oxidation or chemicals. Avoidance of stress cracking calls for greater attention to proper welding temperature, chemical resistances, and material matching, and allowance for thermal expansion and contraction.

11.4.3   Distortion

There are two types of distortion: across the weld and along the length of the weld. Distortion across the weld (Fig. 11-11c) can be due to overheating the joint, welding too slowly or in the wrong sequence, or welding with too small a rod. The problem is inherent in the nature of welding, however, because the base material will expand in response to the heat and pressure of the welding rod. Upon cooling to its original volume, however, it does not shrink back to its original shape. This problem can be solved by going to a double V-weld. For single V-welds, this phenomenon may be counteracted by prebending the pieces as shown in Fig. 11-12.

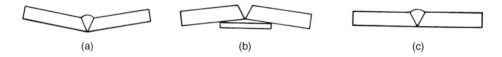

(a)   (b)   (c)

Figure 11-12   Prebending technique: (a) sheet warpage due to shrinkage during the cooling process; (b) prebending prior to welding; (c) sheet pulls itself straight after welding (Diagrams courtesy of Wegener North America)

Distortion along the length of the weld, as in Fig. 11-11d, can also be due to shrinkage of the material. However, it can also be due to overheating and faulty preparation or clamping of the parts. Preheating the material to relieve the stress will help, as will improved clamping and more rapid welding. However, if multilayer welds are used, more time may be needed between welds. Also, the root gap may be too large.

11.4.4   Fusion Problems

Improper fusion (Fig. 11-11e) often is due to a welding temperature that is too cold or a welding rate that is too fast. However, it can also be due to improper surface preparation, rod selection, or welding technique. If the rod material is properly matched and changing the temperature or rate of speed does not produce the desired results, preheating the materials before welding or changing the diameter of the rod (so that a smaller rod is used at the root than for the rest of the welds) may be effective.

11.4.5 Penetration

Poor penetration (Fig. 11-11f) is also likely to be due to too cold a welding temperature or too fast a welding rate. However, it can also be the result of too large a welding rod, too small a root gap, or inadequate surface preparation. Besides adjusting the welding parameters, going to a smaller welding rod, or increasing the root gap, one can opt to increase the angle of the bevel for superior penetration.

11.4.6 Porosity

The porosity evidenced in Fig. 11-11g may be another result of too large a welding rod or too fast a welding rate. It may also be due to improper technique in the form of improper starts and stops, crossing of weld beads, stretching the rod, and imbalance of heat on the rod. Finally, the weld rod itself may be porous. If the rod is the correct size and in proper condition, the emphasis should be on improving welding technique to eliminate this problem.

11.4.7 Scorching

Burning, charring, or scorching of the weld (Fig. 11-11h) is usually the result of too slow a welding rate or too high a welding temperature. However, it can also be due to uneven heating or an attempt to weld a material that is too cold. The correct fanning motion or preheating the part, particularly in cold weather, can resolve the problem if speeding up the welding or lowering the temperature does not produce the desired results.

11.5 Testing the Weld

11.5.1 Nondestructive Testing

11.5.1.1 Visual Examination

The most common test performed by the operator is to examine the weld for the characteristic slight distortion and welding wash along the edge of the welding rod – the signs of a good hot gas weld. There will also be a change in surface appearance along the parts themselves, but this change will not be the same for all materials. Polyolefins will have a dull blush, whereas polyvinyl chloride will be shinier along the weld. A shiny polyolefin, with small splashes of material in a spray pattern away from the bead, indicates that the material was overheated, usually as a result of too low a speed or too great a temperature.

The rod can also give clues to the integrity of the weld. If it has retained its original shape, there was probably not enough pressure exerted during welding or, if there is no discoloration either, the welding was done at too great a speed. Further indications of the integrity of the weld were provided in Section 11.4.1 on appearance problems. Dye that can penetrate the material can be used as an inspection aid, as a proper weld will not permit penetration.

11.5.1.2 Leak Tests

When the product is a vessel, leak tests can be used to establish the integrity of the welds. Depending on its size, the vessel can either be filled with water and observed for leakage or immersed in a tank of water and observed for the bubbles that indicate escaping air. Commercial leakage testers are also available.

11.5.2   Destructive Tests

11.5.2.1 Tensile Test

The tensile test is the traditional method of testing the integrity of a product and the Society of the Plastics Industry has established the following procedure. A 203.2 mm by 152.4 mm (8.00 in. by 6.00 in.) test specimen is constructed of 4.78 mm (0.188 in.) stock. It is cut in half, and a double V-joint is created with a 60° included angle and a 0.40 mm (0.0156 in.) root gap. A 0.40 mm (0.0156 in.) flat is provided at the apex of the bevel. The weld is made, using a small-diameter rod for root welding and a larger diameter rod to complete the weld. The test is performed on at least five specimens cut to 12.7 mm (0.50 in.) width and pulled at a rate of 0.635 mm (0.025 in.) per minute. Weld strengths are typically expressed as a percentage of the original strength of the material. Products that will be exposed to a chemical or elevated temperature environment should be tested after exposure to the same environment.

11.5.2.2 Bending Test

Specimens are prepared in the same manner as for the tensile test. They are bent double along the axis of the weld while they are still hot. The bead should cohere to the beveled surface of the sheet. The same test should be performed on a tensile test specimen 24 hours after welding. It should not break readily when bent 90° by hand. If the base material breaks before the weld, the joint should be adequate.

11.5.2.3 Rod Removal Test

Allow the rod to extend beyond the weld area and pull on it while the rod is still hot. If the filler rod breaks before it pulls away from the base material, it is a good weld. To test a midsection of the rod, pull on it with a knife or scissors or break it out with a hammer and chisel after it has cooled. If the removed section contains bits of the base material, it is a sound weld. For test purposes, the use of colored rod will facilitate the inspection of the weld.

11.5.3 Chemical Test

Immersion of a welded test specimen in acetone for 2 to 4 hours will cause it to swell where residual stresses are present. Faulty welds will separate from the base material. Defective materials will be indicated by delamination and disintegration. Properly welded beads will be difficult to pry out of the joint, even after immersion in acetone.

11.5.4 Spark Test

Vessels can be tested for leaks by means of a spark test. A high voltage electrode is placed on one side of the weld and a conductive material is placed on the other side. When current is applied, a spark will jump through any hole that is present in the weld.

11.6 Applications

The production process whose parts are most often hot-gas-welded together is fabricating, often in combination with extruded pipe. Thermoformed and rotationally molded parts are occasionally hot-gas-welded as well, and hot gas welding is used to join fittings to large vessels produced by these processes. Injection-molded parts are usually sufficiently small in size and high in volume to warrant faster methods with higher investment requirements. Structural foam parts have a relatively thin skin of solid material that does not lend itself well to welding. Thermoset materials cannot be hot gas welded.

Hot gas welding is often used in the construction industry for ductwork, tanks, hoods, and piping, which is probably the largest user of products made by means of the hot gas welding technique. Applications are also found in the marine and automotive industries. The most common repair applications for this technique are found in plumbing, marine, and automotive body work.

11.7 Sources

11.7.1 Welding Rods

V & A Process, Inc., 1230 Colorado Ave., Lorain, OH 44052, (440) 288-8137, fax (440) 288-2323

11.7.2 Welding Equipment

Kamweld Technologies, 90 Access Road, Norwood, MA 02062, (781) 762-6922, fax (781) 762-0052, www.kamweld.com

Laramy Products Co., Inc., P.O. Box 1168, 40 Sandy Lane, Lyndonville, VT 05851, (802) 626-9328, fax (802) 626-5529, www.laramyplasticwelders.com

Leister Technologies LLC, 1253 Hamilton Parkway, Itasca, IL 60143, (630) 760-1000, fax (630) 760-1001, www.leister.com

Wegener North America, Inc., 16W231 S. Frontage Rd., Unit #12, Burr Ridge, IL 60527, (630) 789-0990, fax (630) 789-1380, www.wegenerwelding.com

11.7.3   Welding Rod And Equipment

Seelye, Inc., 333 Enterprise St., Suite C, Ocoee, FL 34761, (800) 258-2936, (407) 656-6677, fax (407) 656-5244, www.seelyeinc.com

12   Induction/Electromagnetic Welding

12.1   Description

The principle of induction/electromagnetic welding is that magnetic materials become hot when subjected to a high frequency, alternating current field. In its most common application, a metal insert or metallic composite preform is placed within the joint and welded into the assembly. Occasionally, the current is used to heat a forming die that is sized, shaped, or located in a fashion difficult to heat by other means. Induction electromagnetic welding is one of only two techniques (The other is laser welding.) that it can achieve noncontact welding in the joint area alone.

This process is most commonly used to join both amorphous and crystalline thermoplastics. In some cases, dissimilar plastics and thermosets can be joined to each other as well as to other materials such as paper, wood, ceramics, and aluminum. Thermoplastic elastomers, as well as foamed and filled materials, can also be used with joints successfully made in materials with as much as 65% filler, a level at which other welding methods do not have enough of the base polymer to make the weld. FDA approval can be obtained for some applications.

Induction/electromagnetic welding is capable of joining very large parts; indeed, bond lines up to 6 m (20 ft) long have been accomplished. This process has also been used experimentally for the use of composites in the repair of steel infrastructure and is widely used for sealing of aseptic food packages. It can also weld joints on contoured surfaces like automobile taillights, where it is competitive with hot plate welding. This technique can produce a hidden joint and can compensate for irregular surfaces. Hermetic seals and very high bond strengths are achievable. Other advantages of induction/electromagnetic welding are precise control of heating, high throughput, and low operating cost. The principal negative factor is the cost of the electromagnetic material (which is sacrificed in each assembly) and of its insertion. This material can also become a contaminant for recycling, although the insert is usually small and the ferromagnetic material rarely exceeds 15% of the insert itself. The process is reversible for disassembly. It is also unsuited to products with other metal inserts or electronic components with metal inserts within the electromagnetic field. This method can join parts that have been injection-molded, rotational-molded, structural foam-molded, blow-molded, extruded, or thermoformed.

12.2   Advantages and Disadvantages

12.2.1   Advantages

1. *High production rates* Induction welding is one of the faster thermoplastic assembly methods, with weld times ranging from 1 to 10 seconds. Rotary tables can speed loading and unloading. Operating costs are low.

2. *Joint strength* Induction welding can provide very high joint strengths.

3. *Entrapment of other parts* Additional parts can be captured between the two parts to be induction-welded, provided they are located such that they do not interfere with the welding.

4. *Permanence and high bond strength* Induction welding can create permanent assemblies with very high bond strengths which cannot be reopened without damaging the parts. As a welded joint, the effects of creep, cold flow, stress relaxation, and other environmental limitations are absent in the joint area, although they may occur elsewhere if other parts of the design are under load. Owing to material limitations, differences in thermal expansion and moisture absorption are rarely of concern once the parts have been welded. Joint strengths equivalent to those of the surrounding walls are possible under ideal conditions.

5. *Reopenability* Induction heating can be used to reopen the joint. This is particularly useful for recycling purposes.

6. *Similar materials with high amounts of filler* Induction welding handles most thermoplastic materials well, including heavily filled resins and crystallines. (Polypropylene is perhaps the easiest material for the process.) Rapid heating of solid sections reduces the likelihood of degradation.

7. *Dissimilar materials* Induction welding can weld dissimilar materials effectively in most cases.

8. *Shape freedom* Induction welds can be made with contoured parts of practically any shape and with fairly thin wall thicknesses as long as there is space for the coils to reach the joint. Hidden joints are possible.

9. *Hermetic seals* Induction welding is capable of creating hermetic seals.

10. *Clean atmosphere* Unlike adhesive and solvent joining systems, no ventilation equipment is necessary for the removal of toxic fumes.

11. *Process freedom* Parts made from virtually all the thermoplastic processes can be induction-welded (when sheet fabrication methods are included).

12. *Large part capability* Part size capability is nearly unlimited when intermittent or continuous welding methods are employed. Parts up to 6 m (20 ft) long have been made in one cycle.

13. *Nondestructive joint verification* X-rays can be used to determine the presence of metal or metal oxide particles in the weld area.

14. *Loose tolerances* Melted electromagnetic material flows into gaps and voids between the parts, thus accommodating irregular surfaces and permitting loose tolerances.

15. *Precise control of heating* The heat flows from the metal source within the joint and can be precisely controlled.

12.2.2 Disadvantages

1. *Additional materials* Induction welding uses additional materials (in the form of preforms, wires, screens, etc.). The cost of such items can be significant and, therefore, this technique is inherently more expensive than methods not requiring such materials and more expensive to recycle as well.

2. *Shape limitations* Induction welding is limited to applications in which the coils can access the joint.

3. *Additional assembly operation* Induction welding requires the placement of three parts in the fixture of the welder: the two parts to be assembled plus the ferromagnetic material.

4. *Stress concentrations* Stress may be caused by metal embedded in plastic at the joint.

5. *Metal inserts* Metal pieces within the electromagnetic field will also heat up.

6. *Initial cost* The cost of the equipment and the development of the optimum coil configuration can be high.

7. *Polymer decomposition* There is a potential for polymer decomposition at the surface of the metal particles. Also, the metal particles can rust if they are ferritic.

8. *Safety* Other metal items of apparel (rings, belt buckles, eyeglass frames, etc.) or utility (pens, pencils, tools) within the induction field can become hot and burn the unwary.

In addition, there is an issue regarding the potential hazard to operators of exposure to electromagnetic radiation. As rf energy passes through the body, it raises its temperature. In the United States, the Occupation Health and Safety Administration (OSHA) has a regulation concerning this type of radiation. The American Conference of Governmental Industrial Hygienists (ACGIH), the American National Standards Institute (ANSI), and the Institute of Electrical and Electronics Engineers (IEEE) have published guidelines for this type of exposure. Prospective purchasers of such equipment should question the manufacturer with regard to these regulations and guidelines.

12.3   The Equipment

As illustrated in Fig. 12-1, the equipment required for electromagnetic welding has five basic components.

1. *Induction generator* This unit converts the 50 to 60 Hz electric current to the output frequency. That frequency can range from 2 to 40 MHz, but the most commonly used frequencies are in the 3 to 8 MHz range. Output power will normally be 2 to 5 kW.

2. *Heat exchanger* Usually combined as a unit with the induction generator, the heat exchanger cools water that is used to prevent heat buildup in both the copper coils and the generator.

3. *Work coil* The work coil is connected directly to the output of the induction generator and conducts the high frequency current. Its function is to provide the magnetic field around the joint. To prevent heat from building up during production, it is normally made of hollow round or square copper tubing so that cooling water can be circulated through it. The design of the work coil is critical to the efficiency of the process as heating cycles of about 0.1 second are possible; however cycle times of 2 to 10 seconds are typical for the process, with larger parts requiring even more time.

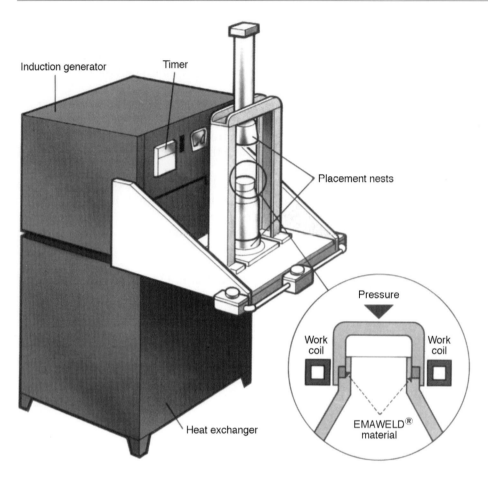

Figure 12-1   Induction/electromagnetic welding equipment (Diagrams courtesy of Emabond Systems)

4. *Pressure ram* The pressure to force the molten insert material into the joint cavity is supplied by a ram attached to a hydraulic or pneumatic cylinder, usually the latter. In most cases, there is an upper, or movable, placement nest attached to the ram. Pressures range from 0.07 to 0.17 MPa (10 to 25 psi) for polyethylene or polypropylene to 0.86 MPa (125 psi) for materials such as nylon, polycarbonate, and polyphenylene oxide.

5. *Placement nests* These are the fixtures that hold the parts during assembly. Depending on the design of the joint, there may be both an upper and a lower nest or just a lower one. Placement nests are made of nonconductive materials such as phenolic or epoxy–fiberglass. To avoid a reduction in the magnetic flux density at the joint, metal components near the work coils must be correctly shielded. It is, therefore, desirable to use the coil to apply the pressure when possible, thereby eliminating the need for the upper nest.

12.4   The Process

The process is illustrated in Fig. 12-2: the electromagnetic insert is shown placed in the part to be joined, located in the lower placement nest. The mating part is located in the upper placement nest, which is movable. The pneumatic cylinder brings the movable nest down to a point at which the insert is trapped in a cavity between the two parts. The joint detail at this stage is shown in the "before welding" portion of Fig. 12-3. Power is then applied to the induction coil, creating the electromagnetic field that melts the insert. The "during welding" illustration depicts the insert being melted by the activated coil. Pressure from above forces the molten insert material to fill in the cavity just as in compression molding. A weld to the joint interface of the mating parts is created. The final joint is shown in the "after welding" illustration.

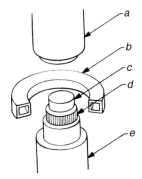

Figure 12-2   The induction/electromagnetic welding process: *a*, upper part inside upper placement nest; *b*, coil; *c*, lower part; *d*, electromagnetic insert; *e*, nest

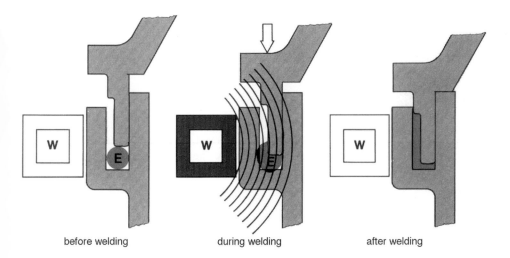

before welding   during welding   after welding

Figure 12-3   Welding stages (Diagrams courtesy of Emabond Systems)

12.5   The Coil

The rate of speed required will, in part, determine the type of coil used. The shape of the joint to be welded will define the coil configuration, inasmuch as the coil must be located as close to the joint as possible for maximum efficiency. Coils are usually made from round, square, or rectangular tubing ranging in sizes from 3.18 to 9.53 mm (0.125–0.375 in.). The larger sizes permit more water to pass through them for better cooling. For this reason, the smallest sizes are not recommended except for extreme circumstances or for small parts and very short heating cycles, where there is little heating effect. Square tubing is preferred because it is desirable for the coil to be as close to the work as possible, since electromagnetic fields become exponentially more concentrated as the coupling distance to the coil is decreased.

There are four basic types of work coil: single-turn, multiturn, hairpin, and split.

12.5.1   Single-Turn Coils

Two varieties of single-turn coils are shown in Fig. 12-4. The simple coil in Fig. 12-4a has a tendency to develop a weak field intensity at the end of the coil leads; however, it requires less space and offers the greatest amount of design shape freedom. A nonactive water-cooled reflector placed on the inside diameter will improve the efficiency of the heating by concentrating the electromagnetic field through the joint. This will help the cold spot at the leads, however the crossover lap coil in Fig. 12-4b will be more effective.

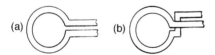

Figure 12-4   Single-turn coils: (a) simple coil and (b) crossover lap coil

12.5.2   Hairpin Coils

The hairpin coil in Fig. 12-5 is a modified (squeezed together) single-turn coil used to join flat sheets or to seal the perimeter of a two-sheet assembly to make mat or tray composites. Such coils are designed to concentrate the magnetic field by reducing the coupling distance between the turns to the thickness of the part. The bond line must be located within the internal diameter of the helical coil because the greatest field strength is within the coil, not on the outside turns. Hairpin coils have been made up to 4.75 m (15 ft) long and in a variety of irregular shapes. They are often used as dies within a press action with movable and fixed platens and are very efficient for joining long flat sheets. They are also cost-effective for creating perimeter seals of glass composite structural components and similar applications.

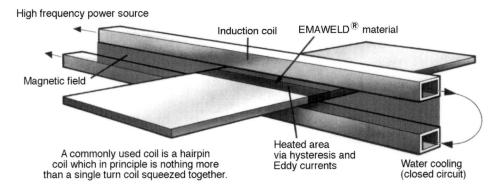

High frequency power source

Induction coil   EMAWELD® material

Magnetic field

A commonly used coil is a hairpin
coil which in principle is nothing more
than a single turn coil squeezed together.

Heated area
via hysteresis and
Eddy currents

Water cooling
(closed circuit)

Figure 12-5   The hairpin coil (Diagrams courtesy of Emabond Systems)

12.5.3   Multi-Turn Coils

The helical or cylindrical coil illustrated in Fig. 12-6a is the most common variety of
multiturn coil, and the number of turns most often used is two. As a general rule,
the length of a multiturn helical coil should not exceed three to four times its diameter.

Round containers are welded with the helical or cylindrical coil in Fig. 12-6a.
Irregular shapes can be welded by coils designed to suit their shapes (coned, square,
rectangular, D-shaped, etc.). Flat surfaces and large surface areas are welded with
the pancake coil (Fig. 12-6b).

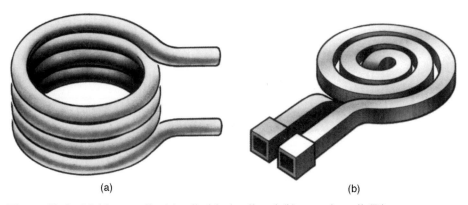

(a)   (b)

Figure 12-6   Multiturn coils: (a) cylindrical coil and (b) pancake coil (Diagrams courtesy of
Emabond Systems)

12.5.4   Split Coils

For most applications, the handling of the parts before and after welding takes more time than the welding itself. The coil can sometimes become an impediment to easy part emplacement and removal. These functions can be eased by splitting the coil into two separate sections, which are moved in and out of position manually or with air cylinders. Large, flat surfaces have been welded using coils made of 1.60 mm (0.063 in.) copper sheeting. They are also effective for large pipe or conduit, or for parts with limited access to the bond line (e.g., flanged fittings).

12.5.5   Other Types of Coils

For some applications, it is difficult to bend hollow copper tubing to the required shape. As shown in Fig. 12-7a, coils can be made of solid copper bar or plate stock with cooling tubes brazed to them. This concept can be used to make multistation coils like the one in Fig. 12-7b. In addition, when induction heating coils are placed inside shoes made of electrically nonconductive materials (phenolic, glass-filled epoxy, etc.), the coils can be used to create the nest and the pressure ram. The shoes illustrated in Fig. 12-7c represent another approach applicable to joining sheet goods and flanges.

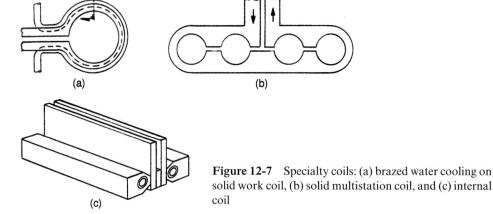

(a)

(b)

(c)

Figure 12-7   Specialty coils: (a) brazed water cooling on solid work coil, (b) solid multistation coil, and (c) internal coil

12.5.6   Coil Positioning

The distance from the coil to the joint, known as the coupling distance, must be as small as possible because the strength of the field varies inversely with the square of the distance between the insert and the coil. Actual contact between the coil and insert is desirable when the coil can provide the downward pressure. Otherwise, a distance of

1.60 mm (0.063 in.) is considered a close coupling. Work coils are custom-designed for each application, with square tubing being preferred because it permits the field to reach closer to the joint. The coil must also be designed to prevent arcing and overloading. An ideal electromagnetic welding application is one in which the coil can be properly located.

12.5.7 Flux Concentrators

The density of the induction field can be controlled and concentrated through the use of ferrite, laminations, or powdered magnetodielectric materials. Configured properly, the heat power density can be increased by a factor of up to 4 over that of a base coil. This can result in a faster cycle, shielding of magnetic fields, heat pattern control, or the use of less expensive equipment that produces fewer kilowatts. While flux concentrators create considerable benefits for heat treating and brazing of metals, they have found few applications for the welding of plastics because the lower temperatures required result in very short cycles even without the use of flux concentrators.

12.6 Materials

12.6.1 Polymers

Induction welding is primarily confined to thermoplastics, most of which can be welded to themselves. In some cases, dissimilar polymers can be induction welded. A compatibility chart is illustrated in Table 12-1.

12.6.2 The Electromagnetic Material

The electromagnetic material can be a metal screen insert or submicrometer-sized particles of magnetic materials dispersed in a liquid or a polymer selected for compatibility with the materials being joined. The electromagnetic material may be special ferrites, graphite powders, stainless steel, or iron oxide. The type of filler and particle size will be determined by the environmental end use and physical properties required. When the materials with magnetic properties are used, their energy losses, which create the heat, are through hysteresis and eddy currents. When the Curie point is reached, magnetic properties cease to exist and eddy current losses are the sole heat source. The induction of heat from nonmagnetic materials is slower because eddy current losses are the only source of heat for them. When mixed in a polymer, the magnetic material can be supplied in a variety of forms such as the molded parts, extruded profiles, die-cut gaskets, and tapes, as illustrated in Fig. 12-8. If the polymer is a hygroscopic material, it may need to be dried before welding. When standard materials are not suitable, custom inserts can be formulated if the volume warrants.

Table 12-1 Material Compatibilities (indicated by •)

	ABS	Acetals	Acrylic	Cellulosics	Ionomer (Surlyn)	Nylon 6/6, 11, 12	Polybutylene	Polycarbonate	Polyethylene	Polyphenylene oxide (Noryl)	Polypropylene	Polystyrene	Polysulfone	Polyvinyl chloride	Polyurethane	SAN	Thermoplastic polyester	Thermoplastic elastomers Copolyester	Styrene block copolymer	Olefin type
ABS	●		●					●								●				
Acetals		●																		
Acrylic	●		●					●				●				●				
Cellulosics				●																
Ionomer (Surlyn)					●															
Nylon 6/6, 11, 12						●														
Polybutylene							●				●									
Polycarbonate	●		●					●					●	●		●				
Polyethylene									●		●									●
Polyphenylene oxide (Noryl)										●		●								
Polypropylene							●		●		●									●
Polystyrene			●					●		●		●				●				
Polysulfone								●					●							
Polyvinyl chloride														●						
Polyurethane															●					
SAN	●		●					●				●				●				
Thermoplastic polyester																	●			
Thermo plastic elastomers Copolyester																		●		●
Styrene block copolymer											●								●	
Olefin type											●									●

Note the indication of which dissimilar thermplastics can be welded (e.g., ABS/acrylic and ABS/polycarbonate). For almost all thermoplastics the Emaweld® process may be used.

Source: Table courtesy of Emabond Systems.

12.6.2.1 Molded-in Pre-Forms

For very high production levels, the preform can actually be molded directly into one of the mating parts. For an explanation of this process, the reader is referred to Chapter 13, "Insert and Multipart Molding."

Figure 12-8 Preform (Photo courtesy of Emabond Systems)

12.6.2.2 Hot Melt Electromagnetic Materials

Hot melt adhesives can be used as preform insert materials provided they have been modified to include electromagnetic particles. The customary technique is to apply the hot melt to one of the parts to be assembled and permit it to cool and harden. The parts are then assembled by electromagnetically reheating the hot melt under pressure. This method can be used in moderate strength applications to join parts made of thermoset materials.

12.6.2.3 Liquid Electromagnetic Materials

Electromagnetic material in liquid form is used for the joining of films or sheets.

12.7   Joint Designs

Besides the coil design, the other principal determinant to the effectiveness of this process is the joint design. There are several basic factors to consider in its determination.

1. *The strength level necessary* Electromagnetic induction welding is capable of providing strengths nearly that of the surrounding material.

2. *Hermetic seal requirement* This process is often used where a hermetic seal is needed.

3. *Joint location and contour* The joint must be located where the coil can be placed close to it and in a shape which lends itself to coil formation.

In designing joints for induction or electromagnetic welding, one must bear in mind that the insert material will be molten and will fill any voids as it follows the path of least resistance. Pressure on the joint will ensure that all the voids and surface irregularities are filled and that the entire surface is wetted. The joint must be designed to contain the molten material and to apply its pressure toward the inside. This pressure is determined by the amount of material available for compression over and beyond the minimum necessary to exactly fill the void. In the equation used to determine A_I, the cross-sectional area of the electromagnetic insert, this additional material is represented by a constant K, which will range from 1.02 to 1.05 depending on the circumstances. The area of the void is represented by A_V.

$$A_I = KA_V \tag{12.1}$$

Equation 12-1 can be used for all joints that have a pocket. However, these joints are not usable in every circumstance. They are particularly difficult to achieve with processes that have only one controlled surface (there is no core side to the mold) such as blow molding, rotational molding, and thermoforming, unless they are double walled. These are also methods in which the type of tolerance needed for such joints is held only with difficulty.

The flat-to-flat joint illustrated in Fig. 12-9a can be used for parts made from the above mentioned processes because the wall section is usually sufficiently uniform for this joint to work. However, for most such applications, it is not suitable for hermetic seals. Also, the edge of the electromagnetic material can be visible unless the flange is sufficiently large to recess it well back from the edge. Long continuous welds for structural applications often use this joint. It requires a flat insert or tape.

An insert with a round cross section can be used to provide better control. It can be placed in a V-groove such as the flat-to-V-groove joint depicted in Fig. 12-9b. More precise location of the insert can be created with the half-round groove shown in Fig. 12-9c. This design can be used for a fitting placed in a hole drilled in a blow- or rotational-molded container. The automotive industry uses a rectangular groove to assemble automobile panels; however, the proportions shown in Fig. 12-9d were taken from an electrically heated food serving tray that needed to withstand repeated dishwasher cycles. The parts were injection-molded from polyphenylene oxide.

The tongue-and-groove design in Fig. 12-9e is well suited to injection-molded parts. It can be used for hermetic seals and pressure joints because it provides the highest weld strength. Since the material is completely contained within the joint area, excellent joint appearance can be attained. The design has been used for medical devices.

Generally, the strongest type of joint is the shear joint. The design in Fig. 12-9g has an external version of this joint. This approach can be used for a blow-molded

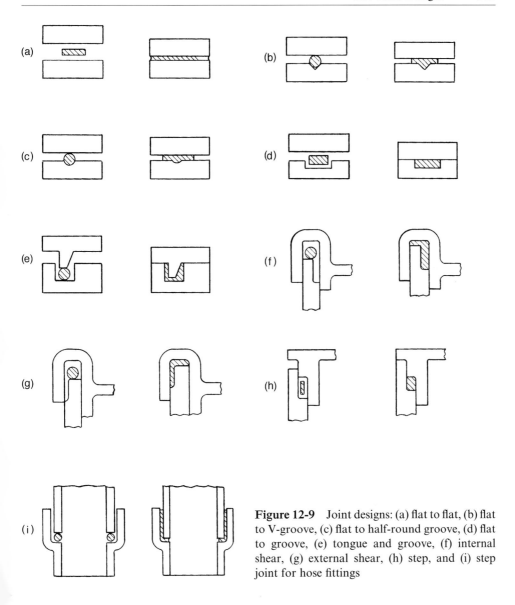

Figure 12-9 Joint designs: (a) flat to flat, (b) flat to V-groove, (c) flat to half-round groove, (d) flat to groove, (e) tongue and groove, (f) internal shear, (g) external shear, (h) step, and (i) step joint for hose fittings

or rotationally molded container; however, it is best suited to external loads. The internal joint in Fig. 12-9f is a superior design for a high pressure seal but cannot be formed on the inside neck of a blow-molded bottle. Equation 12-1 can be applied to these designs.

The step joint in Fig. 12-9h is another form of internal joint that is derived from the shear joint. It can provide excellent pressure seals and is capable of good joints with wide tolerances. It is widely used for small injection-molded parts. Another variety of this concept, used for hose fittings, is shown in Fig. 12-9i. In this case, polyolefin

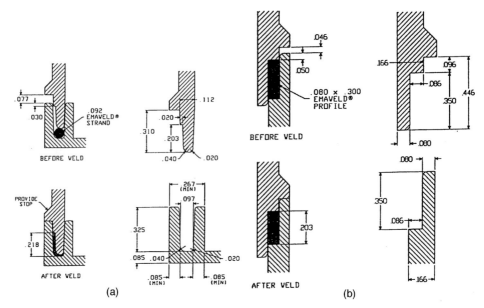

Figure 12-10 Representative joint designs: (a) internal shear joint and (b) external shear joint (dimensions in inches) (Diagrams courtesy of Emabond Systems)

hoses for vacuums and pools were welded to ABS fittings; a mandrel was used to control the inside diameter. Representative joint designs are illustrated in Fig. 12-10.

12.8   Encapsulation

Encapsulation is the method of assembling materials that cannot be welded. Figure 12-11 depicts a metal rivet trapped between two sheets of polyolefin sheeting.

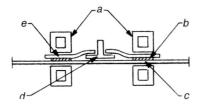

Figure 12-11 Encapsulation: *a*, movable coil; *b*, electromagnetic material; *c*, sheet stock; *d*, metal rivet; *e*, tab stock

12.9   Film and Sheeting

Film and sheet stock ranging in thickness from 0.05 to 5.08 mm (0.002–0.200 in.) can be electromagnetically welded either continuously or intermittently.

12.9.1   Intermittent Sealing

Intermittent sealing of film and sheet is illustrated in Fig. 12-12a. The electromagnetic material is placed between the two sheets and the movable coil is brought down to provide the pressure for sealing. The coil is then lifted, and the next pieces are placed in position for welding. The distortion that can accompany the competitive dielectric sealing method is not present, however the latter method does not incur the extra cost of the electromagnetic material.

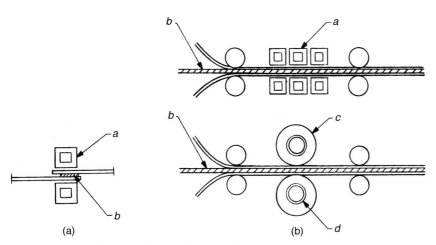

Figure 12-12   Film and sheet welding: (a) intermittent operation (*a*, movable coil; *b*, electromagnetic material) and (b) continuous sealing (*a*, work coil; *b*, electromagnetic material; *c*, electrically conductive pressure wheel; *d*, water-cooled inductor)

12.9.2   Continuous Sealing

Continuous sealing is shown in Fig. 12-12b. It is done with the electromagnetic material formed into a tape or bead of hot melt that is placed between the pieces of film or sheet to be sealed. The combination is run between rollers and then between the coils. It then is pressed between another set of rollers. The total surface area and bonding rate will be determined by the cross-sectional area of the electromagnetic material, however speeds of 30.48 m/min (100 ft/min) have been attained.

12.10   Inserting Metal into Plastic

The use of induction heating as a means of installing threaded metal inserts into thermoplastics was discussed in Chapter 8, "Fasteners and Inserts." There are, however, other applications for this concept. Figure 12-13 illustrates an application in which a steel shaft is to be inserted into a plastic bearing: the shaft is held fixed while

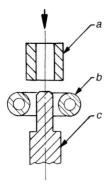

Figure 12-13   Inserting metal shafts into plastic bearings: (a) bearing; (b) coil; (c) metal shaft

the bearing is lowered onto it. Two or more shafts may be heated at once with a multiple-position coil.

Another example of this type of application is the assembly of plastic handles to steel flatware. A flat pancake coil positioned over the tang of the blade is used for this kind of application. Depending on the design, the tang may be heated before or after insertion.

12.11 Sources

Ameritherm Inc., 39 Main St., Scottsville, NY 14546, (800) 456-HEAT, (585) 889-9000, fax (585) 889-4030, www.ameritherm.com

Emabond Systems, Ashland Chemical, Inc., 49 Walnut St., Norwood, NJ 07648, (201) 767-7400, fax (201) 767-3608, www.ashchem.com

Fluxtrol Manufacturing, Inc., 1388 Atlantic Blvd., Auburn Hills, MI 48326, (800) 224-5522, (248) 393-2000, fax (248) 393-0277, www.fluxtrol.com

Hellerbond Technology Co., 3130 Dunlavin Glen Rd., P.O. Box 20156, Columbus, OH 43221-4416, (614) 527-0627, fax (6140) 527-0627

S.P. Industries, 210 Maple Pl., Keyport, NJ 07735, (201) 739-6994

13   Insert and Multipart Molding

13.1   Description

Insert molding is the practice of molding one or more additional parts directly into the host part thereby eliminating further assembly operations. While the greatest number of its applications have historically been to mold threaded metal inserts directly into parts, inserts of many other types (e.g., electrical connectors, metal stampings, decorative panels) have been molded in as well. These are becoming an increasing proportion of the insert molding applications as postmolding insertion of inserts has largely replaced the practice of molding threaded inserts directly into the part. This is particularly true of thermoplastics, where ultrasonic insertion has become the preferred method of insert installation at the time of writing.

Multipart molding can be used to mold together parts of two different materials, such as elastomeric-coated pen barrels, or parts of two colors, as in the case of the numbers in an appliance knob. The term "two-color molding" is often used because the majority of applications have been for the purpose of providing a second color that cannot be removed or for contoured parts that cannot be otherwise decorated. However, the number of parts need not be limited to two, nor need the reason for molding them together be simply to obtain a second color. The only real difference between insert and multipart molding is that in the case of the former, the part to be joined is placed, or inserted, into the mold, whereas in multipart molding it is molded in. These two techniques are not mutually competitive because the choice of which to use depends on the nature of the parts to be joined and the volume required.

13.2   Insert Molding

Insert molding encompasses several categories, each with special requirements. The largest of these application groups comprises round metal threaded inserts, electrical connectors, structural reinforcements, and decorative inserts. The balance consist of a wide assortment whose common thread is that the insert possesses certain elements not available in the host material.

13.2.1   Advantages of Insert Molding

1. *Highest strengths* Molded-in threaded inserts provide the highest levels of torsional and tensile stress resistance of any of the inserts, with the possible exception of helical coil inserts in some cases.

2. *Ability to join dissimilar materials* A wide variety of inserts made of dissimilar materials, typically metals, can be molded into the part. Multipart molding can assemble plastics that cannot be welded or solvent-joined.

3. *Permanence* The joining of the parts is permanent, and separation is possible only by destructive methods.

4. *Low technology requirement* This is a long-established technology that does not require special equipment or highly complex molds for moderate production levels.

5. *Low volume capability* Most inserts do not require very expensive special mold features that would prevent their use at the very lowest production levels.

6. *High volume capability* Very high volume production is available through the use of special equipment.

13.2.2   Disadvantages of Insert Molding

1. *Stress concentrations* Inserts or parts to be molded in must be carefully designed to minimize stress concentrations.

2. *Cost of disassembly for recycling* Since most inserts are metal, they are a contaminant in the recycling waste stream and must be removed by melting or by breaking them out prior to recycling. Even the plastic inserts tend to be sufficiently varied from the host material to make unseparated recycling unlikely.

3. *Cooling rates and thermal contraction* Differences in cooling rates and coefficients of linear thermal expansion between the metal insert and the plastic part create hoop stress within the assembly when the plastic part with the greater CLTE cools and contracts on its insert.

4. *Cost of inserts* Threaded inserts plus machine screws cost more than self-tapping screws without inserts.

5. *Process limitations* Inserts can be readily molded into parts made in the injection, structural foam, and transfer molding processes. They can be molded into parts made in the thermoforming, rotational molding, embediment, and compression molding processes in a more limited fashion. Compression-molded threaded inserts need to be tapped after molding because there is no way to close on the mold to prevent flash.

6. *Insert emplacement* It is difficult and time-consuming to hand-load inserts in volume, a pair of conditions that leads to unpredictably variable cycle times. Automatic feeding, loading, and indexing systems are costly to purchase and install.

7. *Part rejection* A missing or improperly located insert results in a reject part.

8. *Flash* Inserts, particularly threaded inserts, are prone to create flash at their edges, which must be removed as a postmolding operation.

9. *Mold damage* Inserts that have slipped out of position, or "floated," can cause serious mold damage that is expensive to repair and removes the mold from production. Small, light inserts are more prone to this problem than larger ones.

10. *Preheating* To expand large metal inserts and improve flow and cure-in, the inserts must be preheated.

11. *Aging effects* Material relaxation may lead to cracking around the insert on aging. Creep may lead to low insert retention strength, particularly at elevated temperatures, when the plastic expands at a greater rate than the metal insert.

12. *Hermetic seals* Inserts cannot be hermetically sealed in place without extraordinary measures. (The reader is referred to Section 13.2.7.)

13.2.3   Design with Threaded Inserts

Molded-in threaded inserts were the principal means of providing high strength threads in plastic for many years and became the largest single application for the insert molding technique, with the possible exception of electrical connectors. Threaded inserts are still molded in for some applications, particularly with thermosets; however the inherent disadvantages of this method have led molders to seek postmolding means of installing inserts. The method of choice for thermoplastics is ultrasonic installation, whereas the preferred system for thermosets seems to be the use of press-in inserts in combination with adhesives, provided they can supply enough strength. Nonetheless, applications exist for which there is no other viable option, and molded-in inserts provide the highest levels of torsional and tensile stress resistance of any of the inserts, with the possible exception of helical wire inserts for some applications. Thus, the benefits may justify the extra trouble. Molded-in inserts can be used with thermosets or thermoplastics, either with or without glass fibers and other fillers, however the use of such fillers results in an increase in weld line problems.

Inserts complicate the molding process, making it more difficult to mold good parts to tolerance. The melt must flow around the insert and form a good weld line. This can prove to be very difficult if there are many inserts, if there is insufficient space around them for good flow, or if the inserts are located far from the gate. The extra heat required to fill the bosses around the inserts can result in distortion or out-of-tolerance dimensions elsewhere on the part. Also, improperly emplaced inserts or excessive molding pressures can result in plastic flashing into the thread at the tops of the inserts, which then must have their threads chased before they can be used. Finally, a missing or misplaced insert results in a reject part. All reject parts, for whatever reason, must have their inserts removed before the material can be recycled. Some thermoplastic parts are salvageable or, if not, the resin may be reused. The cost of removing the inserts can raise the cost of recycling beyond the value of the reclaimed resin. When inserts are installed after molding, reject moldments can be recycled before the installation of the inserts. For thermoset parts, insert molding can be a more desirable option because ultrasonic/heat installation, the highest strength alternative, is not available. Also, thermoset parts are lost to scrap when they are rejected regardless of whether they have inserts installed.

Insert molding can lead to stress points at the interface between the plastic and the insert. The most common cause of failure of molded-in inserts is probably the delayed effect of material relaxation on radial compressive stress (hoop stress), which is present due to the shrinkage of the plastic boss around the cold insert. This effect is illustrated in Fig. 13-1. It is similar to the shrinkage that occurs around a core pin during molding,

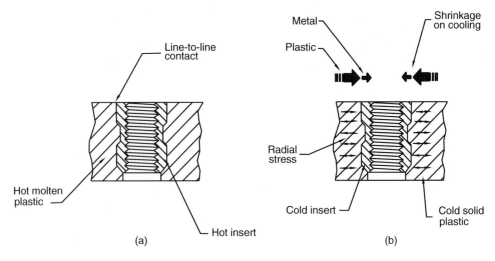

Figure 13-1   Creation of stress on cooling: (a) as molded and (b) radial stress created on cooling

although the core pin is removed from the hole and the insert is not. Heating the insert to 88 to 116 °C (190–240 °F) will help because the insert will then shrink along with the plastic as they cool. However, since the coefficient of linear thermal expansion is usually far greater for the plastic than for the metal insert, this technique only partially eliminates the problem. Also, the heating of the insert must be carefully controlled to maintain consistency. In general, it is best to avoid molded-in inserts with high shrinking plastics, particularly notch-sensitive materials.

The hoop stress may be estimated by the following equation:

$$\sigma = \epsilon \times E_f \tag{13-1}$$

where: σ = stress
  ϵ = strain = mold shrinkage
  E_f = flexural modulus

The strain can be approximated by the mold shrinkage, which increases as the wall thickness increases. If a range of shrinkages is given, the prudent course is to use the high value. The stress can be compared to the tensile strength of the material. However, additional stress will be generated when the plastic contracts more than the metal insert upon cooling of the part. Therefore, the difference in strain must be checked at the low point of the part's service and storage temperature range. The conditions at the height of its temperature range must be investigated as well because, while the plastic part will expand at a greater rate than the metal part, the tensile strength of the material will likely drop as the temperature rises. Material relaxation data should also be examined to determine the likelihood of product failure over its anticipated useful life. Creep can cause the insert to fall out. Bearing in mind the possibility of chemical attack, significant molding stresses, etc., a safety factor should be used when the design is evaluated. It may be that the application is not feasible in the selected material, for some polymers (typically, those having high shrinkage combined with notch sensitivity) are not well suited to inserts.

Sharp outside corners in the insert also become sharp inside corners in the plastic with their inherent high levels of molded-in stress. Blind inserts to be molded in should have no sharp corners and should be rounded on the bottom like the one in Fig. 13-2 to permit the plastic to flow easily around the insert. That caution applies to knurls as well, which should be kept to a minimum, and the edges should not be too sharp. A medium to coarse diamond knurl, with a depth of at least 0.25 mm (0.010 in.) for small inserts and 0.75 mm (0.030 in.) for large inserts, is used to prevent rotation while the recess prevents tensile pullout. The knurl should commence a minimum of 0.75 mm (0.030 in.) from the surface of the part. To avoid a sink mark, the distance (Y) between the bottom of the insert and the edge of the part should be no less than one-sixth the diameter (D) of the insert. Recommended wall thicknesses of bosses surrounding inserts and testing of inserts can be found in Chapter 8, "Fasteners and Inserts."

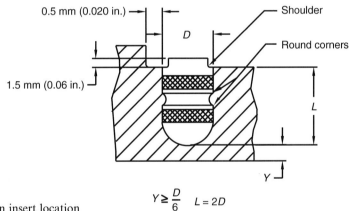

$$Y \geq \frac{D}{6} \qquad L = 2D$$

Figure 13-2   Molded-in insert location

If two inserts are molded opposite each other, the distance between them should be no less than 3.18 mm (0.125 in.). Inserts should absolutely not be located closer than 1.0 mm (0.040 in.) from an inside wall of the part. This is because the insert must be placed in a hole in the core, and there can be no less metal than 1.00 mm or the core will crack. Inserts to be molded into the part can have either closed ends (to prevent the molten plastic from running into the insert) or open ends (such as would be used for a through hole). In the latter case, there needs to be a shut off on both ends to prevent the molten plastic from running into the threads, which would necessitate retapping. Female through inserts should have a minimum length of 0.025 to 0.050 mm (0.001–0.002 in.) greater than the wall thickness of the plastic. For male inserts, that dimension is taken to its shoulder. Ideally, the length (L) of the insert should be twice the outside diameter (D).

Inserts can also be modified for special applications. The one in Fig. 13-3 is made with a vertical sleeve designed to receive a flat part like a metal contact strip. The sleeve is then spun over to lock the contact strip down. If this design were for an electrical contact, the insert should have a shoulder extending off the surface of the plastic part.

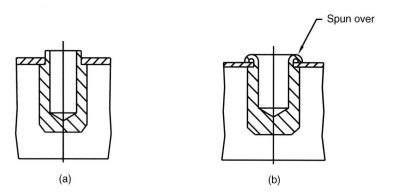

Figure 13-3   Special application inserts: (a) before assembly and (b) after assembly

13.2.4   Mold Considerations for Threaded Inserts

Mold damage is an ongoing concern with insert molding because the mold can close on an improperly placed insert, causing damage to the core and/or cavity. Low pressure closing sensors can be used to prevent this type of damage. Damage can be minimized by building the mold so the pins or holes that hold the inserts before molding are on the stationary half of the mold and cannot be dislodged by the closing of the clamp. For internally threaded through inserts, spring-loaded holding pins from the opposite side of the mold should be used to hold the inserts in place. This also provides support

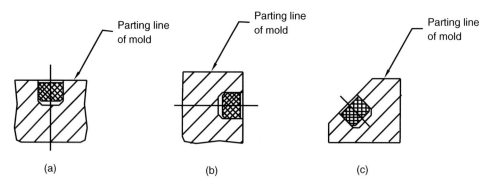

Figure 13-4   Insert location relative to parting line: (a) perpendicular to direction of draw (best design), (b) parallel to direction of draw (fair design), and (c) oblique to direction of draw (poor design, difficult to mold)

for the insert to help it withstand injection pressure and turbulence. However, it is recognized that this cannot be done for blind inserts and is not always feasible for through inserts when they are located on walls that are perpendicular or oblique to the direction of draw, as in Fig. 13-4. Designers should avoid placing molded-in inserts in these positions. Blind inserts can have detents on their pins to help keep them in place. When inserts must be located on the movable side of the mold, there should be an ejector pin underneath each one to counteract the tendency of inserts to hang on their pins.

Gates should be located to ensure that they do not bring the first rush of molten polymer directly to bear on the inserts, as this can break a fragile insert or dislocate it. However, they should be located near enough to the inserts for them to receive the last, and hottest, material to enter the mold, permitting a solid weld line to be formed. (The reader is referred to the section on boss design in Chapter 8, "Fasteners and Inserts.") Ideally, gates should be so located that the insert or inserts are in relatively dead spots in the cavity flow path. Finally, the bottom of the insert should not coincide exactly with the base of its boss.

Some inserts must be placed in positions in the mold that are not feasible for the inserts discussed thus far. In this case, the inserts can be placed in a removable mold segment, which is ejected with the part on each cycle. Since the segment is ejected from the mold with the part, it is usually located on the ejector side of the mold. Multiple mold segments are made to avoid delaying the molding cycle while a given segment is being removed from the molded part. These segments must be keyed to ensure that they cannot be improperly placed in the mold.

The cycle will be extended according to the number of inserts to be placed in the mold. Inserts are typically loaded by hand. This prevents automatic cycling of the molding machine because the operator must be clear of the molding machine before the next cycle can proceed. The more inserts there are to load, the more variable the cycle is likely to be. This operator is pressured to load all the inserts quickly and correctly, since an incorrectly loaded or omitted insert will lead to a reworked part at a minimum, and possibly a reject part or mold damage. Such errors result in inconsistent quality. Manual or automatic loading fixtures can be used to increase the speed of loading a larger number of inserts; however, it may be more cost-effective to use two operators for some molds, since machine time may be the most costly element, the pair might be more capable of keeping up with the molding cycle. Injection molding machines with vertical clamping are designed for insert molding because they permit the inserts to be loaded to a stationary mold positioned in the horizontal plane. Gravity then helps to hold the parts in position while the mold is closing.

For higher volumes, this problem can be alleviated through the use of a vertical shuttle press (Fig. 13-5), which opens vertically and has two lower mold sections. The inserts are loaded into the one that is open (*station* 1) while the other side is molding (*station* 2). The lower sections then shift positions, and the finished parts are ejected from their side and new inserts installed while the opposite side is molding. The cycle is then repeated. This can also be done with a rotary table in place of the shuttle base.

Molds for rotational molding typically have thin walls that do not lend themselves to the techniques described thus far in this section. Inserts for these molds are manually

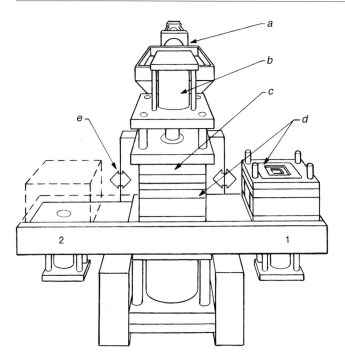

Figure 13-5   Vertical shuttle press for insert molding: (a) injection molding cylinder; (b) clamping cylinder; (c) upper mold half; (d) lower mold halves; (e) lower mold half-travel

screwed to screws or nuts mounted through the mold. This is slow, but the cycle for the process is long enough to allow time to unscrew the inserts manually after molding.

13.2.5   Custom-Designed Inserts

Many inserts of custom design are molded in as well. Metal inserts are often used to provide electrical conductivity, but they have also been used for a variety of custom fastening and other specialty applications like tool blades and steering wheel reinforcements. Many of these applications use inserts that are very similar to standard threaded inserts. The portion that is molded into the plastic can be cold-forged or it can be a screw machine part, the only real difference lying in the functional end that protrudes into the part. Figure 13-6 illustrates several examples of this type of insert.

Some inserts are made of rod having an end that has been modified (by notching, bending, swaging, grooving, etc.) to provide an anchor that protrudes into the moldment. Such inserts are limited to the transfer or injection molding processes and materials that flow readily at melt temperatures. Several of these inserts are illustrated in Fig. 13-7a.

Cylindrical inserts can also be drawn, like those in Fig. 13-7b. Provision for adequate anchorage in the molded part must be provided. For the inserts to seal off and not

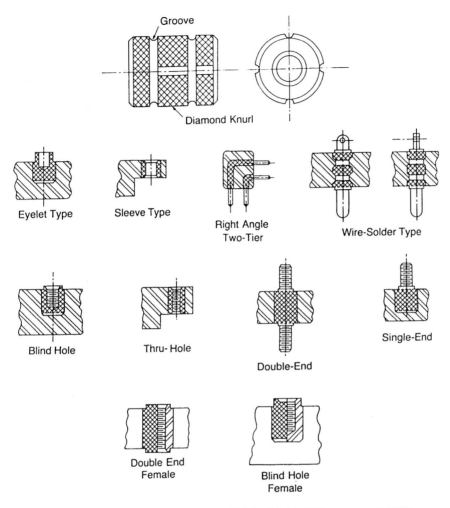

Figure 13-6   Types of insert (From R.E. Wright, Molded Thermosets © 1991)

permit the polymer to flow into the pin, such details should be created by hardened mold pins that form the metal with the tremendous clamping pressure available in the mold when it closes. With transfer molding, very low molding pressures using diallyl phthalate, epoxy, and silicone resins have been used to mold very delicate electronic parts of this type.

When tubular inserts are to be molded to a part, special provisions must be made for anchorage. The easiest solution is to use a perforated metal. However, that is often not acceptable. Internally mounted inserts can be flared like the one in Fig. 13-8a. However, externally located inserts require an inward bead like the one in Fig. 13-8b.

Many inserts are stamped out of metal such as those in Fig. 13-9a. The metal must be selected carefully because a material that is too hard may damage the mold, whereas

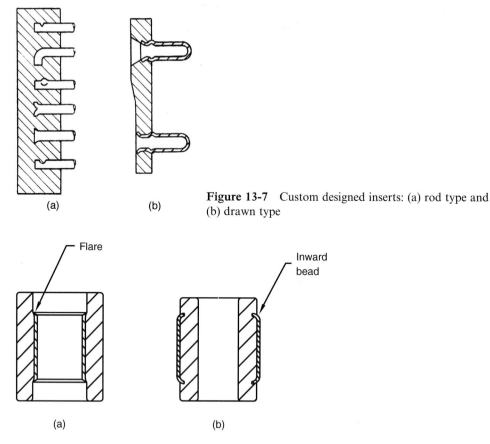

(a)   (b)   **Figure 13-7** Custom designed inserts: (a) rod type and (b) drawn type

(a)   (b)

Figure 13-8 Custom designed tubular inserts: (a) internal – flared and (b) external – inward bead

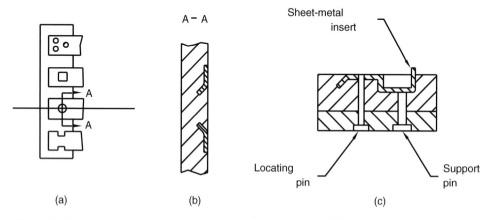

(a)   (b)   (c)

Figure 13-9 Custom-designed stamped inserts: (a) examples, (b) flange improves strength of joining, and (c) stamped inserts with locating pins

one that is too soft can itself be damaged by molding pressures. It is difficult in stamping to hold the tight tolerances required of molded-in inserts. Thus such inserts are best positioned with locating pins, which fit holes in the inserts and are supported with support pins, as shown in Fig. 13-9c, because it is easiest to control dimension on holes. This type of insert should be located on or below the parting line and can be anchored by bent tabs as also shown in Fig. 13-9c. If the function of the part would be impaired by such tabs, lugs can be spot-welded to the insert for this purpose, but at an increased cost.

Care must be taken to ensure that adequate plastic is retained to secure the insert in the host part. The insert illustrated in Fig. 13-10 has a flange for retention. The flare improves this retention, but one must take care to leave enough material between the insert and the edge of the part to avoid breakout.

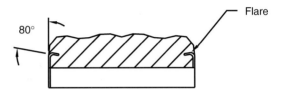

Figure 13-10   Insert security

Large inserts have special problems because of the shrinkage of the plastic and the coefficient of linear thermal expansion for most plastics is so much greater than it is for metals. For example, if the insert is not sturdy enough to be able to withstand the pressures of the shrinkage, the part will warp substantially. If the metal part is on one side, the plastic side will shrink enough to cause its side to be concave. Cooling fixtures will solve the problem in some, but not all, cases. Furthermore, the problem will be exacerbated when the temperature drops and the plastic contracts further. However, at elevated temperatures, the plastic will tend to expand more than the metal, causing the plastic side to become convex if it was straight to begin with. In some cases, the problem can be solved by using a preheated insert.

The most difficult inserts to deal with are irregularly shaped ones. Depending on the contour, the distortion of these parts may be very difficult to predict. This problem can often be resolved by breaking the insert into two or more parts. However, that approach increases the time of loading inserts into the mold.

Improvement in strength, rigidity, safety, and dimensional accuracy can be achieved through the use of inserts for reinforcement. An example of that is the handle in Fig. 13-11. Automobile steering wheels are another such application. If one maintains

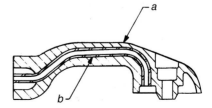

Figure 13-11   Reinforcing insert: (a), molded part; (b), metal reinforcement

a minimum wall thickness of plastic, the rigidity of the reinforcement will overcome the tendency toward warpage of the greater shrinking polymer. Reinforcements are not necessarily metal, as engineering plastics may be used in their place. Besides applications in which a more expensive engineering polymer is used as an insert to be covered by a less costly resin, the outer material may possess other, nonstructural, qualities. An example of this might be a nonslip, rubberlike gripping surface molded over a rigid pen barrel.

13.2.6   Outserts: Inserts Larger than the Moldment

In a number of applications the plastic part to be molded is actually smaller than the part that has been inserted into the mold. An example of this type of application is shown in Fig. 13-12, where several bearings and mounting blocks are molded onto a metal clock frame. These are sometimes referred to as "outserts." The metal is usually aluminum or steel, and thicknesses below 1.0 mm (0.040 in.) can be flimsy and difficult to control in the mold. Normally, any punching or forming operations to be performed on the metal part are done before the molding operation. Using a rigid metal piece as the host part in this fashion is a way to overcome the distortion problems associated with large insert moldments or high-shrinking materials. In addition, closer tolerances can be held with this method. Thermoplastic rejects can be melted off the metal part; thermoset rejects will usually be unsalvageable.

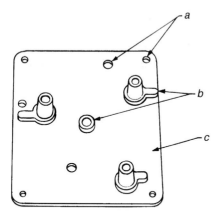

Figure 13-12   Outserts: (a) stamped holes; (b) molded boss and bushing; (c) metal stamping

This type of moldment can be achieved by creating a runner system on one side of the metal part. The part in Fig. 13-13a has a straight runner. When a thin gauge metal part is used, shrinkage of the runners can cause distortion when it overcomes the strength of that piece, as shown in Fig. 13-13b. This condition can be improved by use of curved runners like the one illustrated in Fig. 13-13c. A better method, which is more expensive to tool, calls for the use of a three-plate mold like the one in Fig. 13-13d. This mold permits each plastic piece to be individually gated, thus

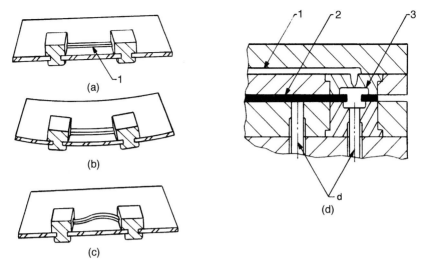

Figure 13-13   Methods of eliminating runner distortion: (a) Part with straight runner (1, runner), (b) distortion resulting from runner shrinking on weak metal part, (c) curved runner reduces effect of shrinkage, and (d) three-plate mold (1, runner; 2, insert; 3, molded part; 4, ejector)

eliminating the problem of the runners. The same principal can be applied with insulated or hot runners.

13.2.7 Hermetic Seals

Because of the postmolding shrinkage of the plastic and the large differentials in coefficient of linear thermal expansion, molded-in inserts cannot provide a hermetic seal without special provisions to accommodate the relative movement between the two materials. Figure 13-14 illustrates a custom-made insert with an O-ring that compresses

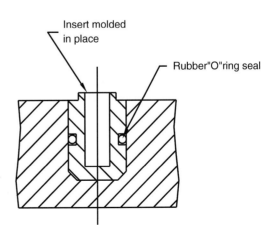

Figure 13-14   Insert for hermetic seal

as the polymer shrinks. When it expands with heat, there is still sufficient compression to create a seal. Another approach is to coat the end of the insert in a rubbery material such as neoprene or polyvinyl chloride–acetate. It is then insert-molded in the usual fashion.

13.2.8   Preparation of Inserts

Mold release agents, dirt, grease, oil, and other contaminants must be removed from inserts before they are placed in the mold for molding. In addition, any loose chips or other debris created during the manufacture of the insert must be removed: such pieces can provide sharp corners, which create stress concentrations, or they may break loose during molding and disturb the finish of the part. In electrical applications they can cause a potentially hazardous short. Finally, the plastic may be attacked by the chemical contaminant.

Chips and oil can be removed by a variety of mechanical techniques including tumbling, blasting (with sand or shot), hand-wiping, and solvent cleaning. Reportedly, however, the best method for metals that are unaffected by alkali is a well-stirred alkali bath followed by a hot water rinse. Metals that are so affected, like aluminum, are best treated with solvents.

When appearance or functional requirements call for a chemically clean insert, silver tarnish and iron rust can be removed from inserts by an acid dip. Nitric acid can be used to remove silver tarnish and to clean brass and bronze. Silver tarnish can also be removed with a diluted solution of one of the cyanides, and a mixture of nitric acid and sulfuric acid can also be used to clean brass and bronze. Trisodium phosphate will remove iron rust.

13.2.9   Decorative Inserts

Nonmetallics have also been molded as inserts, often as decorative items. Examples are wooden inserts in automobile gearshift knobs and doorknobs and decorative trim for automobile armrests, dashboards and center consoles, which have been made this way for many years. The insert is usually thermoformed from a thin plastic sheet around 0.50 mm (0.020 in.) thick with a wood-grain pattern printed on one side of it, as shown in Fig. 13-15a. The part must be stretched over the mold; therefore a minimum radius of 1.58 mm (0.062 in.) with a 2° draft is required, and a maximum of 250% deformation or stretching can be permitted. Then the part is cut to size and placed in the mold, as in Fig. 13-15b. The backing material is then injected to the surface behind the film in Fig. 13-15c. The materials normally used for this process are specialty grades of ABS, polyvinyl chloride, and polystyrene.

Another method of achieving the same objective, illustrated in Fig. 13-16, uses a transparent film 0.08 to 0.25 mm (0.003–0.010 in.) thick, which has had the design printed on it and is made of the same material as the backing. As shown in Fig. 13-16a, the die-cut piece is placed in the mold in much the same fashion as the

Figure 13-15 Decorative insert molding: thermoformed insert. (a) Thermoforming the insert: *a*, sheet heated and lowered to die; *b*, vacuum draw; *c*, formed piece cut on this line to create insert; *d*, softened sheet takes form of die; *e*, thermoforming die. (b) Thermoformed insert emplacement: *a*, insert placed in mold. (c) Molded-in thermoformed insert: *a*, moldment; *b*, insert placed in mold with part molded behind it

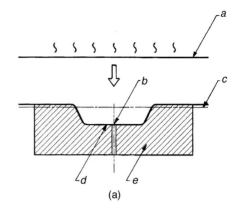

(a)

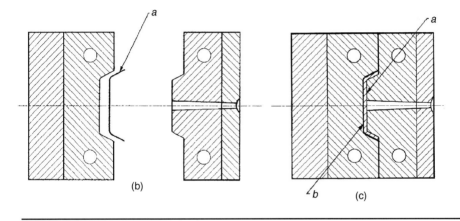

(b)   (c)

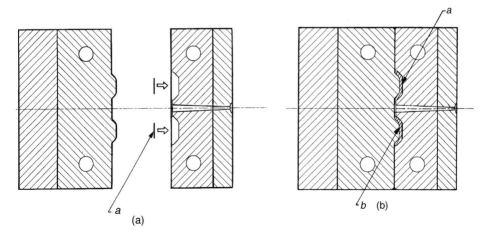

(a)   (b)

Figure 13-16 Decorative insert molding: die cut insert. (a) Die-cut insert placed in mold: *a*, die-cut insert placed in cavity. (b) Molded-in die-cut insert: *a*, insert; *b*, moldment

insert in the example of Fig. 13-15. It must be at least 12.7 mm (0.500 in.) away from the gate and cannot be angled more than 45° from the plane of the parting line. The resin is then injected into the cavity to create the body of the part, as in Fig. 13-16b. A high charge of static electricity is imparted to the insert, which causes it to adhere flat to the surface of the mold. Key caps are one application for this technique.

This method is modified somewhat for compression molding of melamine as used to make products like dinnerware. In this case, the part is first compression-molded to the point of staying in the cavity, and the material is cured enough to be free of blisters. The insert is then placed on the part and the mold is closed to complete the cycle, molding it right into the part.

Another method uses a carrier film or foil, which is on a roll mounted to the mold as in Fig. 13-17. In Fig. 13-17a the mold is shown open so the film can advance to a position where the decoration is still in place. The mold closes on the foil and the part is filled in Fig. 13-17b. When the mold opens, the decoration is molded into the part, which separates from the carrier film. The part is then ejected from the mold and the film moves to the next registration for the following cycle.

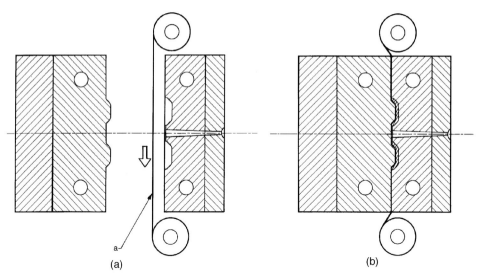

(a)   (b)

Figure 13-17 Decorative insert molding: film transfer. (a) mold open: film advances to next position. (b) mold closed, film molded to part

13.3 Multi-Part Molding

13.3.1 Description

In insert molding, a part or parts manufactured elsewhere are molded into a plastic part. When such parts can be made of an injection-moldable material, they can be molded as the first stage of a multistage process. The concept is simple. First one part is molded and then the next part is molded over or under it, etc. While there are theoretically no limits to the number of stages that can be used, the maximum in commercial use at the time of writing is three. Although the process was developed to create key caps that would withstand years of abrasion without showing signs of wear, it has been used successfully for such applications as multicolor lenses for automotive use and elastomeric grips on rigid pen barrels or tool handles.

By clever mold design and/or by utilizing dissimilar polymers that do not bond to each other, parts can be molded together which can move relative to each other. Examples of in-mold-assembled applications are an HVAC vent register with movable vanes, an air bag connector with a moving hinge, a small crankshaft with a gear that moves, and a toy monkey whose arms and legs rotate.

13.3.2   Advantages Particular to Multipart Molding

1. *Permanence* The parts are permanently joined and can be separated only by destructive methods.
2. *Automatic operation* Joins parts without additional handling.
3. *Ability to join dissimilar materials* Parts made of a wide variety of dissimilar materials can be molded into the part. Multipart molding can assemble plastics that cannot be welded or solvent joined.
4. *Design freedom* Permits the joining of parts that would be difficult or impossible to join by other means.
5. *High volume capability* Very high volume production is available through the use of special equipment.

13.3.3   Disadvantages Particular to Multi-part Molding

1. *Stress concentrations* Parts to be molded in must be carefully designed to minimize stress concentrations.
2. *Cost of disassembly for recycling* Mixed plastic materials can be a contaminant in the recycling waste stream and must be removed by melting or breaking them out prior to recycling. When differences in color are not significant and like materials are used, however, molded-in parts may avoid this disadvantage.
3. *Cooling rates and thermal contraction* Differences between the joined parts in cooling rates and coefficients of linear thermal expansion create hoop stress within the assembly when the plastic part with the greater coefficient cools and contracts on its insert or mating part.
4. *Tooling cost* This is an investment-intensive process, with tooling costs beginning at a level double that of a simple mold.
5. *Equipment cost* As with tooling, the equipment cost is considerably greater for this process. In addition, most custom molders do not possess such equipment.
6. *Process limitation* This technique is limited to the injection molding process.

13.3.4   The Process

There are three principal methods of achieving multipart molding at the time of writing, and there are new ones under development. The three are retractable core, multiple cavity, and rotating core, although two of them are similar in process and differ only in the equipment used. For any of these techniques, the first material is molded in the conventional manner. Such a part is illustrated in Fig. 13-18a. For the retractable core technique, the core shuts off against the cavity wherever the second material is to show through. This core is then retracted sufficiently to permit material from the second injection cylinder to fill the cavity, as illustrated in Fig. 13-18b.

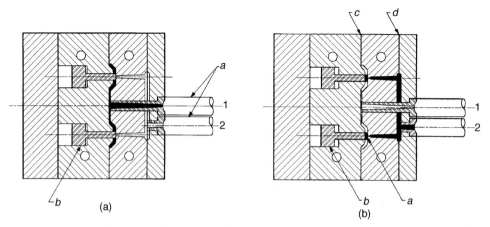

Figure 13-18   Multipart molding – retracting core method. (a) first resin injection: *a*, molding cylinders; *b*, core in forward position. (b) second resin injection: *a*, second resin; *b*, core retracted; *c*, main parting line; *d*, runner parting line

For the other two methods, the second material is molded over or under the first material. Figure 13-19 represents an example of the multiple cavity technique where the second shot is molded over the first. Figure 13-19a shows that part after it has been molded; it then remains on its core and is repositioned to another cavity and its injection molding cylinder. The cavity closes on the first molded part so that any portion of that part which is to show through shuts off against the new cavity. The second material is then injected in the voided area, as illustrated in Fig. 13-19b.

For high production volumes, this process can use a rotary table with two injection cylinders at different positions around the table, with the clamp in the vertical position.

The rotating core method requires a special molding machine with two injection cylinders. The mold is built with a rotating core as shown in Fig. 13-20. The first part is molded in the lower cavity. It is then rotated to the upper position which has the second cavity, which is then filled. The mold then opens and the part is ejected.

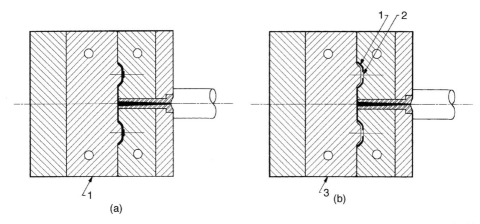

Figure 13-19   Multipart molding: multiple-cavity method. (a) First resin moldment: 1, core half with molded part in place. (b) Second resin moldment: 1, second resin (solid); 2, first resin visible from outside of part at shut-off; 3, core half of mold with first part in place

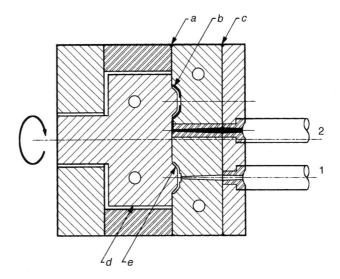

Figure 13-20   Multipart molding: rotating core method: *a*, main parting line; *b*, second resin; *c*, sprue removal parting line (from first resin); *d*, rotating core; *e*, first resin

13.3.5   Materials

Multipart molding is a rapidly developing area, and a great deal of work is in progress at the time of writing. Table 13-1, which compares some plastic materials with respect to bond strength, represents the results of work conducted to date.

Table 13-1 Bonding Strength of Dissimilar Plastic Materials: G, Good Bond; P, Poor Bond; N, No Bond

First Shot	Second Shot															
	Thermoplastics									Hard–Soft Combinations						
										TPR			Elastomers			
	PA6/6	PBT	PC	PMMA	POM	PP	PS	PSU	PVC-W	SEBS	TPU	PP/EPDM	EPDM	NR	SBR	LSR
Thermoplastics																
ABS		G	G	G		P	P		N	G	G					
PA6					N	P	P			G	G	G		N	N	
PA6/6	G	G	P		N	P	P			G	G	G		N	N	
PA6/12													G			
PBT	G	G	G				N				G		G	N	N	P
PC	P	G	G				N	G			G					
PMMA				G		P	N		G							
POM	N				G											
PP	P			P	P		P		P		N	G				
MPPE													G	G	G	
SAN		G	G	G			P		G		G					
TPR: PP/EPDM												G				
Thermoset: BMC																G
Elastomers																
EPDM		G											G			
NR														G		
SBR															G	
LSR																G

Source: Engel machinery Inc.

13.4 Sources

Battenfeld of America, Inc., 31 James P. Murphy Hwy., West Warwick, RI 02893, (800) 248-6015, (401) 823-0700, fax (401) 823-5641 (molding machines), www.battenfeld.com

Caco-Pacific Corp., 813 N. Cummings Rd., Covina, CA 91724-2506, (626) 331-3361, fax (626) 966-4219 (molds), www.cacopacific.com

Engel Machinery Inc., 3740 Board Rd., Rd#5, York, PA 17402, (717) 764-6818, fax (717) 764-0314 (molding machines), www.engelmachinery.com

Ewikon Hot Runner Systems of America, 1051 E. Main St., Unit 101, East Dundee, IL 60118, (800) 980-4815, (847) 844-9351, fax (847) 844-9352, www.ewikon.com

Ferromatik Milacron, 3041 Disney Ave., Oakley OH 45209, (513) 458-8290, fax (513) 841-8593 (molding machines), www.milacron.com

Foboha, 16 Cypress Lane, Hamburg, NJ 07419, (973) 823-9426, fax (973) 823-9427 (molds), www.foboha.de

Hettinga Equipment Inc., 2123 N.W. 111th St., Des Moines, IA 50325, (515) 270-6900, fax (515) 270-1333, www.hettingatechnology.com

Husky Injection Molding Systems Ltd., 500 Queen St. S., Bolton, Ontario, Canada L7E 5S5, (905) 951-5000, fax (905) 951-5384 (molding machines), www.husky.ca

Krauss-Maffei Corp., Injection Molding Div., 7095 Industrial Rd., P.O. Box 6270, Florence, KY 41042-6270, (859) 283-4311, fax (859) 283-0311 (molding machines), www.krauss-muffei.com

Prospect Mold & Die Co., 1100 Main St., Cuyahoga Falls, OH 44221, (330) 929-3311, fax (330) 920-1338 (molds), www.prospectmold.com

Tradesco Mold, Ltd., 1 Paget Rd., Brampton, Ontario, Canada, L6T 5S2, (416) 749-1698, fax (416) 749-2795 (molds), www.tradesco.com

Van Dorn Demag Corp., 11792 Alameda Dr., Strongsville, OH 44149, (440) 876-8960, fax (440) 876-6423, www.vandorndemag.com

14 Press Fits/Force Fits/
Interference Fits/Shrink Fits

14.1 Advantages and Disadvantages

14.1.1 Advantages

Press fits also known as force fits, interference fits, and shrink fits, have long been used to assemble metal parts through the use of interference. The term "shrink fits" refers to plastics applications in which heat is used to expand one of the parts so that it fits readily over the mating part and shrinks onto it as it cools. Sometimes this can be done immediately after molding, which eliminates the cost of reheating the part. Press fits offer several advantages for plastics.

1. *No additional materials* Press fits use no additional materials such as fasteners, inserts, electromagnetic preforms, adhesives, or solvents. Hence, they are inherently lower in cost than methods that do. They also avoid the expensive tooling often associated with the undercuts required for many snap fits.

2. *Ease of assembly* Press fits can be performed without tools in many applications and are well suited to automated assembly. Very high production rates are possible.

3. *Dissimilar materials* Metals can be joined to plastics. Widely differing plastics, as well as similar polymers, can be joined to each other with press fits.

4. *Permanence* Properly executed, press fits can be used for pressure-tight permanent assemblies. Plastic gears and pulleys are often assembled to metal shafts in this manner.

5. *Reopenability* For many applications, press fits can be designed to be reopenable. Pen caps are an example of this type of press fit.

14.1.2 Disadvantages

1. *Shape limitations* Press fits are generally round; however there are some square and rectangular applications involving softer materials and rounded corners. Other shapes have been assembled with several strategically located round studs and bosses.

2. *Process limitations* While tolerances for press fits involving plastics are much looser than those for metal interference fits, they are generally too tight for structural assemblies with many plastic processes. Those that can readily provide parts finished to structural press-fit tolerances are machining plus injection, compression, and transfer molding. The remaining processes are recommended for light-duty press fits only,

and rotational and blow molding are limited to one-cavity volumes and require special measures such as molding both parts as one and cutting them apart.

3. *Hoop stress failure* Excessive interference in combination with high levels of molded-in stress or weak weld lines can result in failure immediately upon assembly. Annealing may be necessary for some applications.

4. *Stress relaxation; creep and cold flow* At a fixed strain, the stresses in a visco-elastic material will relax with time, particularly at elevated temperatures. Excessive strain can result in cracking or crazing of the plastic to relieve the stress. If failure does not occur, the joint can be loosened by these effects as it undergoes stress relaxation until equilibrium is reached.

5. *Environmental limitations* Chemical and ultraviolet light exposure can result in failure due to stress cracking.

6. *Material limitations* Since press fits require the displacement of material through interference fitments, they work best with more ductile plastics.

7. *Thermal expansion* Significant differences in coefficient of linear thermal expansion between the two materials can result in stress cracking or loosening of the joint at the temperature extremes.

8. *Moisture absorption* Some plastics exhibit expansion due to moisture absorption sufficient to result in loosening of the joint.

14.2   Press Fit Engineering

14.2.1   Engineering Notation

The traditional press fit is the one illustrated in Fig. 14-1. In this case, a rigid shaft, either plastic or metal, is forced into a more ductile hub or boss in an interference condition. (The reverse is possible but is less commonly done.) The boss then deforms to the shape of the shaft and is held in place by the pressure of the fitment.

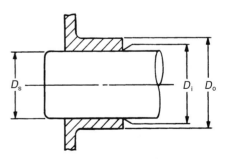

Figure 14-1   Basic press fitment

The circumferential tensile stress exerted on the wall of a cylinder as a result of outward pressure from within is referred to as hoop stress. The interference of the boss with the shaft will create such a stress, and that will be the principal stress on the joint; therefore it must be determined before the proper material can be selected and the appropriate dimensions established.

The following notation will be used for the engineering equations.

C_f = coefficient of Friction
D_I = inside boss diameter (in.)
D_o = outside boss diameter (in.)
D_s = shaft diameter (in.)
D_{si} = inside diameter of sleeve (in.)
ϵ = strain (in.)
E_b = modulus of elasticity for the boss (psi)
E_s = modulus of elasticity (Young's modulus) for the shaft (psi)
F = force of insertion or removal (lb$_f$)
I = diametral Interference $(D_s - D_I)$ (in.)
I_m = maximum allowable interference (in.)
P = pressure at interface between shaft and boss (psi)
R_o = outside boss radius (in.)
R_s = shaft radius (in.)
L = length of press fit surfaces (in.)
σ = design stress (psi) or maximum allowable stress of boss (psi) = yield strength of the material/safety factor
ν_s = Poisson's ratio for the shaft material
ν_b = Poisson's ratio for the boss material

14.2.2   Geometric Factor

There is a geometric factor (G) that is used to reduce the complexity of the computations:

$$G = \frac{1 + (D_s/D_o)^2}{1 - (D_s/D_o)^2} \tag{14-1}$$

14.2.3   Changes Due to Temperature Variations

When determining the diameters to use in these computations, one must also consider the effects of temperature. If both parts are to be made of the same material, they will expand or contract together. If they are different, there will be a difference in the rate of thermal expansion or contraction, and thus a difference in the stress on the joint.

For example, if the shaft were made of metal and the boss of plastic, the joint could weaken at elevated temperatures to the point at which the boss slipped on the shaft. Conversely, at lower temperatures, the boss could contract to the point of failure due to the increase in stress.

Typically, the dimensions of a part are taken to be those at room temperature. Therefore, that is used as the basis. The formula to determine the new diameter at an elevated temperature is:

$$D_2 = D_1 + (t_2 - t_1)C_t D_1 \qquad (14\text{-}2)$$

where D_1 = diameter at room temperature (in.)
$\quad\quad D_2$ = final diameter (in.)
$\quad\quad t_1$ = initial temperature (°F)
$\quad\quad t_2$ = final temperature (°F)
$\quad\quad C_t$ = coefficient of linear thermal expansions (in./in./°F)

If the materials are dissimilar, the new diameters must be found for both. Even a product intended for use at room temperature will encounter temperature extremes during shipment and in storage. Note that equation 14-2 can be used to determine the effect of heating the boss or cooling the shaft to facilitate assembly as well.

One must bear in mind that the strength properties of a material will also vary with temperature. Between 73 and 160 °F, a given grade of acetal may drop in tensile strength from 60.67 MPa to 34.47 MPa (8800 psi to 5000 psi). The design stress may need to be recalculated for high temperature applications according to data available from the resin manufacturer. (Tensile strength increases at lower temperatures.)

If extreme accuracy is required, it should be further noted that the coefficient of linear thermal expansion for a given material can vary according to the direction of flow of the material in filling the part and does itself vary with the change in temperature. For more extreme applications requiring more precise calibrations, the resin manufacturer's laboratory should be consulted.

14.2.4 Hoop Stress

Product performance is defined by the design stress (σ), the highest stress developed in the application the joint is designed to withstand. The equation can be altered, however, to determine the maximum interference (I) the design can endure; maximum interference is the overriding concern whose requirements must be met regardless of all other considerations. If product requirements would be compromised beyond the limits of acceptability, another means of assembly must be selected. The design stress is determined by different equations depending on the rigidity of the shaft.

14.2.4.1 Metal Shaft in Plastic Boss

For a shaft that is far stiffer than the boss (e.g., metal), the following equation applies:

$$\sigma = E_b \frac{I}{D_s} \times \frac{G}{G + \nu_b} \qquad (14\text{-}3)$$

If the maximum allowable interference is required, the equation is altered in the following manner:

$$I_m = D_s \frac{\sigma}{E_b} \times \frac{G + \nu_b}{G} \qquad (14\text{-}4)$$

14.2.4.2 Shaft and Boss of Same Material

For a shaft made of the same material as the boss, the equation is as follows:

$$\sigma = E_b \frac{I}{D_s} \times \frac{G}{G+1} \qquad (14\text{-}5)$$

For the maximum allowable interference in this case, the equation is altered to the following:

$$I_m = D_s \frac{\sigma}{E_b} \times \frac{G+1}{G} \qquad (14\text{-}6)$$

14.2.4.3 Shaft and Boss of Different Plastics

For a shaft and boss made of two different materials, the equation is:

$$\sigma = \frac{GI}{D_s} \left(\frac{G + \nu_b}{E_b} + \frac{1 - \nu_s}{E_s} \right)^{-1} \qquad (14\text{-}7)$$

The equation becomes the following if the maximum interference is required:

$$I_m = \sigma \frac{D_s}{G} \left(\frac{G + \nu_b}{E_b} + \frac{1 - \nu_s}{E_s} \right) \qquad (14\text{-}8)$$

14.2.4.4 Quick Methods

A quick method for determining the proper interferences is to simply refer to the interference versus shaft/boss relationship graph such as the one illustrated in Fig. 14-2. These manufacturers' graphs, which are available only for some of the more common press-fit materials, are resin specific and may not cover the intended

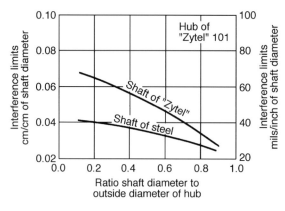

Based on yield point and elastic modulus at room
temperature and average moisture conditions

Figure 14-2   Theoretical interference limits for press fitting (Graph courtesy of DuPont Engineering Polymers)

temperature or humidity range. Therefore this method depends on their availability for the materials under consideration.

If there is no such graph available, there is a relatively inaccurate equation that can be performed quickly and can serve as a screening method to determine whether the parameters of the proposed design are sufficiently close to those desired in the final design to warrant further work. It presumes that the shaft does not distort when pressed into the hub and is most accurate for applications when the hub is metal. In that case, the hoop stress would be:

$$\sigma = \epsilon E_b = \frac{I}{D_i} E_b \qquad (14\text{-}9)$$

The secant modulus, which provides the initial stress, should be used for high strains. The creep or apparent modulus should be used for long-term stresses, where the value that produces creep rupture in the material must not be exceeded by the maximum stress or strain.

14.2.5   Assembly and Disassembly Forces

Press fits require force to accomplish assembly. The level of force required to insert the male part can be determined by the following equation:

$$F = \frac{\pi \sigma C_f D_s L}{G} \qquad (14\text{-}10)$$

This determines the joint strength. Therefore the force required to remove the shaft is the same, but only if it is applied shortly after insertion and before the effects of creep and cold flow can take place.

Equation 14-11 can be used to determine the torsional holding capacity of joint T by multiplying the force F by the shaft radius R_s.

$$T = FR_s \qquad (14\text{-}11)$$

This equation suffers from the fact that an accurate figure for the coefficient of friction is rarely available from the manufacturer for any combination of materials other than the given resin against itself or against a metal. Thus only crude calculations can be done unless that coefficient can be determined by experimenting with the two pertinent materials. Also the coefficient of friction is affected by various factors such as moisture, temperature, stress level, and surface contaminations (grease, mold release etc.).

Assembly of press fits can be eased by enlarging the female part with heat just prior to assembly. In production, this can be accomplished by running a conveyor through a heated tunnel. The reversed condition of cooling the shaft to reduce its size prior to emplacement is a possible alternative. The latter is the more desirable, when feasible, because it is less likely to produce distortion or degradation, which can occur if the boss contains high levels of molded-in stress or is heated to too high a temperature.

14.2.6 Dimensional Changes Due to Assembly

When two parts are press-fitted together, their diameters will change. The majority of the effect will be felt by the softer of the two parts. Thus, for a plastic boss fitted over a metal shaft, the outside diameter of the boss will increase according to the following equation:

$$D_{o1} - D_{o2} = I \frac{2D_s D_o}{D_s^2(1 + \nu_o) + D_o^2(1 - \nu_o)} \qquad (14\text{-}12)$$

In the reverse circumstance, the amount of decrease of the inside diameter of a plastic sleeve fitted into a rigid housing is given by the following equation:

$$D_{si1} - D_{si2} = I \frac{2D_s D_o}{D_i^2(1 - \nu_b) + D_o^2(1 + \nu_b)} \qquad (14\text{-}13)$$

14.2.7 Relationships

From the practical aspect of assembly, the relationships between maximum allowable stress, maximum allowable interference, and the ratio between the shaft diameter and the outside diameter of the boss are of primary interest. It works something like this:

1. If the ratio between the diameters is held constant, the amount of interference permissible will increase as the maximum allowable stress is increased (stronger materials). The increase in permissable interference can be used to permit greater tolerances. The reverse is also true. Thus, the reduction in maximum allowable stress (caused by an increase in the safety factor or a weaker material) will result in a reduction in the allowable interference, leading to a reduction in the amount of tolerance permissible.

2. For a given material (e.g., no change in maximum allowable stress), a decrease in the ratio between the shaft diameter and the outside diameter of the boss (D_s/D_o) resulting in a greater hub wall thickness will permit an increase in the allowable interference. An increase in this ratio can also be used to permit greater tolerances.

3. For a given interference, an increase in maximum allowable stress (stronger material) will permit an increase in the ratio of the shaft diameter to the outside boss diameter, resulting in a thinner boss wall thickness. This could be of particular significance when dealing with the boss sink criteria.

14.2.8 Equation Limitations

A description of the following limitations on the validity of the equations should accompany their presentation.

1. *Stress relaxation* The joint strength will diminish over time as a result of stress relaxation. To determine the holding power after a given period, joint strength must

be determined by means of Equation 14-1 but with the modulus of elasticity after that period of time as a proportion of the original modulus.

$$F' = \frac{\pi C_f D_s L \sigma'}{G} \qquad (14\text{-}14)$$

$$\sigma' = \frac{E_b' \sigma}{E_b} \qquad (14\text{-}15)$$

where F' = force of insertion or removal after time period
σ' = stress after time period
E_b' = modulus of elasticity after time period

The modulus of elasticity over time can be determined from graphs like those in Fig. 14-3. Such graphs are usually available from the manufacturer for engineering resins.

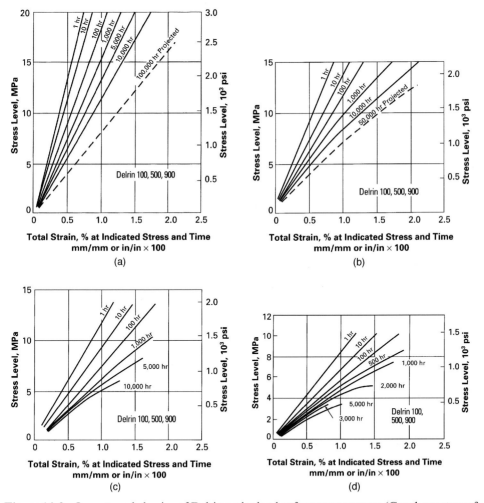

Figure 14-3 Long-term behavior of Delrin under load at four temperatures (Graph courtesy of DuPont Engineering Polymers): (a) 23 °C (73 °F) air, (b) 45 °C (113 °F) air, (c) 85 °C (185 °F) air, and (d) 100 °C (212 °F) air

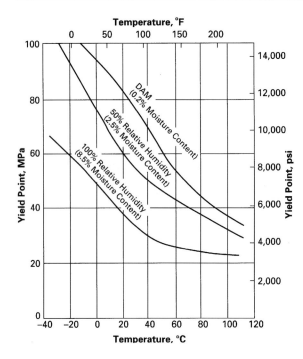

Figure 14-4 Yield point of Zytel 101 versus temperature and moisture content (Graphs courtesy of DuPont Engineering Polymers)

2. *Temperature* Unlike metals, most plastics exhibit considerable variation in tensile strength through the normal operating range. Note the effect of temperature on the moduli in Fig. 14-3. Equations 14-13 and 14-14 can be used with moduli of elasticity from the service extremes to determine the joint strength under those conditions.

3. *Humidity* The tensile strengths of some plastics, nylons particularly, are significantly affected by humidity. The graph in Fig. 14-4 demonstrates the effect of humidity on Zytel 101 (typical for all nylon 6/6 polymers).

Humidity can also alter the dimensions of a part. Figure 14-5 demonstrates the effect of moisture on Delrin, an acetal resin.

4. *Creep and cold flow* These conditions pose a serious problem with press fits because creep rupture or relaxation of the fitment to the point of its slippage can result. To reduce the risk, the apparent or creep modulus should be used for long-term applications. The graphs in Fig. 14-6 illustrate the effect of creep on the stress and strain of Celcon, an acetal resin. A curve such as this may be available from your material supplier.

5. *Environmental factors* Moisture, heat aging, exposure to ultraviolet light, and attack from a variety of chemicals such as lubricants, solvents, and cleaning agents can cause failure in press fits. Unfortunately, the information available from the material manufacturer reflects this kind of exposure according to prescribed test procedures and concentrations that may have little or no relevance to the application under consideration. The long-term testing of the actual product under anticipated exposures is an important consideration in the application of press fits. Figure 14-7 illustrates the effect of outdoor exposure on Celcon, an acetal plastic.

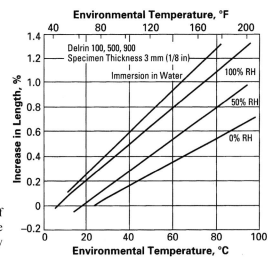

Figure 14-5 Dimensional changes of Delrin with variations of temperature and moisture content (Graph courtesy of DuPont Engineering Polymers)

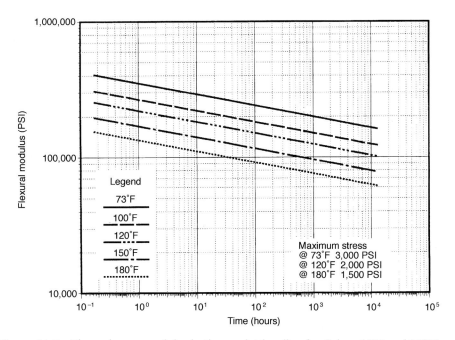

Figure 14-6 Flexural creep modulus in three point bending for Celcon M90 and M270 acetal copolymer; regression line plotted at five temperatures (Graph courtesy of Ticona LLC)

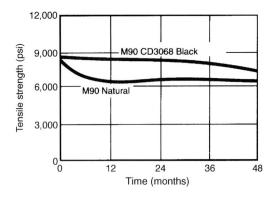

Figure 14-7 Tensile strength versus outdoor exposure (in Arizona and New Jersey) (Graph courtesy of Ticona LLC)

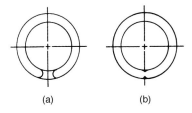

Figure 14-8 Weld or knit lines; (a) molten plastic closing around core pin, and (b) weld or knit line forming

6. *Weld lines* The boss is not formed as an extruded tube in a molded part. It is formed by molten plastic running around the core pin until the edges meet at the other side, as in Fig. 14-8a. Where these edges meet, a weld or knit line is formed (Fig. 14-8b), and it is always weaker than the surrounding material. The best weld line strength tested in the author's experience was 85% of strength of the base polymer for both reinforced and unreinforced resins. A more prudent figure to use in establishing a safety factor would be 50 to 60%. Mold design and molding conditions can significantly affect this condition. Improper molding can result in a weld line with no strength at all.

7. *Out of round* The out-of-round condition typical of many of the more ductile plastics will result in increased stresses on the boss to be taken into account when determining a safety factor. If the material is relatively soft, the boss will take the form of the shaft and negate this effect. Mold construction and molding conditions can significantly affect this condition, particularly with high shrinking materials.

8. *Draft* To permit the removal of the core from the boss, draft is a necessary condition of molded bosses. Unfortunately, it also results in nonuniform stress loading. When the shaft is metal (and depending on the part configuration), the location of the narrowest diameter of the boss can be controlled by the use of split-core pins (Fig. 14-9a) in place of the single-core pin (Fig. 14-9b).

In some cases, the hole may need to be reamed to size – an extra-cost postmolding operation. Do not attempt to resolve this matter by eliminating draft. That can result in serious molding problems and extended molding cycles. When both parts are plastic, it

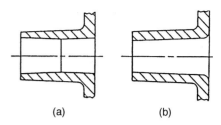

Figure 14-9   Effect of draft on stress loading; (a) split-core pins, and (b) single-core pin

may be possible to resolve the problem by designing the fitment so that both the shaft and boss drafts are in the same direction.

9. *Regrind* Figure 14-10 illustrates the effect of the use of regrind on tensile and impact strength (some resins are less severely affected). If regrind is to be used, this effect must be taken into consideration when a safety factor is established. The author does not recommend the use of regrind for heavy-duty applications, where the strength requirements approach the capabilities of the material at the extremes of the operating conditions.

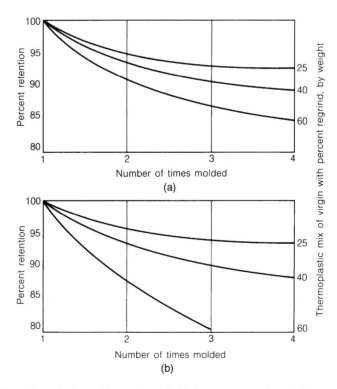

Figure 14-10   Potential effect of regrind on (a) tensile and (b) impact strengths of thermoplastics (From D.V. Rosato, *Rosato's Plastics Encyclopedia and Dictionary* © 1993)

14.3   Safety Factor

All the foregoing criteria must be taken into account when a safety factor is established. Unfortunately, this is not easy: the net effects of increasing the safety factor can be a reduction in the amount of available interference, a decrease in the amount of tolerance available, an enlargement in the outside diameter of the boss (which can result in boss sink), or the need to go to a stronger material (with its increased cost).

Nonetheless, the author recommends a minimum safety factor of 3. That is, the allowable stress to be used in the calculation will be 1/3 the figure supplied by the manufacturer (a factor of 4 would be 1/4, etc.). The author has used a safety factor of 2 on occasion, but not without full knowledge of the risk being taken and certainly not without first prototyping and testing the application. Indeed, regardless of the safety factor used, the application should be thoroughly tested before production is begun.

14.4   Processing

Clearly one cannot utilize bosses with wall sections too thin to fill at their location. This condition can be created with highly viscous resins that must travel a convoluted path (with its resultant pressure drops) to arrive at the boss location. The general rule is that the material should flow from thick to thin section in the mold, never from thin to thick. This is a rule that is commonly violated in the design of parts with bosses with varying degrees of success. In order to flow the melt from thin to thick, the molder must increase the temperature and pressure. This, in turn, makes it more difficult to hold the tight tolerances required to achieve the interferences necessary from press fits and increases the "out-of-round" condition.

14.5   Material Selection

In general, material costs increase as their physical properties improve. The usual practice in material selection is to select the lowest cost material that will meet the product requirements. To do that, the engineer needs to determine the tensile stress requirement. However, the equation to determine stress also requires values for Poisson's ratio and the modulus of elasticity of the material.

To avoid attempting to solve for multiple unknowns, the usual procedure is to select a material and determine either the hoop stress or the interference between the two parts. The quick methods for determining stress or interference were provided for just this type of material screening. Also, some resin manufacturers offer computer programs that can speed up these calculations. Based on the results of those first experimental computations, the decision to substitute another material or to vary the boss dimensions can be made. This process can be repeated until the optimum balance of requirements is achieved.

14.6   Part Design

14.6.1   Heavy-Duty Press Fits

Various design techniques have been used to counteract the tendency of the joint to loosen in response to the various phenomena already discussed. Basically, they take advantage of the natural tendency of the material to cold-flow into crevices. Two of these devices are illustrated in Fig. 14-11. The splined or ribbed shaft (Fig. 14-11a) can be used to help maintain torsional strength. Rings (Fig. 14-11b) will assist in the axial direction; threads will provide the same benefit. These devices can be used in combination to guard against failure in both directions.

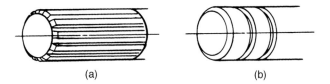

Figure 14-11 Antiloosening design details; (a) splined or ribbed shaft and (b) axial rings
    (a)   (b)

One way to limit axial travel is to create a recess in the shaft for the boss to fit into. This can be accomplished by substituting one large ring for the three small ones in Fig. 14-11b. The final diameter at which the boss rests must be examined for creep effects and the range of environments the design must withstand. The stress created by installation over the initial high shaft diameter must also be checked, however the boss can be enlarged with heat or the shaft cooled in a freezer if it is necessary to avoid that problem.

14.6.2   Light-Duty or Reopenable Press Fits

There are several designs for light-duty or reopenable press fits that reduce costs by permitting looser tolerances for force fits. These concepts take advantage of the ductility of some plastic materials, which permits them to deform or crush under pressure, particularly in thin sections. Several of these approaches were illustrated in Chapter 2. The ribs in Fig. 2-17 crush as the mating part is pressed on. These radial tapered crush ribs provide limited increases in tolerance range with good maintenance of strength and centering. The flexible "bendover" style in Fig. 2-18 is less stiff but can handle more liberal tolerances. Both these rib designs can be placed on either the shaft or the boss.

The stepped design in Fig. 2-20a is typical of traditional laboratory fittings. Hermetic seals would be difficult to attain because a mold with a split cavity would be required, and the mismatch at the parting line would prevent a perfect seal. Such designs are best suited to shafts because they create undercuts in bosses, the cost of which is counterproductive to the objective of cost reduction.

The Luer fitment is illustrated in Figs. 14-12a and 14-12b. It is the standard for needle and intravenous tubing connector fitments in the medical industry. That concept is commonly used with other dimensions for a wide variety of applications. Positive location can be created with devices such as the shoulder in Fig. 14-12c or the ribs in Fig. 14-12d. The angle should be 2° for a firm fit in polypropylene. That angle can be increased to loosen the fitment or decreased to make it even more firm, however 1° should be regarded as the lower limit.

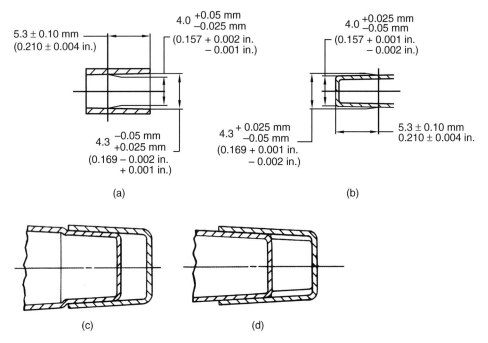

Figure 14-12   Tapered fitments: (a) female Luer fitment dimensions, (b) male Luer fitment dimensions, (c) shoulder design, and (d) ribbed design

14.6.3   Other than Round

The bulk of press fit assemblies, certainly all the highly stressed applications, are done with round parts. However, there are some successful applications, such as food storage containers and cosmetic packages, that have successfully joined square and rectangular shapes.

    The round fitment in Fig. 14-13a illustrates the facility with which two round parts can be press-fitted together. This works particularly well when differences either in wall thickness or in material make one of the parts significantly more rigid than the other and able to force the less rigid member to its contour. This permits greater tolerances and ready accommodation of the ovality of the parts as molded.

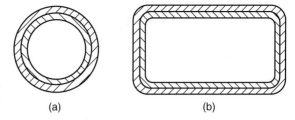

Figure 14-13 Fitment shapes: (a) round fitment and (b) rectangular fitment with rounded corners

(a)   (b)

As the shape is deviated from the round (Fig. 14-13b), this attribute is lost. The corners become stiffer and the sidewall tends to deform inward. The seal then becomes more difficult to maintain. There is a potential for the condition known as a "corner bind," in which the stiffened corners prevent assembly. If a seal is not necessary, corner bind can be avoided by using a larger outside radius for the inner piece than the inside radius of the outer piece (Fig. 2-23b).

15   Snap Fits

15.1   Advantages and Disadvantages

15.1.1   Advantages

Snap fits are similar to press fits and share many of the same advantages and disadvantages. In fact they are competitive for many applications. The principal considerations are the shape of the part, the loadings on the joint, reopenability, mold costs, and whether the joint can withstand the multiple effects of the long-term stress, inherent in press fits, which can be largely avoided in most snap fits. Also, because snap fits are not generally under long-term stress, lower safety factors are required. The advantages of snap fits are as follows.

1. *No additional materials* Snap fits use no additional materials such as fasteners, inserts, electromagnetic preforms, adhesives, or solvents. Therefore, they are inherently lower in cost than methods that do use such materials and are inherently less expensive to disassemble for recycling.

2. *Ease of assembly* Snap fits require only straight insertion; thus they can be performed without tools in many applications and are well suited to automated assembly. Very high production rates are possible.

3. *Dissimilar materials* Metals can be joined to plastics, and widely differing plastics as well as similar polymers can be joined to each other with snap fits. Parts made of metals or plastics that are too brittle for snap fits can have cantilever snap arms attached to them with adhesives or fasteners.

4. *Permanence* Snap fits can be used for permanent assemblies when strong snap fitments that cannot be accessed from the outside of the part are used.

5. *Reopenability* When snap fits are lightly loaded or can be accessed from outside the part, they can be designed to be reopenable. Inexpensive ballpoint or felt-tipped pens are often designed with this type of snap fit.

6. *Shape freedom* While cylindrical or annular snap joints are obviously round and torsion snap joints are essentially fingers, cantilever snap fitments can be designed into a wide variety of shapes.

7. *Entrapment of other parts* Additional parts can be captured between the two parts to be snap-fit together.

8. *Internal joints* Snap fitments can be located internally.

9. *Energy efficiency* Snap fitments are the most energy-efficient assembly method. This also means that there is no excess heat to be removed from the workplace.

10. *Clean atmosphere* Unlike the case of adhesive and solvent joining systems, no ventilation equipment is necessary for the removal of toxic fumes.

11. *Immediate handling* Assembled parts can proceed to other operations at once without waiting for parts to cool and for adhesives or solvents to set. This advantage lends itself well to automated assembly line applications.

12. *High production rates* Production rates of 20 to 60 parts per minute are possible, since the actual joining time is instantaneous.

15.1.2 Disadvantages

There are some disadvantages as well.

1. *Mold limitations* Snap fits usually require mold undercuts. At the least, they can result in limitations in design freedom to allow for the part to be stripped off the core. At the most, expensive collapsing core mechanisms that limit the number of cavities available will be required.

2. *Process limitations* Strippable internal undercuts have a very small window in injection molding during which they can be successfully ejected without distortion. Collapsing core molds are rarely attempted for any process besides injection molding, although some versions are feasible in transfer and compression molding. For low volume applications, internal undercuts can be machined into parts made by most any process, provided the parts are large enough for the tool to access. It is difficult to machine the thin sections necessary for snap arms, however.

3. *Snap failure* Torsion and, particularly, cantilever snap arms are vulnerable to failure through fatigue failure, improper design, and stress from a variety of sources. Cylindrical snap fits are vulnerable to weak spots created by weld lines, gate stress, and voids. Snap failures are impossible to repair and result in complete failure of the assembly unless there is adequate redundancy.

4. *Material limitations* Since snap fits require the flexure of material during the engagement of the snap, they work best with more ductile plastics.

5. *Thermal expansion* Significant differences in coefficient of linear thermal expansion between the two materials can result in stress cracking or loosening of the snap at the temperature extremes.

6. *Moisture absorption* Expansion due to moisture absorption that is sufficient to result in loosening of the snap occurs in some plastics.

7. *Assembly limitations* Exposed snaps arms are vulnerable to breakage during assembly, particularly in automatic assembly equipment. Brittle, filled, and fiber-reinforced materials are most susceptible to this type of failure.

8. *Limited hermetic seals* Hermetic seals are difficult to accomplish with snap fits because of several phenomena that can cause a snap to loosen under load. Some hermetic seals have been accomplished using soft gasketing materials to create the seal.

Ideally, snap fits should be designed to endure stress only during assembly. The remaining disadvantages apply to snap fits under load in the at-rest position.

9. *Stress relaxation* Time diminishes the ability of the material to withstand stress, particularly at elevated temperatures. Stress relaxation can result in cracking or crazing of the plastic.

10. *Creep and cold flow* The joint can be loosened by these effects as it undergoes strain relaxation.

11. *Environmental limitations* Chemical and ultraviolet light exposure can result in failure due to stress cracking.

15.2   General Applications

Snap fits have an ever increasing range of applications spurred largely by cost reduction efforts both in assembly and disassembly for recycling. They can be designed to be easy to reopen, difficult to reopen or impossible to reopen without damage to the part. As a result, designers and engineers have created a wide range of innovative snap fit designs. Applications range from inexpensive ballpoint and felt-tipped pens to office equipment and automotive underhood equipment. There are four basic types of snap fit.

1. *Cantilever snap arms or beams* These are snap hooks attached to the ends of arms, beams, or other appendages. For permanence, they can be designed to be hidden from external view through attachment to internal undercuts. For reopenability, they can be designed to protrude through openings in the side or bottom of the part where they can be accessed for disassembly.

2. *Cylindrical, ring, perimeter, or annular fitments* The felt-tipped pen application is typical of this type of snap fit. It is the variety that most often must be compared to press fits as an alternative.

3. *Ball-and-socket joints* This is the common designation, even though the actual contour is cylindrical as often as it is spherical.

4. *Torsional snap fits* These are designed to be reopenable and to work on the basis of a lever operating about a fulcrum.

15.3   General Engineering Principles

15.3.1   Allowable Dynamic Strain

The snap arm will obviously return to its original position if it is designed to stay below its elastic limit. In many cases, that design provides only a small amount of gripping surface under the hook, not enough to hold the parts together. Based on the use of the initial modulus, the bending stress appears to exceed the yield strength of the material. However, in reality, the yield strength is not exceeded because the initial modulus applies only to very small strains. For large strains, the secant modulus applies. Snap fit design theory is normally based on the allowable dynamic strain limit rather than the yield stress.

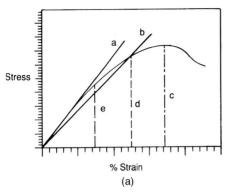 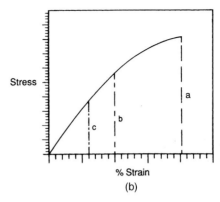

Figure 15-1 Stress–strain curves. (a) Polymer with definite yield point: *a*, modulus of elasticity; *b*, secant modulus; *c*, strain at yield; *d*, 70% of strain at yield for single assembly; *e*, 42% of strain at yield for multiple assemblies. (b) Polymer without definite yield point: *a*, strain at break; *b*, 50% of strain at break for single assembly; *c*, 30% of strain at break for multiple assemblies (Graphs courtesy of Ticona LLC)

The issue then becomes one of establishing a value for the allowable dynamic strain. Figure 15-1a depicts the stress–strain curve for a polymer with a definite yield point, typically an unreinforced thermoplastic. For single assemblies, a value of 70% of the yield strain can be used for amorphous thermoplastics, perhaps as much as 90% for crystallines. If the assembly must be repeatedly reopened, the figures to use are 42% (60% of 70%) of the strain at break for amorphous polymers and 54% (60% of 90%) for the crystallines.

Lower percentages can be used for materials without a definite yield point (Fig. 15-1b), such as glass-reinforced thermoplastics. The value to be used for a single assembly of such a material is 50% of the strain at break; 30% for repeated assemblies.

Note the two moduli illustrated in Fig. 15-1a. The modulus of elasticity is often provided on material specification sheets, however it is obvious that it would misrepresent the true nature of the polymer at the point of 70% yield strain, suggesting that the material is stiffer than it really is. Therefore, the modulus used is the secant modulus drawn to the point of strain intended to be used in order to obtain a value that better represents the actual material. Attention must be paid to the arc of the actual material intended to be used, as the engineer may elect to select another point if it appears to be more representative. The secant modulus is used for all computations in this chapter.

15.3.2 Corner Stress Concentrations

One who is preoccupied with snap fit equations must not lose sight of other engineering phenomena. The stress concentrations created by the inside corners denoted in Fig. 15-2 must not exceed the parameters described in Chapter 2, "Designing for

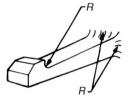

Figure 15-2   Corner stress concentrations

Efficient Assembly." However, 0.5 is a ratio commonly used for inside radii to wall thickness for unreinforced thermoplastics, and it will result in a stress concentration factor of about 1.5. The radii at the base of the arm can be readily set at 0.5, but that is difficult to do under the hook because that radius removes material from the flat portion and, therefore, increases the amount of required deflection. Still, that is a highly vulnerable point in the design and many snap fits fail at that inside radius. Clearly, there is a trade-off to be considered. Therefore the author recommends judgment on the basis of consultation with the resin manufacturer, to determine the appropriate stress concentration factor and strain value for the candidate material.

Glass-reinforced thermoplastics have more latitude in the determination of inside radii. However, with either reinforced or unreinforced polymers, 0.5 mm (0.020 in.) should be regarded as a minimum with a ±0.12 mm (0.005 in.) tolerance so that the radius can never get below 0.38 mm (0.015 in.) at the worst condition.

There is yet one more issue to be resolved with regard to these corners. Referring, once again, to Chapter 2, note that the provision of generous radii to avoid stress concentrations at the inside corners may result in high levels of molded-in stress due to the nonuniform cooling. This would occur if the pocket of material at the base of the snap beam were to be large enough to create a circle whose diameter exceeded the wall thickness by more than 25%.

15.3.3   Engineering Adjustments When Both Parts Are Elastic

The equations in this chapter presume that all the deflection is experienced by one of the two mating parts. That does not necessarily suggest that one need to be metal and the other plastic, as this phenomenon would be largely exhibited by two plastic parts of widely varying stiffness or even those with less variance if the more rigid part had a much heavier wall thickness. Nonetheless, that leaves a great number of applications in which the deflection would be experienced by both parts.

There is a method of dealing with such variations, albeit a somewhat laborious one. Figure 15-3a plots the deflection force P against the deflection for one part, with the 0,0 point in the lower left corner. Figure 15-3b depicts the same plot for the other part, with the 0,0 point in the lower right corner. The two plots are overlaid in Fig. 15-3c. The point of intersection defines the actual deflection force (P) and the portion of deflection (y_1, y_2) experienced by each part. The force required to join the parts and the strains for each can be determined using these data.

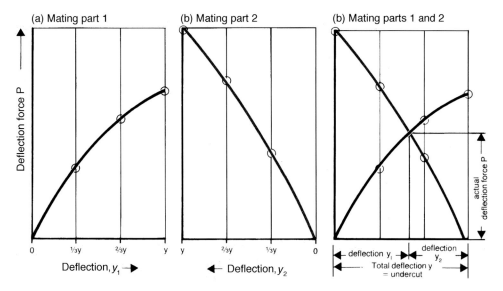

Figure 15-3 Determination of deformation and transverse force when both mating parts are flexible: (a) mating part 1, (b) mating part 2, and (c) mating parts 1 and 2. (Graphs courtesy of Bayer.)

15.3.4 Finite Element Analysis

The design of snap fitments, particularly the cantilever type, lends itself well to finite element analysis (FEA). The classical beam formula used in the design of cantilever snap arms contain inherent errors. They presume that the length-to-height ratio is very high. Shorter snap arms with length-to-height ratios less than 5:1 do not suit these equations well and lead to inaccuracies. The lower the ratio of length to height, the greater the inaccuracy. Consequently, the use of FEA is recommended for these designs in particular, and all cantilever snap fit designs, in general. In addition, FEA can take inside corner stress concentration and adjacent wall deflections into account.

15.4   Cantilever Snap Fits

15.4.1 Cantilever Snap Fit Designs

Cantilever snap fits have a wide variety of applications. They can be designed to be permanent or reopenable with varying degrees of difficulty depending on the requirements of the application. Figure 15-4a illustrates a basic permanent snap fit; the return angle of 90° makes it perpendicular to the cantilever arm and the inner wall of the mating part. That does not mean that this fitment is impossible to reopen as sufficient force can be applied to the joint to cause the hook to fail. However, short

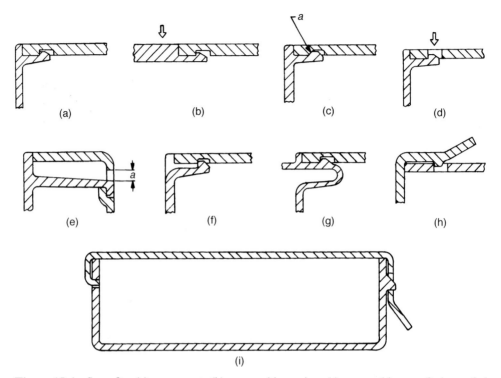

Figure 15-4   Snap fits: (a) permanent, (b) reopenable version, (c) reopenable snap fit (*a*, angled snap), (d) reopenable snap fit – side access, (e) reopenable snap fit – bottom access (*a*, deflection limit), (f) reopenable snap fit – front access, (g) readily reopened snap fit – U-arm, (h) readily reopened snap fit – internal slot, and (i) snap used with hinge lugs to create a reopenable door

of failure it is nonreopenable in this configuration. The snap could be disengaged with a downward pressure on the wall, as shown in Fig. 15-4b. If that angle is changed to one more gradual, as in Fig. 15-4c, the fitment becomes reopenable. The greater the angled surface, the easier it will be to open the fitment and to remove parts from the mold. Therefore, it is possible to control the amount of force required to open the joint by altering the locking angle. This type of snap fitment is more costly to tool, as discussed in Section 15.9 ("Molds for Snap Fits").

For a fitment to be reopenable, there needs to be a way to disengage the hook. If a slot is created instead of a recess as in Figs. 15-4d and 15-4e, the hook can be accessed from the outside of the part, which means that an instrument, such as a screwdriver blade, can be used to disengage it. A tool can also be used to release the hook in Fig. 15-4f. Note that the gap behind the hook in Fig. 15-4e can be set to limit the amount by which the snap arm can be deflected; however, care must be taken to ensure that sufficient space is provided to permit clearance for the snap hook.

If the snap fitment is to be readily opened, such as for use on a battery compartment door, the snap hook can be part of a U-shaped arm (Fig. 15-4g), or it can be external (Fig. 15-4h), as when used for a latch. The hook and the slot in this design can be

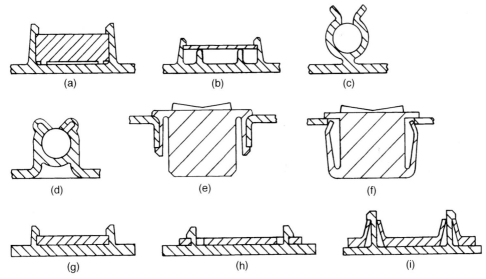

Figure 15-5   Snap fit applications: (a) electrical component lock, (b) printed circuit board mount, (c) reopenable cylindrical lock, (d) semipermanent cylindrical lock, (e) recess locks, (f) panel locks, (g) cantilever hooks with tight tolerance requirements, (h) cantilever hooks that accommodate wide tolerances, and (i) bayonet finger with self-locating feature (accommodates larger tolerances)

reversed, as well (Fig. 15-4i). Figure 15-4i also demonstrates how the snap fitment can be used in conjunction with hinge lugs to create a compartment door.

Figure 15-5 illustrates several snap fit applications, some of which are used to assemble parts that have no integral means of assembly. Printed circuit boards and electrical components can be assembled to plastic parts with the designs in Figs. 15-5a and 15-5b. Rods and wires can be assembled with the types of snap fitments shown in Figs. 15-5c and 15-5d. The design in Fig. 15-5c is intended for removal of the object with approximately the same force as that of assembly, while that in Fig. 15-5d is designed for low installation pressure and high disassembly force. In some cases, locks are molded directly into the component (Figs. 15-5e, 15-5f).

The external cantilever hooks illustrated in Fig. 15-5g are difficult to control dimensionally, and a small deviation in one direction or the other can lead to disengagement or engagement under pressure. The internal hooks in Fig. 15-5h are much more forgiving to dimensional variations, as are the bayonet fingers in Fig. 15-5i. External locators can be used to lessen the vulnerability of the snap fit to dimensional variation. The lugs reduce the number of flexing snap fits required.

Snap hooks are typically molded integral to the parts they are joining. However, that is not always the case, and nonintegral snap hooks that are themselves snapped into a hole can be effectively utilized.

Although the majority of snap fitments are executed in the injection molding process, this method of attachment is not confined to that process. The author has used this method of joining for parts manufactured by the compression molding, rotational molding, blow molding, and thermoforming processes, as well.

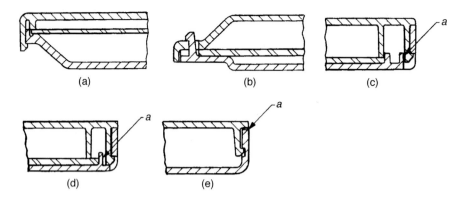

Figure 15-6 Multiple-part assemblies: (a) external snap, (b) bayonet snap, (c) internal snap (*a*, reinforcing rib), (d) limited travel internal snap (*a*, deflection travel limiting rib), and (e) snap used as adhesive clamp or solvent (*a*, adhesive joint)

Multiple-part assemblies can be accomplished with cantilever snap fitments. Two such designs are illustrated in Figs. 15-6a and 15-6b. The design in Fig. 15-6c demonstrates how a rib can be used to reinforce the snap arm and prevent its disengagement. In Fig. 15-6d, the rib is used in a reopenable snap fitment to limit the travel of the snap arm and prevent failure. If the third part in these designs were made of an elastomeric material, the geometry could be configured to accomplish a seal. That would involve the preloading of snap fitments, with the associated creep concerns. Snap fits can also be used as a temporary holding device for handling of the part during assembly or to hold the parts together while an adhesive or solvent is setting up. An example of such a design is shown in Fig. 15-6e.

15.4.2 Cantilever Snap Fit Engineering

The cantilever snap arm is primarily loaded in a flexural manner. The force of installation R is composed of two elements: the normal force N owing to the deflection of the arm and the force F due to the friction, as shown in Fig. 15-7a.

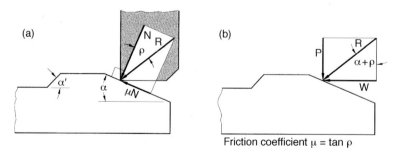

Friction coefficient $\mu = \tan \rho$

Figure 15-7 Cantilever snap arm forces: (a) $F = \mu N$ and (b) friction coefficient $\mu = \tan \rho$ (Diagrams courtesy of Bayer)

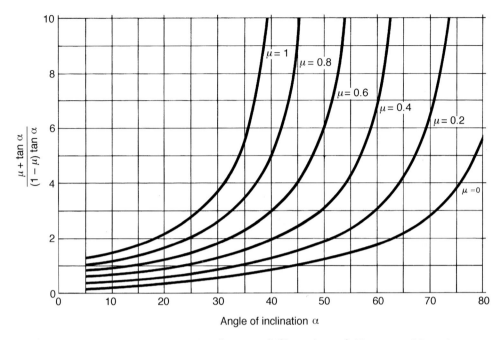

Figure 15-8   Diagram for determining $(\mu + \tan \alpha)/[(1 - \mu) \tan \alpha]$ (Courtesy of Bayer)

Table 15-1   Guide Data for μ, the Coefficient of Friction of Various Plastics on Steel

Material	μ^a
PTFE	0.12–0.22
PE, rigid	0.20–0.25 (×2.0)
PP	0.25–0.30 (×1.5)
POM	0.20–0.35 (×1.5)
PA	0.30–0.40 (×1.5)
PBT	0.35–0.40
PS	0.40–0.50 (×1.2)
SAN	0.45–0.55
PC	0.45–0.55 (×1.2)
PMMA	0.50–0.60 (×1.2)
ABS	0.50–0.65 (×1.2)
PE, flexible	0.55–0.60 (×1.2)
PVC	0.55–0.60 (×1.0)

[a] The figures depend on the relative speed of the mating parts, the pressure applied and on the surface quality. Friction between two different plastic materials gives values equal to or slightly below those shown in Table 15-1. With two components of the same plastic material, the friction coefficient is generally higher. Where the factor is known, it has been indicated in parentheses. *Source*: Bayer.

Translated to its radial (P) and lateral (W) components as in Fig. 15-7b, the equation is expressed as follows:

$$W = P\tan(\alpha + \rho) = P\frac{\mu + \tan\alpha}{(1 - \mu)\tan\alpha} \tag{15-1}$$

where P is the deflection force, μ is the dynamic coefficient of friction between the two materials, and the value for $(\mu + \tan\alpha)/[(1 - \mu)\tan\alpha]$ can be determined from Fig. 15-8. This is to be regarded as an approximate value because the true coefficient of friction is difficult to establish. It can vary with sliding speed, temperature, normal load, atmospheric environment, kinematics, shape and surface topography, and sliding history. Coefficients of friction varying by as much as 100% have been reported for the same materials (see Table 15-1). For most applications, the design should commence with an angle α of 15 to 30°. Depending on the material, it may not be possible to assemble the parts with an angle α greater than 45°.

The retention force is determined in much the same manner by using the back side of the hook. It should be noted, however, that failure can occur even when an angle α' of 90° is used: the loading on the hook is not on the same centerline as the loading on the beam, and this creates a bending moment that causes the hook to rotate back, leading to failure at the inside corner or in the beam. For snaps designed to be reopened, an angle α' of 45 to 60° is recommended unless the design is such that the snap can be accessed without an axial force for purposes of disengagement.

The deflection force P is dependent on the cross section of the cantilever. The equations for determining the value for the deflection force are provided in Table 15-2a with the associated geometric factors in Figs. 15-9 and 15-10. Note that all the equations for cantilever snap arms regard the opposite member with the undercut that receives the snap hook as rigid. An engineering judgment is required in selecting the allowable strain. The symbols and numbered notes used in Table 15-2a are given in Table 15-2b.

The equations for the determination of the permissible deflection or undercut are also provided in Table 15-2. The straight beam with the rectangular cross section is the least expensive to tool. However, it produces the maximum stress concentration at the base of its arm, therefore requiring more material and a longer cycle. Such requirements, in turn, increase the piece part cost; and the more snaps there are, the more significant this becomes.

That stress can be more evenly distributed by going to a tapered beam. The taper reduces the strain on the beam and permits a greater deflection than an equivalent straight snap arm with the same base thickness. Conversely, if the thickness at the hook is maintained, the beam can be made stiffer by increasing the thickness at the base. The beam can be tapered either in thickness or width . The width at the tip may be reduced to one-quarter that of the base; however a good thickness ratio is 1:2. These are the equations provided in Table 15-2.

On occasion, it is necessary to use a different ratio for the thickness. This may be accomplished by means of Equation 15-2, which substitutes a geometry factor K in

Table 15-2a Equations for Designing Across Sections of Snap Fits (Courtesy of Bayer)

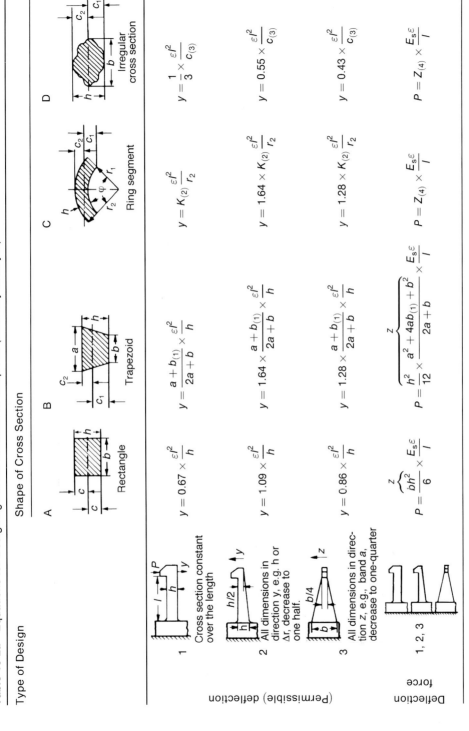

Type of Design	Shape of Cross Section			
	A	B	C	D
	Rectangle	Trapezoid	Ring segment	Irregular cross section
(Permissible) deflection				
1 Cross section constant over the length	$y = 0.67 \times \dfrac{\varepsilon l^2}{h}$	$y = \dfrac{a+b_{(1)}}{2a+b} \times \dfrac{\varepsilon l^2}{h}$	$y = K_{(2)} \dfrac{\varepsilon l^2}{r_2}$	$y = \dfrac{1}{3} \times \dfrac{\varepsilon l^2}{c_{(3)}}$
2 All dimensions in direction y, e.g., h or Δr, decrease to one half.	$y = 1.09 \times \dfrac{\varepsilon l^2}{h}$	$y = 1.64 \times \dfrac{a+b_{(1)}}{2a+b} \times \dfrac{\varepsilon l^2}{h}$	$y = 1.64 \times K_{(2)} \dfrac{\varepsilon l^2}{r_2}$	$y = 0.55 \times \dfrac{\varepsilon l^2}{c_{(3)}}$
3 All dimensions in direction z, e.g., band a, decrease to one-quarter	$y = 0.86 \times \dfrac{\varepsilon l^2}{h}$	$y = 1.28 \times \dfrac{a+b_{(1)}}{2a+b} \times \dfrac{\varepsilon l^2}{h}$	$y = 1.28 \times K_{(2)} \dfrac{\varepsilon l^2}{r_2}$	$y = 0.43 \times \dfrac{\varepsilon l^2}{c_{(3)}}$
Deflection force				
1, 2, 3	$P = \underbrace{\dfrac{bh^2}{6}}_{z} \times \dfrac{E_s \varepsilon}{l}$	$P = \underbrace{\dfrac{h^2}{12} \times \dfrac{a^2 + 4ab_{(1)} + b^2}{2a+b}}_{z} \times \dfrac{E_s \varepsilon}{l}$	$P = Z_{(4)} \times \dfrac{E_s \varepsilon}{l}$	$P = Z_{(4)} \times \dfrac{E_s \varepsilon}{l}$

Table 15-2b   Symbols and Notes for Table 15-2a

Symbols	Notes
y = (permissible) deflection (= undercut) ε = (permissible) strain in the outer fiber at the root; in formulas: ε as absolute value = percentage/100 l = length of arm h = thickness at root b = width at root c = distance between outer fiber and neutral fiber (center of gravity) Z = section modulus $Z = I/c$, where I = axial moment of inertia E_s = secant modulus P = (permissible) deflection force K = geometric factor (see Fig. 15-9) φ = angle	1. These formulas apply when the tensile stress is in the small surface area b. If it occurs in the larger surface area a, however, a and b must be interchanged. 2. If the tensile stress occurs in the convex surface, use K_2 in Fig. 15-9, if it occurs in the concave surface, use K_1 accordingly. 3. c is the distance between the outer fiber and the center of gravity (neutral axis) in the surface subject to tensile stress. 4. The section modulus should be determined for the surface subject to tensile stress. Section moduli for cross section shape type C are given in Fig. 15-10. Section moduli for other basic geometrical shapes are to be found in mechanical engineering manuals. Permissible stresses are usually more affected by temperatures than the associated strains. One preferably determines the strain associated with the permissible stress at room temperature. As a first approximation, the computation may be based on this value regardless of the temperature. Although the equations may appear unfamiliar, they are simple manipulations of the conventional engineering equations to put the analysis in terms of permissible strain levels.

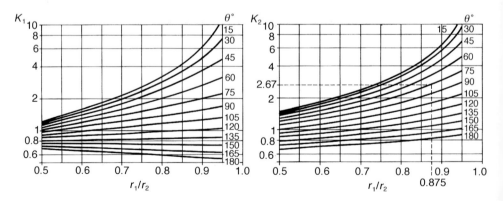

Figure 15-9   Diagrams for determining K_1 and K_2 for cross-sectional shape type C in Table 15-2: K_1, concave side under tensile load; K_2, convex side under tensile load (Diagrams courtesy of Bayer)

$$Z_1 = r_2^3 \times Z/r_2^3 \qquad\qquad\qquad Z_2 = r_2^3 \times Z/r_2^3$$

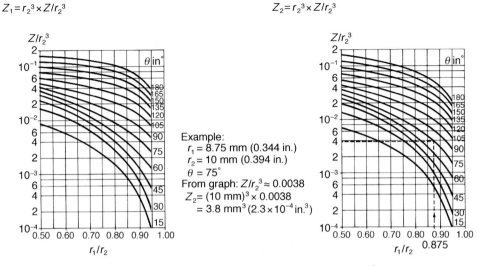

Example:
$r_1 = 8.75$ mm (0.344 in.)
$r_2 = 10$ mm (0.394 in.)
$\theta = 75°$
From graph: $Z/r_2^3 \approx 0.0038$
$Z_2 = (10 \text{ mm})^3 \times 0.0038$
$= 3.8 \text{ mm}^3 (2.3 \times 10^{-4} \text{ in.}^3)$

Figure 15-10 Graphs for determining the dimensionless quantity (Z/r_2^3) used to derive the section modulus (Z) for cross-sectional shape C in Table 15-2: Z_1, concave side under tensile stress; Z_2, convex side under tensile stress (Graphs courtesy of Bayer)

Table 15-3 Values of Geometry Constant K for Use in Equation 15-2

h_L/h_o	K	h_L/h_o	K	h_L/h_o	K	h_L/h_o	K
0.33	2.137	0.53	1.573	0.73	1.259	0.93	1.056
0.34	2.098	0.54	1.553	0.74	1.247	0.94	1.047
0.35	2.060	0.55	1.534	0.75	1.235	0.95	1.039
0.36	2.024	0.56	1.515	0.76	1.223	0.96	1.031
0.37	1.989	0.57	1.497	0.77	1.212	0.97	1.023
0.38	1.956	0.58	1.479	0.78	1.201	0.98	1.015
0.39	1.924	0.59	1.462	0.79	1.190	0.99	1.008
0.40	1.893	0.60	1.445	0.80	1.179	1.00	1.000
0.41	1.863	0.61	1.429	0.81	1.168		
0.42	1.834	0.62	1.413	0.82	1.158		
0.43	1.806	0.63	1.399	0.83	1.148		
0.44	1.780	0.64	1.382	0.84	1.138		
0.45	1.754	0.65	1.367	0.85	1.128		
0.46	1.729	0.66	1.352	0.86	1.118		
0.47	1.704	0.67	1.338	0.87	1.109		
0.48	1.681	0.68	1.324	0.88	1.100		
0.49	1.658	0.69	1.310	0.89	1.091		
0.50	1.636	0.70	1.297	0.90	1.082		
0.51	1.614	0.71	1.284	0.91	1.073		
0.52	1.593	0.72	1.272	0.92	1.064		

h_L = thickness of beam at snap h_o = thickness of beam at base
Source: Ticona LLC.

the straight beam equation:

$$y = K0.67\frac{\varepsilon l^2}{h} \qquad (15\text{-}2)$$

where the constant K can be determined from Table 15-3.

15.5   Cylindrical, Ring, Perimeter, or Annular Snap Fits

15.5.1   Cylindrical Snap Fit Designs

The type of round snap fitment typical of food storage containers, snap caps for packaging containers, and many other applications is designated interchangeably as cylindrical, ring, perimeter, and annular snap fit. Although the concept will work for shapes that are not perfectly round, it becomes more difficult to eject parts from the mold and assemble them, the more square or rectangular they become. Cylindrical fitments too can be designed to be either nonreopenable or reopenable with varying degrees of difficulty. Figure 15-11 illustrates some of the ways in which this type of snap fit can be used.

The cylindrical snap fits typically used for food storage containers and packaging applications and enclosures of other types are shown in Figs. 15-11a and 15-11b.

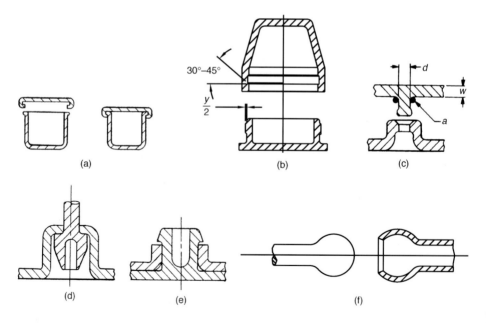

Figure 15-11   Cylindrical snap fits: (a) packaging application, (b) enclosure (courtesy of Bayer), (c) Snap button (*a*, rubber O-ring; $d \leq 0.6w$), (d) and (e) snap locks, and (f) ball and socket

Smaller cylindrical snap fits are often used for applications like pen caps. Even smaller ones can be used for the snap buttons and locks (Figs. 15-11c–15-11e). The solid design can be used when the diameter at the base of the shaft d is equal to or less than 0.6 of the wall thickness. When it is larger than that, the cored designs in Fig. 15-11d or 15-11e should be employed. Note the lead-in angles on these two examples. Experience has shown that 30° is a good starting point as a lead-in angle. Regardless of which design is used, the radius r at the base of the shaft should be no less than $0.25w$, with $0.5w$ being ideal. Finally, there is the ball-and-socket snap fitment, so often used on toys and inexpensive jewelry (Fig. 15-11f).

In some cases, the stresses created on the joint are too great for a continuous snap fitment. For these applications, a discontinuous cylindrical snap joint can be used such as the one in Fig. 15-12. However, it must be pointed out that the slots change the nature of the joint from a continuous snap fitment to a series of cantilever arms. That makes the load on the joint primarily flexural. Therefore, the equations for cantilever snap fits are to be used for discontinuous cylindrical snap fits. Do not use the equations for cylindrical snap fits.

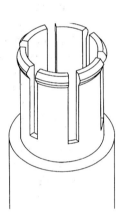

Figure 15-12   Discontinuous cylindrical snap fit (Diagram courtesy of Bayer)

15.5.2   Engineering of Cylindrical, Ring, Perimeter, or Annular Snap Fits

15.5.2.1 *Maximum Permissible Interference*

Cantilever snap fits involve multiaxial stresses because they are rotationally symmetrical. The first step in the execution of cylindrical snap fit design is to determine the maximum permissible interference (I). This means the total interference on the diameter, therefore the interference on one side would be half that value or $I/2$. This is determined by Equation 15-3, which is quite simple (it is an approximation to use for a quick screen):

$$I = \varepsilon d \qquad\qquad (15\text{-}3)$$

where ε is the maximum permissible strain and d is the shaft diameter. This equation presumes that one of the two parts remains absolutely rigid.

When one of the materials is substantially less rigid than the other, it can be expected to exhibit all of the deformation, and therefore all of the strain. Otherwise the strain is proportioned according to the materials' relative rigidity. For example, if the two materials were of equal rigidity, each would be expected to receive half the strain.

A quick screen of dimensional combinations can be made by revising Equation 15-4 to determine whether the allowable strain would be exceeded:

$$\varepsilon = \frac{I}{d}\,100\% \qquad\qquad (15\text{-}4)$$

where I = interference (total) and d = diameter.

15.5.2.2 Transverse and Axial Forces

Cylindrical snap fits are, for the most part, stronger than cantilever snap fits. They also require greater assembly force, although that can be reduced through the use of heat to expand the outer part or cold to reduce the inner part, as with press fits. The computation of the assembly forces for the cylindrical snap joint are the more complex of the equations. The reason is that the stress is distributed over a wide area because the more flexible of the members deflects over a relatively long section of the tubular fitment. Since the length of this stressed area is difficult to predict, the results of these calculations must be regarded as approximate. However, reasonably close approximations have been achieved using equations based on the theory of a beam of infinite length resting on a resilient foundation.

The principal equation will be based on the presumption that the snap fitment is located near the end of the tube. The results of these computations will then be adjusted for the location of the actual fitment. The distribution of this stress is depicted in Fig. 15-13; the distance of the snap fit from the end of the tube is designated by the letter δ.

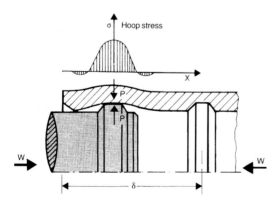

Figure 15-13   Stress distribution during joining of a cylindrical snap fit (Diagram courtesy of Bayer)

The radial force P for a snap fitment located near the end of the tube is determined by the following equation:

$$P = ydE_sX \tag{15-5}$$

where P = radial force

y = undercut

d = diameter at the joint

E_s = secant modulus

X = geometric factor

The geometric rigidity of the snap-fit member is accounted for by the geometric factor X_N. When the shaft is rigid and the outer tube is elastic, the factor is expressed as follows:

$$X_N = 0.62 \frac{\sqrt{(d_o/d - 1)/(d_o/d + 1)}}{[(d_o/d)^2 + 1]/[(d_o/d)^2 - 1] + \nu} \tag{15-6}$$

where d_o = external diameter of the tube

d = diameter at the joint

ν = Poisson's ratio

When the outer tube is rigid and the inner tube is elastic, the factor (X_W) is expressed as follows:

$$X_W = 0.62 \frac{\sqrt{(d/d_i - 1)/(d/d_i + 1)}}{[(d/d_i)^2 + 1]/[(d/d_i)^2 - 1] - \nu} \tag{15-7}$$

where d = diameter at the joint

d_i = internal diameter of the hollow shaft

These equations can be avoided by referring to the diagrams provided in Fig. 15-14.

The transverse force P can then be used to develop the axial or mating force with the equation:

$$W = P \frac{\mu + \tan \alpha}{(1 - \mu) \tan \alpha} \tag{15-8}$$

where μ = coefficient of friction

α = lead angle

The location of the snap joint will affect the accuracy of Equation 15-8. If the distance (δ) exceeds the value in Equation 15-9, the snap joint must be regarded as a remote snap joint:

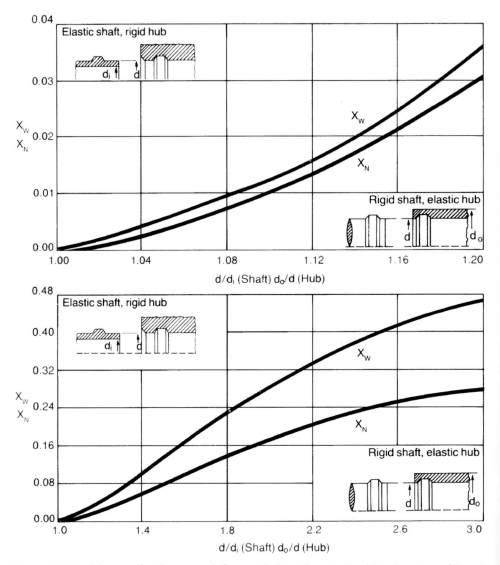

Figure 15-14 Diagrams for the geometric factor X in Equations 15-6 and 15-7 (courtesy of Bayer)

$$\delta_{min} \approx 1.8\sqrt{dt} \tag{15-9}$$

where d = joint diameter

t = wall thickness

For a snap fitment located remote from the end of the tube, both the axial and transverse forces must be increased by a factor of 3:

$$P_{remote} \approx 3P_{near} \tag{15-10}$$

and

$$W_{\text{remote}} \approx 3W_{\text{near}} \qquad (15\text{-}11)$$

These are experimentally derived values (the theoretical values are four times as great). For snap joints located between the end of the tube and the minimum δ for a remote joint, the approximate values for the forces may be determined by an increase proportional to the distance from the end:

$$P_{\text{remote}} = \frac{\delta}{l} P_{\text{near}} \qquad (15\text{-}12)$$

and

$$W_{\text{remote}} = \frac{\delta}{l} W_{\text{near}} \qquad (15\text{-}13)$$

where $l =$ the distance from the end of the tube.

15.6   Torsion Snap Fits

15.6.1   Torsion Snap Fit Designs

Perhaps the least common variety of this type of fitment is the torsion snap joint. It is an excellent approach for reopenable joints and can be used as a release or latch. Figure 15-15 illustrates two examples of snap fitting arms with torsion bars.

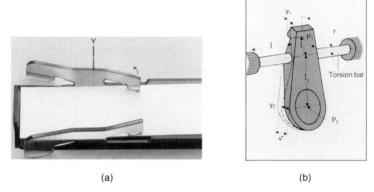

(a)   (b)

Figure 15-15   Torsion snap fits: (a) snap-fitting rocker arm about y axis and (b) snap-fitting arm with torsion bars (Diagrams courtesy of Bayer)

15.6.2   Engineering of Torsion Snap Fits

Shear stresses carry the principal load in torsion snap fits. We are interested in the force P necessary to deflect the arm a given distance y. That force can be exerted either at the hook of the snap joint (position 1) or at the other end of its lever (position 2). There is a proportional relationship between them. Thus:

$$\sin \varphi = \frac{y_1}{l_1} = \frac{y_2}{l_2} \tag{15-14}$$

where $\varphi =$ angle of twist (degrees)

$\quad y_1, y_2 =$ deflections

$\quad l_1, l_2 =$ lengths of lever arm

The maximum angle φ is determined by means of the following equation:

$$\varphi_{pm} = \frac{180}{\pi} \times \frac{\gamma_{pm} l}{r} \tag{15-15}$$

where $\varphi_{pm} =$ permissible total angle of twist in degrees

$\quad \gamma_{pm} =$ permissible shear strain

$\quad l =$ length of torsion bar

$\quad r =$ radius of torsion bar

An approximate value for the maximum permissible shear strain is:

$$\gamma_{pm} \approx (1 + \nu)\varepsilon_{pm} \quad \text{or} \quad \gamma_{pm} \approx 1.35\varepsilon_{pm} \tag{15-16}$$

where $\gamma_{pm} =$ permissible shear strain

$\quad \varepsilon_{pm} =$ permissible strain

$\quad \nu =$ Poisson's ratio (for plastics, approx. 0.35)

The force P is determined by calculating the moment arm, as in Equation 15-17, and dividing it by the length (l_1 or l_2).

$$P_1 l_1 = P_2 l_2 = \frac{\gamma G l_p}{r} \tag{15-17}$$

where $G =$ shearing modulus of elasticity

$\quad \gamma =$ shear strain

$\quad l_p =$ polar moment of inertia $= \pi r^4/2$ for a solid circular cross section

In the case of Fig. 15-15b, there are two torsion bars. Therefore, in that case, the equation becomes:

$$P_1 l_1 = P_2 l_2 = \frac{2\gamma G l_p}{r} \tag{15-18}$$

A fairly accurate value for the shear modulus G can be derived from the secant modulus according to Equation 15-19.

$$G = \frac{E_s}{2(1 + \nu)} = \frac{E_s}{2.7} \tag{15-19}$$

where $E_s =$ secand modulus

$\quad \nu =$ Poisson's ratio $= 0.35$

15.7   Strippable Snap Fits

In the language of injection molding, "strippable" means a part that can be removed directly from the core of the mold without extraordinary mechanisms (which are discussed later in this chapter). This results in a greatly reduced mold cost and a shorter molding cycle. Strippability is, therefore, something to be desired and, consequently, strippable snap fits are a favored technique, particularly in the packaging industry.

When snap fits are employed on the outside walls of the parts, one of the two parts will have its snap fit details on the outside of the part and the other will have its on the inside, or core side, of the part. The latter is the part having snap fit details that must be removed from a mold. Figure 15-16a illustrates a snap that cannot be stripped from a core. Figure 15-16b demonstrates how removal from the core would cause the snap hook to be sheared off. Thus, permanent snap fits, which require a 90° mating surface, are not strippable. To be strippable, the surfaces must be angled and radiused as in Fig. 15-16c. The amount is determined by the depth of the snap, the thickness of the wall and the stiffness of the plastic. Very stiff materials cannot be stripped at all.

Upon ejection from the core (Fig. 15-16d), the snap rides out of the recess in the mold, known as an undercut, and off the core. Unfortunately, that can lead to the deformation of the snap as it is removed (Fig. 15-16e). That, in turn, can lead the engineer to reinforce the back side of the snap, as in Fig. 15-16f. Unfortunately, this course of action may create thick wall sections, which result in external sink (see Chapter 1).

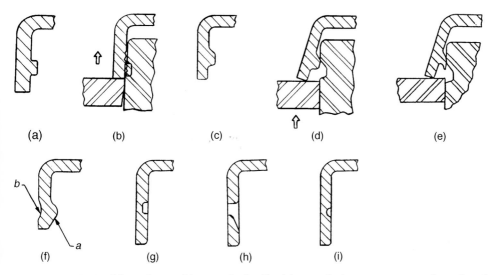

(a)   (b)   (c)   (d)   (e)

(f)   (g)   (h)   (i)

Figure 15-16   Molding of core side snap fit details: (a) poor design – snap cannot be stripped from core, (b) snap sheared off on ejection from mold, (c) better design, (d) snap ejected from mold, (e) snap deformed on ejection, (f) back-filled snap (*a*, back fill; *b*, sink), (g) unstrippable recess, (h) recess strippable in a very soft material, and (i) recess strippable in a soft material

When the detail to be removed from the core is a recess in the part instead of a snap hook, a similar situation occurs. The recess in Fig. 15-16g is not strippable, whereas those in Figs. 15-16h and 15-16i can be stripped, depending on the circumstances.

The rigidity of the plastic at the moment of ejection determines whether the slot or snap hook rides off the core. The component must be soft enough to expand and ride off the core without permanent deformation. (The equation for determining the amount of undercut that can be stripped is found in Section 18.3.4.1). Whether this can be done is dependent on the polymer, the mold construction, and the molding conditions. To fully comprehend the complexities of molding snap fits, it is necessary to understand the injection molding process and the design and construction of its molds.

15.8   The Injection Molding Process

With a few exceptions, snap fits are manufactured by the injection molding process. The fundamentals of the modern injection molding machine are illustrated in Fig. 15-17. The raw resin is poured into the hopper and fed from there into the heated cylinder. The screw in the cylinder turns, moving the plastic forward toward the nozzle. When the balance of the cylinder constitutes the desired shot size, the entire screw moves forward, forcing material through the nozzle into the mold.

The molding cycle is composed of five stages: machine close, injection, cooling or setup, machine open, and part ejection and removal. These are summarized as follows.

1. *Machine close* The beginning of the cycle commences with the closing of the mold to begin the cycle.

2. *Injection* The injection stage is the portion of the cycle during which the molten resin is pushed into the mold.

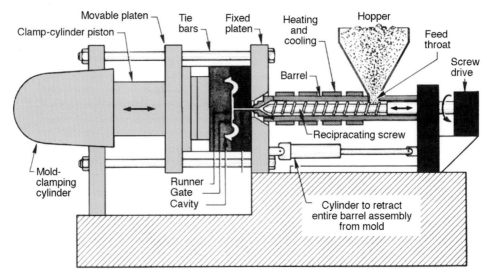

Figure 15-17   In-line screw injection molding machine (From R. E. Wright, *Injection Transfer Molding,* © 1995)

3. *Cooling or setup* During this stage, the plastic cools from the molten state until it is sufficiently rigid to withstand the force of ejection from the mold without distortion. This stage can consume from 70 to 80% of the total cycle and is largely dependent on the wall thickness of the part.

From the consideration of snap fitments, stage 3 is a critical point in the cycle. The cooling time must be long enough for the snap detail to also become rigid enough to withstand ejection without distortion. However, as the angle of the snap hook decreases and its radii increase in order to reduce the forces of ejection, the pocket of solid material at the wall of the snap detail also increases. This, in turn, results in the need for more time to cool. However, the remainder of the part is cooling at the same time and becoming more rigid, perhaps too rigid to expand enough to permit removal from the mold undercut. Thus, there are multiple variables simultaneously in play.

4. *Machine open* When the part has solidified sufficiently, the mold is opened.

5. *Part ejection and removal* As the mold opens, the ejector bars in the molding machine make contact with the ejection mechanism, which pushes the parts out of the mold. Usually, the part is simply pushed off the core. To remove the snap fit from the undercut while the part is still soft enough to expand, however, ejection must be more complex. That raises the issue of the tooling.

15.9   Molds for Snap Fits

15.9.1   The Basics of Injection Mold Construction

To comprehend the tooling complications, one requires a knowledge of injection mold construction and the manner in which such molds operate. First, a basic two-plate mold: the one illustrated in Fig. 15-18 is typical of the majority of molds in use today and is as simple as a production mold gets. (Much simpler tools are made for prototype work.) Item 1 is the sprue bushing into which the sprue is cut. Surrounding it is the locating ring, which aligns the mold to a recess in the stationary platen of the molding machine (item 2). Both bushing and ring fit into the clamp plate (item 3), also known as the top clamp plate because the mold is typically rested on its other end when it is not in the molding machine. The cavity retainer plate (item 4) contains the cavity block and the leader pins (item 5), which fit into the leader pin bushings (item 10). There is a set of these in each corner to guide the two halves together each time the mold is closed. One pair is offset to prevent incorrect reassembly of the mold following maintenance.

Item 6 in Fig. 15-18 is the core retainer plate, also known as the B plate. A pocket to receive the core block is cut in this plate which will also have cooling holes drilled in it. It receives the full impact of the injection pressure and is largely unsupported in the pocket in which the ejection mechanism operates. It is backed up by the support plate (item 7).

Item 9 is the housing that surrounds the ejection pocket. It is illustrated here as one piece, however it is often built as three pieces: a bottom clamp plate and two outer columns. Operating in the pocket are the ejector (item 8) and ejector retainer plates.

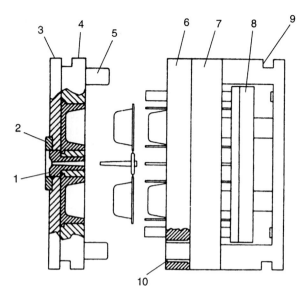

Figure 15-18 Basic two-plate injection mold: 1, sprue bushing; 2, locating ring; 3, clamp plate; 4, cavity retainer plate ("A" plate); 5, leader pin; 6, core retainer plate ("B" plate); 7, support plate; 8, ejector plate; 9, ejector housing; 10, leader pin bushing (From H. Belofsky, *Plastics: Product Design and Processing*, © 1995)

These plates are pushed forward by the ejector bars, which extend from the molding press (not shown). The heads of the ejector pins that eject the parts and the return pins that push the plates back to their original position are captured between the plates. One pin is placed directly under the sprue and is cut to trap some plastic and pull the sprue out. It is referred to as the sprue puller pin . The center of the mold would buckle without additional support; therefore support columns are strategically located in the open space of the ejection pocket.

The core and cavity blocks are placed in their respective pockets. It is customary to place the core in the movable half of the mold (because the part will shrink away from the walls of the cavity) and down on the core (where the ejection mechanism is available to push it off). Occasionally, an unusual design will require these positions to be reversed.

When the mold is filled, the molten plastic runs down the sprue through the runner system to the gate, where it enters the part.

There are significant pressure drops in the system as the plastic makes its way to the part. If a pressure of 137.89 MPa (20,000 psi) is applied to the material in the cylinder, there may be a drop to 103.42 MPa (15,000 psi) at the nozzle. As the plastic travels toward the part it loses further pressure, perhaps dropping to 82.74 MPa (12,000 psi) at the gate. An additional large pressure drop through the gate could reduce pressure to 27.58 MPa (4000 psi) in the part. These are only sample values: the numbers will vary enormously according to the material, the temperature of the cylinder, the contours of the runner system, the temperature of the mold, and the size and type of gate. The actual cylinder pressures will range from a few thousand psi up to 206.84 MPa (30,000 psi) for difficult-to-mold materials. The pressure drops continue in the mold according to the flow path necessitated by the contours of the part.

The drop in pressure is due, in part, to the increase in viscosity of the melt as it cools on its way to the cavity. The condition of the melt when it finally reaches the portion of the cavity that contains the snap fit details is critical to the function of those design elements. Cantilever snap fits are often long thin structures that can be difficult to fill. Thicker versions may result in pockets of solid material exceeding 125% of the surrounding walls, which leads to high levels of molded-in stress.

Cylindrical snap fits will be vulnerable on assembly to a weld line through the snap fitment if the cavity is not center-gated. That is because edge and submarine gates deliver the material from one side of the core. It then travels around the core, and a weld line is formed where it meets on the other side. A center gate delivers the material from the top of the mold and it flows down the sides. Therefore, a weld line through the snap fitment should not be formed. However, where the material meets on the parting line, a small weld line can form if the core is not centered in the cavity (the melt reaches the bottom of the thicker side first and then travels around the core). Therefore, careful mold construction is called for to avoid a weak fitment.

15.9.2 Ejection and Cooling Systems for Stripping Molds

The ejection system in the mold in Fig. 15-18 consists of simple ejector pins. Unfortunately, these pins provide a localized pressure on the plastic part that can cause distortion, as depicted in Fig. 15-19a, unless the part has cooled enough to be rigid. However, if it has cooled that much, it may be too stiff to expand over the core for ejection. The stripper plate shown in Fig.15-19b distributes the ejection pressure over the entire circumference of the part. Therefore, the part can be ejected at a softer state, providing the molder with a great deal more control over the molding conditions. Stripper plates cost more than ejector pins; however, they can return the investment by reducing the molding cycle as much as 35%.

Clearly, the method of cooling the core will have a substantial effect on the cooling of the part. The simple system illustrated in Fig. 15-19a will not provide optimum cooling for the core. It is too far from the heat source, which is the part itself, often referred to as the "moldment."

Note the bubbler in the core of the mold in Fig. 15-19b. This is one of several types of cooling device intended to bring the cooling water up into the core of the mold. If such a device is not used, the core may be hot enough to prevent the molder from establishing a cycle that will eject the part without distorting it. It provides an important means of controlling the molding parameters. Improper cooling is one of the major causes of part failure.

The mold in Fig. 15-19b has snap fits depicted in two different locations. Undercut 1 is near the corner of the part, while undercut 2 is near the lip. If both undercuts are of the same depth, the wall on the right must deflect more than twice as much for undercut 1 (vs. for undercut 2) to achieve ejection because of its relative distance from the corner, which, more or less, becomes the pivot point of the deflection.

The deflection of the part causes the outside diameter to expand, briefly, during ejection. This is no problem for snaps located on the outside of the part because the

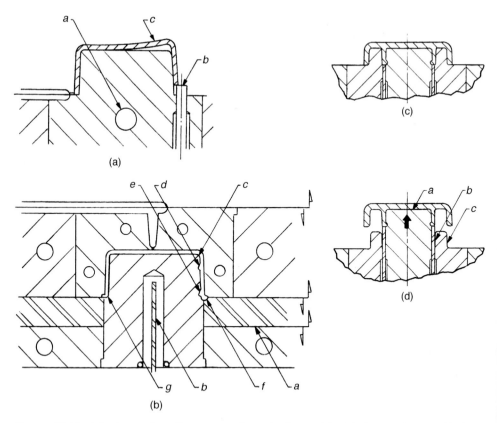

Figure 15-19 Ejection and cooling systems for stripping molds: (a) ejector pins provide localized ejection pressure (*a*, water line; *b*, ejector; *c*, part distortion), (b) stripper plate mold for undercuts (*a*, stripper plate; *b*, "bubbler" core cooling system; *c*, core; *d*, undercut 1; *e*, undercut 2; *f*, full radius lip; *g*, partial radius lip), (c) mold with undercut on internal rib, and (d) riser for undercut (*a*, riser; *b*, ejector sleeve; *c*, core)

cavity separates when the mold is opened. However, for snaps on internal structures such as the one in Fig. 15-19c, coring prevents this expansion. In these cases, a "riser" or "lifter," such as the one shown in Fig. 15-19d, must be employed. This requires a two-stage ejection system, which causes the cored section of the mold to lift the part out of the portion of the mold that is interfering with ejection. Ideally, the ejectors should be placed directly under the undercuts, as shown in Fig. 15-19d. However, other nearby locations have been used successfully. In this case, an ejector sleeve has been used in place of the ejector plate, however ejector blades could be used for parts that are not round. These systems are more costly to build and maintain in the mold.

Because several variables interact as the strippable snap is being removed from the core, it is recommended that an unhardened one-cavity production mold be built for the purpose of experimentally determining the optimum part design, mold design, molding cycle, and material selection. Note the term "one-cavity production mold": it implies that a prototype method that does not duplicate the production cooling, gating, and ejection systems will not provide reliable information on which to base the construction of a costly multicavity production mold.

Bear in mind that the stress–strain data available are very specific to a given set of conditions. Changes in mold and melt temperature, humidity, wall thickness (from the 3 mm standard specimen), gating, and geometric configuration will produce different results. Taking samples from the single-cavity mold will not only permit the testing of the design parameters, it will provide samples for testing of creep and other accelerated aging effects. It is recommended that the trial tool be built "metal safe" (cavities on the low side; cores on the high side) so that the design can be modified by simply removing metal. That eliminates the need to weld metal to the mold and then remove it, a far more costly approach. The revisions are much easier to accomplish with a mold that has not hardened.

15.9.3 Cores for Nonstripping Molds

Many plastic materials are too rigid to permit the stripping of snap details from the core. They require cores with segments, referred to as "collapsing cores," that move out of the way.

However, before we delve further into that subject, it should be noted that there is a simple device for ejecting undercuts. That is to place them on the ejectors themselves. Figure 15-20a shows the part in the mold in the as-molded stage. The ejectors have moved forward in Fig. 15-20b: in this case, the part is not actually ejected from the mold but is left suspended on its ejectors and requires hand or robotic removal. While this method permits the molding of snap fit undercuts at modest initial cost, it is limited in its utility to smaller production quantities because the cost of more sophisticated tooling will be offset by the increased cost of removal beyond one or two cavities. On some designs, it is possible to use a two-stage ejection system to strip the part. The ejector pins are driven by a separate ejector plate, which continues to travel after the pins containing the undercut have stopped, thereby driving the part off those pins.

This method of creating cantilever snap arms can be used for prototype molds to test the design before proceeding to the construction of expensive snap-fit production

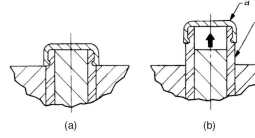

Figure 15-20 Cores for nonstripping molds with undercuts (a) on ejectors – as molded and (b) on ejectors – ejectors forward (*a*, moldment; *b*, ejector)

(a)   (b)

tooling. While desirable to reproduce the same molded-in stresses in a prototype part as in the production part, it is not as important to create all the features of a production mold for a prototype version of a nonstripping mold. A part molded from a hand-operated mold without a production cooling system will tend to have more molded-in stresses, therefore it would tend to err on the safe side, failing at lower stresses or deformations. However, it could also cause a successful design to be discarded because of a prototype mold that was too primitive.

Another way to use mold components driven by the ejection mechanism to create snap fit undercuts is illustrated in Fig. 15-21. It uses an angled core blade mounted to a slide located on the ejector retainer plate. In that way, the vertical opening of the mold translates to a transverse motion as well, causing the snap undercut to come free of the mold to permit ejection. Note that there must be clearance for the transverse movement of the blade, and no other part detail can be placed in a position that would interfere with that motion.

There is another method of collapsing cores that is commercially available; however it is limited in its application to round parts with core diameters between 16.38 and

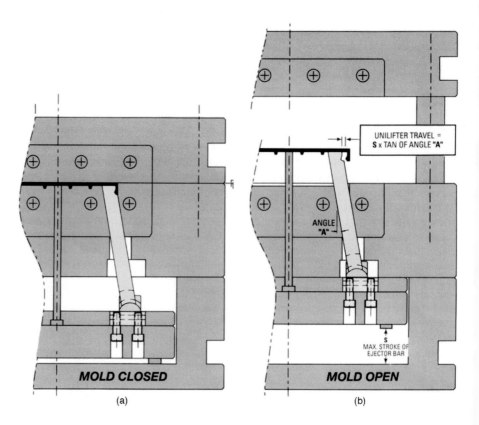

Figure 15-21   Undercut on angled lifter with mold closed (a) and open (b) (Diagrams courtesy of DME Company)

89.78 mm (0.646 and 3.535 in.). In the range between 16.38 and 24.51 mm (0.646 and 0.965 in.), only intermittent undercuts are permissible. This is discussed further in Chapter 18: "Threads: Tapped and Molded In."

Collapsing core molds are expensive to build and maintain. Therefore, the evaluation of the snap fit cost-effectiveness must include these mold cost factors.

15.9.4   Snap Fit Details in the Mold Cavity

Depending on the design, a hole, recess, or snap lug will need to be on the outside of the part. Such a feature would make the part just as difficult to remove from the cavity as an undercut is to eject from the core. In the case of the lug, it may be possible to simply shift the parting line of a simple mold. An example of this approach is shown in Fig. 15-22. Unfortunately, there is a potential for flash at the parting line which would interfere with the operation of the snap. An extra margin of safety would be needed for allowable strain as the flash would add to the deflection of the snap.

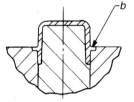

Figure 15-22   Snap fit detail on the parting line: *b*, possible flash on snap lug

If snap lugs cannot be formed on the parting line, the cavities must be split to permit removal of the part. One way to accomplish that is to use the split-cavity version of the spring-loaded collapsing core mold. This is also discussed further in Chapter 18, where a split-cavity mold is illustrated (Fig. 18-6).

A true split-cavity mold is created by cutting the cavity into slides (or side actions). The angled cams force the cavity halves to slide to the sides as the mold opens. They must move only enough to clear the snap hooks. Once that has happened, the part can be ejected in the usual fashion. On closing, the cams force the cavity halves back together and the clamps provide the locking force at the end of the travel.

In some cases, a slot or recess on the cavity side is called for. The same type of cam action used to split the cavity can be used to operate a slide that makes a hole in the side, as illustrated in Fig. 15-23. Such mechanisms are expensive, not merely because of the cost of putting them into the mold, but also because of the space they require in the mold base. This translates into a reduction in the number of cavities that can fit in a given size mold base which, in turn, limits the number of cavities that can fit into a given size molding machine. Looking at it another way, a larger machine with a higher hourly operating cost would be required to accommodate the same number of cavities as a simple mold.

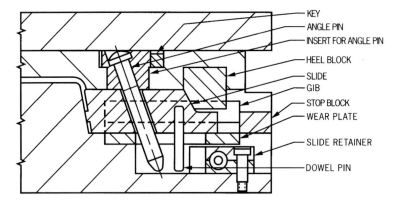

KEY
ANGLE PIN
INSERT FOR ANGLE PIN
HEEL BLOCK
SLIDE
GIB
STOP BLOCK
WEAR PLATE
SLIDE RETAINER
DOWEL PIN

Figure 15-23   Side action mold (Diagram courtesy of DME Company)

The left side of Fig. 15-24 shows a 45° wall. A wall this flat permits the core to shut off against the cavity to create a hole (*b*). This is referred to as a "shutoff." As the angle of the wall is increased toward vertical, it becomes difficult to maintain the wall thickness, as illustrated to the left of that hole. Note that the wall is decreased on the upper part of the cavity and the lower side of the core relative to the left wall. That permits the core and cavity to meet at the hole and create the shutoff (*c*). The wall in the illustration is shown at 80° from horizontal, probably the extreme upper limit for a shutoff of this type. Of course, the wall can be thickened to its nominal weight to each side of the opening.

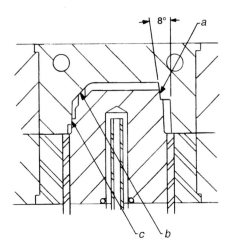

Figure 15-24   Molds for side holes using shutoffs: *a*, corner shutoff hole; *b*, 45° shutoff hole; *c*, maximum angle shutoff with reduced wall thickness

Another type of shutoff is shown on the right side of that illustration, namely, a shutoff at the corner of the part, which in turn permits a shutoff on a normally drafted wall. The angle of the shutoff should be set at 8° because the steel locks at 7° and below. That is not to say that lower angles will not function, but it does mean that the mold must endure greater wear, which will shorten its life and increase its maintenance cost. That concept can be used to create a snap lock in the base of the part, as shown in Fig. 15-25.

Figure 15-25   Moldment with locking feature created by a shutoff

15.10 Conclusions

Creative designers and engineers have found almost limitless ways in which to use snap fits. In most cases, these applications have been successful. However, they usually require a degree of adjustment to achieve the desired level. The problem is that beyond the fundamentals of creating enough holding power and avoiding excessive deformation, the success of the design depends on establishing the right "feel," a property for which no equation has yet been created. The best way to proceed, in principle, is to design to the light side and use a "metal-safe" prototype mold to test the design. If two or more different grades of the resin are chosen for trials, it will be possible to evaluate the effect of a small difference in the properties on the performance of the snap fitment. If a given material is not acceptable for some other reason, such as processing or cost, a similar proportion can be used to establish the revised geometry for the next trial. From the equations, it is clear that geometry plays a large role in the design of snap fitments. Besides experimenting with materials, the designer can add a small rib to the back of a cantilever arm to stiffen it.

It is desirable to build prototype molds with inserts that permit the molding of parts to the limits of the tolerance range. While this is an extra expense, it is a way to test the effects of the tolerance range. When this is not done, engineers tend to specify tighter tolerances than they truly need, thus paying an additional cost for the life of the product instead of just for the prototype mold.

Location can be a critical element in the function of a cantilever snap fitment. The snaps should not, themselves, be used to locate one part to the other. If that is done, an undersize part may disengage despite the presence of properly designed snaps, simply because the part has shifted to one side. Location of the parts relative to each other should be established by separate locating ribs or other such locating devices. The location of the parts is so significant to the effectiveness of the snap fitment that those tolerance extremes should be tested in the mold as well.

The relative height of the snap can also be a problem. If the distance to the snap hook is too small, the snap may not engage. Therefore, the tolerances must be set so that the hook engages at the extreme condition of a short hook length and a long undercut distance. That means that all the tolerances tend toward creating a gap. That gap will result in a loose fitment or rattling if a vibration is present. Snap fits can be designed to be preloaded to prevent loosening due to vibration. However, it is difficult to control the amount of preload without extremely tight tolerances. Furthermore, the preload raises concerns about creep and its related issues similar to those encountered in the design of press fits, although preload stresses are generally at a much lower level than those of press fits.

When evaluating the trial samples, one must be alert to the presence of strain marks (light-colored streaks) or hook deformation. These are telltale signs that the strain limit has been exceeded.

First, check the dimensions of the parts as molded. The most common cause of this type of problem is a part that is out of tolerance. If the source of the difficulty lies elsewhere, the best solution is to look for a variation of the plastic that can accept greater strain. If that fails, the cross section of the arm can be thinned or tapered, or the arm lengthened. However, since those options involve significant mold changes, they represent last resorts.

The good news is that once a designer or engineer has created his or her first snap fitment, the next one is easier. Indeed, after several have been executed, the designer may find that an existing snap fit design can be adapted to the new part, thereby saving a great deal of time.

16 Spin Welding

16.1 Description of Spin Welding

In its crudest form, spin welding, also known as rotary friction welding, can be done with an ordinary lathe or drill press. Indeed, for many years, most spin welding was done with such equipment or with custom equipment fabricated in the manufacturer's own shop.

The principle is very simple. One part is held stationary while the other is spun against it. The only variables are the velocity of the spinning part, the pressure with which it is engaged with the stationary part and the length of time required for the bond to take place. Generally, less than one full turn is required and the elapsed time is under 2 seconds. A brief dwell time allows the joint interface to cool sufficiently for the parts to be removed from the welder. Virtually all the thermoplastics can be spin welded including the crystallines.

16.2 Advantages and Disadvantages of Spin Welding

16.2.1 Advantages

1. *No additional materials* Spin welding uses no additional materials such as fasteners, inserts, electromagnetic preforms, adhesives, or solvents. Therefore, it is inherently less expensive than methods that have such requirements, and costs of disassembly are lower.

2. *Ease of assembly* Spin welding requires only the alignment of the two parts in the fixture of the welder; thus it is well suited to automated assembly. High production rates are possible as the actual welding time rarely exceeds 2 seconds.

3. *Entrapment of other parts* Additional parts can be captured between the two parts to be spin-welded.

4. *Permanence* Spin welding creates permanent assemblies that cannot be reopened without damaging the parts. Since the joints are welded, the effects of creep, cold flow, stress relaxation, and other environmental limitations are absent in the joint area, although they may occur elsewhere if other design elements are under load. Owing to material limitations, differences in thermal expansion and moisture absorption are rarely of concern.

5. *Hermetic seals* Spin welding is capable of creating hermetic seals.

6. *Energy efficiency* Relative to other welding processes, spin welding is highly energy efficient. This also means that there is no excess heat, which must be removed from the workplace, as with hot plate welding.

7. *Clean atmosphere* Unlike adhesive and solvent joining systems, no ventilation equipment is necessary for the removal of toxic fumes.

8. *Immediate handling* Assembled parts can proceed to other operations at once without waiting for parts to cool and for adhesives or solvents to set. This advantage makes the process well suited to automated assembly line applications.

9. *Equipment cost* Inertia spin welders and their tooling are relatively inexpensive in their simplest models. Direct-drive models, which can control orientation, can cost a good deal more.

10. *Size limitations* Theoretically there is no limit to the maximum size of a spin weldment. The largest spin weldment on record is reportedly 1.2 m (48 in.). However, machines of that size are not widely available and most applications are 0.3 m (12 in.) or less.

11. *Far field welding* Spin welders hold an advantage over vibration and ultrasonic welders for assemblies of certain shapes. Tall applications, which would call for far field ultrasonic welding, are better spin-welded, particularly if the shapes are contoured or have openings, or if the materials are crystalline thermoplastics.

16.2.2 Disadvantages

There are some disadvantages as well.

1. *Shape limitations* Spin welding is limited to circular fitments, and most applications involve circular parts as shown in Fig. 16-1a. Sometimes, however, one is of another shape with a circular fitment, as shown in Fig. 16-1b.

2. *Process limitations* Injection molding is the principal process that can create the joint details necessary for spin welding; however, only one of the parts requires the joint details. The other part can be an extrusion that is round in cross section or a part made from any thermoplastic process with a hole drilled in it.

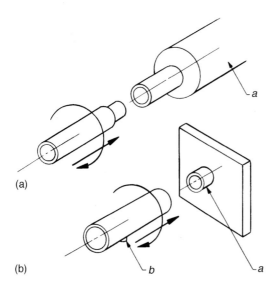

(a)

(b)

Figure 16-1 The spin welding concept: (a) round to round (*a*, stationary part is usually larger) and (b) round to square (*a*, locating boss; *b*, detail can be oriented with difficulty)

3. *Registration difficulty* Basic inertia spin welding cannot reliably orient one part to the other. Models that hold orientation tolerances in the range of ±0.5 to ±2.0° increase the cost by 50 to 300%, the price increasing commensurate with greater precision.

4. *Production rate limitations* Spin welders are capable of welding only one part per cycle.

16.3   Spin Welding Process

Spin welders are classified both by the method by which the rotation is achieved and by the manner in which the parts are handled. Rotation of the driver can be accomplished either by inertia or by direct drive. The parts can be brought into position in a single stage or in two stages.

In a single-stage spin welder, there are two ways to load the parts. One way is for both parts to be placed in the fixture mounted to the base of the spin welder. The driving head then descends, engages the upper part, and spins it to a weld. This technique works best for small parts. With medium to large parts, one is placed in the fixture and the other in the driving head. The drive head then descends spinning and presses that part to the lower part until it welds. Single-stage welders can be either inertia or direct drive.

In two-stage welding, the stationary part is placed in a fixture and the part to be rotated is placed on the first part, as shown in Fig. 16-2a. The drive head then descends to pick up the upper part and begins spinning it, as in Fig. 16-2b. In Fig. 16-2c, the spinning drive head has descended, once again, and is being pressed against the stationary part. Two-stage spin welding is done only with direct drive welders. Driving heads of inertia welders are spinning prior to engagement of the parts. They are then disengaged, and inertia provides the driving force until the weld causes the head to come to a stop. Parts under 38 mm (1.50 in.) require an inertia welder with a brake. Direct-drive welders have a clutch between the driving shaft and the head. When the weld stops the driving head, a clutch disengages it from the shaft. Inertia welders are mostly used for parts under 152 mm (6 in.) in diameter; however, they can be used for parts up to 305 mm (12 in.). Direct-drive welders can do small parts as well, however they provide better control for larger parts.

Regardless of how the head is driven or the parts are engaged, the principle is the same. When the spinning part contacts the stationary part, the friction under pressure causes the surfaces to melt. That leads to friction between the layers. After the polymer has reached its glass transition temperature, it arrives at a steady state at which heat losses through the wall and those due to flash equal the heat being generated. This part of the cycle is application dependent but will usually require less than half a second.

The spinning head is then braked to a halt by friction or a brake (depending on the system) and held in that position until the assembly is cool enough to be removed from the fixture without distortion, as shown in Fig. 16-2d. That usually requires another second, for a total heating and cooling cycle normally under 2 seconds. However, additional time must be allowed for loading and unloading the parts. Thus a typical production rate is on the order of 5 to 15 assemblies per minute. This can be speeded

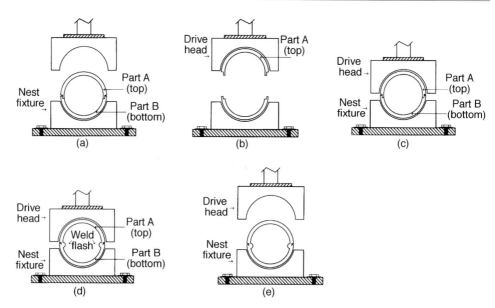

Figure 16-2 The two-stage spin welding process (Diagrams courtesy of Sonics & Materials, Inc.): (a) parts are loaded, (b) drive head picks up upper part and begins spinning, (c) upper head descends and weld takes place, (d) assembly cools until it can be removed, and (e) assembled part ready for removal

with a swivel base that permits one side to be unloaded and reloaded while the other side is spinning. Automated equipment is also feasible.

For inertia spin welding to work, the spin driver (and flywheel, if necessary for the application) must be accelerated to a speed that generates enough kinetic energy to create an effective weld. Once the tool has reached the required speed, it is disengaged from the motor and extended downward to engage the part to be welded. The kinetic energy is converted to heat by the friction under pressure, causing the joint interface to melt and weld as just discussed. There must be sufficient pressure to stop the rotation in less than one second. If rotation extends longer, the solidifying polymer will be sheared and the result will be a reduction in the strength of the weld.

The principal variables are the velocity of the spinning part, the pressure on the joint, and the weld time. The velocity should fall in the range of 9.1 to 15.2 m/s (30–50 ft/s). Speeds as low as 3 m/s (10 ft/s) have been used for small parts with thin wall sections, however the higher velocities are more desirable. The velocity is determined by the following equation:

$$v = \pi DN \tag{16-1}$$

where v = velocity (m/s) (ft/s)

D = average diameter of the joint (m) (ft)

$\pi = 3.1416$

N = angular velocity (rps)

The greater the thickness of the joint, the greater the difference in velocity between its inner and outer diameters. This is not likely to be a problem because the joint thickness is controllable. However, "donut" designs with two weld surfaces at widely different diameters should be avoided. Weld pressures should fall in the range of 207 to 517 N/cm^2 (300–750 psi) and the weld should require 0.1 to 0.3 second to stop rotation.

It is normal to create some flash. To eliminate the flash from view, a flash trap will be required. However, excessive flash is the result of too much energy, and reducing it will not weaken the joint. There are several ways to remove flash. Among them are the following:

1. *Secondary operation* This can be done directly on the spin welder by spinning the part while a cutting tool or rotating brush is brought into contact with the flash. It is sometimes done at a remote location.

2. *Tumbling* The parts can be tumbled in a tumbling barrel with shells, fruit pits, or another medium. In some cases, it may be necessary to apply a blast of nitrogen to reduce the temperature of the parts to −46 °C (−50 °F) to make the flash more brittle. Molding flash will also be removed by this operation.

Weak joints or failure to weld will require an increase in the energy provided. That is achieved by altering the welding conditions. The easiest adjustment is to change the angular velocity. However, the weld pressure is also variable, and a weak joint may be the result of too low a thrust pressure, whereas too great a pressure will stall out the head without creating a weld. The objective is to create a good weld while keeping the weld time as short as possible.

For inertia welders, the weight of the spin driver can be estimated at the rate of 0.70 to 1.41 kg for each square centimeter (10–20 psi) of weld surface. From that point, further adjustments are readily made by altering the rotational speed of the spin driver. Inertia spin welding can produce good, consistent welds as long as good parts with properly designed joints are provided.

16.4 Materials

Any true thermoplastic can be spin welded. In that respect, the process is similar to hot plate and vibration welding. Spin welding is particularly useful for the crystalline thermoplastics that are difficult to join with ultrasonic welding. As with ultrasonic and vibration welding, the amorphous resins are more readily welded than the crystalline resins. However, the crystalline thermoplastics, polyolefins in particular, are more readily spin- or vibration-welded than ultrasonic-welded. As a general statement, most materials spin weld about as well as they vibration weld. However, spin welders are less expensive, particularly if the parts do not require orientation. Table 16-1 will serve as a guide for spin welding. Joining of dissimilar polymers is often unfeasible. When possible, it generally produces lower strength joints. However, greater freedom can be gained when an undercut can be designed into the joint. Then the polymer with the lower melting point will flow into the undercut, creating a mechanical fitment.

Table 16-1   Welding Characteristics of Thermoplastics

Material	Spin Welding Performance
ABS	Good to excellent
ABS/polycarbonate alloy	Good
Acetal	Fair to good
Acrylic	Good
Acrylic multipolymer	Good
Acrylic–styrene–acrylonitrile	Good
Amorphous polyethylene terephthalate	Poor to fair
Butadiene–styrene	Good to excellent
Cellulosics	Good
Fluoropolymers	
Polyvinylidene fluoride (PVDF)	Good
Perfluoro alkoxy alkane (PFA)	Poor
Ionomer	No information
Liquid crystal polymers	Fair to good
Nylon	Good to excellent
PBT/polycarbonate alloy	Good
Polyamide-imide	Fair to good
Polyarylate	Good
Polyaryl sulfone	Good
Polybutylene	Poor to fair
Polycarbonate	Good to excellent
Polyester, thermoplastic	
Polyethylene terephthalate (PET)	Fair to good
Polybutylene terephthalate (PBT)	Good
Polyetheretherketone (PEEK)	No information
Polyetherimide	Good
Polyethersulfone	Good to excellent
Polyethylene	Poor to good (density dependent)
Polyimide	Good
Polymethylpentene	No information
Polyphenylene oxide	Good
Polyphenylene sulfide	Good
Polypropylene	Good
Polystyrene (general purpose)	Good to excellent
Polysulfone	Good
Polyurethane	Poor to fair
PVC (rigid)	Good to excellent
Styrene acrylonitrile	Good to excellent

Filled and reinforced resins, including nylons and acetals with up to 33% glass fill, are readily welded with spin welding. However, only the polymer welds, so the strength of the bond is that of the resin itself and the strength of the joint is further reduced by the percentage given to the filler. Conversely, this process is less affected by mold release agent and other surface contaminants than either hot plate or ultrasonic welding, and it is somewhat less affected by moisture content, although highly hygroscopic polymers must still be handled with care and may require special handling or drying for some applications.

16.5   Design for Spin Welding

16.5.1   Overall Design Considerations

The primary requirement for a part to be spin-welded is that the joint have a circular cross section. It can be spherical, cylindrical, or simply a hole in a part. One of the parts need not itself be round, only the joint must be. For example, the joint could be a boss on a rectangular part. That part would be placed in the stationary fixture. The part to be spun should be round so that it is balanced about its centerline. In addition, the wall below the joint must be strong enough to withstand the welding pressure. The minimum spin welding wall thicknesses are usually found in containers with walls ranging from 0.38 to 1.50 mm (0.015 to 0.060 in.). The largest wall thickness spin welded reached 6.35 mm (0.25 in.), however the initial contact surface of the joint was 1.50 mm (0.060 in.).

Small parts can use friction drivers, such as silicone pads, like the one in Fig. 16-3a. For larger parts, provision must be made to prevent slippage in the drive head during welding. Ribs and bosses, known as "dogs," can be used to grip the parts, and driving lugs such as those shown in Fig. 16-3b can be used in conjunction with a matching driver. In some cases, a flat surface can be driven with a toothed driver as in Fig. 16-3c.

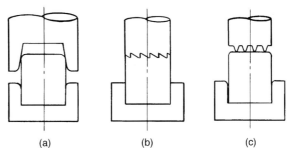

Figure 16-3 Spin welding drivers: (a) 10° driver, friction driver, (b) driving lugs, and (c) tooth driver

(a)   (b)   (c)

16.5.2   Joint Designs

A good spin welding joint must provide alignment between the two parts and a weld area equal or greater than the nominal wall thickness of the part. There are three

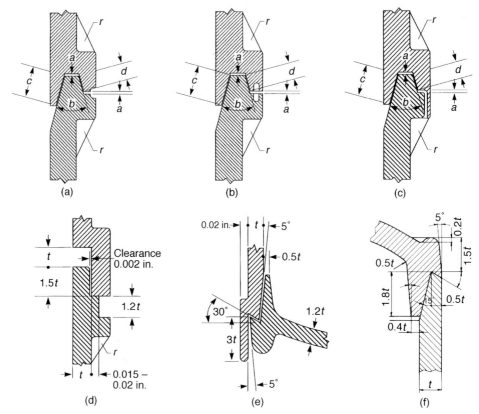

Figure 16-4 Spin driving joint designs: (a) basic tongue-and-groove joint, (b) tongue-and-groove joint with flash trap, (c) tongue-and-groove joint with skirt, (d) basic shear joint, (e) scarf joint with skirt, and (f) basic scarf joint. *a*: Depth of weld; should be 0.5 to 0.8*t*. *b*: Angle of stationary joint interface; must be ≥30° to avoid jamming. *c* + *d*: Weld surface total; should be ≤ 2.5*t*, *r*: Peripheral ribs for gripping purposes. *t*: Wall thickness

basic joints used in spin welding along with a range of variations of each. These are the tongue-and-groove joint, the shear joint, and the scarf joint.

The favored joint is the tongue-and-groove design (Figs. 16-4a–c) because it is the most forgiving. The version in Fig. 16-4a is the basic tongue-and-groove joint. It would provide a secure joint, but it would show flash to both sides. The version in Fig. 16-4b has a flash trap incorporated into the right-hand wall. In this case, flash will be visible to the left side. To conceal it, an additional flash trap would have to be designed into that side. The skirt design in Fig. 16-4c can also be used to hide flash.

Unfortunately, the tongue-and-groove joint is an expensive joint to tool and can lead to a problem of entrapped air in molding. Therefore, the shear joint shown in Fig. 16-4d is preferred for some applications. This is the type of fitment favored for holes and bosses. A shear joint must have sufficient wall rigidity or tooling support to accommodate the effective pressure needed to create a melt. The scarf joint attempts to combine the features of both the preceding joints. Two versions of scarf joints are shown in Figs. 16-4e and 16-4f, the former with a skirt. Note the ribs (*r*) which are provided around the periphery of the parts, to control their relative motion.

16.6   The Equipment for Spin Welding

16.6.1   Drill-Press-Based Spin Welders

A drill press can be equipped for spin welding. While this method of producing a spin weldment has the lowest equipment cost, it is by no means the most cost-effective for large-scale production runs. The drill press is usually operated by an air cylinder for efficiency, however spin welding can be done with a hand-operated model. Even so, the drill press approach is recommended only for prototypes and very short runs. For longer production runs, a commercial spin welder is preferred.

16.6.1.1   Tooling for Drill-Press-Based Inertial Welding

As illustrated in Fig. 16-5, the tool for inertial welding will have a rotating mass and a driving wheel. For the smallest parts, the driving head may be driven by friction

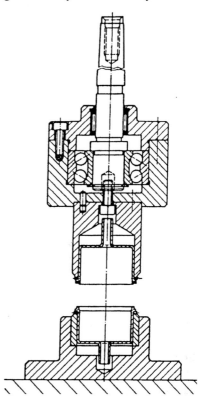

Figure 16-5 Inertia tool for drill press (Diagram courtesy of DuPont Engineering Polymers)

between the shaft and the flywheel through the bearings. For larger parts, a clutch must be provided in the tool. Once the weld has been accomplished, the head will resume spinning after each welding cycle.

16.6.1.2 Tooling for Drill-Press-Based Pivot Tool Welding

The pivot tool shown in Fig. 16-6 is another device for spin welding by means of a drill press. The downward pressure is achieved with a spring acting on the center pivot, which can be adjusted to provide a range of loadings on the part. The pivot is lowered until it finds a recess in the upper part and applies a load. The head, a toothed variety in the illustration, continues to descend until it engages the upper part and drives it until a weld is formed. It then releases immediately, leaving the pivot in place and supplying pressure for 0.5 to 1.0 second so the weld can solidify. The head is then retracted fully, permitting the assembly to be removed and new parts to be placed in the chuck.

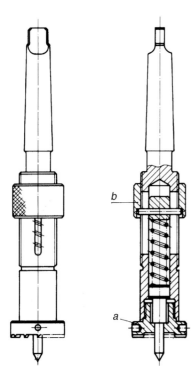

Figure 16-6   Pivot tool for drill press: *a*, toothed crown; *b*, spring loaded adjustment sleeve (Diagram courtesy of DuPont Engineering Polymers)

16.6.2   Commercial Inertia Spin Welders

Beyond the drill press variety, inertia spin welders are the least expensive type of commercially available spin welder. At the time of this writing, these welders can handle parts up to 305 mm (12 in.) in diameter and have a head height adjustment of

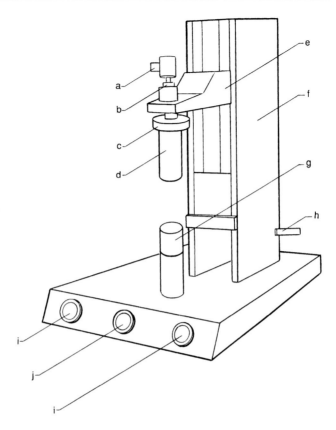

Figure 16-7 Inertia spin welder: *a*, Motor control valve; *b*, pneumatic motor; *c*, kinetic wheel, *d*, driver; *e*, height adjustment locking acoustical stack; *f*, frame; *g*, workpiece positioning anvil; *h*, head down speed control; *i*, dual safety control push-buttons, and *j*, emergency stop (Diagram courtesy of Forward Technology Industries, Inc.)

356 mm (14 in.). However, they cannot control orientation and they have difficulty controlling the weld on larger parts. The commercial spin welder illustrated in Fig. 16-7 is an air-driven system with a control valve (*a*) that regulates the speed of the motor. The air then flows to the compact pneumatic motor (*b*), which drives the flywheel (*c*), also known as the kinetic wheel. The driver (*d*) is attached to the head. Vertical travel is controlled by the air cylinder mounted within the machine frame (*f*) which has its own air supply. It operates an arm (*e*) to which the drive assembly is mounted and is controlled by a speed control (*h*). The part rests in a fixture (*g*). Controls are a set of dual operating buttons (*i*) for safety and an emergency stop button (*j*). A three-stage timer regulates the start time for the flywheel, its rotation time, and the dwell or cooling time after the weld which ensures that there is enough time for the plastic to solidify.

The welder accelerates in the up position and the air supply stops when the flywheel reaches the desired speed. The driver descends to create the weld and comes to a stop when the weld is formed. It is then retracted and remains in the stopped condition while

Figure 16-8 Inertia spin welder (Photograph courtesy of Forward Technology Industries, Inc.)

the part is removed and new parts are placed in position for the next weld. The flywheel and driver are application dependent and must be changed for each job; however, changeover does not require very much time. Higher speed production is achieved with a rotary table that optimizes the machine utilization by separating the loading and unloading functions from the welding cycle.

Inertial spin welders are currently available with motors in three speed ranges. Very small parts between 6 and 38 mm (0.25–1.50 in.) in diameter are best suited to the motor which operates from 0 to 16,000 rpm. The midrange of 38 to 102 mm (1.50–4.00 in.) in diameter is covered by a unit that rotates between 0 and 4800 rpm, while the larger parts, from 102 to 305 mm (4–12 in.) in diameter, require a model that operates in the range from 0 to 2400 rpm. An inertial spin welder is shown in Fig. 16-8.

16.6.3 Commercial Direct-Drive Spin Welders

The majority of spin welders currently available on the market are direct-drive models. These designs provide greater control throughout the range of welding sizes and can be custom-built to practically any size. Standard commercial models handle parts up to 229 mm (9 in.) in diameter and up to 559 mm (22 in.) in height, however custom models are readily available. The bulk of applications are under 305 mm (12 in.), with several mentioned in the 457 to 610 mm (18–24 in.) range. The largest reported application was 1219 mm (48 in.) in diameter. Lower priced models do not control orientation; however, models that can hold orientation from 0.5 to 2° range in cost from 30 to 300% greater than the lowest priced inertia models. Direct-drive spin welders can be either single stage or two stage in their handling of the parts.

A direct-drive spin welder is illustrated in Fig. 16-9. It is quite similar in general layout to the inertial welder shown in Fig. 16-8, however an electric drive motor has replaced the pneumatic motor and its control valve above the driver, and there is no flywheel. The motors range in size from 2 to 3 horsepower in standard models, with custom versions often using motors of 5 horsepower and above. They are rated by the torque they provide. Vertical motion is provided by an air cylinder located within the frame, as with inertia welders. A brake is used to stop rotation.

Figure 16-9 Direct-drive spin welder (Photo courtesy of Sonics & Materials, Inc.)

Dual operating buttons are standard on commercial models. Three-stage timers controlling pickup, spin, and hold periods are found on standard models, but speed must be read manually. However, programmable logic controllers with a key pad and an LED tachometer are available at an extra cost for some models.

16.7   Sources

Branson Ultrasonics Corp., 41 Eagle Rd., P.O. Box 1961, Danbury, CT 06813-1961, (203) 796-0400, fax (203) 796-9838, www.bransonultrasonics.com

Dukane, 2900 Dukane Drive, St. Charles, IL 60174, (630) 584-2300, fax (630) 584-3162, www.dukane.com/us

Forward Technology, 260 Jenks Ave., Cokato, MN 55321, (320) 286-2578, fax (320) 286-2467, www.forwardtech.com

Manufacturing Technology Solutions, 14150 Simone Dr., Shelby Township, MI 48315, (586) 802-0033, fax (586) 802-0034, www.mts-telsonic.com

Mecasonic Div. of Forward Technology Industries, Inc., 13500 County Road 6, Minneapolis, MN 55441, (763) 559-1785, fax (763) 559-3929, www.forwardtech.com

Sonics & Materials Inc., 53 Church Hill Rd., Newtown, CT 06470, (800) 745-1105, (203) 270-4600, fax (203) 270-4610, www.sonics.biz

Ultra Sonic Seal Co., 200 Turner Industrial Way, Aston, PA 19014, (610) 497-5150, fax (610) 497-5195, www.ultrasonicseal.com

17 Staking/Swaging/Peening/ Cold Heading/Cold Forming

17.1 Advantages and Disadvantages of Staking/ Cold Forming

Staking, swaging, peening, and cold forming are very similar in that they are fastening methods that involve the forming of plastics with a tool. Peening and cold forming are performed cold, whereas staking and swaging can be done either hot or cold. When staking is done cold, it is sometimes referred to as cold heading. However, "cold" may be interpreted to mean heated, but to a temperature well below the melting point. Furthermore, there is heat created in cold forming, which can reach temperatures high enough to melt the plastic and even cause degradation. While these are permanent fastening methods principally used for thermoplastics, some thermosets have been occasionally peened and cold-formed. All the thermoplastic processes can make parts that can be swaged and cold formed to some degree, only injection molding can readily create the studs necessary for staking, peening, and cold heading in most cases, however.

17.1.1 Advantages

1. *Dissimilar materials* These processes can assemble dissimilar materials, such as glass, metals, or thermoset plastics, provided that one of the parts is of a formable polymer. High stresses and distortion created by wide differences in coefficient of linear thermal expansion may be associated with dissimilar materials, however, and the reader is cautioned to be wary.

2. *Speed of assembly* These methods are performed very rapidly, assembly times of 1 or 2 seconds are quite common.

3. *Elimination of assembly devices* The need for fasteners, adhesives, solvents or electromagnetic preforms is eliminated along with the costs of such items.

4. *Loose tolerances* While close dimensional control is required for the highest strength applications, low value assemblies can be accomplished with fairly loose tolerances.

5. *Disassembly for recycling* Although these assemblies are not reopenable, they can be readily disassembled for recycling, particularly if heat is employed.

6. *Low investment* Staking equipment is low in cost, and very little in the way of fixturing is required.

17.1.2   Disadvantages

1. *Permanent assembly* Parts joined by these methods are not reopenable. Therefore, their application is limited to products that will not require servicing.

2. *Recovery* The material tends to recover some of its original shape after forming While this leads to loosening of the assembly, the use of heat can minimize this problem.

3. *Appearance* The nature of these processes, generally prevents the achievement of high cosmetic levels. Obviously, this is a subjective statement that might be arguable by those who point out that smooth, consistent stake heads can be achieved with the use of heat for many materials. The counter-position could be that even a smooth stake head is not attractive and that the weaker flat head designs are the only ones that are even passable. If applied with care, the use of heat can improve the appearance. However, the strength of the joint may decline when attention must be paid to appearance, since the reduction in packing pressure to prevent flashing may result in voids under the head.

4. *Filler materials* The use of these materials will affect the formation of the joining. Depending on the location of the gate relative to the joint, the filler material may not flow into the stud as readily, thus leaving it "resin rich" without the same properties as the balance of the part.

5. *Tool clearance* It is not always possible to position the forming tool such that it can function. Handheld heat and ultrasonic devices are available which can create crude heads in some difficult-to-access locations.

6. *Process limitations* The nature of stud formation limits these methods largely to parts made by the injection molding process.

17.2   Staking

The principal reason to apply all the processes discussed in this chapter except swaging, is to form stakes. A stake is basically a post that has had a head formed on it much like a rivet to capture two or more parts. Unlike a rivet, the post is molded integral to one of the parts, thus saving the cost of the rivet. The die that forms the head can be cold, or it can be heated by hot air, electrical resistance, or ultrasonic vibration. If the distance and height differential between stakes is not too great, the heated methods can be used to weld multiple stakes with one die.

Several criteria are used to determine which method of forming a stake is best suited to a given application. They are as follows.

1. *Strength requirements* The type of loading, such as shear or tensile, and its level must be established. In addition to the size and type of stake to be used, the number of stakes required must be determined. Ultrasonic horns can be made which form multiple stakes at one time, provided their spacing and height differentials are not too great.

2. *Horn or die accessibility and support* Both hot and cold forming equipment require direct access for reciprocal die action, but the hot-formed stakes can be

created with handheld devices having more limited capabilities. However, the different processes also require varying levels of pressure, and thus the wall below the stake may need to be supported to avoid distortion.

3. *Material limitations.* The material will determine the effectiveness of a given method and establish its design parameters. Ultrasonic techniques are particularly material sensitive.

17.2.1   Cold Forming of Stakes

Cold forming can theoretically be applied to all thermoplastics. However, it is particularly suited to acetal, nylon, and other materials with good impact resistance. Some thermoplastics are too brittle to be cold-formed. Thermosets can be cold-formed provided they are in the B, or partially cured, stage of cure.

Essentially, these processes rely on the application of compressive pressure beyond the yield strength of the plastic to cause it to deform into the shape of the head or some other configuration. To avoid fracture, the load must be applied gradually to carefully controlled levels.

Cold forming of plastics is quite similar to cold forming of metals; however, properties of plastic deformation and strain recovery, also known as "springback," introduce a time dependence aspect. Indeed, some recovery will take place in cold-formed parts regardless of the time factor, particularly if the assembly is exposed to heat in excess of that created in the forming process. However, holding the deformed segment under pressure for a period of time following forming will reduce the amount of strain recovery that takes place and will cause it to occur more slowly. In addition, molecular orientation in the direction of principal strain will take place. This has the effect of improving strength in the direction of the molecular orientation and reducing it in the cross direction. Temperature also has an effect on recovery, inasmuch as exposure to elevated temperatures will cause the recovery to take place. For applications in which such exposures are likely, the use of heat in the formation of the head is recommended since the head will not recover appreciably unless the temperature at which it was formed is exceeded. Hot forming should be used if a relaxation-free head is required. In addition to the recovery issue, cold-formed parts are more prone to crazing and chemical attack than hot-formed parts.

Depending on the application, there will be some heat associated with this process. Cold forming of thermoplastics is also done at temperatures that are elevated, but to a level well below the melting temperature. In both cases, however, temperatures can rise to the point of degrading the material. Since such degradation can lead to failure, it is wise to take care to prevent it from occurring. Simple cold forming can be done with a range of devices ranging from an arbor press to a multiple-heading automatic machine.

Cold heading of a plastic stake is illustrated in Fig. 17-1. The spring-loaded pilot sleeve holds the two parts together while the heading tool descends on the stud. Yielding and deformation will take place at the point where the cross section of the stud is smallest and is unsupported by the wall of the mating part. The walls of the rivet in Fig. 17-1 are uniform, however tapered or tubular designs are possible.

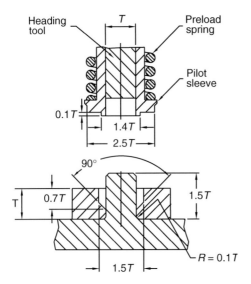

Figure 17-1   Cold heading of a plastic stake

A plastic rivet joining two parts is shown in Fig. 17-2. Here the portion of the rivet to be deformed is cored out to create a tubular shape, thereby reducing the cross-sectional area to be deformed. A reduction in cross section will require less force for deformation, which reduces the amount of elastic deformation on the portion of the shaft supported by the mating part. That in turn results in less elastic recovery, which loosens the joint. The stud in this design also has a head contoured to provide a spring effect when it is compressed. That feature maintains pressure on the assembly, preventing it from loosening due to recovery.

The dimensions of the stud are determined by a number of factors. Once the amount and direction of the principal load on the assembly have been established,

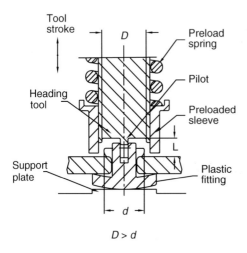

Figure 17-2   Cold heading of a plastic rivet

the number of stakes and their diameters can be determined from the strength equations. Cold-formed heads will exhibit tensile pull strengths that are approximately 50% of the material's shear strength. The stud diameter determines its length, as its unsupported length may not exceed twice the diameter without incurring buckling during head formation. When this length-to-diameter ratio must exceed 2, the forming operation can be broken into two stages, the first of which blowforms a cone shape and reduces the height to an acceptable level. Experimentation with the forming conditions can also solve a buckling problem, since the proper assembly parameters vary somewhat with the material, speed of assembly, and material temperature.

17.2.2   Hot Air/Cold Staking

When heat is used, the principal method of warming the material to be formed is to use hot air. This is illustrated in Fig. 17-3a, where a stud to be formed into a stake is being preheated by a jet of heated air from a tube. The heating of the air is performed by an in-line heater. In Fig. 17-3b, the cold forming die, normally driven by a pneumatic

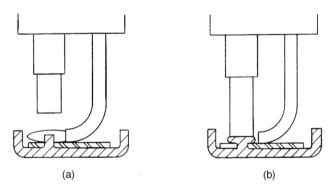

(a)   (b)

Figure 17-3   Hot air/cold staking: (a) preheating of stud and (b) cold heading (Diagrams courtesy of Orbitform, Inc.)

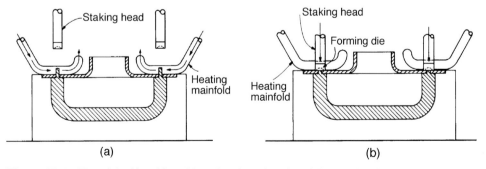

(a)   (b)

Figure 17-4   Hot air/cold staking: (a) preheating of stud and (b) cold heading (Diagrams courtesy of Forward Technology Industries, Inc.)

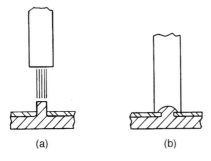

Figure 17-5 Hot air/cold staking: (a) preheating of stud and (b) cold heading (Diagrams courtesy of Service Tectonics Inc.)

(a)   (b)

cylinder, has descended on the softened stud to form the stake. It remains in that position until the plastic solidifies. Care must be taken to apply the load gradually and with only enough force to achieve the forming, since excessive load and speed will cause fracture of the stud. Figures 17-4 and 17-5 illustrate competing approaches. This process has a typical cycle of 8 to 20 seconds, however it lends itself well to multiple stations, permitting several heads to be formed at one time. The ability of the various thermoplastics to use this process is shown in Table 17-1.

Peening is a form of cold staking that uses a heated die similar in shape to the one illustrated in Fig. 17-3. In this case, however, the plastic is not heated. Peening is limited to resilient thermoplastic materials capable of withstanding the impact of the die without breaking.

17.2.3   Ultrasonic Cold Forming

Ultrasonics can also be used for cold forming. Referred to as the "high pressure method," this method uses ultrasonic vibrations to melt the material to a point at which the surface is softened but not melted. A high bearing force is exerted on the flat-faced horn to press the softened plastic into the mushroom shape illustrated in Fig. 17-6. As long as the amount of downward travel is controlled, the parts may be securely joined or they may be set to permit relative motion. Resilient materials like ABS, high impact styrene, and the polyolefins are best suited to this process, however it is also effective with acetal and polycarbonate.

Staking is not the only application for cold forming: this process can also be used to swage, crimp, fold, or otherwise form the heated plastic.

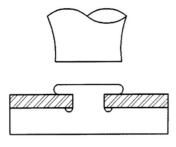

Figure 17-6 Ultrasonic cold forming (Diagram courtesy of Branson Ultrasonics Corporation, Danbury, CT)

Table 17-1 Staking and Swaging Characteristics of Thermoplastics

Material	Hot Air/Cold Staking	Thermal	Ultrasonic
ABS	Excellent	Excellent	Excellent
ABS/Polycarbonate	Excellent	Excellent	Good
ABS/PVC	No information	Good	Good
Acetal homopolymer and copolymer	Poor	Good	Fair to good
Acrylic	Good	Good	Fair
Acrylic multipolymer-XT polymer	No information	Good	Good
Acrylic/PVC	No information	Good	Good
Acrylic – impact modified	No information	Good	Fair
Butadiene–styrene (BDS)	Good	Good	Good
Cellulosics: CA, CAB, CAP	Fair	Good	Good
Fluoropolymers	Poor to good	Not suitable	Not suitable
Liquid crystal polymer	Good	No information	Fair to good
Nylon	Fair	Good	Fair to good
PC–PET	No information	Good	Good
Polyarylate	No information	Good	Fair
Polycarbonate	Excellent	Good	Fair
Polyester			
(PBT)	Good	Good	Fair
(PET)	Fair to good	Good	Fair
Polyetheretherketone	No information	Good	Good
Polyetherimide	No information	Good	Good
Polyethylene			
Low and high density	Good	Excellent	Fair to good
Ultrahigh molecular weight	No information	Not suitable	Not suitable
Polymethylpentene	No information	Excellent	Fair to good
Polyphenylene oxide	Good	Excellent	Good to excellent
Polyphenylene sulfide	Good	Excellent	Poor
Polypropylene	Good	Excellent	Excellent
Polystyrene			
General purpose	Excellent	Good	Fair
Impact modified	Good	Good	Fair to excellent
PVC			
Flexible	No information	Not suitable	Not suitable
Rigid	Good	Good	Good
SAN–NAS–ASA	Good	Good	Fair
Styrene–maleic–anhydride	No information	Good	Excellent
Sulfone polymers	Good	Good	Fair

17.2.4   Hot Die Forming of Stakes (Thermal Staking)

Heat can be used to raise the temperature of the plastic above its melting point. That helps to create tight assemblies and reduces the tendency toward recovery to no more than the normal relaxation of the material. The heat is normally provided by an ordinary heated die or by ultrasonic vibrations, although electromagnetic/induction heating has also been used for this purpose.

Hot dies work in a very simple fashion. The die is heated to a temperature sufficient to heat the surface of the plastic as pressure is applied to form the head of the stake. Care must be taken to provide adequate heat for forming without causing degradation. Hot dies cannot be turned on and off like ultrasonics equipment; however, they are better suited to some materials – particularly those with heavy glass loadings. In addition, they do not damage sensitive electronic components as ultrasonics can, although they do have some elastic recovery. The ability of various plastics to accept thermal staking is also listed in Table 17-1.

17.2.5   Ultrasonic Hot Forming of Stakes

High frequency ultrasonic vibrations can be used to melt the top of the stud and re-form it into the head of a stake. The shape of the head of the stake is machined into the tip of the horn much like a hot die. As the vibrations melt the plastic, it gradually creates the head of the stake. The reader is referred to Chapter 19, "Ultrasonic Welding," for a more detailed explanation of the melting process and ultrasonic welding equipment.

The principal advantages of using ultrasonics to create stakes are as follows: unlike cold and hot die forming ultrasonic joining does not produce elastic recovery; moreover, the process provides greater control over the formation of the stake head than the other methods. Ultrasonics can offer cycle times under one second, excellent repeatability, and process control. These advantages are present because only the top of the plastic stud, is melted, a refinement often accomplished by starting the ultrasonic horn before it contacts the stud. Ultrasonic welds are usually performed using high amplitude and low pressure to permit the material to melt and flow into the cavity in the horn. The rate of speed of the downward travel must be carefully controlled so that it does not exceed the melting rate of the plastic; otherwise, the stud might crack or deform. Such damage can also occur if the downward pressure is too great or the amplitude is too low. Commencing the ultrasonic vibrations before contact with the stud, known as "pretriggering," can help resolve this problem.

As the plastic melts, it fills the contour of the die, forming the head of the stake. When the limit of travel is reached, the horn is held in that position, providing downward pressure on the joint until the head has cooled enough for the horn to be removed. If the horn is removed too quickly, the formed head may remain in the staking cavity.

Materials are significant in the use of ultrasonics to form stakes. Table 17-1 present a partial rating of their ability to form ultrasonic stakes, along with the thermal and hot air/cold staking techniques.

17.2.6 Laser Staking

Laser energy can also be used to melt the plastic stud so the plastic can be formed by a die. The laser beam is directed through the transparent die down on the plastic stud to be welded until the polymer is soft enough to be formed in a fashion similar to that shown in Fig. 17-5. The die then presses down on the stud to shape the head of the stake. To avoid the loss of light transmission due to abrasion of the die, the plastic stud cannot be made of glass fiber filled resins. Such applications require a metal die, which is applied as a second stage. Laser welding can be used for most polymers; however, only those particular compounds that absorb laser energy can be used for laser staking. The source for this type of equipment is Extol, Inc. For more information on laser welding, the reader is referred to Chapter 21.

17.3 Stake Design

17.3.1 The Stud

The stud determines many of the design parameters. Its diameter will determine the strength of the joint and which type of head to use. The first issue is the matter of nonuniform cooling resulting when wall thickness variations at the base of the stake exceed 25%, which can lead to high levels of molded-in stress as evidenced by voids and sink marks. The stress will diminish the strength of the stake, and the sink can also detract from the quality of an appearance part. As a general guide, this condition can be avoided by limiting the diameter of the stud to 50% of the nominal wall thickness. When this is not possible, another alternative is the use of the hollow profile head, discussed later in the chapter.

Two basic stud designs are illustrated in Fig. 17-7. The base of the stud is subjected to stress that can lead to failure during both the formation of the stake and the life of the product. Note the radius at the base of the stud in Fig. 17-7a. This is done on studs in order to meet the stress factor criteria for inside radii in plastics. The same objective is met by the radius in the recess at the base of the stake in Fig. 17-7b. Without recessing into the surface of the wall, however, a chamfer must be created in the mating part to clear that radius. This adds some cost to the tooling for that moldment, the amount depending on which is the core side of the part. If the chamfer is on the core side, it can be readily achieved by turning down a larger core pin. Otherwise, an extra set of core pins must be used, with a shutoff taking place where they meet.

(a)

(b)

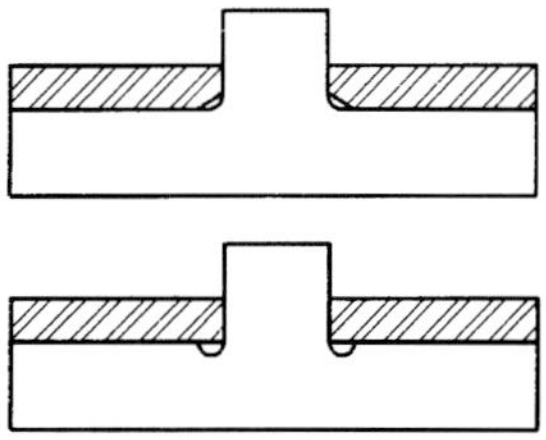

Figure 17-7 Basic stud design: (a) surface stud radius and (b) recessed stud radius (Diagrams courtesy of Branson Ultrasonics Corporation, Danbury, CT)

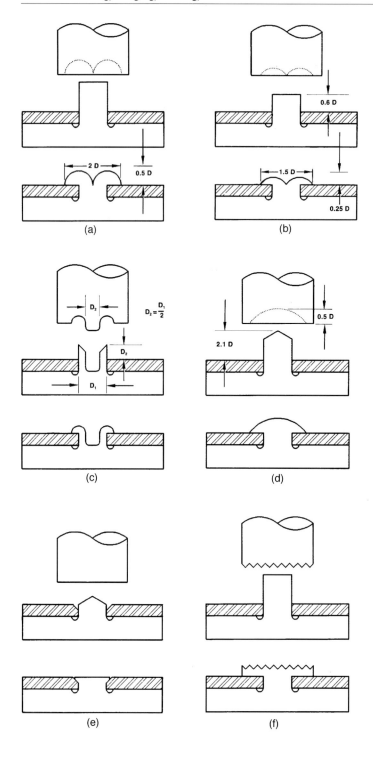

(a)

(b)

(c)

(d)

(e)

(f)

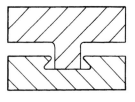

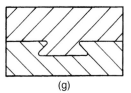

(g)

Figure 17-8 Stake configurations: (a) standard rosette head, (b) low profile rosette head, (c) hollow profile head, (d) dome stake, (e) flush stake, (f) knurled stake and (g) blind stake (Diagrams courtesy of Branson Ultrasonics Corporation, Danbury, CT)

A shutoff may be necessary in any case to create a radius at the top of the mating part. That radius would create a radius at the base of the stake head as it is formed, once again to reduce the stress concentration factor at an inside corner. If this feature is too costly for a given design, much of the benefit can be gained by a 45° chamfer in its place. These details add some cost to the molds, however they create a significant increase in the strength of the completed stakes.

17.3.2 Stake Heads

Tolerances have a cost/strength effect, as is the case with most of the assembly methods. In cold staking, the hole can be oversize within reason with a relatively little effect on the strength of the joint. In hot staking, the plastic melts and can ooze down the gap around the stud, thus depriving the head of its material and weakening the joint. However, the melted material in hot staking results in fewer internal stresses, which makes for an inherently stronger joint. The strongest stake is achieved by hot staking to close tolerances, since loose tolerances create less contact area under the head and can actually result in a loose fitment. Although that problem may also be the result of insufficient pressure on the head or for too short a period of time. When tolerances are the principal cost driver, the most economic strong stake would be a hot stake made to looser tolerances, but with an experimentally derived amount of extra material in the stake (which can be established once the amount of material being lost down the sides of the stud is determined).

The dimensions of the head of the stake are designed to provide the proper amount of material to form the head of the stake. Too large a stake cavity will result in a ragged or irregularly shaped head, whereas one that is too small will cause excessive flash around the stake head. The same result can occur if the stud is improperly sized. Too large a stud will produce flash, while too small a stud can cause a misshapen head.

Several of the most common varieties of stake heads are illustrated in Fig. 17-8. For the most part, these are designed for ultrasonics; most, however, have been successfully formed by other methods as well. The amount of material used to form the head determines its dimensions and those of the stud. Basically, there has to be enough material available to form a solid head. These dimensions should be regarded as a starting point, as additional stud height may be required to avoid the displacement of material in the surface of the mating part or to create a more solid head through packing. The additional plastic may, however, result in unsightly flash.

Figure 17-8a depicts the standard profile rosette stake design. The point in the center of the rosette head starts the melt and serves to reduce the amount of energy required to form the stake. It must be carefully centered over the center of the flat top of the stud, as misalignment will result in a lopsided head, flash around the head, or a collapse at the base of the stud. When forming is completed, the head fills out to a diameter twice that of the original stud.

The rosette stake is an excellent design for most thermoplastics, including the less rigid types; it is not used for compounds with abrasive fillers such as glass and minerals, however, because these prematurely wear off the point in the head. Crystalline, high melt thermoplastics formed under high pressure, high amplitude, and high trigger pressure can sometimes avoid brittleness through the use of the standard profile rosette stakes. This design is commonly used for stakes with stud diameters of 1.6 to 4.0 mm (0.063–0.156 in.), although standard ultrasonic horn tips are available from 0.8 to 6.35 mm (0.031–0.250 in.). For diameters greater than 4.0 mm (0.156 in.) and limited space applications, a lower profile configuration such as that shown in Fig. 17-8b can be used. It requires less energy and a shorter cycle than the standard profile stake, but there is also less material in the head, resulting in a weaker stake. Since the top of the stud is flat for either profile rosette stake, rosette studs are more economical to put in the mold than studs for dome stakes that have a coned top.

The preferred design for diameters greater than 4.0 mm (0.156 in.) is the hollow stake design depicted in Fig. 17-8c. It can create a large, strong head with less material, which results in a lower energy requirement for head formation. When this configuration is used, the stud diameter must be at least 4.0 mm (0.156 in.), to assure enough material for the stake while retaining a minimum core diameter of 1.5 mm (0.059 in.) for the inside hollow. That recess can be designed so that the walls meet the sink and void avoidance criteria. The walls of the recess must be drafted to ease removal from the mold, and the inside corner must be radiused to minimize the stress concentration factor. While stakes are regarded as a nonreopenable form of assembly, the hollow stake can be drilled out and replaced with a self-tapping screw to provide reopenability of a sort. In most cases, hollow studs are more costly to tool than the other varieties.

The dome stake in Fig. 17-8d is used for very small diameter studs [<3.18 mm (0.125 in.)], multiple stakes, and for larger diameters made of glass- or mineral-filled materials that are too abrasive for profile stakes. They also are particularly recommended for crystalline thermoplastics and those that degrade readily. In place of a point in the head, the stud is cone shaped. This mainly eliminates the need to replace dies because of worn-down points, however it does result in an additional cost in the mold, which is modest unless the number of stakes is very large. The curve of the dome will tend to align itself with the center of the stud, resulting in a less precise die location requirement and looser tolerances. The pan head and flat head are two other versions of dome heads that provide more material in the head for greater strength. All three designs will result in a head height that is one-half the diameter. The pan and flat head diameters will be twice the stake diameter; the simple dome diameter will be 2.23 times the stake diameter.

For some appearance-sensitive applications, the flush stake shown in Fig. 17-8e can be used. It utilizes a flat horn to press the excess material into a countersunk hole. As with the dome stake, a cone-shaped stud is used. While pleasing to the eye, it is a much weaker design, since there is far less material available to form the head and a thin wedge of material is created at the base of the head. The thinner the thickness of the mating part, the less polymer available for the head; in the worst case, there may not be enough for adequate strength. In addition to the lack of plastic, this configuration can result in a stress concentration at the base of the stake due to a sharp inside corner. Flush heads can be used with all thermoplastics but should be limited to light-duty applications.

The knurled stake in Fig. 17-8f is probably the easiest to use in that it can be performed quickly, with very crude alignment; a handheld ultrasonic tool and a stud of virtually any diameter will do. The deficiencies of the knurled stake lie in unpleasant appearance and questionable strength. It can be useful as a means of assembling prototypes or for low volume production, as it is readily accomplished with a handheld ultrasonic welder; however its lack of precision location requirements lends it to high volume, high speed applications as well. A variety of standard tips are available with both positive and negative patterns ranging from fine to coarse.

The inverse stake, also known as the blind stake or mechanical interlock, is illustrated in Fig. 17-8g. The stud is placed in a mating hole with an undercut, and an inexpensive flat-faced horn is placed over the stud. Alignment is not critical. A permanent blind stake or mechanical interlock is created when the stud is melted into the cavity by ultrasonic vibrations.

17.4   Swaging

Swaging is the term applied to either hot or cold forming, where an entire wall, or portion of a wall, is displaced with a tool to entrap another part or parts, thereby creating an assembly without a molecular bond. The entrapped part is nearly always made of a dissimilar material (such as a nonplastic, thermoset, or thermoplastic that cannot be welded to the original part). Hot forming with ultrasonics has also been used to form the ends of tubes. Swaging's principal advantage is that mold costs are reduced because there are none of the costly undercuts or studs normally associated with entrapment of dissimilar materials. In addition, it provides a tight assembly without fasteners or adhesives and can be accomplished with fast cycle times This procedure is quick and has a minimal stress buildup. Swaging can be continuous or segmented to reduce the energy cost. The concept is illustrated in Fig. 17-9.

In Fig. 17-9a, the assembly is shown before swaging has taken place. Without swaging, the vertical wall would require an undercut. The assembly is created when the die in Fig. 17-9b descends to displace material over the part or parts to be entrapped. The principal process parameters are the temperature of the swaging tool,

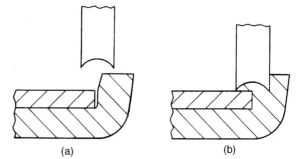

Figure 17-9 Swaging: (a) before swaging and (b) swaged

the temperature of the plastic, the amount of pressure applied, and the length of time it is applied. Ultrasonic swaging is quicker than that done with a hot die, and there is no stress buildup, as with cold forming. When swaging is done with ultrasonics, specially designed tooling is needed. The ultrasonic horn applies pressure with controlled velocity to the wall, and a cold flow of plastics begins. Pretriggering is used to enable the melting of the plastic to begin the instant the die meets the wall.

It then melts and is plowed over the mating part by the shape of the recess in the horn. A heavy hold pressure is then used to prevent elastic memory before the melted section is cooled. When it cools, the mating part is trapped in place.

Low to medium stiffness polymers lend themselves best to ultrasonic swaging because they permit the cold forming of the material to begin prior to the application of the ultrasonic vibrations. These include ABS, the cellulosics, impact polystyrene, polyethylene, polymethylpentene, and polypropylene. However, rigid thermoplastics can also be swaged with greater difficulty.

Cold swaging can be performed with the softer materials, such as crystalline or the tougher amorphous thermoplastics, which can withstand the large strains resulting from the process. However, they are far more likely to exhibit elastic recovery than swaging with heat.

17.5   Sources

17.5.1   Thermal Staking

Forward Technology, 260 Jenks Ave., Cokato, MN 55321, (320) 286-2578, fax (320) 286-2467, www.forwardtech.com

Heraeus Noblelight, Inc., 2150 Northmont Parkway, Duluth, GA 30096, (770) 418-0707, (770) 418-0688, www.noblelight.net

Manufacturing Technology Solutions, 14150 Simone Dr., Shelby Township, MI 48315, (586) 802-0033, fax (586) 802-0034, www.mts-telsonic.com

Sonitek - Sonic & Thermal Technologies, Inc., 84 Research Dr., Milford, CT 06460, (203) 878-9321, fax (203) 878-6786, www.sonitek.com

17.5.2   Hot Air/Cold Staking

Hydra Sealer Div. of Forward Technologies, Inc., 13500 County Road 6, Minneapolis, MN 55441, (763) 559-1785, fax (763) 559-3929

Orbitform, Inc., 1600 Executive Dr., Jackson, MI 49203, (517) 787-9447, fax (517) 787-6609, www.orbitform.com

Service Tectonics Inc., 2827 Treat St., Adrian, MI 49221, (517) 263-0758, fax (517) 263-4145, www.padprinting.net

17.5.3   Laser Staking

Extol, Inc., 651 Case Karsten Dr., Zeeland, MI 49464, (616) 292-1771, fax (616) 748-0555, www.extolinc.com

17.5.4   Ultrasonic

Refer to Chapter 19 "Ultrasonic Welding."

18 Threads: Tapped and Molded-in

18.1 Advantages and Disadvantages of Integral Threads

18.1.1 Advantages Common to Threads of Both Types

1. *No additional materials* Besides the screw itself, integral threads use no additional materials such as inserts, electromagnetic preforms, adhesives, or solvents. Therefore, these threads are inherently lower in cost than methods that do have such requirements and are less expensive to disassemble for recycling, although not necessarily as low in cost as those that do not have a screw either.

2. *Ease of assembly* The insertion of screws into threaded holes is a long-established assembly method that requires little special training. In addition, automatic equipment for performing this function is widely available, making it suitable for automated assembly.

3. *Dissimilar materials* Metals can be joined to plastics, and widely differing as well as similar plastics can be joined to each other by means of screws with integral screw threads.

4. *Reopenability* Integral threads are often selected for their reopenability. They are superior to those created by self-tapping screws in the respect that those screws may destroy the hole on repeated reopenings. Molded-in threads are superior to tapped threads in this respect.

18.1.2 Disadvantages Common to Threads of Both Types

1. *Slow assembly* Relative to other assembly methods, screws are emplaced at a slow rate of speed.

2. *Thermal expansion* Significant differences in coefficient of linear thermal expansion between the two materials can result in stress cracking of the thread or the boss, or loosening of the screw.

3. *Moisture absorption* Some plastics exhibit expansion due to moisture absorption that is sufficient to result in loosening of the screw.

4. *Limited hermetic seals* Gasketing is required to accomplish hermetic seals.

5. *Stress relaxation* Time diminishes the ability of the material to withstand stress, particularly at elevated temperatures. Stress relaxation can result in cracking or crazing of the plastic threads.

6. *Creep and cold flow* The screw can be loosened by these effects as it undergoes strain relaxation.

7. *Environmental limitations* Chemical and ultraviolet light exposure can result in failure of the thread due to stress cracking.

18.2   Drilled and Tapped Holes in Plastics

The drilling and tapping of holes in plastics is normally thought of as a technique limited to prototyping, short runs, and parts that are very large, awkward in shape, or both. That is because the process is very slow in comparison to molding the threads into the part. However, molding in the threads has a relatively high tooling cost, which tends to increase the number of parts that can be drilled and tapped before the volume can justify such a tooling investment. In some cases, it is worthwhile to use a hybrid technique whereby the hole is molded and then tapped. However, depending on the quality of thread required, it may be necessary to drill and ream the hole anyway to remove the draft. Still, there is a time saving over drilling the full diameter.

Drilling and tapping can sometimes be justified on the basis of hole location accuracy (with the proper equipment), however tapped threads cannot be held to as high a class of fit as molded threads. Location accuracy can be increased by molding into the surface of the part a tiny depression that acts as a start for the drill. Drilling fixtures with drill bushings are often used for higher accuracy, and there are commercial fixtures available with prepositioned drill bits installed. In addition, the time required to move the drilling head between hole sites can be reduced to one-tenth that of manual operations through the use of microprocessor-controlled equipment. Nonetheless, it may be more cost-effective to place threaded inserts in the drilled or molded holes, since less time is needed to do this than is required to tap holes. In addition, such inserts provide a much stronger metal thread. The reader is referred to Chapter 8, "Fasteners and Inserts," for a further discussion of threaded inserts as well as self-tapping screws.

18.2.1   Advantages Unique to Tapped Threads

1. *Shape freedom* Tapped threads can be placed in most parts regardless of shape and are not shape-limited in the way that most of the welding methods are. Tapped threads require no assembly details to be placed in the mold (if the pilot holes are not molded in).

2. *Process freedom* Holes can be drilled and tapped in parts made from any process, provided the wall thickness of the part is adequate.

3. *Precision* The most precise hole location possible is attainable with precision machine tools.

18.2.2   Disadvantages Unique to Tapped Threads

1. *Production rates* Manual drilling and tapping of screws is much slower than most other means of assembly. Production rates are dependent on the pitch of the threads,

the length of the screw, and the number of holes that can be gang-drilled and tapped at one time.

2. *Strength*   Tapped threads are weaker than molded-in threads or the metal threads available through the use of inserts.

3. *Stress cracking*   Whether ISO or ANSI taps are used, tapped threads are typically V-threads, which create stress concentrations that can lead to stress cracking.

4. *Cleanliness*   The creation of chips and the use of lubricants require the cleaning of tapped holes.

5. *Material limitations*   Flexible plastics are difficult, if not impossible, to tap.

6. *Wall thickness limitations*   Thin walls cannot be drilled and tapped. The minimum wall thickness for which this technique can be employed varies with the material, with stiffer polymers being able to handle thinner walls.

18.2.3   Drilling Holes in Plastics

Experienced machinists familiar in working with metals will discover some differences in drilling holes in plastics. The first problem they are likely to encounter is melting and deformation of the hole resulting from heat buildup due to friction. This is partly because plastics conduct heat at a much lower rate than metals do. Thus, the heat created by the friction of the drill cutting the plastic will result in a dramatic rise in temperature of the drill and, consequently, the material at the interface. The higher (than metals) thermal expansion exhibited by the plastic further increases friction and exacerbates this condition. When these changes are combined with the lower melting point of the plastic, melting occurs and a misshapen hole results. Even when the diameter of the hole is true, the finish may be affected. Dull drill bits increase the heat due to friction and the likelihood of problems.

Holes drilled in plastic can be distorted in other ways as well. Plastics are, for the most part, much softer than metals. Therefore, they tend to deform under the pressure of the drill in two ways. The pressure of the drill bit can cause the entire structure to deform, resulting in a distorted hole, or it can cause the wall of the hole to compress under the load. When the drill is removed and the wall exhibits elastic recovery, the hole is a smaller diameter than the drill that created it.

In general, standard drill presses with wood and metal drill bits can be used with plastics; however, they vary in their machinability. The greater the hardness, rigidity, and lubricity of a particular polymer and the higher its melting point, the better it is suited to drilling and tapping. Proper procedure, drill bit configuration, and the use of coolant will reduce this temperature buildup. These factors vary with different plastics, and recommendations will be found for many individual resins in Chapter 5, "Assembly Method Selection by Material." For the remainder, the general parameters are a lip clearance of 12 to 18° and a point with an included angle of 60 to 90°, although large-diameter drills can go as high as 120°. Best results are obtained with special drills for plastics, which have one or two highly polished, honed or chrome-plated flutes, narrow lands, and large helix angles to minimize frictional heat and to quickly expel the chips. Rake angles large enough to create continuous chips but not large enough to cause brittle failure and discontinuous chips should be used. An unequal angle on the length of the cutting edge will cause the hole to be oversize, and an improperly ground drill will leave a rough or burred hole. Improper heat treatment after regrinding will result in chipping of the drill.

General practice is to start low and increase the speeds and feeds on a test piece until discoloration, a rough or burred hole, melting, or burning signal that the limit has been reached. Then back off a safe amount and proceed with the drilling operation on actual parts. Feed rates should range between 30.5 and 61.0 m/min (100–200 ft/min). Feed rate varies with the diameter of the hole according to Table 18-1. Too coarse a feed can also cause chipping of the drill.

Drilling speeds increase as the hardness of the material increases and the size of the holes decreases. Operators must be alert to a rough or burred hole, which can result from dulling of the drill bit. Besides reject work, continued use of a dull drill will likely result in a broken bit. A burr on the drill bit can cause an oversize hole. The likelihood of problems like this can be reduced through the use of carbide tipped drills, which can be used at high speeds and may not require cooling. Frequent removal of the tool may be required to clear the chips from deep holes.

Chilled compressed air is the ideal coolant for drilling plastics, as it blows the chips away in addition to providing cooling. Most important, it does not contaminate the part and require cleaning the way water, metal cutting fluids, and oils do. The latter two fluids are particularly undesirable because they can degrade or otherwise attack some polymers. Fluids can also cause the part to slip from its clamp and grab and spin.

18.2.4   Reaming Holes in Plastics

Reaming may be required to provide the accuracy or finish needed for a given application. As with drilling, it is important that the reamers be sharp enough to cut the plastic and not deform it. Straight flute reamers can be used with through holes, but helically fluted reamers may be used in blind holes as well, and they provide a smoother cut and finer finish. Standard unaltered straight or fluted high speed steel reamers can be used with a normal chamfer angle of 45° and rake angle of 5°. As with drilling, cooling will improve the finish. Recommended reaming conditions for selected plastics are provided in Table 18-2.

18.2.5   Tapping Holes in Plastics

Provided the same heat, chip removal, tool maintenance, and cooling parameters are maintained as for drilling of plastics, standard taps and dies designed for metal-cutting can be used for hard plastics that exhibit little compression and recovery. Otherwise, oversize taps are recommended. Up to 3.2 mm (0.125 in.), two-fluted taps are recommended. For larger holes, three-fluted taps are preferred over four-fluted taps because the chips can be removed more easily. The flutes should be finish-ground to reduce friction and the resultant heat. Carbide taps are recommended, particularly for filled

Table 18-1 Drilling Thermoplastics and Thermosets

Material	Brinnell Hardness Number	Condition	Speed (ft/min)	Feed (in./rev) Nominal Hole Diameter (in.)								High Speed Steel Tool Material
				$\frac{1}{16}$	$\frac{1}{8}$	$\frac{1}{4}$	$\frac{1}{2}$	$\frac{3}{4}$	1	$1\frac{1}{2}$	2	
Thermoplastics												
Polyethylene Polypropylene TFE-fluorocarbon Butyrate	$31R_R$ to $116R_R$	Extruded, molded, or cast	150–200	0.002	0.003	0.005	0.010	0.015	0.020	0.025	0.030	M10 M7 M1
High impact styrene Acrylonitrile–butadiene–styrene Modified acrylic	$83R_R$ to $107R_R$	Extruded, molded, or cast	150–200	0.002	0.004	0.005	0.006	0.006	0.008	0.008	0.010	M10 M7 M1
Nylon Acetals Polycarbonate	$79R_M$ to $100R_M$	Molded	150–200	0.002	0.003	0.005	0.008	0.010	0.012	0.015	0.015	M10 M7 M1
Acrylics	$80R_M$ to $103R_M$	Extruded, molded, or cast	150–200	0.001	0.002	0.004	0.008	0.010	0.012	0.015	0.015	M10 M7 M1
Polystyrenes	$70R_M$ to $95R_M$	Molded or extruded	150–200	0.001	0.002	0.003	0.004	0.005	0.006	0.007	0.008	M10 M7 M1
Thermosets												
Paper or cotton base	$50R_M$ to $125R_M$	Cast, molded, or filled	200–400	0.002	0.003	0.005	0.006	0.010	0.012	0.015	0.015	M10 M7 M1
Fiber glass, graphitized, and asbestos base	$50R_M$ to $125R_M$	Cast, molded, or filled	200–250	0.002	0.003	0.005	0.008	0.010	0.012	0.015	0.015	M10 M7 M1

Source: Machining Data Handbook, 2nd ed., Machinability Data Center, Metcut Research Associates, Inc., Cincinnatti, OH., 1972.

Table 18-2 Reaming Thermoplastics and Thermosets

Material	Brinell Hardness Number	Condition	High-Speed Steel Tool Speed (ft/min)	Feed (in./rev)[a] Reamer diameter (in.) $\frac{1}{8}$	$\frac{1}{4}$	$\frac{1}{2}$	1	$1\frac{1}{2}$	2	Tool Metal	Carbide Tool Speed (ft/min)	Feed (in./rev)[a] Reamer diameter (in.) $\frac{1}{8}$	$\frac{1}{4}$	$\frac{1}{2}$	1	$1\frac{1}{2}$	2	Tool Metal
Thermoplastics																		
Polyethylene Polypropylene Fluorocarbons Butyrates	31R$_R$ to 116R$_R$	Extruded, molded, or cast	250 to 300	0.006	0.008	0.010	0.010	0.012	0.015	M2 M7 M1	500 to 600	0.006	0.008	0.010	0.010	0.012	0.015	C-2
Nylon Acetals Polycarbonates	79R$_M$ to 100R$_M$	Molded	250 to 300	0.004	0.006	0.008	0.010	0.012	0.015	M2 M7 M1	350 to 450	0.004	0.006	0.008	0.010	0.012	0.015	C-2
Acrylics	80R$_M$ to 103R$_M$	Extruded, molded, or cast	200 to 300	0.006	0.008	0.010	0.010	0.012	0.015	M2 M7 M1	300 to 400	0.006	0.008	0.010	0.010	0.012	0.015	C-2
Thermosets																		
Paper and cotton base reinforced	50R$_M$ to 125R$_M$	Cast, molded, or filled	200 to 250	0.003	0.003	0.004	0.005	0.005	0.005	M2 M7 M1	250 to 300	0.003	0.003	0.004	0.005	0.005	0.005	C-2
Fiber glass and graphitized base, reinforced	50R$_M$ to 125R$_M$	Cast, molded, or filled	100 to 150	0.002	0.002	0.002	0.003	0.005	0.005	M2 M7 M1	150 to 200	0.002	0.002	0.002	0.003	0.005	0.005	C-2

[a] Based on a six-flute reamer.
Source: *Machining Data Handbook.* 2nd ed., Machinability Data Center, Metcut Research Associates, Inc., Cincinnatti, OH, 1972.

polymers or production runs beyond a very few. Threads finer than a 1.0 mm pitch or 28 threads per inch are not recommended. Coarse threads are preferred because there is more material in the thread to provide greater strength, and the increased lead makes chip removal easier, although the latter point is not an issue for through holes. To avoid cracking or chipping at the hole entrance, holes to be tapped should be countersunk or chamfered.

The diameter of a hole to be tapped in plastic should be oversize to prevent peeling or breakage of the threads. The amount is dependent on the size of the hole and the elastic recovery properties of the plastic. A general rule, however, is that the hole should be large enough to result in a thread height 70% of a full thread.

18.3   Molded Threads in Plastics

18.3.1   Advantages Unique to Molded-in Threads

1. *Efficiency* Threads can be molded directly into plastic parts, thereby completely eliminating the drilling and tapping operations.
2. *Thin walls* Threads can be molded into walls that would not withstand the stress of drilling and tapping.
3. *Material freedom* Threads can be molded in materials that are too soft to withstand drilling and tapping or cannot withstand the heat associated with those techniques.
4. *Strength* Molded-in threads are stronger than tapped threads.

18.3.2   Disadvantages Unique to Molded-in Threads

1. *Mold limitations* Molded-in threads have severe mold limitations.
2. *Mold costs* The cost of molds that can create integral threads is much greater than those for parts that use threads created by other means.
3. *Process limitations* Stripping or unscrewing of molds is rarely attempted for any process besides injection molding, although some versions are feasible in transfer and compression molding.

18.3.3   Thread Design

Molded threads are stronger than tapped threads and are, therefore, preferable for integral thread applications for which the additional tooling cost can be justified. Furthermore, as molten polymer fills the mold, a "skin" of compacted plastic forms on the outer surface. It is denser than the core material and forms a moisture barrier as well. Also, molded threads can usually be held to tighter tolerances than tapped threads, provided they are not made of difficult-to-control materials with high shrinkages. However, there are limits to the efficiency of molded threads, and those that are greater in diameter or longer than 12.7 mm (0.50 in.) should be tapped instead of molded. Also, those molded of filled material may have a shortage of filler at the crest of the thread, resulting in lower strength at that point.

Tapped holes are generally made to standard thread forms in order to accommodate common screws, whereas molded threads are more often used for applications like closures, where they must fit standard bottle threads or can be readily made to a custom thread form. This property is significant with integral threads because plastics are notoriously notch sensitive, and the V-grooves of standard threads make perfect notches. Thus, they are vulnerable to the stresses generated by the screws when they are tightened.

Vulnerability to the type of stress just described is a particular problem when the threaded hole is placed in a thin-walled structure such as a boss or a bottle cap. The high hoop stresses created by such threads, when combined with the impossibility of molding the threaded bosses without weld or knit lines, makes the use of V-shaped thread forms a very risky venture. Pipe threads, which have V-forms and develop even higher forces, are that much worse. When V-threads must be used, their roots must be rounded to minimize the stress concentration and their crests flatted to avoid a bind with the radiused roots of the mating part. For a further discussion regarding the forces developed by screw threads on bosses, the reader is referred to Chapter 8, "Fasteners and Inserts." Regardless of the computed values, every application must be tested thoroughly over a significant time period, and some mold modifications must be anticipated. The exact design of the various standard screw threads can be found in the *Machinery's Handbook 24th Edition* (E. Oberg, F. D. Jones, H. L. Horton, H. H. Ryffel and R. E. Green, Ed., Industrial Press Inc., New York, NY, 1992).

Very fine threads are not well suited to molded threads. A lead of 0.80 mm or 32 threads per inch should be regarded as the minimum thread size, and molded threads should not exceed 75% of full depth. Also, a class 2 fit, which is classified as a moderate to free fit for interchangeable parts, is described in *National Institute of Standard and Technology Handbook H-28* (formerly *National Bureau of Standards Handbook H-28*) and is the finest that can be expected on a production basis from materials with low shrinkages. A class 3 semiprecision or medium fit should be attempted only for low volumes that can be tightly controlled. Materials that have lower dimensional stability should have the coarser class 1 fit, described as a loose fit for quick and easy assembly. Threads made from such materials should also be kept as short as possible to avoid the accumulated effect of a lead that is out of tolerance.

The V-shaped thread form depicted in Fig. 18-1a is sometimes recommended when the forces are light and there is concern over the appearance of sink on the outer surface of the part. In this case, the circle drawn through the largest plastic area is readily held to a point where its diameter (D) is equal or less than 1.25 the nominal wall thickness (W), thus preventing the occurrence of sink opposite the thread.

The round thread in Fig. 18-1b is a stronger thread. Unfortunately, the largest pocket of material as described by a circle drawn through its' thickest area has a diameter that easily exceeds 1.25 times its nominal wall thickness.

The modified buttress thread form in Fig. 18-1c is a compromise design that strives to achieve the maximum available force and thread strength consistent with a limited

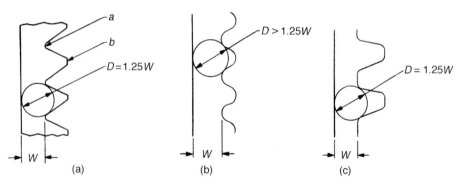

Figure 18-1 Molded thread forms: (a) V-thread (*a*, root must be radiused, *b*, crest must be flatted), (b) round thread, and (c) modified buttress thread

accumulation of plastic. Thus, the diameter drawn through its thickest section can be kept close to 1.25 times its nominal wall thickness. It also fits the standard bottle thread form, as depicted in the same illustration. The standards for bottle threads necessary to develope a thread form for a closure can be obtained from the following organization:
Glass Packaging Institute
1627 K St. NW, Suite 800
Washington, D.C. 20006
Telephone no. (202) 887-4850
Facsimile no. (202) 785-5377

The Closure Manufacturers' Association, an adjunct committee of the Glass Packaging Institute, can be found at the same location.

The dimensions for the starting point of a modified buttress thread design are found in Fig. 18-2a. From this point, the design must be adjusted for the particular application. In Fig. 18-2b, the thread is shown to be blended out short of the end. If it is permitted to run out, a thin wedge of steel will be formed in the mold. This will break off, leaving a damaged thread that must be repaired, only to break again in short order. Therefore, the thread is ended 0.75 mm (0.030 in.) from the end to avoid this problem. This condition exists at the base of the thread, as well, when it runs up to a shoulder. If it is necessary to run the thread to the shoulder, it may be necessary to create a recess for the thread to end short of the wall, as shown in Fig. 18-2c.

18.3.4 Molds for Threads

As might be expected, a special tool is required to mold threads. Internal threads can be made in three ways. They can be stripped off the core while still hot; they can have a core that collapses; or they can be unscrewed off the core. In very small quantities, the parts can be unscrewed manually from a core that has been removed from the mold. Two cores are used, so that one can be molding while the part is being removed from the other.

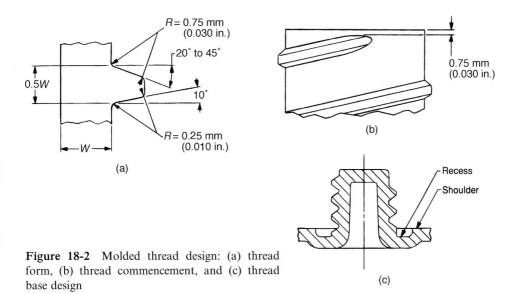

Figure 18-2 Molded thread design: (a) thread form, (b) thread commencement, and (c) thread base design

Whichever method is used, it is desirable to keep the internal threads as short as possible. In the case of the stripping mold, the shorter the thread, the less strain created. The collapsing core mold has a limited amount of space available for the thread, and the unscrewing mold must wait for all the threads to be unscrewed before the cycle can proceed. In some cases, double or triple threads with their longer leads are used to shorten the cycle.

External threads can also be made with mold components that are removed from the tool for part retrieval except, in this case, the component would be a cavity insert. The concept for external threads, known as an expandable cavity, is similar to the collapsing core used for internal threads. However, external threads are usually made in volume with a split cavity.

18.3.4.1 Stripping Molds for Internal Threads

Shallow internal threads or those in reasonably flexible materials can be stripped off the core while they are still hot and relatively soft. In doing so, the part must actually be stretched over the threads of the core. This ability of a thread to be stripped off the core is determined by the amount of strain the part can withstand. This is determined according to the following equation:

$$\frac{D_M - D_m}{D_m} \times 100 = \%S \tag{18-1}$$

where D_M = major diameter

D_m = minor diameter

S = strain

Table 18-3 Recommended Allowable Strain for Closure Materials

Material	% Strain at 66 °C (150 °F)
ABS	8
Acetal	5
Acrylic	4
Nylon	9
Polythylene	
Low density	21
High density	6
Polypropylene	5

Source: R. D. Beck, Plastic Product Design, Van Nostrand Reinhold Co. Inc., New York, 1980.

The recommended allowable strain for some common closure materials is provided in Table 18-3.

Other materials have been stripped besides the ones listed in Table 18-3. For example, there are special grades of urea and phenolic designed for this purpose. Besides the ability to withstand strain, the key element in determining the ability of a part to be stripped is its stiffness at the time it is ejected from the mold. That is dependent on the temperature of the part at the time of ejection. The objective is to allow the part to cool just enough to be ejected without incurring a permanent set. The relevant numbers can be determined experimentally with a prototype mold. A strain of 4% is recommended as a starting point, and the mold should have a stripper plate and cooling equivalent to that of a production mold. It is advisable to keep the number of turns of thread to a minimum: with many applications one full turn of thread is enough, and most have less than two full turns of thread. The stress on the part can be reduced by using intermittent threads such as those found on many bottle caps. While such threads have been on the market for many years, the reader is advised to proceed with caution in using them, since some versions of intermittent threads may be covered by active patents.

18.3.4.2 Collapsing Core Molds for Internal Threads

Closures and other round parts with core diameters between 16.4 and 89.8 mm (0.645 and 3.535 in.) can be made with commercially made collapsing cores. In the range between 16.4 and 24.5 mm (0.645–0.965 in.) only intermittent undercuts are permissible. Figure 18-3a illustrates the core in the uncollapsed position. Its main components are a solid center pin and a core sleeve that has been cut in wedges as shown in the plan view. These wedges are made of spring steel, and the center pin is forcing them into the diameter of the core. When the ejector plates move forward, the core sleeve is pushed off the center pin and the wedge segments collapse inward,

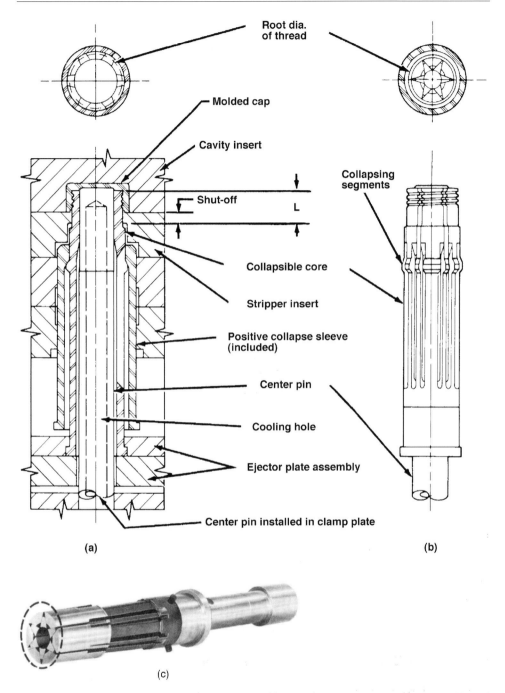

Figure 18-3 Collapsing core mold: (a) uncollapsed, (b) collapsed, and (c) core shown collapsed before threads have been cut in (Diagrams and photograph courtesy of D-M-E Company)

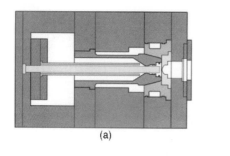

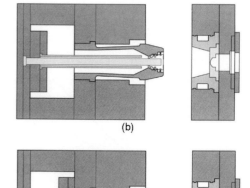

Figure 18-4 Expandable cavity mold: (a) mold closed, (b) mold open, cavity expanded, and (c) ejector forward, part ejected (Diagrams and photograph courtesy of D-M-E Company)

as in Fig. 18-3b. The positive collapse sleeve ensures that the wedge segments are forced inward. As it moves forward, it also activates the stripper insert, forcing the part off the core sleeve. The illustration in Fig. 18-3c provides a better view of the wedge segments. Although expensive to build and maintain, collapsing core molds are less costly than unscrewing molds for low to medium volume applications for which stripping molds are inappropriate.

18.3.4.3 Expandable Cavity Molds for External Threads

For external threads, the cavity must be split to permit removal of the part. One way to accomplish that is to use the expandable cavity cousin of the spring-loaded collapsing core mold. There are two versions of this concept. Figure 18-4 depicts a version that has its strikers on the cavity, or "A" side of the mold. As illustrated in Fig. 18-4a, they act on the spring-loaded cavity to keep it closed when the mold is closed. When the mold is opened, they spring apart to permit the moldment to be ejected, as in Fig. 18-4b. Part ejection is illustrated in Fig. 18-4c. This design allows the cavity to remain stationary, but a striker is required on the cavity side of the mold. The striker can also be located on the core or "B" side of the mold, as in Fig. 18-5; however, that approach requires the cavity to move forward to permit the part to be ejected.

18.3.4.4 Split-Cavity Molds for External Threads

The mold illustrated in Fig. 18-6 is designed to make a threaded container on the left with a split cavity. The external threads on the container would lock in the cavity and prevent the removal of the part; therefore the cavity must split for part removal.

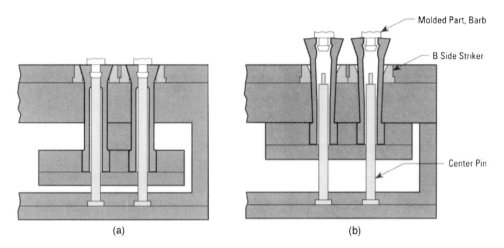

Figure 18-5   Expandable cavity mold — "B" side striker insert application: (a) mold closed and (b) mold open (Diagrams courtesy of D-M-E Company)

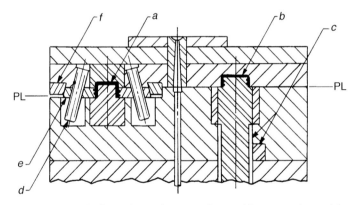

Figure 18-6   Split cavity and unscrewing mold: *a*, container with external threads; *b*, cap with internal threads; *c*, rack and pinion; *d*, cam; *e*, slide with thread details; *f*, clamp or lock

This is accomplished by the angled cams mounted to the "A" plate. They operate slides into which the cavity thread details are cut. The slides move outward when there is relative motion between them and the cams. They have to move only enough to clear the threads. Once they are clear of the part and there is sufficient space, the container can be ejected by the large pin acting on its base. When the mold closes, the cams force the slides to close and the clamps lock them in place. It is important that the cavity halves be built with precision and be sufficiently robust to withstand the molding and locking forces. Otherwise there will be flash on the threads, which may require an additional chasing operation for removal.

Such mechanisms are expensive, not merely because of the cost of putting them into the mold, but also because of the space they occupy in the mold base. This translates into a reduction in the number of cavities that can fit in a given size mold base, which, in turn, limits the number of cavities that can fit into a molding machine of a given size. Looking at it another way, a larger machine with a higher hourly operating cost would be required for the same number of cavities as a simple mold.

18.3.4.5 Unscrewing Molds for Internal Threads

The tool used for high volume applications for parts that cannot be stripped off the core is the unscrewing mold. The right side of the tool in Fig. 18-6 is the mold for a cap. It is mounted in the traditional fashion with the cavity in the "A" plate and the core in the "B" plate. In this case, the cap is held by the cavity while the core is unscrewed by one of two methods. The one illustrated on the right-hand side of Fig. 18-6 is the rack-and-pinion system in which an air cylinder operates a rack, which, in turn, rotates a pinion attached to the core. In the other method of operating unscrewing cores, electric motors operate gear-driven cores.

18.3.4.6 Unscrewing Chuck Plate Mold

Another method of unscrewing the parts from the mold is the unscrewing chuck mold. In this mold, the cores remain stationary and a large plate with unscrewing chucks is positioned over the parts and unscrews them. To maintain production rates, this type of mold is built with two core halves that rotate into molding position so that one can be molding while the parts are removed from the other. Although the initial investment is greater, this system offers improved cooling, reduced maintenance, less part contamination from oil or grease, and the elimination of wear from the rotation of the cores over the unscrewing core mold.

18.3.4.7 Molds for Parts with Less than One Turn of Thread

Whether the thread is external or internal, inexpensive tools can be used when there is less than one full turn of thread. This is accomplished through the use of shutoffs. The internal thread is formed where the core from one side meets the core from the other side in the section taken along the curve. Where the thread ends, the steel meets, creating what is referred to as a shutoff. The angle between the ends of the thread should be at least 8° for long tool life. As long as the thread ends do not overlap, no undercut is created. Unfortunately, this approach is not applicable for the internal threads on bottle caps because the core must enter from both ends.

For external threads, the shutoff occurs in the cavity. When applicable, this permits the use of a simple mold without split cavities and the expensive mechanism described for that system.

18.4 Sources

18.4.1 Collapsing Cores and Cavities

D.M.E. Company, 29111 Stephenson Hwy., Madison Heights, MI 48071, (800) 626-6653, (248) 398-6000, fax (888) 808-4363, www.dme.net

Roehr Tool Corporation, 14 South St., Hudson, MA 01749, (877) 563-1912, (978) 562-4488, fax (978) 562-3660, www.roehrtool.com

18.4.2 Unscrewing Chuck

Husky Injection Molding Systems Ltd., 500 Queen St. S., Bolton, Ontario, Canada L7E 5S5, (905) 951-5000, fax (905) 951-5384, www.husky.ca

19   Ultrasonic Welding

19.1   Advantages and Disadvantages of Ultrasonic Welding

19.1.1   Advantages

Ultrasonic welding is one of the thermoplastic welding processes and shares many of the same advantages and disadvantages. Indeed, the thermoplastic welding processes are competitive for many applications. The principal considerations in making the decision are the geometry of the part, the loadings on the joint, size, mold costs, and required rate of assembly.

There are, however, a variety of applications for ultrasonic techniques beyond the welding of two plastic parts together with energy directors. They can be used for spot welding, stud welding, fabric and film sealing, and to plug holes. The use of ultrasonics in the insertion of metal parts into plastics is covered in Chapter 8, "Fasteners and Inserts" and its use in staking and swaging is discussed in Chapter 17, "Staking/ Swaging/Peening/Cold Heading/Cold Forming." Beyond the scope of this book, it is also used for slitting and the ultrasonic welding of metals. The advantages of ultrasonic welding are as follows.

1. *No additional materials* Ultrasonic welding uses no additional materials such as fasteners, inserts, electromagnetic preforms, adhesives, or solvents. Therefore, it is inherently lower in cost than methods that have such requirements, and items so joined are less expensive to disassemble for recycling.

2. *Ease of assembly* Ultrasonic welding requires only the alignment of the two parts in the fixture of the ultrasonic welder, thus it is well suited to automated assembly.

3. *Entrapment of other parts* Additional parts can be captured between the two parts to be ultrasonically welded. However, care must be taken to avoid fragile parts, which would be damaged by the ultrasonic vibrations.

4. *Permanence* Ultrasonic welding creates permanent assemblies that cannot be reopened without damaging the parts. Since a welded joint is produced, the effects of creep, cold flow, stress relaxation, and other environmental limitations are absent in the joint area. Owing to material limitations, differences in thermal expansion and moisture absorption are rarely of concern (except for nylons) once the parts have been welded. Joint strengths equivalent to those of the surrounding walls are possible under ideal conditions.

5. *Internal joints* With small, flat-surfaced parts, internal ribs can be ultrasonically welded.

6. *Contour freedom* Unlike spin and vibration welding, ultrasonic welding can be done with parts of practically any surface contour as long as the joining surfaces can

be created within the required size and shape limitations. Holes can be plugged ultrasonically.

7. *Hermetic seals* Ultrasonic welding is capable of creating hermetic seals.

8. *Energy efficiency* Relative to other welding processes, ultrasonic welding is highly energy efficient. This also means that there is no excess heat, which must be removed from the workplace when other processes are used.

9. *Clean atmosphere* Unlike adhesive and solvent joining systems, no ventilation equipment is necessary for the removal of toxic fumes (except for polyvinyl chloride).

10. *Immediate handling* Assembled parts can proceed to other operations at once without waiting for parts to cool and for adhesives or solvents to set. This advantage lends the process well to automated assembly line applications.

11. *High production rates* Production rates of 20 to 60 parts per minute are possible, since the actual welding time rarely exceeds one second.

19.1.2 Disadvantages

There are some disadvantages as well.

1. *Shape limitations* There must be a flat or nearly flat joining surface between the parts, as well as a surface reasonably close to the joint and a constant distance from it that can be accessible for the horn.

2. *Process limitations* Injection molding is the principal process that can create the joint details necessary for ultrasonic welding; although pressure thermoforming can also produce energy directors. Tight process control is required to create the necessary tolerances and flatness. Some extruded, rotationally molded, thermoformed parts, and those fabricated from sheet goods have also been ultrasonically welded, most often with spot welds, which do not require energy directors. Textiles and films are welded with special anvils and equipment.

3. *Damage to electronic components* Ultrasonic vibrations can damage some electronic components or their assemblies. The use of higher frequency equipment that operates at lower levels of vibration may solve this problem.

4. *Size limitations* The largest ultrasonic welders can create welds in the range of 0.23 m × 0.3 m (9 in. × 12 in.). However, machines of that size were not widely available at the time of writing, and most applications are much smaller. Often, multiple welders are used to weld large parts.

5. *Consistency* Ultrasonic welding dates from the mid-1960s. Historically, there have been problems of consistency due to power variations, surface irregularities, material differences, and multicavity part dimensional variations. However, the advent of microprocessor-controlled equipment and improvements in the design of ultrasonic joints have largely overcome these issues.

6. *Material limitations* Materials for ultrasonic welding are limited to compatible thermoplastics. This is an involved subject that requires elaboration. Therefore, the reader is referred to Section 19.4, which deals with this topic.

7. *Sound concerns* Ultrasonic welders usually emit a shrill, high-pitched whistlelike sound that can be annoying to operators, who may require ear protection. Sound

enclosures are available, and the higher frequency equipment operates at lower noise levels.

8. *Equipment cost* In their simplest form, ultrasonic welders are not very expensive. However, the price increases rapidly as features such as microprocessor controllers are added. This can result in a significant investment if multiple welders are required.

19.2   General Applications

Spurred largely by cost reduction efforts and new molding technology, ultrasonic welding has enjoyed an ever increasing range of applications. Many companies regard it as the welding method of choice because of its speed, efficiency, and lack of environmental problems. In short, for relatively flat parts of modest size and compatible materials, it is a quick and clean way to permanently join thermoplastic parts.

Designers and engineers have created a wide range of innovative ultrasonic welding applications. In addition to the direct welding of two parts, ultrasonic techniques are used for staking and swaging (see Chapter 17: "Staking/Swaging/Peening/Cold Heading/Cold Forming), the installation of threaded inserts (see Chapter 8, "Fasteners and Inserts"), spot welding, and the welding of textiles and films. The medical industry uses ultrasonics for the assembly of many small devices used in the packaging of medicines and their delivery to the human body. The automotive industry uses ultrasonics to weld automobile lamp assemblies, instrument panels, air ducts, upholstery, body components, and small engine components. Many small to medium parts for the toy, appliance, and electronics industries are assembled this way. The cosmetic and food industries use ultrasonic welding to create a vast range of packaging applications.

19.3   The Principal of Ultrasonic Welding

All vibrations make a sound; however not all sounds are audible. Ultrasonic, which literally means "beyond sound," is a term used to refer to that portion of the mechanical vibration frequency range which is beyond the audible spectrum and cannot be heard. That is regarded as anything above 18,000 Hz, the approximate limit of human hearing.

In ultrasonic welding, vibrations produced by a horn create intermolecular and boundary friction at the joint interface of the parts to be welded. The polymer begins to soften locally and, as the dampening factor increases, the reaction accelerates itself, causing a greater proportion of the vibration energy to be translated into heat. That raises the temperatures of the plastic parts until their melting point is achieved. After the vibrations are halted, the pressure is maintained until a molecular bond is created and the joint is sufficiently cooled.

Logically, the closer the source of the vibrations to the joint, the fewer vibrations lost to absorption by the part on their way to the joint. When the horn can be located within 6.25 mm (0.25 in.) of the joint, the process is referred to as near-field welding. Excellent results can be obtained with near-field welding, even with thin walls and crystalline or low stiffness polymers. Near-field welding is shown in Fig. 19-1a.

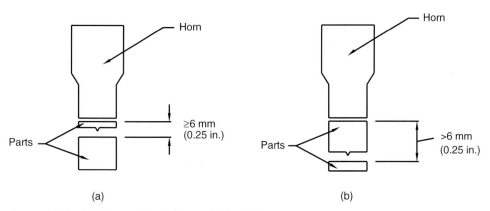

(a)   (b)

Figure 19-1   (a) Near field and (b) far-field welding

Far-field welding (Fig. 19-1b) is used for greater distances, which require thicker walls of amorphous or high stiffness thermoplastics. Under the optimum conditions, far-field welds have been successfully accomplished with horn contacts as far as 250 mm (10 in.) from the joint interface. However, far-field welds require longer weld times, higher than normal amplitudes, and higher weld force to achieve the same quality obtainable with near-field welds.

19.4   Materials for Ultrasonic Welding

The resin is one of the key factors determining energy levels for ultrasonic welding. The stiffer a material is, the more successfully it transmits ultrasonic vibrations from the horn to the joint interface. The rigid amorphous thermoplastics have a random molecular structure and transmit ultrasonic vibrations very efficiently with little attenuation. They also have softening temperatures that cover a broad range because they do not have a defined melting point. As their temperature rises, they soften gradually as they pass from the rigid state through their glass transition temperature into a rubbery condition, which leads to the molten state. Cooling follows the same pattern of transitions in reverse, and thus the material is permitted to flow easily without solidifying prematurely. Amorphous polymers exhibit very low levels of postmolding shrinkage, permitting them to be welded immediately after molding without incurring stresses due to shrinkage. It should be noted that the less rigid the amorphous thermoplastic, the greater the amplitude and energy required to weld. At some point, the polymer may be so soft that the ultrasonic vibrations are dampened so much that welding is not possible. This condition can also be created by the decrease in density that accompanies the foaming of the material, as in structural foam.

Crystalline thermoplastics have a very orderly molecular structure, which absorbs mechanical energy in the solid state. This makes it much more difficult to transmit vibrations from the horn to the joint interface without considerable attenuation. Consequently, far-field welding is very difficult to accomplish, if it can be done at all.

Excessive distances from horn to joint can result in melting of the plastic at the horn, which affects appearance. Crystallines also have a high heat of fusion and a sharp melting point, and a very high level of heat energy is required to break down the crystalline structure before melting can occur. Thus, the polymer remains rigid until its melt temperature is attained. Then it becomes molten very abruptly. The net result is that very high energy is required to break down the crystalline structure and permit the polymer to melt and flow. Thus, high amplitudes and shear joint designs are usually necessary to weld these materials, particularly if a hermetic seal is required. Cooling is also sudden because the molecules recrystallize abruptly. Crystalline thermoplastics experience a high level of postmolding shrinkage, which can create stresses in the joint if welded within 24 hours of molding. The differences in melt characteristics of amorphous and crystalline thermoplastics are illustrated in Fig. 19-2.

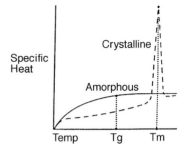

Figure 19-2   Specific heat versus temperature (Graph courtesy of Branson Ultrasonics Corp., Danbury, CT)

Ideally, both parts should be made of the same polymer to ensure that both sides of the joint interface will melt at the same temperature. In some circumstances different materials are desirable for the two parts. In general, dissimilar materials can be welded provided their glass transition temperatures are within 22 °C (40 °F) of each other and their molecular structure is similar. (If the temperature differential is greater than this, the material with the lower melt temperature will become too soft to create frictional heat with the other polymer before it can adequately soften.) Examples of combinations that meet this criterion are polycarbonate to acrylic and polystyrene to modified polyphenylene oxide. However, the two materials must be chemically compatible as well. Polyethylene and polypropylene meet the temperature criterion, but are not chemically compatible and cannot be welded together. Blends and alloys that share a common component can often be welded to each other as well. For example, ABS can be welded to acrylic or polystyrene.

The compatibility of the various thermoplastics for ultrasonic welding is demonstrated in Table 19-1. The dark squares indicate compatibility. In most cases, these materials are completely compatible only with themselves; however there are grade differences within resins that are sufficiently significant to prevent weldability. As a general guide, the two grades should be similar in molecular weight, and their melt temperatures should be within 22 °C (40 °F) of each other. The circles in Table 19-1 indicate partial compatibility, meaning that not all grades and compositions are compatible.

Table 19-1   Compatibility of Thermoplastics[a]

	Amorphous Resins	ABS	ABS/polycarbonate alloy (Cycoloy 800)	Acrylic	Acrylic multipolymer	Butadiene–styrene	Phenylene-oxide based resins (Noryl)	Polyamide-imide (Torlon)	Polyarylate	Polycarbonate	Polyetherimide	Polyethersulfone	Polystyrene (general purpose)	Polystyrene (rubber modified)	Polysulfone	PVC (rigid)	SAN-NAS-ASA	Semicrystalline Resins	Acetal	Cellulosics (CA, CAB, CAP)	Fluoropolymers	Ionomer	Liquid crystal polymers	Nylon	Polybutylene terephthalate (PBT)	Polyetheretherketone (PEEK)	Polyethylene terephthalate (PET)	Polyethylene	Polymethylpentene	Polyphenylene sulfide	Polypropylene
Amorphous Resins																															
ABS		█	○	○	○											○	○														
ABS/polycarbonate alloy (Cycoloy 800)		○	█		○					█																					
Acrylic		○		█	○		○						○				○														
Acrylic multipolymer		○	○	○	█								○				○														
Butadiene–styrene						█							○																		
Phenylene-oxide based resins (Noryl)				○			█			○				○			○														
Polyamide-imide (Torlon)								█																							
Polyarylate									█																						
Polycarbonate			█				○			█					○																
Polyetherimide											█				○																
Polyethersulfone												█																			
Polystyrene (general purpose)				○	○	○							█				○														
Polystyrene (rubber modified)							○							█																	
Polysulfone										○	○				█																
PVC (rigid)		○														█															
SAN-NAS-ASA		○		○	○		○						○				█														
Semicrystalline Resins																															
Acetal																			█												
Cellulosics (CA, CAB, CAP)																				█											
Fluoropolymers																					█										
Ionomer																						█									
Liquid crystal polymers																							█								
Nylon																								█							
Polybutylene terephthalate (PBT)																									█						
Polyetheretherketone (PEEK)																										█					
Polyethylene terephthalate (PET)																											█				
Polyethylene																												█			
Polymethylpentene																													█		
Polyphenylene sulfide																														█	
Polypropylene																															█

[a] Solid squares, compatibility; open circles, compatibility in some cases (usually blends).
Source: Branson Ultrasonics Corp., Danbury, CT

Some of the materials are hygroscopic. The presence of moisture makes it more difficult to accomplish a good ultrasonic weldment because the water boils off when the melt reaches 100 °C (212 °F) and creates porosity in the form of a foamy condition or voids at the joint interface. This weakens the joint, causes degradation, creates a frosty appearance, and makes it very difficult to create a hermetic seal. Moisture-absorbing plastics such as these should be dried before welding unless they have been stored in sealed polyethylene bags with a desiccant or are welded immediately after molding. The principal hygroscopic materials are nylon, polysulfone, polycarbonate, polyester, and alloys that contain these polymers.

The ability of the various resins to be ultrasonically welded varies with the type of ultrasonic welding and whether it is near- or far-field welding, as illustrated in Table 19-2. The codes in this table indicate relative ease of welding for the more common thermoplastics. In addition to the material factors, ease of welding is a function of joint design, part geometry, energy requirements, amplitude, and fixturing. These ratings, which were promulgated by an equipment manufacturer, do not relate to the strength obtainable. The manufacturer, Branson Ultrasonics Corp., points out that since variations in resins may produce results slightly different from those indicated here, the ratings in Table 19-2 should be used only as a guide.

19.4.1 Additives and Contaminants

19.4.1.1 Colorants

Colorants are either dyes or pigments. Except for some whites and blacks, they have little effect on ultrasonic weldability. Titanium dioxide, the primary white pigment, is inorganic and chemically inert; at levels exceeding 5%, however, it can act as a lubricant and inhibit weldability. Carbon black, commonly used as a black colorant, can also inhibit weldability. As a precaution, testing should be performed on pigmented samples to determine effect. It should be noted that organic pigments are replacing metal-based pigments for many applications.

19.4.1.2 Fillers, Extenders, and Fibrous Reinforcements

The presence of fillers and extenders, such as glass fibers, talc, calcium carbonate, kaolin, wood, aluminum, titanium, and carbon fibers can adversely affect weldability. The addition of 10 to 20% glass can improve the weldability of crystalline thermoplastics considerably by making them stiffer and more efficient at transmitting mechanical vibrations. However, fillers replace polymer with materials that will not weld. When they reach levels of 30 to 40%, they can inhibit weldability significantly, with the result that strong welds cannot be assured because there is insufficient polymer at the joint interface and the energy director simply embeds itself into the mating part. Even at lower levels they can cause marking of the part sufficient to require lowering the filler content.

It is the percentage of some fillers in the joint that is significant, and that can be higher than the overall percentage in the part. That is because fillers such as long fibers of glass can collect at the gate area as the mold fills and then be swept up into the cavity in clumps to create concentrations of glass fibers at the joint surface. This is known as agglomeration or filler enrichment. Uneven welding can result from nonuniform distribution of filler in the joint area. Substitution of short-fiber glass reinforcement, relocation of the gates, changes in molding conditions, and reduction in filler content can resolve this problem.

Table 19-2   Weldability of Resins: E, Excellent; G, Good; F, Fair; P, Poor

Material	Ease of welding		Swaging and staking	Insertion	Spot welding	Vibration welding
	Near field[a]	Far field[b]				
Amorphous resins						
ABS	E	G	E	E	E	E
ABS/polycarbonate alloy	E-G	G	G	E-G	G	E
Acrylic[c]	G	G-F	F	G	G	E
Acrylic multipolymer	G	F	G	G	G	E
Butadiene-styrene	G	F	G	G	G	G
Phenylene-oxide based resins	G	G	G-E	E	G	E-F
Polyamide-imide	G	F				G
Polyarylate	G	F				
Polycarbonate[d]	G	G	G-F	G	G	E
Polyetherimide	G	F				
Polyethersulfone	G	F				
Polystyrene						
(general purpose)	E	E	F	G-E	F	E
(rubber modified)	G	G-F	E	E	E	E
Polysulfone[d]	G	F	G-F	G	F	E
PVC (rigid)	F-P	P	G	E	G-F	G
SAN-NAS-ASA	E	E	F	G	G-F	E
Xenoy (PBT/ polycarbonate alloy)	G	F	F	G	G	E
Semicrystalline resins[e]						
Acetal	G	F	G-F	G	F	E
Cellulosics	F-P	P	G	E	F-P	E
Fluoropolymers	P					F
Ionomer	F	P				
Liquid crystal polymers	F	P	G-F			
Nylon[d]	G	F	G-F	G	F	E
Polyester, thermoplastic						
Polyethylene terephthalate (PET)	G-F	P				
Polybutylene terephthalate (PBT)		P				
Polyetheretherketone (PEEK)	F	P				G
Polyethylene	F-P	P	G-F	G	G	G-F
Polymethylpentene	F	F-P	G-F	E	G	E
Polyphenylene sulfide	G	F	P	G	F	G
Polypropylene	F	P	E	G	E	E

[a] Near-field welding refers to a joint 6.35 mm (0.25 in.) or less from the horn contact surface.
[b] Far-field welding refers to a joint more than 6.35 mm (0.25 in.) from the horn contact surface.
[c] Cast grades are more difficult to weld due to higher molecular weight.
[d] Moisture will inhibit welds.
[e] Semicrystalline resins in general require higher amplitudes due to polymer structure and higher energy levels due to higher melt temperatures and heat of fusion
(*Source*: Branson Ultrasonics Corp., Danbury, CT)

Up to a level of 10%, fillers have little effect on the weldability of amorphous thermoplastics and can be welded in the normal manner. At that level, they begin to affect costs because glass and mineral fillers act as an abrasive and increase the rate of wear on the horn surface and can damage the fixture as well. At levels beyond 20%, this effect becomes significant. Fillers can require special tooling, such as hardened steel or carbide-coated titanium horns, and higher powered equipment.

19.4.1.3 Flame Retardants

Inorganic oxides or halogenated organic elements, such as aluminum, antimony, boron, chlorine, bromine, sulfur, nitrogen, and phosphorus, are added to a resin to alter its burning characteristics or to inhibit ignition. The amount used can range from a small percentage to over 50%. Use of flame retardants at these higher levels reduces the amount of resin available for welding in much the same fashion as fillers. This can reduce the strength of the joint and can also require higher powered equipment operating at higher than normal amplitudes to achieve a weld of adequate strength.

19.4.1.4 Foaming Agents

Foamed parts have a cellular structure beginning just below the surface, with the cells becoming increasingly large as they progress toward the center of the wall. Voids in these cells interrupt the transmission of mechanical vibrations and reduce the amount of energy available at the joint interface. This effect is dependent on the density of the foam. The greater the decrease in density, the more difficult it is to weld the parts ultrasonically. Furthermore, any unreacted chemical blowing agent still present in the welding area may result in the creation of foam.

19.4.1.5 Impact Modifiers

Typically, rubber impact modifiers also reduce the resin's ability to transmit mechanical vibrations. Thus the material's ability to permit far-field welding is reduced and greater amplitudes are required to produce an acceptable weld. In addition, the amount of resin available for welding at the joint interface is reduced by the amount of impact modifier used.

19.4.1.6 Lubricants

Lubricants affect the weldability of the plastic parts because they lower the coefficient of friction between the two materials, making it difficult to generate heat. Additives used to improve flow characteristics, such as waxes, zinc stearate, aluminum stearate, stearic acid, and fatty acids must be avoided; however small amounts of these additives will have minimal effect on weldability.

19.4.1.7 Mold Releases

Silicones, fluorocarbons, zinc stearate, and aluminum stearate have an effect similar to that of lubricants. In addition, the release agent can chemically contaminate the resin and, thereby, inhibit the formation of a proper bond. The best way to ensure release of the parts from the core without the use of mold release is to employ greater draft angles of 2° or more. If that is not possible and mold release must be used, it should be used sparingly and limited to paintable and printable grades, which interfere with ultrasonic welding the least and can sometimes be used without solvent cleaning before assembly. (Refer to Chapter 7, "Adhesive and Solvent Joining," for specific solvent recommendations.) Some mold releases migrate, and proximity to a nearby molding machine where they are being used liberally can be the source of a problem. As with most assembly techniques, cleanliness remains a problem, and to avoid a cleaning operation, parts to be ultrasonically welded must be kept clean and free of contaminants such as grease.

19.4.1.8 Painted Parts

The requirement for maintenance of color uniformity has led to a need to assemble painted parts. Experiments with painted PC/ABS and PP specimens welded to unpainted pieces of the same material have led to the conclusion that this can be done with both amorphous and crystalline thermoplastics. However, a drop in weld strength is associated with the ultrasonic welding of painted samples.

19.4.1.9 Plasticizers

Principally used in vinyls, plasticizers are additives that make a resin more flexible, decreasing its efficiency at transmitting mechanical vibrations. This diminution in efficiency requires the use of higher amplitudes and makes far-field welding difficult. The additives the FDA has approved for use in products in contact with substances transmitted within the human body have no metals and are preferred to metallic plasticizers.

19.4.1.10 Regrind

The use of regrind affects the physical properties of the plastic. Consequently, it has an adverse effect on its welding characteristics and can lead to uneven welding. Very low levels of uncontaminated regrind that has not been degraded will be insignificant for most applications. Samples using the intended amount of regrind must be tested before regrind can be accepted for a given application, since the results can differ from those obtained with 100% virgin plastic. In some cases, it may be necessary to exclude the use of regrind.

19.5   Part Design for Ultrasonic Welding

19.5.1   Overall Ultrasonic Welding Considerations

By this point, it is presumed that the basic design parameters of the assembly have clearly indicated that ultrasonic welding is the assembly method of choice. Now it becomes necessary to further define these requirements such that the details of the design can be developed.

19.5.1.1 Strength Requirements

First we need to establish what the strength of the assembly must be. Properly executed, ultrasonic welding is capable of producing a joint nearly as strong as the surrounding walls in many polymers. A key concern in this process is the question of whether a hermetic seal will be required. Once the required strength and seal have been established, the material and wall thickness can be selected.

19.5.1.2 Appearance Requirements

Another key issue is the appearance requirement for the assembly. The need to fixture one part and the requirement for proper horn placement on the other may have a pronounced effect on the shape of the parts. If a visually attractive appearance is needed and flash will be a concern, these issues will largely determine the design of the joint.

For a weld to take place, it is essential that sufficient vibrations be delivered to the joint interface to create the friction necessary to melt the plastic. It is desirable to do so in the most efficient manner possible.

19.5.1.3 Rigidity Considerations

Rigid walls deliver mechanical vibrations more efficiently than do those that are less rigid. Therefore, the walls need to be stiff enough to permit such delivery. The thickness required is dependent on the material composition and the distance between the horn and the joint interface. However, the vibrating side is not the only concern. For the weld to take place, the stationary side must be as stationary as possible, since the vibrations transmitted as vibrations are not used to create the weld. However, the wall can be reinforced by the fixture in which it rests during welding. Thus, in designing the joint area of the assembly, the following considerations must be taken into account:

a. *Appendages*   Damage can occur to tabs, fingers, or other appendages that protrude from the side of the part beyond the joint as depicted in Fig. 19-3. Remedies include increases in the inside radii at the base of the appendage to reduce stress concentrations, altering the molding conditions, increasing the wall thickness, reducing the weld time, reducing the amplitude, dampening the vibrations with external fixturing, and switching to a higher frequency welder.

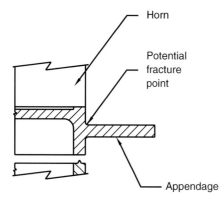

Figure 19-3   Part design for ultrasonic welding – appendages

b. Bends   When bends occur in the path of the vibrations as they travel from the horn to the joint interface as illustrated in Fig. 19-4a, a dampening effect occurs, resulting in a weakened joint below the bend. Crystalline materials suffer from this effect more than amorphous polymers.

c. Holes or Voids   Openings in the vibrated part will disturb the path of the ultrasonic vibrations en route from the horn to the joint interface because they stop when they reach the hole, as demonstrated in Fig. 19-4b. If the hole is very small and the material is an amorphous polymer, the vibration pattern may suffice to allow for a continuous joint below the hole. That will be the weakest portion of the joint, and a hermetic seal would be very difficult. For crystalline polymers or larger holes, there will be no weld below the opening. It should be noted that poor weld lines will have a similar effect, and the joint will be weakest where such a weld line exists.

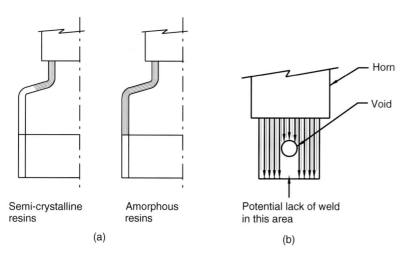

Figure 19-4   Part design for ultrasonic welding: effects of (a) bends and (b) voids (Diagrams courtesy of Branson Ultrasonics Corp., Danbury, CT)

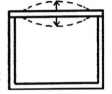

Figure 19-5 Part design for ultrasonic welding – diaphragmming (Sketch courtesy of Branson Ultrasonics Corp., Danbury, CT)

d. Diaphragmming Figure 19-5 illustrates the effect of diaphragmming, sometimes referred to as "oil canning," which occurs when the horn is placed in contact with the thin-walled section of a flat, circular part. This can result in burn-through. Design remedies are internal support ribs or thicker wall sections under the gate area. Sometimes a relocation of the gate can be of help. Processing remedies include a higher or lower amplitude (usually lower), shorter weld time, or a nodal plunger on the horn. Finally, a change to higher frequency equipment can be the solution.

e. Horn Location The distance between the horn and the joint interface determines the amount of energy that will be lost en route to the joint. Furthermore, the part should be designed so that the horn can bear directly over the joint area, as shown in Fig. 19-6a. This helps prevent marking the part by directing the mechanical vibrations and will likely require the horn to overhang the part. The horn-to-part contact area must be larger than the total weld area, also to prevent surface marking. Figure 19-6a shows a raised section on the part to provide a better coupling surface. Figure 19-6b demonstrates how a flange can be added for an energy director joint so the horn can be placed directly over the joint.

f. Inside Sharp Corners The stress concentrations related to inside sharp corners may result in damage in the form of fracturing or melting on subjection to ultrasonic mechanical vibrations. (This topic is discussed in further detail in Chapter 2,

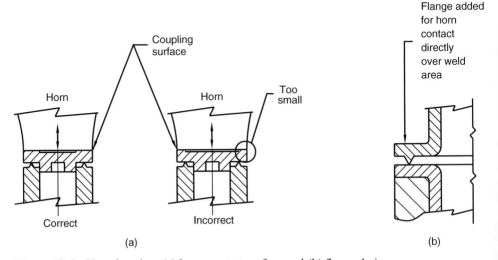

Figure 19-6 Horn location: (a) horn contact surface and (b) flange design

"Designing for Efficient Assembly.") If sharp corners cannot be eliminated from the design, it may be necessary to reduce the amplitude, change to higher frequency equipment, or find some other way to dampen the vibrations. It may be possible to reduce the molded-in stress by altering the molding conditions.

g. *Internal Components*   If components placed internally must be in direct contact with the part to be welded, they should be made of materials that cannot weld together. If that is not possible, they can be lubricated to prevent such welding together. This problem is similar to that of internal parts fragile enough to break from ultrasonic vibrations. Solutions that benefit both circumstances are the relocation of the parts further from the weld areas, the introduction of a dampening material, or a reduction in amplitude, pressure, or weld time. A change to a higher frequency welder with a more limited vibrational pattern is another solution.

h. *Material*   The part material or materials will determine the strength of the weld and the efficiency of the process, as discussed earlier in this chapter.

i. *Rigidity of the Vibrated Part*   The vibrated part must be sufficiently rigid to transmit the energy from the horn. Otherwise, the result may be nonuniform welds or no welds at all. When the mechanical vibrations are being inadequately transmitted to the joint interface, remedies include increasing the wall thickness, changing to a stiffer material, adding stiffening ribs, and revising the part to reduce the distance the vibrations must travel.

j. *Rigidity of the Stationary Part*   The rigidity of this part in its fixture will determine the amount of energy available to create friction. If the part, which is supposed to be stationary, is vibrating along with the vibrated part, failure to secure an acceptable weld may result. Design solutions include increasing the wall thickness, changing to a stiffer material, and adding stiffening ribs. However, the problem is often attributed to improper fixturing. While low temperature, amorphous plastics can be supported by a fixture made of a resilient material such as cork or poured urethane, high temperature, high energy amorphous materials and crystalline polymers should be supported by rigid steel or aluminum fixtures. Side support up the wall as close to the joint area as possible is required for shear joints, and energy directors are needed on very thin walls to prevent outward deflection during welding. Flanged energy directors should have support directly under the flange, as shown in Fig. 19-6b.

19.5.2   Joint Fundamentals

The three fundamental elements of ultrasonic joint design are part alignment, uniform vibration travel distance, and minimal contact area. We discuss each of these in turn.

19.5.2.1 *Part Alignment*

The ultrasonically welded joint and tooling should not be relied on for the location of the mating parts. Flexing of the walls of the part can lead to misalignment, and extension of the fixture high enough to align the parts can lead to surface marking. Locating

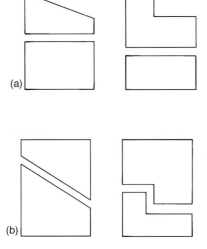

Figure 19-7 Uniform vibration travel distance: variations in (a) horn contact height (surface of horn contact should be on a single plane parallel to the joint interface) and (b) joint interface height (joint interface should be on a single plane parallel to the surface of horn contact).

pins, ribs, tongue-and-groove fitments, or other design features integral to the parts should always be used to ensure correct alignment.

19.5.2.2 Uniform Vibration Travel Distance

Mating surfaces are best kept in a single plane that can be mounted horizontally in the welder. To maintain a weld of uniform strength, the distance the ultrasonic vibrations must travel between the horn and the joint interface should be held constant: see the near- and far-field weld examples in Fig. 19-1.

Neither variations in the horn contact surface height such as those in Fig. 19-7a nor variations in the joint interface height as in Fig. 19-7b are desirable. High strength welds and hermetic seals become extremely difficult as the joint surface deviates from the horizontal plane. Weld uniformity can also be affected by unintentional joint surface deviations resulting from cavity-to-cavity variations, part distortion in molding, or a lack of parallelism between horn, part, and fixture. Ejector pins should not be located near an energy director, since there is often a slight depression in the surface near an ejector pin. Some deviation can be tolerated if weld uniformity is not important, and angles up to 45° have been welded. However, significant variations from the horizontal are difficult and require special tooling. Figure 19-8 illustrates a method of using two separate horns to ultrasonically weld a joint that is not on a horizontal plane.

19.5.2.3 Minimal Initial Contact Area

Regardless of which type of joint is used, the basic principal is to focus the ultrasonic vibrations in a small, uniform initial contact area. That means the point of the energy director must be sharp. Since mold makers tend to leave small radii on the order of 0.1 to 0.2 mm (0.004–0.008 in.) where no radius is specified and may not recognize the significance of a sharp corner on an energy director, the author has found it

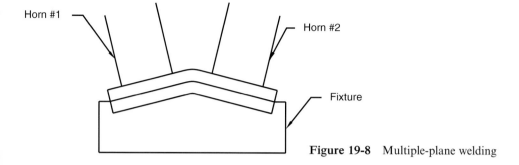

Figure 19-8 Multiple-plane welding

worthwhile to emphasize this detail with the notation "SHARP EDGE – ENERGY DIRECTOR" and an arrow pointing to the tip of the energy director. This will concentrate the energy to achieve rapid, localized energy dissipation. Even so, the nature of plastic flow in the mold is such that the tip will never be absolutely sharp. In addition to reducing the time and energy required to accomplish the weld, the sharp tip will result in less scuffing and flash.

19.5.3 Energy Director Joints

19.5.3.1 Butt Joint

The object of all the various joint designs is to minimize contact area. The practical effect of this objective is demonstrated in Fig. 19-9. The surface of the flat butt joint in Fig. 19-9a cannot be expected to be absolutely flat; therefore there will be voids in the welded surface. The welding of the large area associated with a simple butt joint

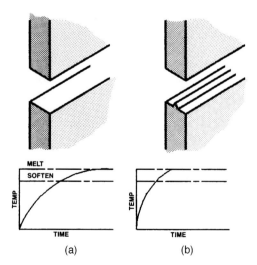

Figure 19-9 (a) Butt joint versus (b) energy director (Diagrams courtesy of Branson Ultrasonics Corp., Danbury, CT)

requires an excessive amount of time and energy. There is also an unsightly melt-out resulting from this type of joint. As a result of these negatives, this unmodified version of the butt joint is generally regarded as obsolete and is rarely used.

The energy director is the V-shaped rib protruding from one of the parts to be assembled (Fig. 19-9b). The point of the V forms the joint interface with the flat of the mating part. In this way, the energy is concentrated so as to rapidly initiate the softening and melting of the joint interface. In a state of equilibrium, the material at the point of the V will be at, or very nearly at, its elastic limit. The high forces at high frequency modulation increase the stress beyond the elastic limit, causing a permanent deformation. As a result, hysteresis losses (heat) are generated, which are part of the internal molecular friction used to elevate the temperature at the part interface, and significantly shorten assembly time. Note how much less time and energy are required to soften and melt the plastic when an energy director is used. The dimensions of the basic energy director joint are provided in Table 19-3.

Generally speaking, smaller parts call for smaller energy directors. As a rough guide, the width of the energy director (W_j) should be no more than 20 to 25% of the wall thickness (W). Stiff, amorphous polymers melt across the joint surface, and those that melt at moderate temperatures can do with an absolute minimum energy director height as low as 0.13 mm (0.005 in.); however, a recommended minimum would be 0.25 mm (0.010 in.). Higher melt temperature amorphous plastics, like polycarbonate, and crystalline polymers require a minimum height of 0.5 mm (0.020 in.), although an absolute minimum would be 0.3 mm (0.013 in.). When the height of the energy director must exceed 0.75 mm (0.030 in.), consideration should be given to the use of two energy directors in place of one large one. In that case, the total of their heights should equal 10% of the joint width. Too large an energy director can result in excessive melt-out or flash, and variations in the height in the energy director will result in inconsistent welds.

Crystalline polymers achieve maximum joint strength only across the width of the base of the energy director. Therefore, shear joints are preferred for crystalline materials requiring a strong weld or a hermetic seal. When they must be used, an energy director with a 60° included angle, which will raise the height of the energy director and give it a more pronounced point, is recommended. That point penetrates deeper into the surface of the mating part during the initial stage of the weld. It, therefore, has less exposure to the air, which causes degradation and premature solidification. A stronger joint will result, along with improved prospects for a hermetic seal except when a very soft crystalline polymer is used. In that case, the tip will not have sufficient rigidity and a 90° included angle is recommended. Glass-filled materials also perform better with the 90° included angle.

It does not matter which of the two parts gets the energy director when both are made of the same material. When the materials are different, the one with the highest melt temperature and stiffness gets the energy director. When a copolymer or terpolymer (e.g., ABS) is to be welded to a homopolymer (e.g., acrylic), the greatest strength is attained when the energy director is on the homopolymer part. Since butt joints do not provide location and alignment, use of the fixture can result in scuffing. Some other locating device, such as pins and sockets, must be found.

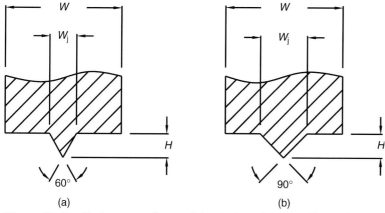

Figure 19-10 Basic energy director joint designs: (a) crystalline polymers and (b) amorphous polymers

Table 19-3 Dimensions for Basic Energy Director Joints for Crystalline and Amorphous Polymers

Dimension	Crystalline Polymers[a]			Amorphous Polymers[b]	
	in.	mm		in.	mm
W_j					
Minimum	0.015	0.4	Minimum	0.010	0.25
Maximum	0.050	1.3	Maximum	0.050	1.3
Absolute minimum	0.013	0.3	Absolute minimum	0.005	0.13
H					
Recommended minimum	0.020	0.5	Recommended minimum	0.010	0.25
Maximum	0.043	1.1	Maximum	0.025	0.6

[a] $W = 4W_j$ to $5W_j$; $H = 0.5W_j{}^c$ to $0.866W_j$.
[b] $W = 4W_j$ to $5W_j H = 0.5W_j$.
[c] For softest crystallines.

19.5.3.2 Joint Layout

Generally, unless a bond is needed on an internal wall, ultrasonic joints are placed around the perimeter of the part. When a hermetic seal is required, the joint must follow the contour that must be sealed. However, sharp corners, such as those illustrated in Fig. 19-11a, can lead to blistering, accumulations of material, and the formation of stresses with their resultant cracks. Figure 19-11b depicts the corners rounded with a centerline radius (R_2) equal to its distance from the inner wall (W) plus the inside radius (R_1). When there is no need for a hermetic seal, the corner can be eliminated altogether, as in Fig. 19-11c. In such cases, an intermittent or interrupted joint, such as that in Fig. 19-12a, can be used. This will reduce the required energy or

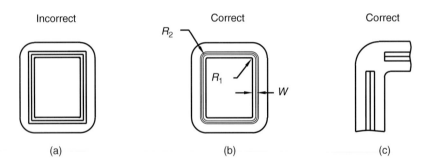

Incorrect   Correct   Correct

(a)   (b)   (c)

Figure 19-11   Joint layout: (a) sharp corners (not desirable), (b) corners with radii, and (c) interrupted corners

power level. When there is a need to resist peeling forces or reduce flash visible from the outside, the energy directors can be turned perpendicular to the wall, as in Fig. 19-12b.

This concept can be combined with an energy director that follows the contour of the wall to produce the crisscross design illustrated in Fig. 19-12c. That provides a larger volume of material for the joint; the result is increased strength at the joint, yet the initial contact created at the interface is minimal. For these designs, the energy director dimensions should be reduced to 60% of those of a standard energy director. The sawtooth configuration with energy directors incorporating a 60° included angle is recommended for air- or liquid-tight seals. Because of the high level of flash associated with this design, a configuration that controls the flash, such as a tongue-in-groove or step joint, is preferred.

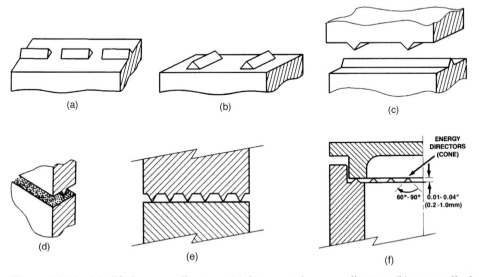

Figure 19-12   Modified energy directors: (a) interrupted energy director, (b) perpendicular energy director, (c) crisscross energy director, (d) textured surface, and (e), (f) cone energy directors (Diagrams courtesy of Branson Ultrasonics Corp., Danbury, CT)

Table 19-4   Textured Depth and Energy Director Height

Texture	Pattern Depth		Energy Director Height	
	in.	mm	in.	mm
300	0.003	0.08	0.005–0.011	0.13–0.28
450	0.0045	0.11	0.012–0.017	0.30–0.43
600	0.006	0.15	≥0.018	≥0.46

19.5.3.3 Textured Surface

Branson Ultrasonics Corporation has patented the concept of the textured surface, as illustrated in Fig. 19-12d, for which license to use is granted with the purchase of the company's equipment. The theory of the textured surface is based on the concept that the point of the energy director comes in contact with a mating surface of numerous small projections, which themselves act as miniature energy directors. They prevent the energy director from side to side "skidding" and that improves the efficiency of the heating process. In addition, these projections form a barrier to the molten material and prevent it from flowing out of the joint area. Finally, they help to retain heat, reduce flash and particulate matter, and generate a greater surface area for bonding, which in turn increases bond strength.

The increased efficiency of the textured surface is claimed to result in a reduction in the total energy required and an increase in strength of up to three times. Branson has worked with a company that creates mold textures (Mold-Tech) to create three finish depths with accompanying energy director heights as shown in Table 19-4. Further information may be obtained from Branson, whose address is listed in the "Sources" section at the end of this chapter.

For light-duty joints requiring lower structural strength, the cone-shaped energy directors illustrated in Fig. 19-12e can be used. This design reduces the overall area to be welded, and thus welds are completed more quickly, with less energy usage. A cone-shaped energy director also generates less heat, resulting in less shrinkage and material degradation.

19.5.3.4 Step Joint

The butt joint is a strong effective joint, however it does nothing to hide the flash or melt-out that often accompanies an ultrasonic weld. An outer wall can do that and provide location and an improvement in shear resistance as well. This type of joint, known as a step joint and illustrated in Fig. 19-13a, provides additional distance for the flash to travel and keeps it from becoming visible to the outside. Furthermore, the weld extends up the wall, providing additional strength. Although a slip fit is specified [0.05–0.10 mm (0.002–0.004 in.) is good], the step wall also provides lateral location when used in conjunction with a similar wall on the opposite side of the part. When external melt-out is not a significant concern, the step can provide location as an internal locating wall or

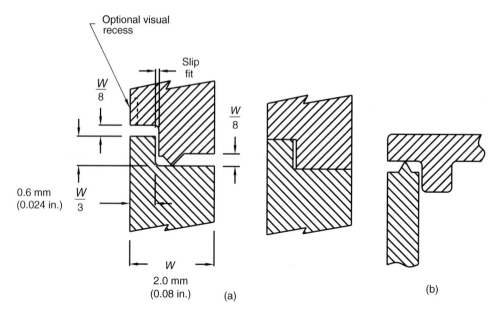

Figure 19-13   Step joints able to: (a) locate and hide flash and (b) locate internally

rib, as in Fig. 19-13b. The joint details described for the butt joint are applicable to the step joint as well.

The principal of visual separation of mating surfaces to diminish the effect of imperfections, discussed in Chapter 2, "Designing for Efficient Assembly," finds useful employment in step joints. In addition to the recesses described in Chapter 2, a separation can be designed into these joints. The amount of desired "shadow line," usually 0.25 to 1.5 mm (0.010–0.060 in.), is added to the $W/8$ gap between the step and the mating recess shown on the left of Fig. 19-13a. The shadow line should not be used with the version in Fig. 19-13b. The step joint can be applied in a variety of configurations as long as the basic principals are maintained.

Although the step joint uses the same energy director used for the butt joint, the space available for a weld is restricted by the vertical step because the flat is only about two-thirds that of a butt weld. That means that the joint will not be quite as strong as a butt weld of the same wall thickness for amorphous thermoplastics (crystallines will be about the same).

19.5.3.5 Tongue-and-Groove Joint

The tongue-and-groove design contains the material on both sides of the joint, thereby eliminating flash and providing location from either or both sides. Since only half the wall thickness is available for the weld, joint strength would seem to be lower than that of a full butt joint. However, some of the weld will run up the sides of the groove, thus creating a weld in both shear and tension. The tongue-and-groove joint is an excellent design for hermetic seals.

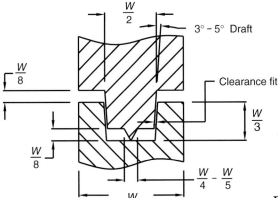

Figure 19-14   Tongue-and-groove joint

The joint details shown in Fig. 19-14 will apply to the tongue-and-groove joint. A slip fit of 0.05 to 0.10 mm (0.002–0.004 in.) per side is good. This type of fitment has an additional cost for the molds and moldments.

19.5.3.6 Thin-Walled Joint

Figure 19-15 illustrates an energy director joint used in welding thin-walled parts. When at least one of the parts can be injection-molded to provide the energy director detail, the other could be thermoformed as in this illustration. Spot welding, stud welding, and ultrasonic stitching, as described later in this chapter, are other means of using ultrasonic welding to assemble thin-walled parts, sheet, and fabric.

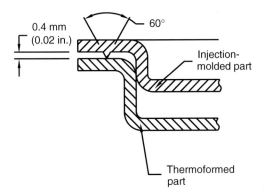

Figure 19-15   Thin-walled joint

19.5.4 Shear Joints

The energy director is the joint of choice for amorphous thermoplastics because it requires less energy than the shear joint. It also requires one-fourth to one-third the weld time of the shear joint. However, it is not necessarily the best material for crystalline resins because the energy director can either degrade or recrystallize before it flows across the joint to form a weld. When this happens, weld strength is reduced, making it difficult to obtain a strong weld or a hermetic seal. The open energy director joint permits the exposure of the joint to more air during welding. This accelerates the crystallization and, with nylon, may cause oxidative degradation. Shear joint designs are usually preferred for crystalline thermoplastics and joints requiring very high strength.

The basic shear joint is illustrated in Fig. 19-16. Essentially, it is a miniature butt joint. (Note that this is a change from earlier angled designs, which led to wedging in many cases.) A lead-in is often provided to guide the two parts together. The shear joint is designed to eliminate exposure of the weld area to the air, thus preventing premature solidification and degradation due to oxidation. It is well suited to the crystalline materials, which have a relatively narrow temperature range in which they are molten.

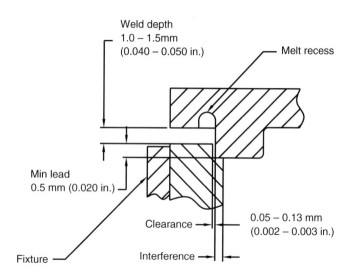

Figure 19-16
Basic shear joint

The strength of the joint is determined by the depth of weld more than by the interference. Optimum weld depths fall in the range of 1.5 to 2.0 mm (0.060–0.079 in.). A weld depth 1.25 to 1.5 times the wall thickness of the part is sufficient to provide a weld nearly as strong as the surrounding wall. Note the melt recess, also known as a flash trap, provided in the part above the weld. This optional design detail encapsulates the melt for appearance applications, where any melt-out would be objectionable. Its size is determined by estimating the amount of material to be melted and adding a safety margin of 10%. Care must be taken to avoid overwelding, which can defeat this device and lead to leveling and alignment problems as well.

Table 19-5   Shear Joint Interference Guidelines

Maximum Part Dimension	Range of Interference per Side	Part Dimension Tolerance
<18 mm (0.75 in.)	0.2–0.3 mm (0.008–0.012 in.)	±0.025 mm (±0.001 in.)
18–35 mm (0.75–1.50 in.)	0.3–0.4 mm (0.012–0.016 in.)	±0.050 mm (±0.002 in.)
>35 mm (1.50 in.)	0.4–0.5 mm (0.016–0.020 in.)	±0.075 mm (±0.003 in.)

Source: Branson Ultrasonics Corp., Danbury, CT.

The parts are guided together by a lead-in, and the melt is controlled by the amount of interference between the two parts. The tolerances for those interferences are provided in Table 19-5. Excessive interference can result in excessive melt-out or flash. The tolerances for shear joints are very tight, particularly for crystalline materials. This limits the utility of such joints with these materials to parts that do not exceed 90 mm (3.54 in.), however parts made of amorphous materials can be larger. Note that the fixture in Fig. 19-16 reaches up the wall to the top of the stationary part. This is to provide rigidity during welding, since this type of joint can result in an outward force that can cause the wall to deflect without support from the fixture.

Several additional variations of the shear joint are illustrated in Fig. 19-17. The designs in Figs. 19-17a, 19-17d, and 19-17e are designed to melt a small area on initial contact. Figure 19-17a shows an outside wall that provides alignment and hides melt-out, whereas the designs in Figs. 19-17b and 19-17c control the vibrations with a recess in the surface meeting the horn and a stress factor reducing radius at the base of the energy director wall. The flat joint in Fig. 19-17b provides a strong weld but uses a high level of energy, while the joint in Fig. 19-17c uses less energy but is not as strong.

The joint in Fig. 19-17d is sometimes referred to as an "energy director shear joint" because it combines characteristics of both. In this design, the energy director's principal force components are downward and inward. The outward component is much reduced. The version in Fig. 19-17e, also referred to as a "mash joint," produces a stronger joint by increasing the length of the weld. A hermetic seal is created because no melt is squeezed out of the bonding area.

The inside/outside fitment has been used successfully to overcome the tight tolerances associated with shear joints. The joint illustrated in Fig. 19-18a is designed such that interference lost on one wall is picked up on the other. When the draft is 2° or more, a wedging effect is created as the weld takes place. The concept works because it is much easier to control the tolerances of the slot and the energy director wall than it is to control diameters and keep them round. Diametrical tolerances two to three times as great as those indicated in Table 19-5 have been successfully employed in this fashion. Since, however, this design can lead to higher energy usage, it is not to be used indiscriminately.

Figure 19-17 (opposite)   Additional shear joints: (a) shear joint with skirt to hide melt-out, (b) shear joint with radius at base of energy director wall – heavy duty, (c) shear joint with radius at base of energy director wall – light duty, (d) energy director shear joint, and (e) mash joint – before welding, and (f) after welding

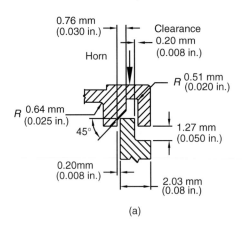

0.76 mm
(0.030 in.)

Clearance
0.20 mm
(0.008 in.)

Horn

R 0.51 mm
(0.020 in.)

R 0.64 mm
(0.025 in.)

45°

1.27 mm
(0.050 in.)

0.20mm
(0.008 in.)

2.03 mm
(0.08 in.)

(a)

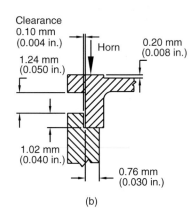

Clearance
0.10 mm
(0.004 in.)

Horn

0.20 mm
(0.008 in.)

1.24 mm
(0.050 in.)

1.02 mm
(0.040 in.)

0.76 mm
(0.030 in.)

(b)

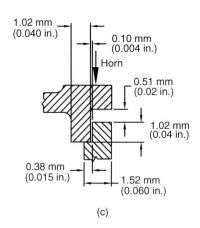

1.02 mm
(0.040 in.)

0.10 mm
(0.004 in.)

Horn

0.51 mm
(0.02 in.)

1.02 mm
(0.04 in.)

0.38 mm
(0.015 in.)

1.52 mm
(0.060 in.)

(c)

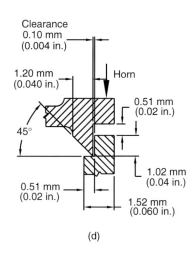

Clearance
0.10 mm
(0.004 in.)

1.20 mm
(0.040 in.)

Horn

0.51 mm
(0.02 in.)

45°

0.51 mm
(0.02 in.)

1.02 mm
(0.04 in.)

1.52 mm
(0.060 in.)

(d)

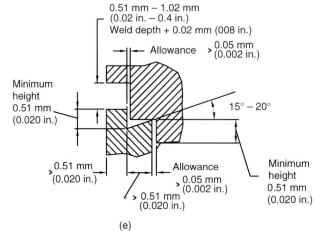

0.51 mm − 1.02 mm
(0.02 in. − 0.4 in.)
Weld depth + 0.02 mm (008 in.)

Allowance > 0.05 mm
(0.002 in.)

Minimum
height
0.51 mm
(0.020 in.)

15° − 20°

0.51 mm
> (0.020 in.)

Allowance
> 0.05 mm
(0.002 in.)

> 0.51 mm
(0.020 in.)

Minimum
height
0.51 mm
(0.020 in.)

(e)

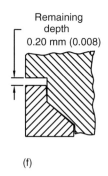

Remaining
depth
0.20 mm (0.008)

(f)

The design in Fig. 19-18a is shown on an outside wall. However it could also be used on an inside wall, such as the one in Fig. 19-18b, which demonstrates a single-weld version similar in appearance to the inside/outside joint in Fig. 19-18a but with the shear joint on one side only.

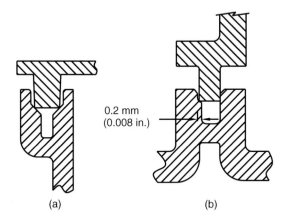

0.2 mm
(0.008 in.)

(a)   (b)

Figure 19-18   Inside/outside shear joints: (a) dual shear and (b) single shear

19.5.5   Hermetic Seals

When a hermetic seal is required, a design must be created that permits a continuous joint to be established in a flat plane. The distance between the horn and the joint interface must be constant and as short as possible. There can be no openings in the part between the horn and the joint, and the walls should be straight and free of curves. If there must be contours, they should be as soft as is feasible.

Material selection is critical in the design of hermetic seals. Hygroscopic resins must be welded immediately after molding or the parts must be placed in bags with a desiccant. Since moisture creates frothing, which destroys hermetic seals, parts may need to be dried before sealing. Rigid, amorphous plastics are best. Energy director joints can be reliably used with them. If a crystalline material must be used, the recommended joint is a shear joint. If there is no way a shear joint can be used, an energy director joint in a tongue-and-groove configuration such as the one in Fig. 19-14 can be a possibility. However, consultation with the equipment manufacturer is recommended before proceeding. If the concept is deemed feasible, the manufacturer will probably recommend an energy director with a 90° included angle. However, when the material is polycarbonate, a 60° included angle energy director is used with a width equal to 25 to 30% of the joint width.

If the part contours preclude the creation of a continuous weld, or if the ability of the design or material to meet the requirement for a hermetic seal is questionable, a gasket or O-ring such as the one illustrated in Fig. 19-19 can be used. The O-ring should be compressed only at the end of the weld.

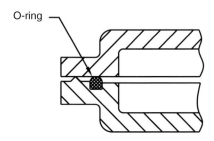

Figure 19-19   Hermetic seal energy director joint

19.5.6   Scan Welding

Scan welding is a high speed welding technique that is used for large parts and small parts with very high production rates. It can be employed when at least one of the joint surfaces is flat and the materials are rigid, although it can be used with some film and fabric applications, provided their nonpolymer fiber content does not exceed 35%. In this technique, the horn is stationary and the parts are moved in relation to it.

Standard scan welding is illustrated in Fig. 19-20a. The parts are carried by a conveyor belt to power-driven rollers that pass them under the welder at a uniform rate of speed. They also facilitate horn entry, minimize drag, and prevent part "kick-up." When rigid parts are to be welded, a self-aligning joint such as a step joint or tongue-and-groove design should be used. While production rates are application dependent, speeds greater than 10.6 m/min (35 ft/min) have been achieved. The shapes of some products are not well suited to the flat-bed table used in standard scan welding. Round parts can be mounted on a rotary table as in Fig. 19-20b, a technique known as rotary scan welding.

Radial scan welding is a near-field welding technique that can be used for parts that require a seal around their periphery. For this technique, a part is slowly rotated in the horizontal plane and receives vibrational energy from a welder mounted perpendicular to its circumference. Parts do not need to be perfectly round for radial scan welding; however they may not have any sharp edges for the horn to have to pass over. In the application shown in Fig. 19-20c, a hermetic seal is achieved. The surface speed for this application was 9.14 m/min (30 ft/min) and the weld time was 1.2 seconds. Size is not a determining factor in the use of scan welding.

19.5.7   Stud Welding, Staking, Swaging, and Spot Welding

Stud welding, staking, swaging, and spot welding are similar in the respect that a series of localized weld sites are used in place of a continuous weld. In addition, they all require that the horn be placed directly over the weldment, although stud welding can tolerate some moderate displacement. The methods differ in the way they are used. Stud welding can be used only with parts made of like or compatible resins, and the studs must either be close enough to be welded by a single horn or far enough apart to permit sufficient

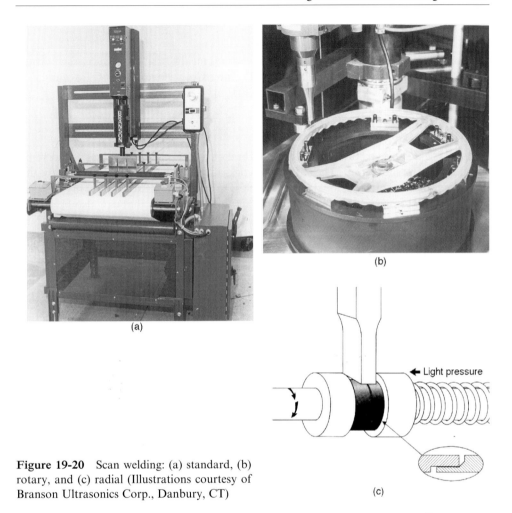

Figure 19-20   Scan welding: (a) standard, (b) rotary, and (c) radial (Illustrations courtesy of Branson Ultrasonics Corp., Danbury, CT)

flexure between a welded stud and one that has not yet been welded and is resting atop its hole. Stud welding also requires tighter tolerances than the other three. Staking and swaging are principally employed when dissimilar materials are used, since all the welding takes place on only one of the mating parts. Staking, swaging, and stud welding can be used to trap another part of a dissimilar material between them. Spot welding is used for like or compatible plastics when the wall thicknesses are very thin or the process will not produce the type of precise detail and tolerances that stud welding requires. Examples of such processes are extrusion, thermoforming, rotational molding, and fabrication of parts made of sheet or film.

The advantage of these methods is that they lend themselves to applications involving parts too large to be joined with continuous welds or contours that do not permit these welds. The disadvantages are that localized stress points are created, and even with the use of an O-ring, the methods are not recommended for hermetic seals.

19.5.7.1 Staking and Swaging

Staking and swaging can be accomplished by other means besides ultrasonics and were treated in Chapter 17, "Staking/Swaging/Peening/Cold Heading/Cold Forming," to which the reader is referred for further discussion.

19.5.7.2 Stud Welding

Figure 19-21a shows a stud-welded joint before, during, and after welding. Observe the similarity to the shear joint. The stud weld works in much the same manner, except that it is much smaller and it is repeated many times. In addition, the stud is usually designed with a lead-in to provide its own means of alignment; however that can also be accomplished with a chamfer on the top of the hole. The 0.2 to 0.3 mm (0.008–0.012 in.) interference specified for a stud 12.5 mm (0.50 in.) or less in diameter provides the small initial contact area, which concentrates the vibrations in that area and helps keep the cycle time short. Larger diameters can use an interference up to 0.4 mm

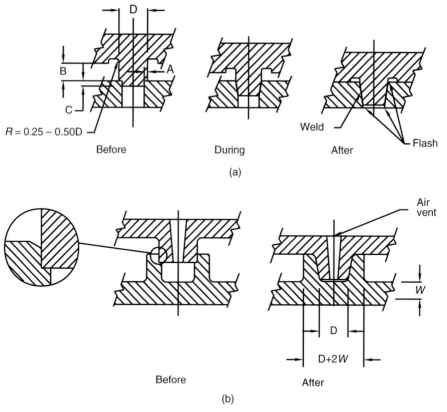

Figure 19-21   Ultrasonic stud welding: (a) ultrasonic stud welding joint and (b) stud in boss or blind hole

(0.016 in.). The strength of the weld is determined by the total area welded, which is a function of the diameter of the weld and its depth. Note that a melt recess to permit a flush fit has been provided at the base of the stud, which has been radiused to reduce stress concentrations.

The version depicted in Fig. 19-21b can be used when the walls must be separated or flush walls are needed on both parts. An air vent must be provided in one of the two parts to prevent an air trap, which will create an air pressure buildup. That is preferred to a break in the wall of the boss, which would substantially weaken the joint. The outside diameter of the boss must be at least the diameter of the stud plus twice the nominal wall thickness. To prevent deflection or breakout, the edge of the hole in the part or the boss must be at least 3 mm (0.125 in.) from the edge of the part.

a. Entrapment of Additional Parts   Stud welding can be used to trap additional parts of dissimilar materials between the external parts. Examples of this type of application are shown in Fig. 19-22. In Fig. 19-22a a third part was placed over a stud and the mating part was welded over it. The recess at the base of the stud is there to provide space for a radius. To accommodate the flash and permit the parts to lie flat, a flash trap has been provided at the base of the hole in the mating part. The same goals can be accomplished with molded plastic rivets, as in Fig. 19-22b. This concept has an advantage for applications featuring a wide temperature range, and the differences in the coefficient of linear thermal expansion between the plastic parts and metal rivets or screws would lead to excessive stresses or loosening of the fasteners. Since the molded rivet is of the same material as the mating plastic part, there is no relative differential. Location, however, must be provided by some other means.

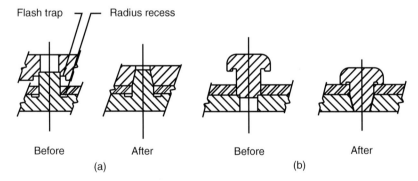

Figure 19-22   Entrapment of additional parts: (a) integral stud and (b) molded rivets

b. Reduced Energy Studs   Because they are essentially shear joints, stud welds can use significant amounts of energy. When this becomes an issue or when there is a need to reduce the welding cycle, energy director studs that minimize the power requirements and reduce the cycle time can be employed. An example of such a joint using energy directors in the hole is shown in Fig. 19-23a. A version using a conical tip appears in Fig. 19-23b. Both versions would require tight tolerance control to maintain a consistent weld.

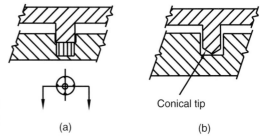

Figure 19-23 Energy director studs: (a) stud weld with vertical energy director and (b) stud with energy director

(a)

Conical tip

(b)

c. Thin-Walled Parts   Parts with a nominal wall thickness of 1.5 mm (0.060 in.) or less are considered to be thin-walled parts. As shown in Fig. 19-24a, they may require a two-stepped stud in place of a normal stud. This provides the same strength of weld while reducing the travel by 50%. This technique can also be used where space restrictions imposed for other reasons limit the amount of travel available for the weld.

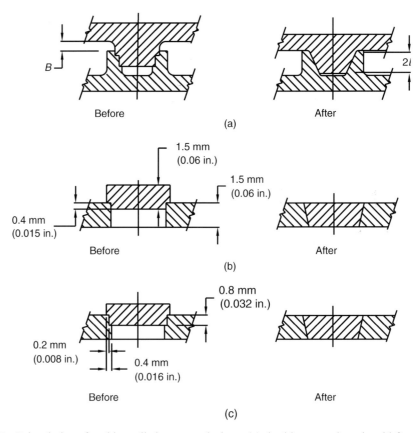

Figure 19-24 Joint designs for thin-walled parts and plugs: (a) double-stepped stud weld for thin walls, (b) single-step plug, and (c) double-step plug

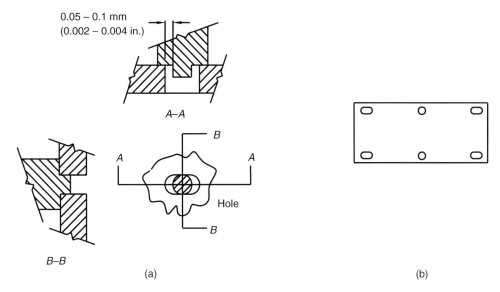

Figure 19-25   Stud placement in elongated holes: (a) design details and (b) slot layout (center holes used for location)

d. Plugs   Holes in plastic parts can be closed by means of a plug emplaced with the stud welding technique, as illustrated in Fig. 19-24b. When the walls are thin, the stepped technique in Fig. 19-24c may be employed for this application as well.

e. Stud Placement   Studs that are placed within 76 mm (3 in.) of each other can be welded by a single horn. Those located beyond that distance will need to be welded individually. As the distance between welds grows larger, it becomes increasingly difficult to hold the necessary tolerances. This can be accommodated with studs welding into elongated holes, as shown in Fig. 19-25a. A layout using center holes for location and slots for the outer welds to permit greater tolerances is shown in Fig. 19-25b.

19.5.7.3 Spot Welding

Spot welding provides quick (<1 second), high strength welded joints with one side unmarked for appearance purposes. Standard tips are available for material thicknesses ranging from 0.80 to 7.14 mm (0.031–0.281 in.) with tip diameters of 12.7, 19.05, and 25.4 mm (0.50, 0.75, and 1.00 in.). That, plus the simple requirement of flat, uniform surfaces, makes spot welding applicable to parts made from processes that cannot produce the details required of other types of ultrasonic joint, such as extrusion, thermoforming, rotational molding, and parts fabricated of sheet or film. It can be used for parts made of corrugated plastic board and those that have been thin-gauge thermoformed, for which there are few other cost-effective options. Spot welding lends itself well to most thermoplastics (see Table 19-3) and can be accomplished with hand guns, which makes it applicable to large parts with complicated contours.

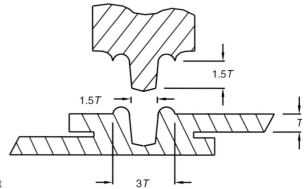

Figure 19-26   Spot welding joint

Referring to Fig. 19-26, we see that the pilot of the spot welding tip melts through the top part and partially through the wall of the mating part. The displaced material forms a ring in the surface of first contact defined by the contour of the welding head. At the joint interface, frictional heat is produced, and the molten polymer that was displaced by the welding tip is forced into the interface to create a permanent molecular bond. Tips are designed so that the head diameter is three times the thickness (T) of the top layer and the length is one and a half times the thickness of the top layer.

For applications in which the horn must access the appearance side and a weld mark is undesirable, the concept can be reversed, with a flat horn used over an anvil shaped to the contour of the spot welding tip. In that way, the spot welding mark is confined to the back side and the outer surface is relatively unblemished; this is referred to as "inverse" or "blind" spot welding. (A mark may still be caused by a flat-faced horn if the weld time is long.)

19.6   Fabric and Film Sealing

Ultrasonics competes with thermal bonding techniques for fabric and film sealing materials because the energy usage is much lower and rates of speed are quite high. Branson produces a model specifically suited to this application that simultaneously cuts and sews at a rate of 22.9 m/min (75 ft/min), and higher speeds can be achieved with modifications to the equipment.

The principal of ultrasonic bonding is illustrated in Fig. 19-27. The fabrics to be bonded pass between a rotary drum or anvil and a vibrating horn. The rotary drum will have a raised pattern on its surface and is normally made of steel. The mechanical vibrations of the horn and the compressive force between it and the drum surface create frictional heat at their point of contact. The pattern that will be created is the pattern that has been machined into the drum surface. A high level of softness, breathability, and absorption in the fabrics is maintained because bonding occurs only at the contact points.

Applications for ultrasonic welding of films and fabrics include automotive components, bedding, carpets, clothing, curtains, footwear, luggage, rainwear, and tents, plus

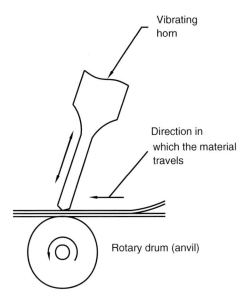

Vibrating horn

Direction in which the material travels

Rotary drum (anvil)

Figure 19-27   Fabric and film sealing

disposable garments and diapers of many kinds. There are a variety of standard pattern wheels available and custom ones are readily created. Weldability varies with the type of material and the form to be welded, according to Table 19-6. These ratings differ from those of similar materials to be rigid plastic welded because films and fabrics are welded with the ultrasonic horn in immediate proximity to the material. Additional brief descriptions of ultrasonically weldable fabrics and films are as follows.

- *Woven.* These are textiles formed by the regular interweaving of filaments and yarns. Thread density, thermoplastic content, tightness of weave, and uniformity of thickness will influence weldability. Weld strengths can vary with yarn or filament orientation.
- *Nonwoven.* These are textiles formed by bonding and/or interlocking fibers, yarns, or filaments through mechanical, thermal, or chemical means. Uniformity of thickness and thermoplastic content will affect weldability. Excellent strength is created through random orientation of fibers.
- *Knitted.* Knitteds are textiles formed by interconnecting continuous loops of filaments or yarns. Style of knit, thermoplastic content, and elasticity of construction affect weldability.
- *Coated materials.* When textiles and films are covered with a thermoplastic layer like urethane or polypropylene, they are referred to as "coated materials." The base material may be a nonthermoplastic, such as paper or coated cardboard. Coating thickness, material, and substrate characteristics affect a coated material's weldability.
- *Laminates.* Any combination of two or more layers of textiles and films constitutes a laminate. For weldability, the mating surface must have a lower melt temperature than the other layers.
- *Films.* Whether blown, cast, or extruded, thermoplastic sheet goods having a wall thickness of 0.25 mm (0.010 in.) or less are regarded as "film." To be weldable,

Table 19-6 Fabric and Film Sealability of Materials: E, Excellent; G, Good; F, Fair; P, Poor

Material	Woven	Nonwoven	Knitted	Coated materials	Laminates	Films
Polyester	E-G	E	G	E	E	E
Nylon	E-G	E-G	G	G	G	G
Polypropylene	E-G	E	G	G	G	E-G
Polyethylene		E		E-G	G	F-P[a]
PVC	G-P			G-P	G-P	G-P[b]
Acrylic	F-P					—[c]
Urethane				E-G		E-G[d]
Saran				E-G		
EVA				E-G		
Surlyn				E-G		

[a] Thin polyethylene film [<0.13 mm (0.005 in.)] is generally considered poor for ultrasonic welding because it has a low coefficient of friction and tends to break down or degrade during welding.
[b] PVC sheet or fiber is difficult to predict due to the broad range of additives used in its manufacture. Plasticizers are often added to rigid PVC to increase flexibility. As the content of plasticizer increases, the ability to ultrasonically bond PVC can be inhibited.
[c] Acrylics can be ultrasonically tacked or cut. Continuous bonding is generally unsatisfactory due to embrittlement and low strength.
[d] Thermoplastic urethane (ester base) coated materials exhibit excellent strength when bonded ultrasonically. Thermosetting urethanes (ether base) will degrade when subjected to ultrasonic energy.
Source: Branson Ultrasonics Corp., Danbury, CT.

however, the materials must be at least 0.013 mm (0.0005 in.) thick. Other factors affecting weldability are the film's density and the type of polymer it is made from.
- *Blends*. Blends of thermoplastics with cotton, linen, wool, silk, rayon, acetate, and triacetate can be ultrasonically welded provided there is a minimum 65% thermoplastic content. The greater the thermoplastic content, the greater the bond strength attainable.
- *Modacrylics*. Copolymers of acrylonitrile and other resins, like PVC, can achieve good bonds depending on fabric construction and seam selection.

Generally, polyester is considered to be the best material for ultrasonic applications. However, both nylon 6 and nylon 6/6 can produce strong, neat stitches with ultrasonics. Of the polyolefins, polypropylene is preferred over polyethylene for general textile applications; however most polyolefins have good ultrasonic welding characteristics. In addition to being the lightest weight materials, polyolefin fibers possess a unique wicking characteristic that permits them to draw body moisture from the skin up through the fabric to the outer surface. Ultrasonic welding is the method of choice for joining Mylar films. Polycarbonate film can also be ultrasonically welded, as well as some other thermoplastics that are unlisted because they are rarely encountered. Some of these films require a buffer sheet for optimum results.

19.7   The Ultrasonic Equipment

19.7.1   The Basic Principles

There are seven essential components required to accomplish ultrasonic welding. They are the stand, the power supply, the converter, the booster, the horn, the fixture, and the controls. The base, the frame, and the column constitute the stand, which supports the actuator, a unit comprising in itself the converter, booster, and horn. An air cylinder mounted above the actuator applies pneumatic pressure through the actuator to the parts. The actuator brings the horn down into contact with the vibrated part to be welded, applies the required force, and retracts the horn when the weld is completed. The power supply, or ultrasonic generator, supplies high frequency electrical energy to the converter. The converter changes electrical energy into mechanical vibrations. The booster, also known as the booster horn, is an amplitude-modifying device that can either increase or decrease the amplitude of the vibration supplied to the horn. The horn is the tool that transmits the vibrations to the part. The fixture rigidly holds the stationary part to be welded. Figure 19-28a shows an integrated ultrasonic welder.

19.7.2   The Power Supply or Generator

The power supply, or generator, is shown in Fig. 19-28b as a separate component, but the actuator is mounted in a stand. This is known as a "modular" design, since the actuator can also be purchased without a stand, as it might be if it were to be incorporated into a multifunction assembly line. Some models, however, have the actuator, power supply, and stand incorporated into one unit. These are known as "integrated" units. Basically, the generator transforms input voltage to high frequency electrical energy, which it transmits to the converter. The standard 120 or 240 V alternating current at 50/60 Hz is fed to transistors in the power amplifier, which are switched on and off at the welding frequency (e.g., 20,000 cycles per second for a 20 kHz system) (see Section 19.7.8). These high-powered signals are transmitted to the converter or transducer, which begins to vibrate. Variations in supply voltage have traditionally been a problem for ultrasonic generators, however a generator with line load regulation can resolve this matter.

19.7.3   The Converter or Transducer

The converter or transducer, shown at the top of the stack in Fig. 19-28c, converts the electrical energy received from the generator into high frequency mechanical vibrations. When an alternating voltage is applied to the opposite sides of piezoelectric ceramic elements, they expand and contract as the electric polarity changes from positive to negative. Therefore, they physically vibrate at the same frequency as the current. These elements are sandwiched between two metal pieces. The ceramic elements are brittle and can shatter if dropped or subjected to excessive running stress.

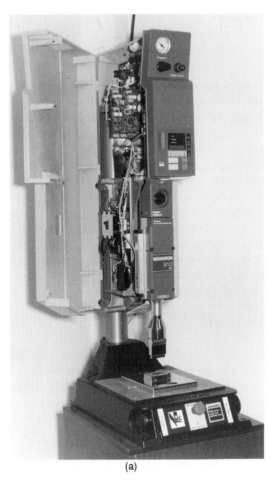

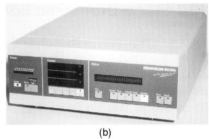

(b)

(a)   (c)

Figure 19-28 Integrated ultrasonic welder (a), an independent power supply (b), and a "stack" consisting of a converter, booster, and horn (c: top, middle, and bottom, respectively) (Illustrations courtesy of Branson Ultrasonics Corp., Danbury, CT)

19.7.4 The Booster

Also known as the "booster horn," the "impedance transformer," or the "amplitude transformer," the booster is a half-wavelength resonant element whose function is to increase or decrease the amplitude of the vibrations and, thereby, create the desired amount of vibrations at the joint interface. As illustrated in Fig. 19-28c, the booster is located between the converter and the horn. To be able to modify the amplitude, the booster must have different diameters at either side of its center or nodal point (the point at which there is little or no motion). A booster with a smaller diameter at the end connected to the horn will increase the amplitude, while a decrease will occur with a booster that has a larger diameter connected to the horn. A "coupling bar" is a booster that does not alter the amplitude.

Standard boosters will multiply the vibrations from 0.5 to 2.5 times the converter output amplitude (referred to as "gain"). Custom boosters with higher gain can be made, but that must be done with care lest the horn's ability to withstand the greater amplitude without fracture be exceeded. Boosters are readily interchangeable, and an underwelding condition can be countered with a booster of higher gain; the reverse is done for overwelding. Boosters typically are made of either aluminum or titanium. For optimum performance, the horn–booster combination must be carefully tailored to the application.

19.7.5 The Horn

The last component in the chain is the horn, which is a tuned half-wavelength resonant element that transmits the vibratory energy and the application force from the booster to the parts to be welded through direct contact. It is usually a metal section 0.5 wavelength long and made of titanium, alloy steel, or aluminum. Titanium is the material of choice for high amplitude applications because it has the best acoustical properties of the high strength alloys and is the horn material most resistant to stress. It is also the most expensive to buy and machine. Titanium horns can be faced with carbide, nickel, or Teflon to improve their wear resistance. Low amplitude applications like inserting and those requiring materials with high concentrations of abrasive fillers are appropriate for alloy steel horns, although they lose more ultrasonic energy. Simple, inexpensive horns are often made with aluminum, which has the best acoustical properties of the lower strength alloys and is least costly to purchase and machine. Such horns can be chrome-plated, carbide-coated, or anodized to extend their service life. Chrome plating and a layer of polyethylene film between the horn and the part are also used to prevent marking of the part due to the transfer of oxide from the horn to the part.

The horn is expanding and contracting at the frequency of the welder (e.g., 20,000 times per second for a 20 kHz unit). The term "horn displacement amplitude" refers to the full range of the travel of the tip of the horn from full extension to full compression, or "peak to peak." Therefore, a horn moving over a peak to peak distance of 0.13 mm (0.005 in.) would have a horn displacement amplitude of 0.13 mm (0.005 in.). The typical range is 0.013 to 0.064 mm (0.0005–0.0025 in.) for amorphous thermoplastics; crystalline may go as high as 0.13 mm (0.005 in.). At rest, the horn will be at the midpoint of the horn displacement or peak-to-peak amplitude. The rate of motion of the horn face is referred to as the "horn velocity."

The fitment of the horn to the part is very important to the efficient transmission of the vibrations, and an improper fit can lead to uneven welding. A poor fit can be caused by wear to the horn, equipment misalignment, part distortion, cavity-to-cavity variations, or raised lettering. It may be necessary to relieve the surface of the horn to place the vibrations in the most effective location.

The shape of the horn in cross section will create gain. A straight-sided horn like that in Fig. 19-29a provides no gain at all. Straight horns are available in square and

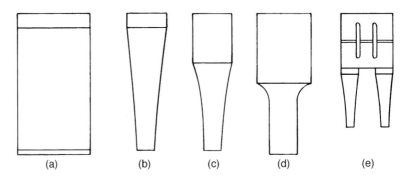

Figure 19-29 Ultrasonic horn types: (a) straight-sided horn, (b) exponential horn, (c) catenoidal horn, (d) step horn, and (e) full-wave composite horn

rectangular configurations. These horns generally range in size from 3 to 254 mm (0.120–10 in.), the longest slotted bar horn made being about 600 mm (24 in.) long; horns over 254 mm (10 in.) produce uneven welding, however. Horns over 90 mm (3.50 in.) use slots to reduce lateral stress. This is accomplished by breaking up critical dimensions that produce unwanted lateral motion and other modes of vibration into individual members, each of which is vibrating in a longitudinal mode, with side action minimized.

Straight horns are also made in circular configurations, which are available solid or hollowed out, as appropriate. They also require slotting to reduce cross-coupled or radial stresses when their diameter exceeds 90 mm (3.50 in.). Circular horns have been made up to 280 mm (11 in.).

Low gain and nodal stress will be created by the exponential horn in Fig. 19-29b. Its gradual taper follows an exponential curve, which distributes internal stress over a large area.

The catenoidal horn in Fig. 19-29c, whose length follows a catenoidal curve, provides medium gain and is a compromise solution between the exponential horn and the high gain step horn. It can achieve moderately high amplitudes with a limited amount of stress. This horn is well suited to staking or welding small parts.

The highest gain for a given input-to-output ratio, with its resulting high levels of nodal stress, is provided by the step horn (Fig. 19-29d). A step horn has two sections, each of which has a different but uniform cross section. The maximum stress occurs at the radius between the two sections, the point at which the horn is most likely to fail when it is driven at an excessive amplitude by an improper booster. Gain factors up to 9:1 can be attained with step horns.

An application of unusual shape or size may prevent a half-wavelength horn from contacting or developing enough amplitude to weld the parts. In such cases, a composite horn (Fig. 19-29e), also known as a compound horn, can sometimes provide the solution. Such a horn can provide a higher amplitude without creating excessive stress. It is composed of individual horns, known as "horn tips," which are attached to a large coupling horn. Together, they form a single, tuned, full-wavelength unit whose amplitude at the face of the horn is considerably greater than what a single horn could deliver. The coupling horn may be made of aluminum or titanium, while the individual horns are usually titanium or steel. Composite horns have been used as a solution for amplitude or wear problems with large, multiple insertion, staking, and stud welding applications. They can sometimes eliminate the need for multiple welders by providing greater part coverage.

19.7.6   The Fixture

The fixture performs two functions. It aligns the mating part to the vibrated part, and it provides support to the joint area to withstand the applied pressure and permit the ultrasonic energy to be transmitted efficiently. If the lower part can also vibrate while the upper part transmits vibrations to the joint from the horn, part marking may occur and the required heat buildup from friction will be reduced.

Figure 19-30a illustrates a fixture for an energy director joint. Note that it is designed to reach a point just below the top of the lower part. A fixture for a shear joint (Fig. 19-30b) is extended to the highest point on the part to the provide support that will prevent deflection to the side. However, other factors, such as the material to be welded, the part geometry and its symmetry, and wall thickness, will also affect the design of the fixture. Production requirements such as ease of loading and quick release mechanisms also dictate the fixture design.

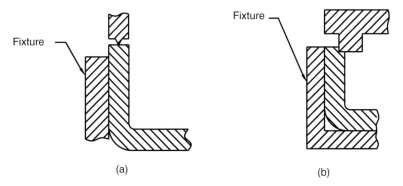

(a)   (b)

Figure 19-30   Ultrasonic fixtures for: (a) energy director joint and (b) shear joint

A fixture used for rigid, amorphous thermoplastics would likely be made, at least in part, of a resilient material such as poured or cast urethane, neoprene, cork, or Teflon. These materials absorb more energy but minimize part marking and inconsistent welds; also, construction is less expensive than for a rigid fixture. A rigid material, such as aluminum or stainless steel, is used when the fixture must hold a part made of a flexible material like a crystalline thermoplastic. Insufficient support in critical areas can lead to uneven welds through excessive vibration or nonuniform horn contact. Rigid fixtures are also used for insertion, staking, swaging, and spot welding. They are sometimes chrome-plated to increase resistance to wear and to prevent part marking. Prototype and low volume fixtures are sometimes made of wood, epoxy, or plaster.

19.7.7 The Controls

The process elements to be controlled are the weld distance, weld time, weld energy, weld pressure, hold time, stack gain (converter, booster, and horn) and the power draw. Basic, inexpensive ultrasonic welders have a control system that is able to control the time elements only. The welder is actuated and descends to the part, where vibration takes place for a preset period of time. These elements are important because improper weld speed, time, or pressure can result in damage to internal parts and uneven welding, or over- or underwelding. After a hold period, the actuator retracts. The timer can control the time before, during, and after the weld. Other elements, such as travel and pressure, are controlled manually. This "open-loop" type of system represents the type of control expected on a basic ultrasonic welder. It is perfectly adequate for many noncritical applications.

Microprocessor-controlled equipment provides much greater regulation of the welding parameters. These systems are referred to as "closed-loop" systems because the feedback permits the resetting of the elements according to the predetermined parameters (actually, they are only partially closed-loop setups). The capabilities and mode of operation vary according to the specific equipment manufacturer. However, there are five control modes that can be set to predetermined limits available with this type of equipment:

- *Continuous ultrasonics*. Ultrasonics are set to be either on or off, with the use of constant sonics permitted by controlled triggering.
- *Distance*. The weld can be controlled to melt either to a preset height or to meltdown.
- *Energy*. Constant energy levels can be preset to accomplish acceptable assemblies.
- *Time*. Time is still the primary controlling factor, and the same timing elements can be regulated with microprocessor-controlled equipment.
- *Time/energy compensation*. When the controls are set for energy priority regulation, the time will be automatically extended until the minimum energy level is achieved if that does not occur within the preset time.

The equipment also provides feedback to the operator, which can be used for precision adjustment of the assembly conditions. Once a set of acceptable parameters has been established, it can be stored in the memory to save time the next time the job is set up. In addition, a "process window" can be created. An alarm is thereby triggered whenever a given weldment fails to fall within this range. In addition to the controlled variable, the display will also monitor several others. When this equipment is connected to a printer, a complete printout of the parameters of each weldment can be obtained, for use in statistical analysis. Such data, in turn, can be used to monitor the process and determine when it has gone statistically out of control. The continuing development of microprocessor-controlled ultrasonic welders constitutes a quantum leap in the controllability of the process and has permitted the successful welding of difficult-to-weld engineering thermoplastics.

Sophisticated equipment provides the capability to do "profiling," that is, to vary the vibrational amplitude and weld force during the welding. Recent work with polycarbonate and polyamide suggests that significant improvements in performance can be achieved with this refinement. Amplitude profiling promotes stronger welds and does not reduce polymer degradation. Force profiling can reduce cycle times by 28 to 43% and can increase weld strength with relatively low modulus materials such as nylon. The combination of amplitude and force profiling can increase weld strength and decrease cycle times.

19.7.8 Equipment Frequency

Originally, virtually all ultrasonic welders operated at 20 kHz. Equipment that operated at the higher frequencies of 35 to 40 kHz was developed when the need to address some of the weaknesses of 20 kHz equipment was recognized. These higher frequency units are smaller and quieter; they also have a gentler action, along with greater part protection, material reaction, and speed. The converter, booster, and horn are smaller because the wavelength at 35 or 40 kHz is approximately one-half that at 20 kHz. Thus the components can be positioned closely, so that the units fit into the tight spaces available in automated equipment, and there is less mass to be moved for units that are part of moving mechanisms. A cost reduction for the equipment of about 10% results, as well.

The high-pitched squeal of ultrasonics is the noise of the part itself being vibrated. Since the amplitude of the higher frequencies also is approximately half that at 20 kHz, there is a corresponding reduction in the amount of energy transmitted to the part and its subsequent noise emission. This also translates to a gentler action on the parts as they are welded, which reduces cyclic stressing, heating beyond the joint interface, and the damage to other parts such as delicate electronic components. With this reduced energy, however, the ability to transmit ultrasonic energy through the part is also limited. Thus, the distance from the horn to the joint interface for 35 or 40 kHz equipment is limited to about 6.35 mm (0.25 in.).

The use of lower force levels and the enhanced ability to control the mechanical vibrations transmitted to the part result in improved process control and weld quality. The drop in energy dissipated within the part minimizes stress and degradation of the material, and a structurally sounder and better-looking finished assembly is produced. The improvement in energy control results in faster welding cycles in some cases and better melt characteristics of some materials, such as acrylic, vinyl, and glass-filled thermoplastics. The use of higher frequency equipment can be more effective for spot welding and staking.

There are significant size limits to the use of 40 kHz equipment. The horn sizes of 40 kHz horns are about half those of 20 kHz units; therefore, the maximum size of the parts that can be welded is similarly limited. The smaller size of the tooling does permit the ganging of units to weld larger assemblies. However, the size of the parts that can be welded is also limited by the 1000 W maximum capacity of the high frequency equipment (due to the smaller size of the crystals inside the converter). By comparison, 20 kHz welders have a maximum power supply capacity of 5000 W (the majority of welders are rated at only 1000 W). The greater power requirement for the 20 kHz welder results in a higher operating cost for energy.

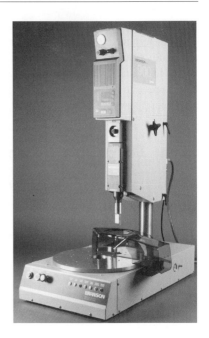

Figure 19-31 Rotary ultrasonic welding system (Photograph courtesy of Branson Ultrasonics Corp., Danbury, CT)

In addition to the principal frequencies of 20 and 35 or 40 kHz discussed in this section, welders that operate at 15, 30, and 70 kHz are available and are intended to provide optimum results for specialized applications. The lower frequency equipment permits even greater amplitudes and the welding of larger parts and/or from longer distances. The higher frequency equipment allows greater precision in welding fragile sections without damaging nearby areas or components, and the intermediate equipment permits the optimization of characteristics.

19.7.9   Automation of Ultrasonic Welding

Ultrasonic welding itself requires so little time that a large portion of the cycle is lost to handling of the parts. This can be substantially reduced through the use of a rotary indexing table such as that shown in Fig. 19-31.

Modular welding heads can also be integrated into automated production lines.

19.8   Sources

Branson Ultrasonics Corp., 41 Eagle Rd., P.O. Box 1961, Danbury, CT 06813-1961, (203) 796-0400, fax (203) 796-9838, www.bransonultrasonics.com

Dukane, 2900 Dukane Drive, St. Charles, IL 60174, (630) 584-2300, fax (630) 584-3162, www.dukane.com/us

Forward Technology, 260 Jenks Ave., Cokato, MN 55321, (320) 286-2578, fax (320) 286-2467, www.forwardtech.com

Herrmann Ultrasonics, Inc., 620 Estes Ave., Schaumburg, IL 60193, (847) 985-7344, fax (847) 985-1470, www.herrmannultrasonics.com

Manufacturing Technology Solutions, 14150 Simone Dr., Shelby Township, MI 48315, (586) 802-0033, fax (586) 802-0034, www.mts-telsonic.com

Mastersonics, Inc., 12877 Industrial Dr., Granger, IN 46530, (219) 277-0210, fax (219) 277-0210

Mecasonic, Div. Forward Technology Industries, Inc., 13500 County Rd. 6, Minneapolis, MN 55441, (763) 559-1785, fax (763) 559-3929, www.mecasonic.com

MS Plastic Welders, Inc., 37732 Hills Tech Dr., Farmington Hills, MI 48331, (248) 553-8330, fax (248) 553-8490, www.msplasticwelders.com

Sonic Ease, Inc., 210 Maple Pl., P.O. Box 87, Keyport, NJ 07735, (908) 739-9230, fax (908) 739-4540

Sonics & Materials Inc., 53 Church Hill Rd., Newtown, CT 06470, (800) 745-1105, (203) 270-4600, fax (203) 270-4610, www.sonics.biz

Sonitek - Sonic & Thermal Technologies, Inc., 84 Research Dr., Milford, CT 06460, (203) 878-9321, fax (203) 878-6786, www.sonitek.com

Sonobond Ultrasonics, Inc., 1191 McDermott Dr., West Chester, PA 19380, (800) 323-1269, (610) 696-4710, fax (610) 692-0674, www.sonobondultrasonic.com

Ultra Sonic Seal Co., 200 Turner Industrial Way, Aston, PA 19014, (610) 497-5150, fax (610) 497-5195, www.ultrasonicseal.com

20　Vibration Welding

20.1　Advantages and Disadvantages

20.1.1　Comparison with Ultrasonic Welding

Vibration welding can be described as the process that picks up where ultrasonic welding leaves off, particularly in terms of part size. That it is somewhat similar in principle is true. Consequently, it shares many of the same attributes. In addition, the equipment is made by manufacturers who also produce ultrasonic welding equipment, so the two processes complement each other nicely. Because of the considerable cost of the equipment, however, one normally uses vibration welding only for applications unsuited to ultrasonic welding. There are some significant differences between the two techniques.

While both processes create heat by means of high frequency vibrations, they apply them in a different manner and at much different frequencies. Ultrasonic welding vibrations are vertical, range from 15,000 to 72,000 Hz, and create heat through molecular friction. Vibration welding is horizontal, with frequencies between 120 and 300 Hz, and creates heat through surface friction as it physically moves the whole part. The net result is that vibration welding can handle much larger parts than ultrasonic welding and does not require the energy directors that largely limit ultrasonic welding to parts made by the injection molding process. While thermoset parts cannot be vibration-welded, parts made from virtually all the thermoplastic processes can be joined by this process.

Hot plate (or fusion) welding can also be considered to be a competitor to vibration welding because its joint strengths and material compatibilities are approximately equivalent. However, while it has much lower equipment cost and a greater range in size and shape, it has a much longer cycle time and higher equipment, fixture, energy, and maintenance costs than vibration welding.

20.1.2　Advantages of Vibration Welding

1. *No additional materials* Vibration welding uses no additional materials such as fasteners, inserts, electromagnetic preforms, adhesives or solvents. Therefore, it is inherently lower in cost than methods that do and is less expensive to disassemble for recycling.

2. *Low surface preparation* Vibration welding is relatively insensitive to poor surface preparation.

3. *Ease of assembly* Vibration welding requires only the placement of the two parts in the fixture of the vibration welder.

4. *Entrapment of other parts* Additional parts can be captured between the two parts to be vibration-welded provided the additional parts are located such that they do not interfere with the welding.

5. *Permanence* Vibration welding creates permanent assemblies that cannot be reopened without damaging the parts. Since joints are welded, the effects of creep, cold flow, stress relaxation, and other environmental limitations are absent in the joint area, although they may occur elsewhere if other parts of the design are under load. Owing to material limitations, differences in thermal expansion and moisture absorption are rarely of concern once the parts have been welded. Joint strengths equivalent to those of the surrounding walls are possible under ideal conditions.

6. *Internal joints* In some cases, internal walls that meet at the welding plane can be vibration-welded.

7. *Shape freedom* Vibration welds can be made with parts of practically any shape as long as the horizontal joining surfaces can be created within the required limits. Vibration welding is not dependent on the vibration transmission qualities of the plastic; therefore it can weld parts that are not well suited to ultrasonic welding because they are too flimsy, contoured, have holes in the walls, or do not have a surface well suited to ultrasonic horns.

8. *Hermetic seals* Vibration welding is capable of creating hermetic seals.

9. *Energy efficiency* Relative to other welding processes, vibration welding is highly energy efficient. This also means that there is no excess heat that must be removed from the workplace.

10. *Clean atmosphere* Unlike adhesive and solvent joining systems, no ventilation equipment is necessary for the removal of toxic fumes.

11. *Immediate handling* Assembled parts can proceed to other operations at once without waiting for parts to cool and for adhesives or solvents to set.

12. *High production rates* Depending on the application, vibration welding is capable of production rates of 4 to 30 parts per minute based on a single weldment per cycle and not taking into account the part handling time, which varies with the application. Higher rates are possible if multiple parts are welded with each cycle.

13. *Process freedom* Parts made from virtually all the thermoplastic processes can be vibration welded.

14. *Large part capability* Equipment is available that can weld parts up to 1016 mm × 2032 mm (40 in. × 80 in.). A 1524 mm (60 in.) automobile bumper has been successfully vibration-welded.

15. *Precision control* Vibration welding permits precision control of process variables.

16. *Quick changeover* Vibration welding equipment can be quickly changed from one job to the next.

17. *Controllability* This process is readily controlled and is not likely to result in surface degradation due to overheating.

20.1.3   Disadvantages of Vibration Welding

1. *Shape limitations* There must be a flat, horizontal welding surface.

2. *Damage to electronic components* Vibrations can damage some electronic components or their assemblies.

3. *Alignment* Locating pins or other devices cannot be molded into the part. Alignment between the two parts is set by the final resting place of the two parts.

4. *Material limitations* Materials for vibration welding are limited to compatible thermoplastics.

5. *Sound concerns* The foghorn noise associated with vibration welding (90–95 dB) makes the use of sound enclosures commonplace. These reduce the sound level to approximately 80 dB.

6. *Equipment cost* Vibration welders are more expensive than hot plate welders and cost considerably more than ultrasonic welders or spin welders, a consideration that tends to limit their use to applications too large or poorly configured for the less expensive techniques.

20.2   The Process of Vibration Welding

Outdoor camping manuals often described a method for creating a fire by rubbing two pieces of dry wood together. The author was never successful at this technique; however, this particular former Boy Scout now realizes that the problem was that he never achieved speeds of 120 cycles per second in rubbing the two pieces of wood together. Crudely, that in effect, is the concept behind vibration welding: pressure plus friction combining to create heat. In this case, however, the parts being rubbed together are made of plastic.

The cycle begins with the loading of one of the parts to be welded in the nest or fixture of the lower clamp plate of the machine illustrated in Fig. 20-1a. The mating part is then placed in position on the lower part. Next the lower clamp plate is raised to the vibrating platen in which rests the upper fixture, as in Fig. 20-1b. It fits the mating part closely and fixes its location. The two parts are clamped together under a controlled load, and the part in the vibrating platen is vibrated through an amplitude ranging from 0.75 to

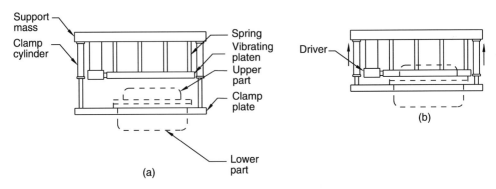

Figure 20-1   The vibration welding process: (a) machine open and (b) vibrating position

5.0 mm (0.030–0.200 in.) at a frequency from 120 to 300 Hz to create frictional heat at the interface between them. The pressure is important because, without the pressure, no heat is generated. Low weld pressures result in increased cycle times. Sophisticated systems can optimize the process control by varying the pressure during the cycle.

Basically, there are four phases to vibration welding. In the first phase, the vibration of the rigid members creates Coulomb friction, which generates heat at the joint interface. No penetration (movement of the parts toward each other) takes place in this phase. When the glass transition temperature is reached and viscous flow occurs, the second phase begins, wherein heat is generated by viscous dissipation in the molten polymer. Lateral flow of the polymer permits the penetration to take place. In the third phase, both melt and flow have reached a steady state at which heat losses through the wall due to flash are equal to the heat being generated. At this point melt is flowing laterally and weld penetration increases linearly with time. This part of the cycle is application dependent and will require 0.5 to 10 seconds, about two-thirds of the welding cycle. The penetration required to reach steady state condition increases with the wall thickness of the part, but decreases with increasing weld pressure.

When sufficient heat has been generated to melt the material at the interface, the vibrations are halted and the fourth phase commences. Weld penetration continues because the clamping pressure causes the molten polymer to flow until it solidifies. The parts are held clamped in the desired final position while they cool sufficiently to withstand handling, a period ranging from 0.5 to 5.0 seconds. Clamping is critical because there is no means of controlling lateral location of the two parts to each other beyond the final position of the vibrating platen. When the cooling period is over, the lower clamp plate is lowered to its original position and the completed assembly is removed. The equipment is now ready to commence a new cycle.

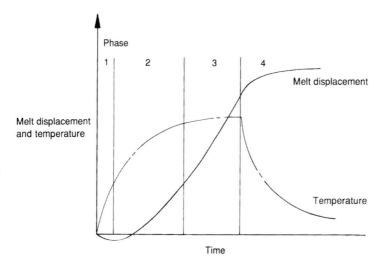

Figure 20-2   Relationship of melt displacement to melt temperature through the four welding phases (Potente, H. and Uebbing, M., Computer-Aided Layout of the Vibration Welding Process, SPE ANTEC Tech. Papers, 1992)

The relationship between melt displacement and temperature through the four phases is illustrated in Fig. 20-2. With weld parameters held constant, cycle time increases with the wall thickness. Depending on the part configuration, the material displaced will range from 0.38 to 0.75 mm (0.015–0.030 in.). The less material that must be displaced, the shorter the cycle time will be, since the extra material requires more time for both heating and cooling. The amount of displacement required is affected by the flatness of the welding interface. The greater the warpage, the more material there is that needs to be displaced and the longer the cycle must be. However, since this process is usually used for medium to large parts, a larger portion of the cycle is devoted to loading and unloading the machine than for most other processes.

20.2.1 Linear Vibration Welding

The vibration can be applied in either linear or rotational fashion. Linear welding is by far the oldest and most commonly used variety, with thousands of machines in use. It provides the best value and a wide range of sizes. The manner in which the parts are driven is dependent on their shape and rigidity. The driving force is applied on the external shoulders in Fig. 20-3a. However, it can also be applied from a recess in the part, as in Fig. 20-3b. The welding of internal ribs or midwalls calls for a fixture such as the one in Fig. 20-3c, in which pressure is applied over the internal joint.

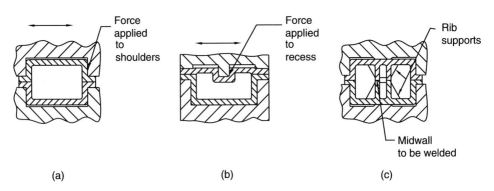

Figure 20-3 Fixtures for linear vibration welding: (a) external shoulder fixture, (b) recessed fixture, and (c) multiple internal rib fixture

20.2.2 Orbital Vibration Welding

The principal rotational type of vibration welding, orbital vibration welding, was created to fill a gap in the market for a midsize machine that can handle parts up to 305 mm (12 in.) in diameter and to accommodate some parts that cannot be made with ordinary vibration welding. With this technique, the two platens rotate in a

Figure 20-4   Orbital welding

circular pattern relative to each other, as illustrated in Figure 20-4. Whereas linear vibration operates best at resonance frequency, orbital vibration welding is used at off-resonance frequency. Unlike linear vibration welding, which has a nonuniform welding velocity because it must start and stop at each end of its cycle, orbital welding is continuous. This reduces the time necessary to create the weld, which lessens the amount of energy required. It also requires less weld amplitude, thus reducing the clearance required for the weld motion and providing better control of the flash. It has less effect on flexible walls and can produce stronger bonds. Unfortunately, orbital vibration is application sensitive and cannot be used for many products.

20.2.3   Angular Vibration Welding

Angular welding is that which has a rotational action around a pivot. This technique is reported to be well suited to round parts or those having a joint interface located at the same radius. However, the author's research has failed to locate an angular vibration welder currently being manufactured commercially.

20.3   Materials

Virtually all the thermoplastics can be vibration-welded. In that respect, the process is similar to hot plate and spin welding. Vibration welding is particularly useful for the crystalline thermoplastics, which are difficult to join with ultrasonic welding. This is clear in the rating of vibration welding characteristics of common thermoplastics in the center column of Table 20-1. As with ultrasonic welding, the amorphous resins are more readily welded than the crystalline resins. However, the crystalline thermoplastics are more readily vibration-welded than ultrasonic-welded. Materials with low coefficients of friction may require higher vibration frequencies for optimum weld quality. The process is claimed to have been successfully employed with thermoplastic rubber and elastomers. High temperature engineering thermoplastics have been successfully vibration-welded. The right-hand column shows the dissimilar materials

Table 20-1   Welding Characteristics of Thermoplastics

Material	Vibration Welding	Dissimilar Materials Having Compatibility
ABS	Excellent	ABS/PC, acrylic, SAN, PEI, PET, PPO, PC
ABS/polycarbonate alloy	Excellent	ABS, PC, acrylic
Acetal	Excellent	
Acrylic	Excellent	ABS, ABS/PC, acrylic multipolymer, P-AM-IM, PC, SAN
Acrylic multipolymer	Excellent	ABS, acrylic, PS, SAN
ASA	Fair	
Butadiene–styrene	Good	
Cellulosics	Excellent	
Fluoropolymers	Fair	
Ionomer	Good	
Liquid crystal polymers	Good	
Nylon	Excellent	
PBT/polycarbonate alloy	Excellent	PC
Polyamide	Fair to good	
Polyamide-imide	Good	
Polyarylate	Excellent	
Polycarbonate	Excellent	ABS, acrylic, PPO, PC/PBT, PBT, polysulfone, polyetherimide, PPO/PA, PC/ABS
Polycarbonate/polyester alloy	Excellent	
Polyesters (PET, PBT)	Excellent	ABS, PC, PEI
Polyetheretherketone (PEEK)	Good	
Polyetherimide	Excellent	Polycarbonate, ABS, PBT
Polyethersulfone	Excellent	
Polyethylene	Good to fair	
Polymethylpentene	Excellent	
Polyphenylene oxide (modified)	Excellent to fair	PS, acrylic, PC, SAN, PPO/PA, ABS
Polyphenylene oxide/polyamide	Excellent	PPO, PC
Polyphenylene sulfide	Good	
Polypropylene	Excellent	
Polystyrene		
General purpose	Excellent	PPO, acrylic multipolymer, SAN
Rubber modified	Excellent	
Polysulfone	Excellent	
PVC (rigid)	Good	
SAN-NAS-ASA	Excellent	ABS, ACRYLIC, acrylic multipolymer, PPO, PS

that can be joined by vibration welding. Joining of dissimilar materials is limited to amorphous thermoplastics as with ultrasonic welding.

Filled and reinforced resins are readily welded with vibration welding. However, only the polymer welds, so the strength of the bond is that of the resin itself, and the strength of the joint is further reduced by the percentage given to the filler. In particular, glass fibers may protrude through the surface of the weld at its center. Conversely, this process is less affected by mold release and other surface contaminants than either hot plate or ultrasonic welding, and it is somewhat less affected by moisture content, although highly hygroscopic polymers like nylon must still be welded with care. The presence of moisture can lead to bubbles in the joint due to the formation of water vapor. Predrying the parts can reduce bubble formation and welding time. A high joining pressure can prevent the formation of bubbles in the weld. At 50% relative humidity, nylon 6 weld strengths can drop to around half those of dry welds. Nylon 6/6 fares somewhat better, dropping less than one-third.

Dr. V. K. Stokes has done extensive work in the field of vibration welding. In May 1995 he published the data in Table 20-2 based on his experiments. These figures represent the proportion of original strength (1.0 = 100%) and the nominal strain to failure Stokes was able to achieve for each combination of nine materials. Thus, the ABS-to-ABS combination in the upper left-hand corner is read as follows: a grade of ABS with a tensile strength of 44 MPa (6.4 ksi) and a strain at failure of 2.2% was used. A weld strength of 0.9 or (0.9 × 44 = 39.6) 39.6 MPa (5.76 ksi) was achieved with a nominal strain to failure of 2.1%. When two different materials are used, the proportion applies to the weaker of the two materials.

20.4   Vibration Welding Part Design

20.4.1   Basic Considerations

If two parts are to be vibration-welded, nothing can be permitted that could interfere with the relative motion of the parts in the welding plane. A boss or a change in wall thickness could prevent the travel of the parts relative to each other. Furthermore, the joint must be supported during welding. Finally, the absolute minimum width of the joint surface cannot be less than 80% of the stroke or amplitude.

It should be noted that thin-walled thermoformed parts with wall thicknesses as low as 0.75 mm (0.030 in.) have been successfully vibration-welded (on their flanges) because it is possible to weld with an amplitude as low as 0.75 to 1.50 mm (0.030–0.060 in.) using high frequency vibration welding (in the range of 250–300 Hz). With their low amplitudes, these frequencies permit tight-fitting designs, like those in Fig. 20-5a and 20-5b, and are more suitable to small parts. The clearance for each of the lenses in these housings was only 0.64 mm (0.025 in.) per side.

Vibration welding can be used for joints up to 15° off the horizontal plane, as in Fig. 20-5c. Note that the parts in this illustration are contoured in the cross direction. Studies indicate that excellent strength can be attained in both directions.

Table 20-2 Achievable Strengths of Vibration Welds of Nine Thermoplastics[a]

Material	ABS	ASA	M-PPO	M-PPO/PA	PC	PBT	PC/ABS	PC/PBT	PEI
MPa (ksi)	44 (6.4)	32.5 (4.7)	45.5 (6.6)	58 (8.5)	68 (9.9)	65 (9.5)	60 (8.7)	50 (7.3)	119 (17.3)
Strain at Failure	2.2%	2.9%	2.5%	>18%	6%	3.5%	4.5%		6%
ABS 44 (6.4) 2.2%	0.9 / 2.1%		0.76 / 1.45%		0.83 / 1.7%	0.8 / 1.6%	0.85[b] / 1.8%		0.65 / 1.14%
ASA 32.5 (4.7) 2.9%		0.46 / 0.9%							
M-PPO 45.5 (6.6) 2.5%	0.76 / 1.45%		1.0 / 2.4%	0.22 / 0.35%	0.24 / 0.4%				0
M-PPO/PA 58 (8.5) >18%			0.22 / 0.35%	1.0 / >10%	0.29 / 0.75%				
PC 68 (9.9) 6%	0.83 / 1.7%		0.24 / 0.4%	0.29 / 0.75%	1.0 / 6%	1.0 / 1.7%	0.7[b] / 1.8%	1.0 / 4.9%	0.95 / 2.75%
PBT 65 (9.5) 3.5%	0.8 / 1.6%				1.0 / 1.7%	0.96 / 3.5%			0.95 / 4.1%
PC/ABS 60 (8.7) 4.5%	0.85[b] / 1.8%				0.7[b] / 1.8%		0.85 / 2.3%		
PC/PBT 50 (7.3)					1.0 / 4.9%			1.0 / >15%	
PEI 119 (17.3) 6%	0.65 / 1.14%		0		0.95 / 2.75%	0.95 / 4.1%			1.0[c] / 6%

[a] Except as noted, data obtained through room temperature tensile tests on welded specimens 6.3 mm (0.25 in.) thick at a strain rate of 10^{-2} s^{-1}. Most data are for 120 Hz welds. See text on page 508 for key to reading this chart.
[b] Data obtained through tests on specimens 3.2 mm (0.125 in.) thick.
[c] High strength can be achieved only through high frequency (250 and 400 Hz) welds.

Source: V. K. Stokes, Toward a Weld Strength Data Base for Vibration Welding of Thermoplastics, SPE ANTEC Technical Papers, 1995.

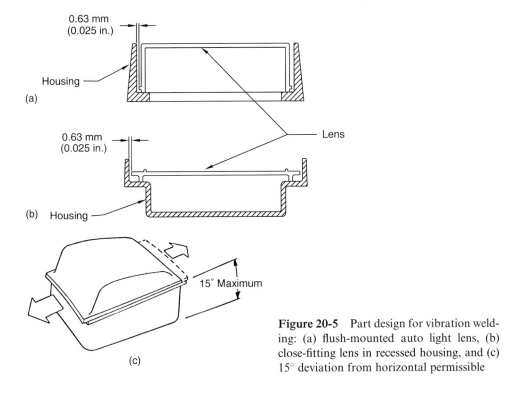

Figure 20-5 Part design for vibration welding: (a) flush-mounted auto light lens, (b) close-fitting lens in recessed housing, and (c) 15° deviation from horizontal permissible

20.4.2   Joint Designs for Linear Vibration Welding

The basic butt joint shown in Fig. 20-6 is the earliest joint used in vibration welding. The flange on the design in Fig. 20-6a provides more weld area and a better grip for the clamp, it also permits the application of pressure closer to the joint. For very thin walls or those that are unsupported, stiffness can be imparted by the addition of

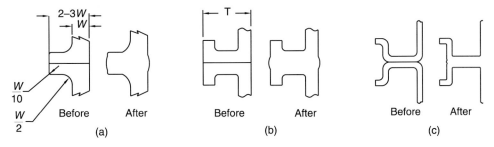

Figure 20-6 Butt joint designs for vibration welding: (a) flange butt joint, (b) ribbed flange butt joint (T, weld surface; M, stroke; $T_{min} = 0.8M$), and (c) thermoformed butt joint with return

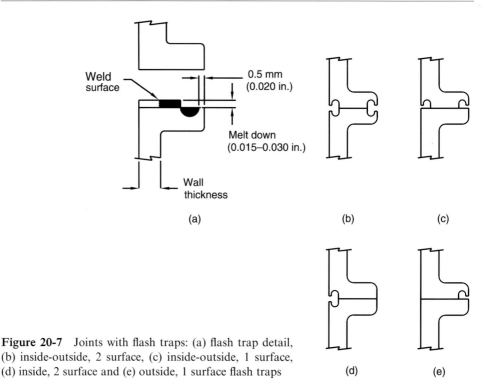

Figure 20-7   Joints with flash traps: (a) flash trap detail,
(b) inside-outside, 2 surface, (c) inside-outside, 1 surface,
(d) inside, 2 surface and (e) outside, 1 surface flash traps

a rib on the outer edge of the flange, as in Fig. 20-6b. For thermoformed parts, it is more cost-effective to add a horizontal return, as shown in Fig. 20-6c, so that parts can be inexpensively die-cut.

With a meltdown of 0.38 to 0.75 mm (0.015–0.030 in.), a considerable amount of flash can be generated around the surface. To some extent, this can be controlled by the amplitude and frequency. A fine edge of flash along the joint interface is formed when large amplitudes are used because they move the material further than small amplitudes. It is sometimes possible to achieve an acceptable level of flash with a basic butt joint simply by going to a higher frequency and smaller amplitude. The smaller amplitude results in heat being confined to a smaller area, with less flash generation.

When the approach just described will not suffice, a joint with a flash trap or melt recess will be required. Joints with flash traps are illustrated in Fig. 20-7. The size of the flash trap should be at least 30% greater than the calculated flash, are shown in the design of Fig. 20-7a.

When aesthetic requirements are more demanding, the tongue-and-groove design in Fig. 20-8a can be used. This variety of tongue-and-groove joint varies from the type used for other joining methods in that a minimum of 0.75 mm (0.030 in.) clearance must be provided for high frequency vibration motion or 1.50 mm (0.060 in.) for low frequency. Thus, it can be used to locate the parts to each other prior to

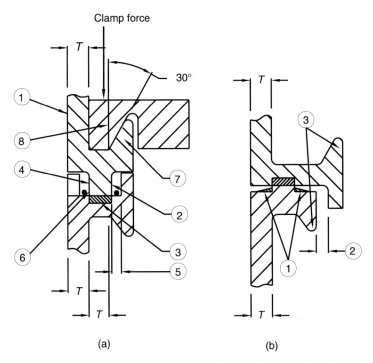

Figure 20-8   Tongue-and-groove joints: (a) basic joint and (b) skirted tongue-and-groove joint

joining; however it does not provide a very precise location. This illustration shows the assembly in the welded position with the tool still in place. Item 1, the 2.50 mm (0.100 in.) nominal wall thickness, is there to provide sufficient rigidity to ensure the transmission of the vibratory motion through the part. The tongue (item 2) is the same thickness, and it is molded 0.50 to 0.75 mm (0.020–0.030 in.) longer than the joint is intended to be. In the weld area (item 3) half of that thickness of material will be welded. The tongue height after molding (item 4) should be equal to the tongue thickness, and the space (item 5) between the tongue and the lip of the groove must be sufficient for the vibratory motion to take place. Flash (item 6) will be found to each side of the tongue. This is the portion of the tongue excess that did not become part of the weld.

A simple butt joint does not compensate for warpage. Note the angle portion of the joint shown as item 7 in Fig. 20-8a. This is the gripping tab, or reverse flange, which the tool locks onto to help overcome distortion. When this detail is not present, the portion of the tool (item 8) positioned just over the tongue may be knurled to hold the part down. This can leave a slight pattern on the finished part. The 30° angle on the top of that joint will help to pull the parts to their proper position.

A skirted version of this joint is shown in Fig. 20-8b. Item 1 illustrates the flash, which is hidden by the skirt. A clearance of 1.30 mm (0.05 in.) is required (item 2). The gripping tabs are indicated as item 3.

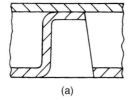 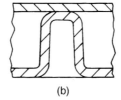 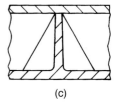

(a)   (b)   (c)

Figure 20-9   Internal vibration welding joints: (a) cored wall, (b) double wall, and (c) ribbed wall with gussets

Internal joints like those in Fig. 20-9 cannot be accessed for coring to support flanges without creating openings in the outer wall like the one in Fig. 20-9a. Alternative solutions are the double-wall configuration in Fig. 20-9b or the gussets in Fig. 20-9c.

20.5   The Equipment

Construction varies somewhat among competitive models; however, a vibration welder like the one in Fig. 20-10 is composed of five basic components.

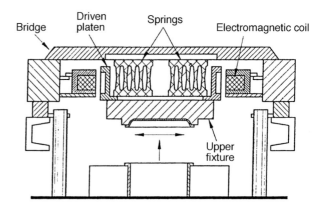

Figure 20-10   Basic vibration welder

1. *Vibrator* The vibrator is composed of a vibrating platen, springs, and two electromagnets. The vibrating platen is suspended from springs, which are also the resonating members and return the vibrating platen to alignment after each cycle. The vibrating platen grips the part to be vibrated and vibrates it when the electromagnets are energized. Some models operate the vibrating platen by mechanical or hydraulic means.

2. *Clamp* The clamping mechanism has a lower clamp plate, which is operated by a hydraulic cylinder that lifts it to meet the vibrating platen and supplies the pressure for welding. The clamp loading is variable.

3. *Fixture* The lower part to be welded rests in a fixture that is custom-made for each application. Fixtures for vibration welding are usually simple cutouts to match

the part contours, which are made from aluminum or urethane castings taken from the part itself. Chrome-plated steel fixtures are sometimes used for high wear applications. These fixtures are easily replaced, making changeover from one job to the next possible in a very short time. Some fixtures are made in which one part is loaded into the top and the other into the bottom. The entire plate surface is available for welding with each cycle; therefore it is possible to weld as many parts as will fit at one time to optimize the operation.

4. *Controls* There are six variables to be controlled on a vibration welder. These are the weld time, hold time, joint pressure, amplitude, frequency, and voltage. Closed-loop and programmable controllers are now available which permit a number of preset job programs to be recalled quickly for fast job changeover. Equipment offering multiple or variable pressures is now widely available.

5. *Machine frame* The machine frame provides the structure that supports the other components. Most machines are sold with sound enclosures as part of the package.

20.6   Sources

Bielomatik, Inc., 55397 Lyon Industrial Dr., New Hudson, MI 48165, (248) 446-9910, fax (248) 446-6244, www.bielomatikinc.com

Branson Plastic Joining Inc., 1001 Lehigh Station Road, Henrietta, NY 14467-9389, (716) 359-3100, fax (716) 359-1189

Branson Ultrasonics Corp., 41 Eagle Rd., P.O. Box 1961, Danbury, CT 06813-1961, (203) 796-0400, fax (203) 796-9838, www.bransonultrasonics.com

Dukane, 2900 Dukane Drive, St. Charles, IL 60174, (630) 584-2300, fax (630) 584-3162, www.dukane.com/us

Forward Technology, 260 Jenks Ave., Cokato, MN 55321, (320) 286-2578, fax (320) 286-2467, www.forwardtech.com

Manufacturing Technology Solutions, 14150 Simone Dr., Shelby Township, MI 48315, (586) 802-0033, fax (586) 802-0034, www.mts-telsonic.com

Mecasonic, Div. Forward Technology Industries, Inc., 13500 County Rd. 6, Minneapolis, MN 55441, (763) 559-1785, Fax (763) 559-3929, www.mecasonic.com

Sonics & Materials Inc., 53 Church Hill Rd., Newtown, CT 06470, (800) 745-1105, (203) 270-4600, fax (203) 270-4610, www.sonics.biz

Ultra Sonic Seal Co., 200 Turner Industrial Way, Aston, PA 19014, (610) 497.5150, fax (610) 497-5195, www.ultrasonicseal.com

21   Welding with Lasers

21.1   Advantages and Disadvantages

21.1.1   Non-Contact, Surface, Direct, or Butt Laser Welding

Laser welding is infrared welding using a laser as the energy source. Infrared welding using halogen and parabolic lamps predates the use of the laser and first appeared in the 1970s. These early systems did not offer outstanding advantages over other welding techniques and created problems of heat build-up in fixtures, critical focusing, and limited lamp life. Lasers were introduced as a light source in the early 1990s and found a number of applications by the end of the century.

The laser can be used to weld plastics either by directing the laser beam on the surface of a laser-absorbing plastic and welding by fusion or by transmitting a laser beam through a laser-transparent material and welding at the interface with the laser-absorbing material. The fusion welding method is also known as "non-contact welding", "surface welding", "direct welding," or "butt welding" and is very similar to hot plate welding, except that a laser is used in place of the hot plate. In fact, some machines for this type of laser welding have been built using a hot plate welder as the basis. Angled mirrors are placed between the two parts to be welded as illustrated in Fig. 21-1. The laser melts the plastic on the weld surface of each of the parts to be welded by absorption of the laser beam. The mirrors are removed from between the parts and the parts are then pressed together as in hot plate welding. This process offers the advantage of simultaneous welding of both parts and any degraded material is buried in the weld. Unfortunately, it is a relatively slow process that leaves a large weld bead and expensive process engineering is required.

Non-contact infra-red welding is best accomplished when the parts to be joined are made of the same thermoplastic. Melting of both parts will occur at the same rate, at the same temperature, and with the same amount of collapse. This process can be used to join dissimilar plastics with varying degrees of success. The differences in thermal properties will result in one part melting at a different temperature than the other and to a different depth. Consequently, the inter-diffusion of the polymers at the melt interface will be limited and the joint strength will be reduced. The greater the difference in thermal properties between the two plastics, the poorer the weld will be. Some polymers cannot be welded together.

This type of welding originated using infrared lamps and they are still used for some applications such as butt welding of pipes. Infra-red lamps offer the advantages of a low cost, mass-produced light source, low maintenance, high efficiency, high power density and, for medium or long wave applications, they are not sensitive to the type of plastic used. Unfortunately, infra-red lamps suffer from short lamp life and heat build-up in

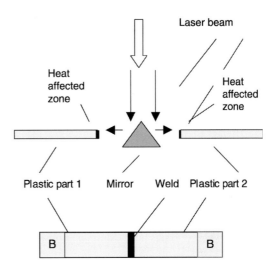

Figure 21-1 Non-contact laser welding (courtesy SAE)

fixtures. In addition, they are notoriously difficult to focus and that is their great disadvantage. Today, lasers are used in place of infra-red lamps for surface or direct welding. The equipment for surface or direct welding requires either a beam splitter or a laser for each part and is very expensive. Consequently, surface or direct welding is used only when none of the other welding techniques will suit the application.

21.1.2 Laser Staking

Laser energy can also be used to melt a plastic stud so it can be formed by a die. The laser beam is directed through the transparent die down on the plastic stud to be welded until the polymer is soft enough to be formed. The die then presses down on the stud to shape the head of the stake. To avoid loss of light transmission due to abrasion of the die, the plastic stud should not be made of glass fiber filled resins. Such applications require a metal die which is applied as a second stage. The source for this type of equipment is Extol, Inc.

21.1.3 Through Transmission Infra-Red Laser Welding

This section will focus on through transmission infra-red (TTIr) welding as this is the principle laser welding process. It is of most interest to design engineers because of its unique characteristics. From this point on, this discussion will apply only to that variety of laser welding.

   With parts for which it is well-suited, laser welding can create a weld that is closer to perfection than the welds made with any other method. Laser welding is unique in the respect that it is a non-contact thermoplastic welding process, which leaves no marks or particulates on the finished product. It is the most precise of all welding processes, with a very small area affected by heat. It is also capable of welding in recesses that are inaccessible by the other welding methods. Unfortunately, the cost of the equipment is still high, although it is becoming competitive with other types of welding equipment. Laser welding is not exactly cost competitive with the other welding processes, but rather its combination of unique properties provides the engineer with the capability to perform welds or achieve quality levels otherwise impossible.

21.1.4   Advantages of Through Transmission Laser Welding

1. *Non-contact welding* Laser welding uses a smooth glass plate to hold the parts to be welded with a light clamping force of, generally, 100 pounds or less and there is no contact with the welding instrument. Consequently, there is nothing to mar the surface of the welded parts and there is no welding pressure required that could damage delicate parts.

2. *Sub-surface welding* The weld occurs at the interface between the two parts. The weld is nearly invisible with transparent parts and completely invisible with opaque parts. The weld can be just below the surface or deep in the parts

3. *Precise welding* The size and location of the laser weld can be very precise and there is no vibration transmitted to the part to be welded. For those versions of the process that require relative motion between the laser and the weldments, it is the laser that moves. Consequently, laser welding is the most precise of the welding processes.

4. *Minimal heat-affected area* The weld spot is very small and the heat-affected area is quite contained. As a consequence, it is possible to weld very close to other components without affecting them, thus permitting welding of sub-assemblies with sensitive and fragile parts previously emplaced.

In addition, distortion and degradation due to heat is extremely limited and there is reduced residual stress.

5. *Ability to weld dissimilar plastics* Dissimilar plastics which meet the necessary criteria can be welded to each other.

6. *Reduced mold cost* Laser welding requires no special joint configuration thereby reducing the cost of tooling.

7. *No flash* In laser welding, the weld is totally enclosed and, in most cases, no flash is created to be trimmed off. There are no loose particulates as well.

8. *Welding of elastomers and difficult materials* Laser welding has the unique ability to weld elastomers. It can also weld PEEK (polyetheretherketone), LCP (liquid crystal polymer), COC (cyclo-olefin copolymer), and PEI (polyetherimide).

9. *Flexible to flexible* In addition to welding elastomers, laser welding can also be used to weld films to each other.

10. *Flexible to rigid welding* Film and elastomers can be welded to rigid extruded and molded parts.

11. *No additional materials* Laser welding uses no additional materials, such as fasteners, inserts, electromagnetic preforms, adhesives, or solvents for material combinations in which one of the two materials is transparent to laser light and the other

absorbs it. Therefore, except for initial capital investment, it can be inherently lower in cost than methods which do use additional materials and less expensive to disassemble for recycling.

12. *Low surface preparation* Laser welding is extremely forgiving of poor surface preparation and surface deviations of up to ±1 mm (0.039 in.) can be tolerated depending on the particular laser welding technique employed. Overcoming surface deviations does, however, result in flash.

13. *Entrapment of other parts* Additional parts can be captured between the two parts to be laser welded and they can be located closer to the weld than with other processes.

14. *Permanence* Laser welding creates permanent assemblies, which cannot be reopened without damaging the parts unless a laser is used to remelt the weld interface. As a welded joint, the effects of creep, cold flow, stress relaxation, and other environmental limitations are absent in the joint area, although they may occur elsewhere if other parts of the design are under load. Owing to material limitations, differences in thermal expansion and moisture absorption are rarely of concern once the parts are welded. Joint strengths equivalent to those of the surrounding walls are possible under ideal conditions.

15. *Internal joints* Internal walls can be laser welded provided there is clear access for the laser beam.

16. *Shape freedom* The laser is capable of creating virtually any pattern, therefore laser welds can be made with parts of practically any shape so long as the horizontal joining surfaces can be created within the required height limits. Depending on the variety of laser welding to be used, permissible height variations can range from 1 mm (0.039 in.) to 40 mm (1.575 in.). If a continuous weld is not required, much greater vertical freedom is available. Laser welding is not dependent on the vibration transmission qualities of the plastic. Therefore, it can weld parts which are not well suited to ultrasonic welding because they are too flimsy, contoured, or do not have a surface well suited to ultrasonic horns.

17. *Hermetic seals* Laser welding is capable of creating a hermetically sealed, gas-tight weld.

18. *Energy efficiency* Compared to the hot plate welding technique, laser welding is highly energy efficient. This also means that there is no excess heat which must be removed from the workplace.

19. *Clean atmosphere* No loose particulates are created and, unlike adhesive and solvent joining systems, no ventilation equipment is necessary for the removal of toxic fumes.

20. *Immediate handling* Assembled parts can proceed on to other operations at once without waiting for parts to cool and for adhesives or solvents to set.

21. *High production rates* Depending on the application, laser welding is capable of extremely high production rates for small parts such as 30 to 60 parts per minute based on a single weldment per cycle and not taking into account the part handling time, which varies with each application. Higher rates are possible if multiple parts are welded with each cycle. Larger parts require more time, however the technique is still fast compared to competing processes.

22. *Process freedom* Parts made by virtually every thermoplastic process can be laser welded.

23. *Small part capability* Parts as small as 2 mm (0.079 in.) by 5 mm (0.197 in.) have been laser welded.

24. *Large part capability* Equipment is available that can weld parts up to 250 mm (9.843 in.) by 400 mm (15.748 in.). Larger parts can be laser welded if a continuous weld is not required.

25. *Quick changeover* Except for the simultaneous welding technique, laser welding equipment can be quickly changed from one job to the next.

21.1.5   Disadvantages of Through Transmission Laser Welding

1. *Material limitations* The limited availability of materials for laser welding has probably been its most serious disadvantage to date. Materials for laser welding have been limited to compatible thermoplastics where one of the two materials is transparent to laser light and the other absorbs laser energy. Recent development of coatings that can be applied at the weld interface have considerably expanded the range of feasible materials, however at an increased cost per unit.

2. *Shape limitations* Although there are some versions of the process that can handle height variations, most successful applications have had a flat, horizontal welding surface at the time of this writing.

3. *Equipment cost* Depending on the power of the laser in question and the degree of control sophistication of the competing welding equipment in question, laser welding equipment ranges from comparable in cost to considerably more expensive than other processes. On a size to size comparison, the latter is most often the case and laser welding becomes the technique of choice for those applications requiring its special characteristics and no other method will do.

21.2   The Process of Laser Welding

21.2.1   The Laser

The term laser is actually an acronym for the process known as Light Amplification by Simulated Emission of Radiation. Basically, a light source is reflected between two reflective surfaces and passes through a "laseractive medium". The light source is repeatedly reinforced in such a fashion to produce a strong concentration of the light beam.

Three types of lasers are normally used for laser welding: diode lasers, carbon dioxide (CO_2) lasers, and neodyniumyttrium-aluminum-garnet (Nd:YAG) lasers. As shown in Fig. 21-2, diode and Nd:YAG lasers operate in the near infra-red (NIR) wavelength range while CO_2 lasers function in the infra-red (IR) range. The curve represents a typical thermoplastic's light absorption characteristics. High power diode lasers (HPDL) produce laser light in the wavelengths range from 600 nm to

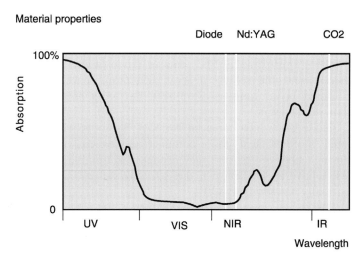

Figure 21-2 Absorption v. wavelength of a typical thermoplastic (courtesy Leister Technologies)

1600 nm, however, those used for laser welding are normally in the range of 800 to 980 nm. The wavelength used is dependent on the crystal structure of the diode.

As shown in Table 21-1, diode lasers offer several characteristics, which make them the preferred laser for many plastics applications. They have low maintenance cost, a high efficiency (in the range of 30 to 50%), and are the smallest of the lasers used for laser welding. These lasers are composed of Gallium (Ga), Indium (In), and Aluminum (Al) on one side and Phosphorous (P), Arsenic (As), and Antimony (Sb) on the other side.

Table 21-1 Lasers Used for Welding Plastics

	CO_2-Laser	Nd:YAG-Laser	High power diode laser
Type of active media	Gas	Solid state	semiconductor
Wavelength [nm]	10600	1064	~800–980
max.Pwr Output cw [kW]	upto 50	upto 5	0.06 up to > 1 (modular)
Focus-Intensity [W/cm^2]	10^6–10^8	10^5–10^8	10^3–10^5
Efficiency [%]	5–10	3–5 (8 DP*)	30–50
Size (Head) [cm^3/W]	~200	~20–50	~1
Price [$/W] Source: Rofin Sinar	70	180 (270 DP*)	70
Service interval [h]	1000	800 (Lamps)	servicefree

DP* = Diode Pumped Nd:YAG-Laser
(Courtesy: Leister Technologies)

The neodyniumyttrium-aluminum-garnet (Nd:YAG) laser operates at 1060 nm and is the next most commonly used laser for plastics applications. The carbon dioxide

(CO_2) laser operates at 10600 nm, which is impractical for most plastics applications. Therefore, it is the least commonly used for laser welding. CO_2 lasers are not very efficient for thick parts, but are used to join plastics films. In terms of cost per watt, they are approximately the same as diode lasers, however they are far larger than diode lasers and require a great deal more space. Carbon dioxide lasers have an efficiency in the range of 5 to 10%, but are capable of much greater output power than diode lasers.

Diode lasers have replaced Nd:YAG lasers in many applications because they are much more efficient, more compact, less expensive, their beam can be transmitted through fiber optics, and they are more adaptable to different geometries. Nd:YAG lasers have an efficiency in the 3 to 5% range. Laser power for plastics welding is normally less than 100 watts.

Lasers have been used to weld metals for some time. However, laser welding of metals requires highly focused laser beams of much greater power, in the range of 150 to 1000 watts, in order to achieve the much higher metal melting temperatures. In addition, the keyhole through the material commonly found in metal welding is to be avoided in plastics welding in order to prevent degradation of the polymer. In some cases, it is possible to weld plastics to metal.

21.2.2   Basic Through Transmission Laser Welding Methods

The term "through transmission" welding derives from the fact that the laser passes through the laser-transparent upper part to the surface of the laser absorbent lower part, as depicted in Fig. 21-3. There are three stages involved in the laser welding of plastics. During the heating stage, the laser energy heats the laser-absorbent polymer at the interface causing the part to expand, which increases the weld pressure. At the focus of the beam the laser power is at its maximum and the part begins to melt at that point creating a melt zone. This is the start of the second stage; the melting stage. Further expansion of the parts takes place creating additional expansion with a corresponding increase in pressure. At this point, a small amount of welding has taken place. The welding stage is the final stage. Sufficient molten material is being generated to create a real weld.

Typically, the plastic melts to a depth of 0.2 mm (0.008 in.) and the depth of the weld is approximately the same for both parts. The greater the depth of the weld, the more slowly the weld center cools down. This results in an increase in internal/ residual stresses. The longer the parts are held in this stage, the wider the weld bead becomes. If it is held too long, degradation can take place. Mixing is accomplished as the melted areas are pressed together resulting from the pressure created by the expansion of the material, which is constrained by an external pressure of a clamping force of, generally, 20 to 100 pounds. Sufficient external pressure must be supplied to reduce the gaps due to shrinkage or warpage. The weld is created as the material cools.

As with hot plate or fusion welding, there is a collapse of the melted surface. In this case, it is in the range of 0.13 mm (0.005 in.) to 1 mm (0.039 in.) for all but a small

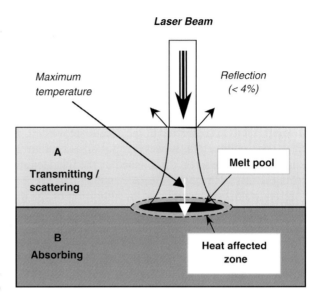

Figure 21-3 Through transmission laser welding (courtesy SAE)

number of applications. Typical weld times are on the order of one to eight seconds for most of its applications, although larger parts can run longer.

The two parts to be welded must have different optical absorption properties at a given wavelength. The top, laser-transparent part must transmit as much of the wavelength as possible while the bottom, laser-absorbent part must absorb as much as possible in a thin layer. This results in the most energy efficient combination because a high amount of energy is absorbed at the interface. If energy absorption of the top part is too high, the amount of energy required will be increased and it can degrade the top part before the bottom part is soft enough to weld. Excessive reflectivity also causes an increase in the amount of energy that is needed for welding.

For use in welding of plastics, the beam can be used in four different ways: spot or contour welding, quasi-simultaneous welding, simultaneous line welding, and a Leister patented process called mask welding.

21.2.3   Spot or Contour Welding

Spot or contour welding is the most versatile of all modes of laser welding. Either term describes this method because the process utilizes a spot of energy, which follows the contour of the weld surface according to a preprogrammed path. Nearly any path can be programmed, which gives this process considerable flexibility, one of the principal advantages of this technique. A second advantage is that the equipment can be quickly changed from one application to another, making spot or contour welding well suited to short runs or "just in time" manufacturing. In addition, weld bead width is readily modified and power requirements are low (usually less than

60 watts, but can reach 500 watts) for this type of welding. Power requirements are determined by the size of the weld bead, the speed of the laser, the type of laser, and the efficiency of the optical system. The equipment operator can adjust the power according to the needs of the application by controlling the input amperage.

For a diode laser, the laser light beam begins to diverge as it is emitted from the diode. It is then collimated, shaped and focused into a circular spot using mirrors and lenses. A fiber optic delivers the light to the work area where it is again refocused through a lens and directed at the area to be welded. The spot varies in size from 0.6 mm to several millimeters; however, spot sizes of 1 to 2 mm are recommended for most applications. The size is established by the focal length of the focus lens and the distance from the lens to the weld interface as shown in Fig. 21-4. The size of the spot, and therefore the weld, can be adjusted by alteration of the distance from the focal point of the lens to the weld interface. If the laser output power is held constant, an increase in the diameter of the spot will result in a reduction in intensity because the energy is spread over a wider area. A longer weld time will be required in order to melt the plastic sufficiently for welding to take place. Alternatively, the output power of the laser can be increased. The amount of energy that reaches the weld area is determined by the amount of energy supplied by the laser, less the optics and coupling efficiency losses, and the travel speed of the laser, in addition to the size of the spot.

The weld bead will be wider than the diameter of the spot. The amount varies with the material and amorphous polymers will have smaller beads than semi-crystalline materials. In the case of the latter, the bead can be as much as 50% larger than the diameter of the spot.

The term "contour" as used in the description of this process refers to two-dimensional contours as this technique's ability to handle interruptions in the weld surface, such as ejector pin recesses, and vertical dimensional deviations is presently

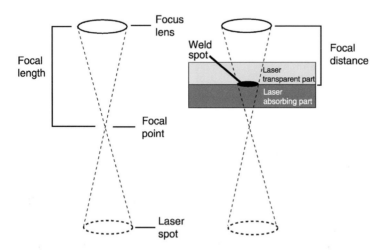

Figure 21-4. Weld spot size determination (courtesy Leister Technologies)

very limited. However, for applications where a continuous weld or a hermitic seal is not required, Bielomatik and Leister do offer equipment, which has the laser mounted to a robot arm that can follow the three-dimensional contours of the part in a manner similar to a robotic paint spray nozzle. Periodic spot welds are created just as in traditional metal spot welding. This process permits true three dimensional laser welding.

21.2.4   Simultaneous Through Transmission Infra-Red (STTIr) Laser Welding (Also Known as Simultaneous Line or Flash Welding)

Simultaneous is the key word in describing the difference between this type of laser welding and the spot or contour welding described previously. For STTIr, the entire weld surface is heated, melted, and welded simultaneously. This produces a faster weld (1 to 10 seconds depending on size) with no moving parts and fewer residual stresses. In addition, it is capable of producing higher weld strengths through greater surface deviations. Whereas spot or contour welding weld strengths suffer from surface deviations well under 1 mm (0.039 in.), STTIr can maintain its weld strength at deviation levels of 2 mm (0.079 in.) or more. The greater the surface deviations, however, the greater the pressure that will be required and the greater the flash that will be created. Unsupported internal walls with complex curvatures can also be welded. Set-up, however, is generally longer with STTIr than with spot or contour welding.

For simultaneous through transmission infra-red laser welding, the light beams emitted from the diodes are also collimated, shaped, and focused as in spot or contour welding. However, in this case, they are focused into a line or circle, instead of a spot, using mirrors and lenses as shown in Fig. 21-5.

The infra-red light source may be more than one diode. The laser line may be composed of arrays of diodes turned on simultaneously as depicted in Fig. 21-6.

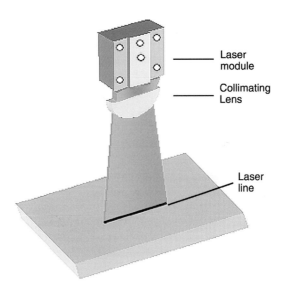

Laser module

Collimating Lens

Laser line

Figure 21-5   Laser line creation (courtesy SPE)

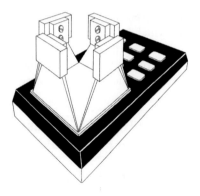

Figure 21-6   Simultaneous welding (courtesy Leister Technologies)

This type of equipment can weld applications with complicated weld surfaces. Squares, rectangles, and circles ranging from 2 to 50 mm can be line-welded. The typical STTIr weld line width is 1 to 2 mm.

21.2.5   Quasi-Simultaneous Laser Welding

Quasi-simultaneous laser welding (QSLW) is a variant of spot or contour welding performed with a single laser traveling at the very high speed of 10 m/s. At this rate, it competes with simultaneous line welding. However, it can also be slowed to the speed that would produce spot or contour welding.

As illustrated in Fig. 21-7, a high speed scanning mirror system guides the laser beam through a lens over the weld length repeatedly to achieve a weld. The laser beam is moved by the mirror system. In this manner, the weld interface is quasi-simultaneously and homogeneously heated and melted. This allows distorted parts to be welded with appropriate pressure. The quasi-simultaneous welding technique can produce hermetic seals.

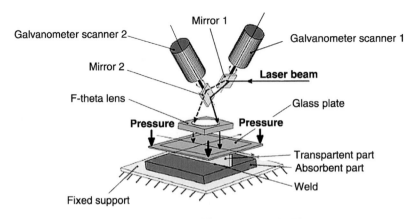

Figure 21-7   Quasi-simultaneous welding (courtesy SPE)

A solid-state Nd:YAG laser with a maximum power output in continuous-wave operation of approximately 250 W is employed. Extremely fine welds only 0.1 mm (0.004 in.) wide have been achieved. The system can weld two-dimensional contours as well as vertical contours up to $+/-$ 40 mm. Parts as small as 2×5 mm and as large as 250×560 mm have been welded by this process, the latter using twin lasers. Welding time is 2 to 8 seconds depending on length of the weld. Bielomatik is the principal source of quasi-simultaneous welding equipment and has pioneered development of the technique. Leister offers a similar process using high-power diode lasers.

21.2.6   Mask Welding (Leister Patented Process)

Mask welding combines laser welding with the masking technique common to coating technology. It is the process of choice when extremely fine or precise weld lines or welds of varying sizes and shapes are required. Fully automated mask welding can be achieved with a resolution of $2 \mu m$ (0.00008 in.). Microfluidic devices that require high positioning and welding accuracy in the micrometer range can be joined using mask welding. Such a device with a channel width of 100 μm has been successfully welded with this technique. Mask welding is both fast and flexible and is ideally suited for area bonds with complex and fine structures. In one example, the total cycle time for a 100 mm^2 (0.155 in.^{2}) product is in the range of 12 seconds. The actual welding time for this applications is less than a second.

With the mask welding process, the entire surface of the part is scanned with a laser line and the areas that are not to be welded are masked so that the light does not reach them, as illustrated in Fig. 21-8. The mask is located over the glass plate and fixed to a three-axis plate that permits precise alignment of the mask relative to the parts vertically or in either horizontal axis. Micro mask welding can be controlled with an accuracy of 2 μm (0.00008 in.) and can weld a curtain of laser light 14 mm (0.551 in.) wide at one time. Macro mask welding can weld a curtain of laser light 120 mm (4.72 in.) wide and can hold accuracy to 0.05 mm (0.002 in.). If greater widths are to be welded, multiple passes with slight overlap must be performed, leaving a seam. As with the other TTIr welding methods, the parts to be welded are placed beneath a transparent glass plate and pressure is applied. The laser is a high-powered laser with an emission wavelength between 808 nm and 980 nm. Power in the range of 80-120 W permits rapid heating and cooling of the polymer.

The mask must be constructed of a material that either reflects or absorbs laser energy. It is produced from a metallic coated glass plate from which the coating has been removed by a photolithographic process wherever the weld is to be created. This provides the designer with considerable design freedom in creating the weld contour as needed.

Both Diode and Nd:YAG lasers are used for multiple welding methods and each technique has its benefits as illustrated in Table 21-2. This information is accurate, but general in nature at the time of this writing. For any given application, the relativity of each technique for a given attribute could vary.

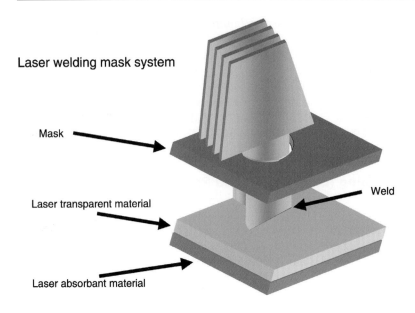

Laser welding mask system

Mask

Weld

Laser transparent material

Laser absorbant material

Figure 21-8   Mask welding (courtesy Leister Technologies)

Table 21-2   Basic Advantages and Limitations of the Through Transmission Laser Welding Techniques

Parameter/Method		C	SI		QS	MA
Laser type	Diode	+	+		+	+
	Nd: YAG	+			+	
Relative cost$_{9)}$		M	M/high		M/high	M
Weld shape		3D		2D		3D
Weld cycle		M/L	S			S/M
Size of plastic part		Any	L			
Lenght of weld		Any				Any

C-Contour Welding, SI-Simultaneous Welding, QS-Quasi-simultaneous Welding, MA-Mask welding, L-large, M-medium, S-small, Any-no limitation. (Val A. Kagan, Society of Automotive Engineers, 2002-01-2011)

21.3   Materials for Laser Welding

21.3.1   Material Properties Affecting Laser Weldability

Material suitability is a more sensitive issue with laser welding than with any other joining process. Laser welding can theoretically weld all thermoplastics that are transparent to the laser beam. However, each polymer's ability to transmit light from lasers of the three principle wavelengths used in laser welding is different. In addition, the response of a given polymer to laser beams will be altered by the presence of fillers, additives, and pigments.

The properties of plastics that affect their ability to transmit laser beams are transmission, absorption, and reflectance. Figure 21-9a shows that, of the amount of light that first reaches the top surface of the plastic, a small amount (about 4 to 5%) is immediately reflected off the surface. For the amorphous polymer depicted, the balance of the light will be transmitted through the material with minimal internal reflection, provided there are no fillers, additives, or pigments present.

Semi-crystalline materials have crystalline structures that cause scattering of light as shown in Fig. 21-9b. The amount of light lost in this fashion varies with the degree of crystallinity and the crystalline size. The greater the crystallization rate and the size of the spherulitic microstructure, the lower the transmittance of laser energy. Crystalline materials can also produce heating at or near the surface due to light scattering.

The principle properties of a polymer that affect its laser weldability are its laser energy transmittance, reflectance, and absorption. These properties are generally defined in percentages as follows:

$$\text{Transmittance}(\%): \quad T = 100\frac{I_t}{I_0} \tag{21-1}$$

$$\text{Reflectance}(\%): \quad R = 100\frac{I_r}{I_0} \tag{21-2}$$

$$\text{Absorption}(\%): \quad A = 100\frac{I_0 - I_t - I_r}{I_0} \tag{21-3}$$

where: I_t = laser beam transmittance intensity
I_0 = laser beam incident intensity
I_r = laser beam reflection intensity

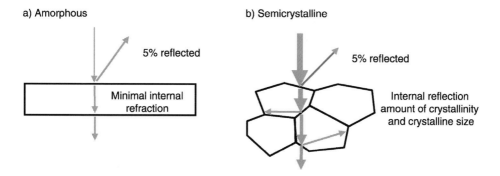

Figure 21-9   Light transmission characteristics; a) amorphous plastics; b) semi-crystalline plastics (courtesy Leister Technologies)

Reflectance is not generally a problem, as it is in the 4 to 5% range for most of the thermoplastics that are typically laser welded. Materials with a high transmittance rate are ideal for the upper part in TTIr welding, because they allow most of the laser energy to reach the weld interface. Transmittance is high for amorphous thermoplastics and varies for semi-crystalline thermoplastics. This characteristic limits the weldable thickness of semi-crystalline materials. Semi-crystalline polymers can be modified to varying degrees by increasing or decreasing the crystallinity and the number of spherules. Increasing these elements results in improved absorbancy, whereas decreasing them improves transmittance.

Materials with a high absorption rate are ideal for the lower part in TTIr welding, because they absorb most of the energy at the weld interface and create a thin heated surface layer that is ideal for laser welding. Materials with absorption rates in the middle ranges can be a problem, because they absorb the laser energy through the thickness of the material and that results in heating of the whole material thickness. These materials allow very little energy through the material, therefore they are poor candidates for the top material in TTIr laser welding. In addition, they do not absorb very much energy at their top edge and, consequently, make a poor selection for the lower material.

21.3.2   Effects of Refraction Properties on Material Selection

Refraction does not effect any laser application in which the laser beam is delivered perpendicular to the work surface unless it must pass through several layers of plastic to reach the weld interface. Refraction is of principle concern with amorphous plastics as semi-crystalline materials have crystalline structures that cause scattering of light as shown in Fig. 21-9b. Refraction is also a concern when using techniques in which the laser beam is not delivered perpendicular to the work area, such as in quasi-simultaneous laser welding. That system uses mirrors to deflect the beam around the weld surface. Consequently, the beam strikes the work surface at an angle. The larger the overall weld surface, the greater the angle will be.

Refraction of the light beam occurs when it passes from a medium of one density to a medium of another density. This process is governed by Snell's Law (see Fig. 21-10). It is of importance with respect to the upper material through which light is transmitted, because Snell's Law will affect where the light contacts the interface of the light beam with the lower (absorbent) material.

$$\text{Snell's Law:} \quad \frac{\sin\phi_A}{\sin\phi_B} = \frac{n\theta_A}{n\theta_B} \tag{21-4}$$

where: $n\theta_A$ = Index of refraction of medium A
$n\theta_B$ = Index of refraction of medium B
θ = Angles of incidence and reflection
$\theta_A = 90° - \phi$

$$\text{Angle of refraction} \quad \beta = \theta_A - \theta_B \tag{21-5}$$

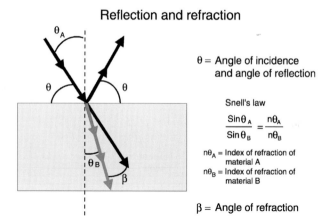

Figure 21-10   Reflection and refraction (courtesy Leister Technologies)

21.3.3   Effects of Pigments, Fillers, and Additives on Light Transmission

Fillers such as glass reinforcements similarly reduce transmittance. Dr. Val Kagan, Robert Bray, and Al Chambers have performed considerable work to determine the effect of glass fibers and pigments on the light absorption of polyamide. In Fig. 21-11 they show that the transmittance property of nylon 6 is reduced by light scattering as the percentage of glass fibers is increased. The effective path length increases by approximately 35% as the glass percentage is increased from 0% to 45%. Correspondingly, transmittance of laser energy drops from nearly 70% with no glass content to only 20% with glass fiber content at 65%. The intrinsic absorption remains essentially unaffected and any change in absorption is the result of diffuse scattering of laser energy. Filled materials have been welded with up to 50% glass content, but hermetic seals could not be achieved.

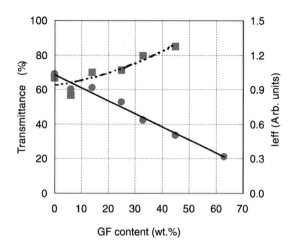

Figure 21-11   Transmittance and effective wavelength v. glass fiber content (courtesy SPE)

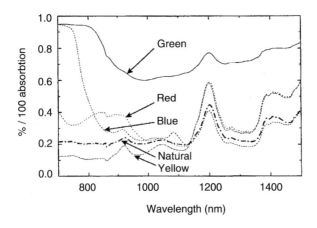

Figure 21-12 Effect of wavelength on absorption (courtesy SPE)

Pigments vary considerably in their effect on the transmission of light. As a broad statement, carbon black and titanium oxide (white) absorb laser energy. Therefore, colorants that contain these pigments will cause an increase in laser energy absorbance and a decrease in transmission. It should be noted that many colors other than black and white contain these pigments to some degree.

Figure 21-12 illustrates the absorption characteristics of nylon 6 pigmented in four colors, along with those of uncolored nylon 6, through a range of wavelengths from 600 nm to 1500 nm. Except for the green colorant, all colors and the uncolored sample were in the vicinity of 20% laser energy absorbancy in the range of wavelengths (around 1000 nm) in which they could be welded with diode or Nd:YAG lasers. The curve for the green colorant was higher across the board and shows approx. 60% laser energy absorbancy in the welding range of wavelengths.

The color requirement for a given application will affect the suitability of an application to laser welding and must be considered as one of the determining factors. It should also be noted that many plastics have an amber cast to them in their natural state. Resin manufacturers often add a small amount of blue colorant to give the plastic a "water clear" appearance. Dyes are composed of much finer particles than pigments and have much less effect on light transmission characteristics. Black or red dyes that transmit light can be used to create all-black products.

21.3.4 Laser Welding Transmitting Materials

In order to be absorbent to laser energy, one part must be black, white, or contain a high amount of absorbent additives. This restriction has severely limited the utility of through transmission laser welding applications for plastics applications, particularly in the medical fields. Recent successful developments have demonstrated that a thin film or coating containing laser absorbent dye can be placed between two transmitting materials to produce a weldable interface. The coating acts as a focal point that absorbs the laser energy creating sufficient heat to melt the surfaces of the two mating parts at the weld interface as shown in Fig. 21-13.

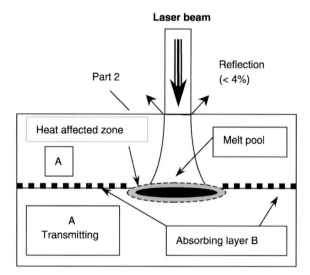

Figure 21-13 Through transmission laser welding of two transmissive plastics (courtesy SAE)

The coating is available commercially in the form of an ink marketed by Gentex Corp. under the name "ClearWeld™". Clearweld™ is capable of absorbing near-infrared light in the wavelength range used by diode and Nd:YAG lasers. Essentially, it is a solvent-based system, which acts as a carrier for the laser absorbent material. The coating is deposited on the surface of one of the two parts and the solvent quickly (1 to 2 s) evaporates. It is applied in a thin, microns-thick layer with ink-jet or needle dispensing equipment. The ink has a green color dye that absorbs the laser energy. The dye has the secondary benefit of identifying the location of the coating on the work piece. On welding, the green disappears and the joint is optically clear as shown in Fig. 21-14. The ink is non-toxic according to USP cytotoxicity tests and additional compatibility tests are under way.

Lap joints such as the one depicted in Fig. 21-14 have been tested by Gentex to strengths that match or exceed that of the surrounding material. These joints have been tested to 7705 psi in polycarbonate, 2119 psi in PMMA and 1929 psi in polyethylene film with failures occurring in the surrounding material, not in the weld area. At the time of this writing, Clearweld™ has been successfully used with all materials listed in Table 21-3. Work is in progress on other plastics as well, however,

Figure 21-14 Clearweld lap joint (courtesy Gentex Corporation)

the ink needs to be individually formulated for each plastic material, because different carrier solvents are used to avoid chemical attacks on the plastics being used.

Table 21-3   Polymers Successfully Welded with Clearweld™ (Courtesy: Gentex Corporation)

ABS*	Polythylene terephthalate*
Acetal*	PETG
Acrylic	PCTG
Cellulose acetate	Polyester
Cylclic olefin copolymer	Polyetherimide
Ethylene vinyl acetate	Polyether block amide
PFA	Polyethylene - LDPE, MDPE, HDPE, UHMWPE*
ECTFE	Polyimide*
ETFE	Polyphenylene oxide, modified*
THV	Polystyrene
FEP	Polyurethane
Ionomer - Surlyn	PVDC
PEEK*	PVDF
Polyamide*	Polysulfone
Polyamide - glass reinforced*	TPE
Polycarbonate	PVC
Polybutylene terephthalate*	

*Top substrate thickness may be limited

21.3.5   Compatibility of Plastics for Laser Welding

In addition to the polymers listed in Table 21-3, polyketone, polypropylene, polysulfone, and SAN have been laser welded. Dissimilar materials have been successfully laser welded in some cases. In order to be welded, the two plastics must have a similar polymer structure and overlapping softening ranges. PET mesh, PBT, and acrylic all have been successfully welded to polycarbonate; nylon has been welded to PC/ASA, and different types of nylon have been combined. There have been experiments with various combinations of polycarbonate, acrylic, ABS, polystyrene, and SAN. Figure 21-15 represents the laser welding compatibility of various plastics.

The reader should be cautioned that only specific formulations of a given polymer may be laser weldable and that thickness of the top layer may be restricted owing to limited laser energy transmittancy. There is a great deal of development activity in this field at the time of this writing; consequently, the chart will be in need of an up-date by the time it reaches print. It is recommended to regard it as incomplete and to contact the polymer manufacturer for the current status with regard to laser weldability of a given resin.

21.4   Joint Designs

One of the best features of laser welding is the elimination of precisely detailed joint features requiring tight tolerancing. Most joints are flat lap joints, such as the one

Material combinations

transp. absorb.	ABS	PMMA	PA6	PA66	PA12	PA11	PC	SAN	PS	PE	PP	POM	PET	PBT	TPU
ABS	w	w	n	n	n	n	-	p	p??	n	n	?		-	v
PMMA	p	w	n	n	n	n	p??	p	n??	n	n	n		n	
PA6	n	n	w	p	p	p	n	n	n	n	n	n	n	n	n
PA66	n	n	p	w	p	p	n	n	n	n	n	n	n	n	n
PA12	n	n	p	p	w	p	n	n	n	n	n	n	n	n	n
PA11	n	n	p	p	p	w	n	n	n	n	n	n	n	n	n
PC	-	p	n	n	n	n	w	p??	p??	n	n	n		-	
SAN	p	p	n	n	n	n	p??	w	p??	n	n	n		n	
PS	p??	n?	n	n	n	n	p??	p??	w	n	n	n		n	
PE	n	n	n	n	n	n	n	n	n	(w)	p	n		n	
PP	n	n	n	n	n	n	n	n	n	p	(w)	n		n	
POM	n	n	n	n	n	n	n	n	n	n	n	(w)		n	
PET			n	n	n	n							(w)	n	
PBT	-	n	n	n	n	n	-	n	n	n	n	n	n	(w)	
TPU	p		n	n	n	n	p?					p?			?

w = weldable
p = possible
n = not weldable or not tested
- = poor weld strength

Figure 21-15   Material compatibility for laser welding (courtesy Leister Technologies)

illustrated in Fig. 21-14. Surface deviations up to +/- 1 mm can be tolerated. That means the only tight tolerances to be concerned with are those related to the location of the two parts to be joined - which is common to all joining methods.

Most of the hot plate welding joint designs, such as those shown in Fig. 10-5, can also be used for laser welding; however, their use is limited to applications where the distance the light must travel through the laser transparent part is short to avoid diffusion of the beam. Therefore, the butt joint shown in Fig. 10-5 could not be used and one such as the one depicted in Fig. 21-16 should be employed instead.

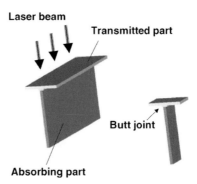

Laser beam

Transmitted part

Butt joint

Absorbing part

Figure 21-16   Butt weld (courtesy SPE)

21.5   Equipment

Construction varies somewhat between competitive models, however a through transmission laser welder such as the one in Fig. 21-17 is composed of six basic components.

1. *Laser* The laser will be one of the three types (diode, carbon dioxide and Nd:YAG) described in Section 21.2. The power is supplied by a generator that converts line voltage and frequency to the requirements of the particular laser.

2. *Clamp* The clamping mechanism has an upper clamp plate, which is operated by a hydraulic cylinder, which lowers it to meet the lower clamp plate and supplies the pressure for welding. The clamp loading is variable.

3. *Fixtures* The lower part to be welded rests in a fixture, which is custom-made for each application. Lower fixtures for laser welding are usually simple cutouts to match

a)   b)   c)   d)

Figure 21-17 Through transmission laser welders; a) simultaneous welder (courtesy Branson Ultrasonics); b) contour welder (courtesy Leister Technologies); c) quasi-simultaneous welder (courtesy Bielomatik); d) mask welder (courtesy Leister Technologies)

the part contours. Upper fixtures vary according to the type of system. For diode or Nd:YAG systems, the part is usually held stationary and the laser moves, although the reverse may be true for small parts. For carbon dioxide lasers, the laser is held steady and the part moves. The particular configuration of the upper fixture is application sensitive. Motion is controlled by robotics.

4. *Controls* The machine status and process parameters are usually controlled by personal computers. Typically, the computer will control weld time, joint pressure, melt collapse, and laser power. For other than simultaneous welding, the computer will also control the robotic path and travel speed.

5. *Machine frame* The machine frame provides the structure that supports the other components.

6. *Enclosure* Laser welding equipment is enclosed in order to protect the operator from radiation and intense light. On most equipment, a camera is provided to permit the operator to observe the process.

21.6   Applications

Each of the various welding techniques is best suited for specific applications. Ultrasonic welding is the least expensive, consequently, it is the process the designer usually thinks of first, unless the parts are round and lend themselves well to spin welding. If the part is too large for ultrasonics, vibration or hot plate welding gets the next look. If vibration equipment is available or the budget will tolerate the cost of a new machine, vibration welding would be the next choice, because the energy cost is lower than for hot plate welding. However, part contours or high weld strength requirements could force the designer to use hot plate welding despite its high energy cost.

Applications for which laser welding becomes the method of choice are the ones requiring high levels of precision or those that cannot tolerate surface marks or particulates. Other exceptional characteristics would be the ability to weld into recesses that would not accept the tooling required of the other processes, the capability of welding elastomers, and the absence of heat or vibrations, allowing the joining of sub-assemblies with sensitive components.

The majority of laser welding applications to date have been either in the automotive or in the medical industries. Automotive applications include steering fluid reservoirs, brake fluid reservoirs, tail light assemblies, fuel line components, and display housings. The two parts of the power steering fluid reservoir for the VW New Beetle are hermetically sealed with through transmission laser welding (Fig. 21-18a). Some medical applications are valves, housings, lenses, and intravenous equipment in which loose particulates are unacceptable. Other applications include cell phone components, gas analyzer, and light sensors. The hermetically sealed gas analyzer is illustrated in Fig. 21-18b. Both parts are made of ABS, the upper part being transparent.

Development of through transmission laser welding applications have been shrouded in secrecy as manufacturers seek to gain a competitive edge. However, the technique has been on the market for several years and applications that have been

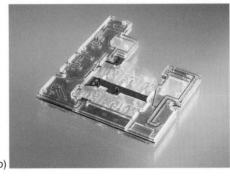

(a)   (b)

Figure 21-18.   Laser welding applications; a) power steering fluid reservoir for VW new beetle (courtesy Bielomatik); b) gas analyzer (courtesy Leister Technologies)

in development are surfacing. Laser welding is a valuable new weapon in the designer's arsenal that will find many applications in plastics joining.

21.7   Sources

Bielomatik Inc. USA, 55397 Lyon Industrial Drive, New Hudson, MI 48165, (248) 446-9910, fax (248) 446-6244, www.bielomatikinc.com

Branson Ultrasonics Corp., 41 Eagle Rd., P.O. Box 1961, Danbury, CT 06813-1961, (203) 796-0400, fax (203) 796-9838, www.bransonultrasonics.com

Dukane, 2900 Dukane Drive, St. Charles, IL 60174, (630) 584-2300, fax (630) 584-3162, www.dukane.com/us

Extol, Inc., 651 Case Karsten Dr., Zeeland, MI 49464, (616) 292-1771, fax (616) 748-0555, www.extolinc.com

Forward Technology, 260 Jenks Ave., Cokato, MN 55321, (320) 286-2578, fax (320) 286-2467, www.forwardtech.com

Gentex Corporation, 324 Main Street, Simpson, PA 18407, (570) 282-8631, fax (570) 282-8555, www.gentexcorp.com

Heraeus Noblelight, Inc., 2150 Northmont Parkway, Duluth, GA 30096, (770) 418-0707, (770) 418-0688, www.noblelight.net

LaserQuipment, 28220 Southwest Boberg Rd., Wilsonville, OR 97070, (503) 454-4200, (503) 682-7151, www.laserquipment.com

Leister Technologies LLC, 1253 Hamilton Parkway, Itasca, IL 60143, (630) 760-1000, fax (630) 760-1001, www.leister.com

Rofin-Sinar Inc., 40984 Concept Dr., Plymouth, MI, 48170, (734)455-5400. fax (734)455-2471, info@rofin.com

Sonotronic, Inc., 3741 Venture Drive, Suite 335, Duluth, GA 30096, (770) 232-4539, www.sonotronic.com

References

Amodel PPA Resins Engineering Data, Amoco Polymers Inc., Tech. Bulletin AM-F-50060

Application Design, DuPont Polymers, Tech. Bulletin A-23013-20M (1961)

Assembly Magazine, Sept. 1982

Baker, A., *Threaded Fasteners to the Rescue*, Design News, Cahners Publishing Co. (Feb. 1996)

Basdekis, C. H., *ABS Plastics*, Reinhold Publishing Corp., New York (1964)

Bauman, J. and Park, J., *Optimization of the Spin Welding Parameters for an Air Induction Part*, SPE ANTEC Tech Papers (1996)

Beall, G. L., *Designer's Guide to Pressure Forming*, Plastics Design Forum Magazine (Sept./Oct. 1985)

Beall, G. L., *Product Design for Rotational Molding* (Nov./Dec. 1984)

Beall, G. L., *Remove Speculation from Specification of Molded Parts*, Plastics Engineering Magazine (Sept. 1980)

Beck, R. D., *Plastic Product Design*, Van Nostrand Reinhold Co. Inc., New York (1980)

Behnfeldt, M. J. and Bouyoucos, J. V., *Variable Frequenty Technique for Vibration Welding of Plastics Parts*, Plastics Machinery and Equipment Magazine (1978)

Belofsky, H., *Plastics: Product Design and Process Engineering,* Hanser Gardner Publications, Cincinnati (1995)

Berins, M. L., editor, *Plastics Engineering Handbook* , Fifth Edition, The Society of the Plastics Industry, Inc., Chapman & Hall, New York (1991)

Blitshteyn, M., *Surface Treatment of Plastics*, Tantec, Inc. (Nov. 1992)

Bodnar, M. J., *Bonding Plastics*

Botelle, J. D., *Post Molding VS Molded-In Inserts*, University of Wisconsin

Buchman, A., Sidess, A., Dodink, H., *Fatigue of Adhesively Bonded Thermoplastics*, SPE ANTEC Tech Papers (1996)

Carpenter, R., *Design Opportunities Utilizing In-Mold Techniques*, SPE ANTEC Tech Papers (1988)

Celcon-Acetal Polymer, Hoechst-Celanese, Tech. Bulletin CE-1A (Dec. 1995)

Characteristics and Compatibility of Thermoplastics for Ultrasonic Assembly, Branson Ultrasonics Corp. Tech. Bulletin PW-1

Characteristics of Thermoplastics for Ultrasonic Assembly Applications, Ultra Sonic Seal (1984)

Chen, S.C., Hsu, K. S., Jeng, M. C., Hsu, K. F., *Numerical Simulations and Experimentals Studies of the Co-Injection Molding Process*, SPE ANTEC Tech Papers (1993)

Chookazian, S. M., *Electromagnetic Fusion Welding for Thermoplastic Assembly*, SPE ANTEC Tech Papers (1990)

Coldforming Plastic Studs, Appliance Manufacturing Magazine (June 1994)

Crawford, R. J., editor, *Rotational Moulding of Plastics*, Research Studies Press Ltd., Taunton, Somerset, England (1992)

Cycolac Design Guide, General Electric Company Tech. Bulletin CDC-540A (2/94) RTB10M

Cycoloy Design Guide, General Electric Company Tech. Bulletin CLY-400 (9/92) RTB

Dalton, S., *The Story of Hot Air Staking*, SPE ANTEC Tech Papers (1988)

Design Handbook for DuPont Engineering Polymers — Module I General Design Principles — DuPont Polymers Tech Bulletin 201742B (9/92)

Designing for the Environment – A Design Guide for Information and Technology Equipment, American Plastics Council (1994)

Designing Joints for Ultrasonic Welding, Dukane Corporation/Ultrasonics Div. Tech. Bulletin 9125-A-73

Designing Premature Failure Out of Injection Molded Parts – Miles Inc. Tech. Bulletin 516 (4M) 4/93

Designing With Fiberglass Reinforced Plastics, Molded Fiber Glass Company, Ashtabula, OH (1989)

Designing With Plastic – The Fundamentals – Hoechst Celanese Tech. Bulletin 93-320/10M/0294

Design Issues For Living Hinges, Montell Polyolefins (Dec. 1995)

Desmond, W. X., *Emaweld*^{TM} *Applications in the Automotive Industry*, SPE ANTEC Tech Papers (1985)

Dratschmidt, F. and Ehrenstein, G. W., *Polymer Thread Joints Under Dynamic Load*, SPE ANTEC Tech Papers (1995)

DuBois, J. H. and John, F. W., *Plastics*, Reinhold Publishing Corp., New York (1967)

DuBois, J. H. and Levy, S., *Plastics Product Design Engineering Handbook*, Van Nostrand Reinhold Co., New York (1977)

DuBois, J. H. and Pribble, W. I., *Plastics Mold Engineering*, Revised Edition, Reinhold Publishing Corp., New York (1965)

DuPont Zytel Design Handbook, DuPont Polymers Tech. Bulletin E-44971

Dym, J. B., *Injection Molds and Molding*, Van Nostrand Reinhold Co., New York (1979)

Ellsworth Adhesive Systems Catalog

Emaweld^{TM}, Emabond Systems, Ashland Chemical Co. Tech. Bulletin 1688

Facts For Spin Welding, Mecasonics Div. of Forward Technologies Tech. Bulletin 389032

Ford, R. L. and Driscoll, S. B., *Ultrasonic Welding with Amplitude Profiling*, SPE ANTEC Tech. Papers (1995)

Froment, I. D., *Vibration Welding Nylon 6 and Nylon 6/6 – A Comparative Study*, SPE ANTEC Tech. Papers (1995)

Gallagan, S. T., *Material and Additive Factors in Ultrasonic Assembly*, SPE ANTEC Tech. Papers (1985)

Genc, S., Messler, Dr. R. W., Jr., Gabrielle, Dr. G. A., *Issues in the Selection of Locking Features in Integral Attachment Design Using Snap Fits*, SPE ANTEC Tech. Papers (1997)

Giese, M. and Ehrenstein, G. W., *Identification of Phase 3 in the Vibration Welding Process*, SPE ANTEC Tech. Papers (1993)

Giese, M. and Ehrenstein, G. W., *Vibration Welding of Random Glass Mat Reinforced Thermoplastic Composites*, SPE ANTEC Tech. Papers (1992)

Gomes, K., *The Development and History of Taptite Thread Rolling Screws*, University of Wisconsin (1991)

Goodman, S. H. and Schwartz, S. S., *Plastics Materials and Processes*, Van Nostrand Reinhold Co., New York (1982)

Green, R. E., *Machinery's Handbook*, Industrial Press Inc. (1992)

Grewell, D. A., *Amplitude and Force Profiling: Studies in Ultrasonic Welding of Thermoplastics*, SPE ANTEC Tech Papers (1996)

Grewell, D. A., *Weldability of ABS and Testing of Weld Strength at Various Strain Rates; A Study in Ultrasonic Welding*, SPE ANTEC Tech. Papers (1997)

Grewell, D. A., Benatar, A., Park, J. B., *Plastics and Composites Welding Handbook*, Hanser Publishers, Munich (2003)

He, F. and Benatar, A., *Effect of Amplitude and Pressure Control on the Strength of Ultrasonically Welded Thermoplastics*, SPE ANTEC Tech. Papers (1996)

Hebert, L. P., Girard, P., Gaudreault, M., Salloum, G., *Effects of Inserts on Shrinkage in Injection Molded HDPE Disks*, SPE ANTEC Tech Papers (1991)

Helms, J. E., Yang, C., Pang, S. S., *Analysis of a Taper – Taper Adhesive Bonded Joint in a Composite Flat Plate*, SPE ANTEC Tech Papers (1994)

Herrmann, T., *Seam Design in Ultrasonic Welding*, Kunststoffe German Plastics (1987/7)

How to Bond Plastics, Loctite Corp., Newington, CT

Inserting Metal into Plastic, Ameritherm Inc. Tech.Bulletin Vol. 2, No. 1 (Sept. 1996)

Joint Designs For Hot Plate Welding, Plastics Design Forum Magazine (Nov.–Dec. 1992)

Joint Design for Ultrasonic Welding, Branson Ultrasonics Corp. Tech. Bulletin PW-3

Joint Designs for Ultrasonic Welding, Sonics and Materials, Inc. Tech. Bulletin (1989)

Joint Designs for Ultrasonic Welding, Ultra Sonic Seal Tech. Bulletin (1987)

Kagan, V. A., *Innovations in Laser Welding of Thermoplastics: This Advanced Technology is Ready to be Commercialized*, Society of Automotive Engineers, 2002-01-2011

Kagan, V. A., *Laser Transmission Welding of Semi-Crystalline Thermoplastics – Part II: Analysis of Mechanical Performance of Welded Nylon*, SPE ANTEC Tech. Papers (2000)

Kagan, V. A., Bray, R., Chambers, A., *Forward To Better Understanding of Optical Characterization and Development of Colored Polyamides for the Infra-red/Laser Welding: Part I – Efficiency of Polyamides for Infra-red Welding*, SPE ANTEC Tech. Papers (2001)

Kagan, V. A., Woosman, N. M., *Efficiency of Clear-Welding Technology for Polyamides*, SPE ANTEC Tech. Papers (2002)

Kaminsky, S., *Plastic Welding Techniques for Equipment Fabrication, Repair*, Plant Engineering (Jan. 23, 1969)

Karamuk, E., Wetzel, E. D., Gillespie, J. W., Jr., *Modeling and Design of Induction Bonding Process for Infrastructure Rehabilitation with Composite Materials*, SPE ANTEC Tech Papers (1995)

Kirkland, C., *Automolding 2000: Four World Class Processes For World Cars*, Injection Molding Magazine (Feb. 1997)

Kocheny, S. A., Zybko, J., *Three Approaches in Utilizing High Power Diode Laser to Join Thermoplastics*, SPE ANTEC Tech. Papers (2002)

LaBounty, T. J., *Spin Welding, Up-Dating an Old Technique*, SPE ANTEC Tech Papers (1985)

Lamb, D., *Fastener Strategy Considerations*, Appliance Manufacturer Magazine (Aug. 1996)

Landrock, A. H., *3M Adhesive Answer Book for Product Assembly*, 3M

Larson, R. F. and Larsen, G. C., *The Next Generation of PC Software for Traditional and FEA Snap-Fit Design*, SPE ANTEC Tech Papers (1994)

Lexan Design Guide, G. E. Plastics Tech. Bulletin CDC-5361 (2/94) RTB-15M

Lincoln, B., Gomes, K., Braden, J. F., *Mechanical Fastening of Plastics*, Marcel Dekker, Inc., New York (1984)

Lokensgard, E., *Effects of Process Conditions and Internal Lubricants on the Pull-Out Resistance of Threaded Inserts in Polypropylene*, SPE ANTEC Tech Papers (1994)

Lokensgard, E., *Time Lapse Effects On Pull-Out Resistance of Threaded Inserts in Polypropylene*, SPE ANTEC Tech Papers (1995)

Luscher, A. F., Bonenberger, P. R., Gabrielle, Dr. G. A., *Attachment Strategy Guidelines for Integral Attachment Features*, SPE ANTEC Tech Papers (1995)

Luscher, A. F., Bonenberger, P. R., Gabrielle, Dr. G. A., Messler, Dr. R. W., Jr., *A Classification Scheme for Integral Attachment Features*, SPE ANTEC Tech Papers (1995)

Macosko, C. W., *RIM Fundamentals of Reaction Injection Molding*, Carl Hanser Verlag, Munich (1989)

Machining, Finishing and Fastening, Monsanto Plastics Div., 1013A

Malloy, R. A., *Molding Options for Snap Fit Assembly*, Plastics World Magazine (May 1997)

Malloy, R. A., *Plastic Part Design for Injection Molding*, Hanser Gardner, Carl Hanser Verlag, Munich (1994)

Malloy, R. A., *What's the Best Assembly Method*, Plastics World Magazine (March, 1997)

Malloy, R. A., Orroth, S. A., Arnold, E. S., *Self Threading Screw Boss Design*, SPE ANTEC Tech Papers (1985)

Maniscalco, M., *SnapFit Software Closes the Loop*, Injection Molding Magazine (Jan. 1997)

Massey, F. L., *Boss and Fastener Design Specifications for SMC Materials*, 32nd Annual Technical Conference, Reinforced Plastics/Composites Institute, The Society of the Plastics Industry, Inc. (1977)

Materials Compatible with Ultrasonic Fabric and Film Sealing, Branson Ultrasonics Corp. Tech. Bulletin FF-5

McChesney, C. E. and Groteau, G. P., *Knit Line Strength of Reinforced Polyphenylene Sulfide Injection Moldings*, SPE ANTEC Tech Papers (1993)

Michaeli, W. and Brinkman, S., *Multi-Component Injection Molding – Influence of Process Parameters on the Adhesion of Rigid-Flexible Combinations*, SPE ANTEC Tech Papers (1995)

Monsanto Plastics Design Guide

National Institute of Standards and Technology Handbook

Nieh, J. Y. and Lee, L. J., *Morphological Characterization of the Heat Affected Zone in Hot Plate Welding*, SPE ANTEC Tech Papers (1993)

Nordgren, J. A., *Joining with Adhesives*, Adhesive Systems, Industrial Tape and Specialties Div., 3M (1994)

Noryl Injection Molding, General Electric Co. Tech. Bulletin CDX-81C (6/88) RTB

Oberg, E., Jones, F. D, Horton, H. L., Ryffel, H. H., Green, R.E. Ed., *Machinery's Handbook*, 24th Edition, Industrial Press, Inc., New York (1992)

Plastics Are Recyclable, Society of Plastics Engineers, Inc. (1994)

Plastics Design Manual, Pub. No. 7138A, Monsanto Company, St. Louis (1994)

Plastics Joining Techniques, A Design Guide, Miles Inc. Tech. Bulletin KU-F3023 (10) 1993

Plastic Snap Fit Joints, Miles Inc. Tech. Bulletin 505.KU-F-2049 (10) L (1992)

Poslinski, A. J. and Stokes, V. K., *Analysis of the Hot Tool Welding Process*, SPE ANTEC Tech Papers (1992)

Potente, H. and BruBel, A., *The Problem of Stress Cracking in Heated Tool Butt Welded Tail Lights Made of PMMR and ABS*, SPE ANTEC Tech Papers (1996)

Potente, H. and Uebbing, M., *Computer-Aided Layout of the Vibration Welding Process*, SPE ANTEC Tech. Papers (1992)

Potente, H. and Uebbing, M., *The Effects of Moisture on Vibration Welding*, SPE ANTEC Tech. Papers (1994)

Potente, H. and Uebbing, M., *Vibration Welding of High Temperature Plastics*, SPE ANTEC Tech. Papers (1993)

Potente, H., Fiegler, G., Becker, F., and Korte, J., *Comparative Investigations on Quasi-simultaneous Welding on the Basis of the Materials PEEK and PC*, SPE ANTEC Tech. Papers (2002)

Rahut, Dr. H., *Assembling Plastic Parts with Adhesives*, Plastics Design and Processing (April, 1971)

Rantz, L. E., *Proper Surface Preparation: Bonding's Critical First Step*

Rao, N. S., *Design Formulas for Plastics Engineers*, Carl Hanser Verlag, Munich (1991)

Rogers, J. K., *Why Instrument Panel Redesign is Essential*, Modern Plastics Magazine (Feb. 1991)

Romani, K. and Zhao, W., *The Development of Residual Stresses During Thermoplastic Adhesive Bonding to Metals*, SPE ANTEC Tech Papers (1997)

Rosato, D. V., *Rosato's Plastics Encyclopedia and Dictionary*, Carl Hanser Verlag, Munich (1993)

Rose, P.W. and Liston, E., *Gas Plasma Technology and Surface Treatment of Polymers Prior to Adhesive Bonding*, SPE ANTEC Tech Papers (1985)

Rotheiser, J. I., *The Bigger Picture*, Plastics Engineering Magazine (Jan. 1997)

Ruffini, R. S. and Nemkov, V. S., *Materials for Effective Magnetic Flux Control and Concentration in Induction Heating Processes*, Industrial Heating Magazine (Nov. 1996)

Rubin, I. I., editor, *Handbook of Plastic Materials and Technology*, John Wiley & Sons, Inc., New York (1990)

Rubin, I. I., *Injection Molding Theory and Practice*, Wiley – Interscience, NY (1972)

Sawyer, W. G., Blanchet, T. A., Knapp, K. N. II, Lee, D., *Friction Modeling and Experimentation for Integral Fasteners*, SPE ANTEC Tech Papers (1997)

Simo, J., Ellwood, P. A., Taylor, H. J., *Pollutants From Laser Cutting and Hot Gas Welding of Polymer Based Materials*, SPE ANTEC Tech Papers (1994)

Sirkin, A. L., *Mira Spring Polypropylene Hinge*, Enjay Chemical Co., Tech. Bulletin ELD-60633

Skeist, I., *Handbook of Adhesives*, Von Nostrand Reinhold Co., New York (1962)

SMC Design Manual – Exterior Body Panels, SMC Automotive Alliance, Bloomfield Hills, MI (1991)

SMC Design Manual, The Composites Institute, Society of the Plastics Industry, Inc. (1991)

Snyder, M. R., *Design Flexibility Spurs Molding of Multicomponent Parts*, Modern Plastics Magazine (June 1997)

Spahr, T., *Snap-Fits for Assembly and Disassembly*, University of Wisconsin (1991)

Staking Configurations, Dukane Corporation/Ultrasonics Div., Tech. Bulletin 9839-B-78

St. John, M. and Park, J. B., *Ultrasonic Weldability of Painted Plastics*, SPE ANTEC Tech. Papers (1997)

Stokes, V. K., *Cross-Thickness Vibration Welding of Thermoplastics*, SPE ANTEC Tech. Papers (1992)

Stokes, V. K., *Experiments on the Hot-Tool Welding of Dissimilar Thermoplastics*, SPE ANTEC Tech. Papers (1993)

Stokes, V. K., *Joining Methods for Plastics and Plastic Composites: An Overview*, SPE ANTEC Tech. Papers (1989)

Stokes, V. K., *Strength and Bonding Mechanisms in Vibration Welded Polycarbonate*, SPE ANTEC Tech. Papers (1989)

Stokes, V. K., *Thickness Effects in the Vibration Welding of Polycarbonate*, SPE ANTEC Tech. Papers (1989)

Stokes, V. K., *Toward a Weld-Strength Data Base for Vibration Welding of Thermoplastics*, SPE ANTEC Tech. Papers (1995)

Stokes, V. K., *Vibration Welding of Thermoplastics, Part I: Phenomenology of the Welding Process*, Polymer Engineering and Science (1988)

Stokes, V. K., *Vibration Welding of Thermoplastics, Part II: Analysis of the Welding Process*, Polymer Engineering and Science (1988)

Structural Adhesives Guide for Industrial Product Design and Assembly, 3M Industrial Specialties Div. (1990)

Tarahoni, S., *Collapsible Core Analysis and Part Design Feasibility for Injection Molding with Collapsible Cores*, SPE ANTEC Tech Papers (1991)

Taylor, N., *The Ultrasonic Welding of Short Glass Fibre Reinforced Thermoplastics*, SPE ANTEC Tech. Papers (1991)

Textured Surface Technology, Branson Ultrasonics Corp. Tech. Bulletin #59 (May 1992)

Temin, S. C. *Adhesion and Solvent Bonding, Handbook of Plastic Materials and Technology*, Rubin, I. I. (Ed.), John Wiley & Sons, Inc., New York (1990)

The Integral Hinge, Enjay Chemical Co., Tech. Bulletin PLA-63-850

Thomas, D. W. and Potter, J. P., *Making Even Better Plastic Welds*, Laramy Products Co., Inc. (1993)

Thompson, R., *Processor Rules for Good Adhesive Assembly*, Plastics World Magazine (May 1985)

Topping, M., *Electromagnetic Welding of Thermoplastics and Specific Design Criteria*, Ashland Chemical Co.

Tres, P. A., *Designing Plastic Parts for Assembly*, Hanser Gardner, Carl Hanser Verlag, Munich (1994)

Ultrasonic and Vibration Welding Characteristics and Compatibility of Thermoplastics, Branson Ultrasonics Corp. Tech. Bulletin PW-1

Ultrasonic Insertion, Branson Ultrasonics Corp. (Jan. 1991)

Ultrasonic Installation of Inserts in Thermoplastic Components, Ultra Sonic Seal Co. Tech. Bulletin (July 1991)

Ultrasonic Installation of Inserts in Thermoplastic Components , Sonics and Materials, Inc., Tech. Bulletin

Ultrasonic Radial Scan Welding, Branson Ultrasonics Corp. Tech. Bulletin #17

Ultrasonic Spot Welding, Branson Ultrasonics Corp. Tech. Bulletin PW-9

Ultrasonic Staking, Branson Ultrasonics Corp. Tech. Bulletin PW-6 (Jan. 1989)

Ultrasonic Staking and Spot Welding of Thermoplastic Assemblies, Sonics and Materials, Inc. Tech. Bulletin (April 1993)

Ultrasonic Staking and Spot Welding of Thermoplastic Assemblies, Ultra Sonic Seal Co. Tech. Bulletin (1985)

Ultrasonic Staking — Troubleshooting, Branson Ultrasonics Corp. Tech. Bulletin #8 (Nov. 1989)

Ultrasonic Stud Welding, Branson Ultrasonics Corp. Tech. Bulletin PW-5

Ultrasonic Welding Characteristics of Textiles and Films, Branson Ultrasonics Corp. Tech. Bulletin P10-7

Understanding Thermoplastic Part Warpage, LNP Engineering Plastics Tech. Bulletin (Spring 1995)

Valox Design Guide, General Electric Co. Tech. Bulletin VAL-500 (6/88) RTB

Varadan, V. K., Varadan, V. V., Schaffer, T. L., *Microwave Enhanced Solvent Welding*, SPE ANTEC Tech Papers (1992)

Viscio, D. P., *A Survey of Metal Threaded Inserts for Insertion into Plastics*, Heli-Coil Products, Div. of Mite Corp.

Watson, M. N., *Improvements to Productivity With the Butt Fusion Process for Joining Plastic Pipes*, SPE ANTEC Tech Papers (1990)

Weick, P., *Threaded Closure Molding: An Alternative Approach*, SPE ANTEC Tech Papers (1989)

Wegener, M. and Michaels, W., *Influence of Thermal Extension Mismatch in the Adhesive Joining of Different Materials*, SPE ANTEC Tech. Papers (1991)

Wetzel, E. D., Don, R. C., Gillespie, J. W., Jr., *Modeling Thermal Degradation During Thermoplastic Fusion Bonding of Thermoset Composites*, SPE ANTEC Tech Papers (1994)

Wolcott, J. S., *Designing Parts for Ultrasonic Assembly*, SPE ANTEC Tech Papers (1990)

Wright, R. E., *Injection/Transfer Molding of Thermosetting Plastics*, Hanser Gardner, Carl Hanser Verlag, Munich (1995)

Wright, R. E., *Molded Thermosets*, Hanser Gardner, Carl Hanser Verlag, Munich (1991)

Yeh, H. J., Schott, C. L., Park, J. B., *Experimental Study on Hot-Air Cold Staking of ABS and Polypropylene Samples*, SPE ANTEC Tech Papers (1997)

Index

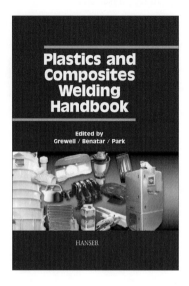

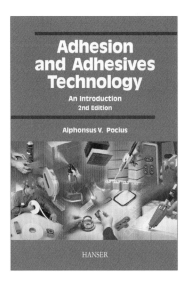

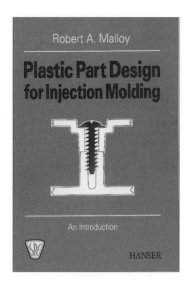